P9-DKF-044

GENERALIZED METHODS OF VIBRATION ANALYSIS

GENERALIZED METHODS OF VIBRATION ANALYSIS

RALPH J. HARKER

University of Wisconsin, Madison

A Wiley-Interscience Publication

JOHN WILEY & SONS

New York Chichester Brisbane Toronto Singapore

Library of Congress Cataloging in Publication Data:

Harker, Ralph J., 1915-
Generalized methods of vibration analysis.

"A Wiley-Interscience publication."
Bibliography: p.
Includes index.
1. Vibration. 2. Engineering mathematics.
I. Title.
TA355.H34 1983 620.3 82-17440
ISBN 0-471-86735-7

Printed in the United States of America

10 9 8 7 6 5 4 3 2 1

To Enes

PREFACE

The subject of mechanical vibration has spawned numerous texts and books and many are of excellent quality. Although considerably diverse in detail, they share much common ground in topical coverage and in development of subject matter.

This book is innovative in several respects, particularly in the consistent use of *normalized parameters* to generalize analytical results. System mass and stiffness elements are all referenced to a single mass and spring, providing dimensionless variables and relationships for natural frequencies, normal modes, and forced response. This approach makes it possible to group various types of fundamental systems in tabulated format to display generalized results.

Normalization also makes possible the use of personal calculators in a number of computations, decreasing reliance on software and computers. For instance, a normalized Holzer solution becomes tractable numerically when performed on a personal calculator with only one memory. The data presented, coupled with the use of the calculator, greatly enhance individual problem-solving capability. As will be shown, natural modes can be calculated directly from equations for the normalized system, and it is not necessary to use any absolute values of spring or mass.

Normalizing also reduces difficulties related to the dual system of units now in use. The ratios provided can be applied in the generalized equations without dimensional hazards. After considering the nature and characteristics of a system we can obtain absolute results with specific actual dimensional data, applied to the previous results in a single calculation.

Considerable emphasis is placed quantifying vibratory response, on both a total and a modal component basis. In the former, numerous tables are provided for system combinations with several masses. In the latter, a single-degree analog system is introduced to simulate the modal vibratory characteristics of more complex systems.

To provide an in-depth treatment of fundamental vibratory behavior, the analysis has been restricted to classical linear systems, primarily in steady state. However, the result has been a comprehensive overview of basic system mechanisms, which should be valuable in developing vibrational concepts and in approximating more exotic models.

The material in this book has its roots in an elective course I have taught in mechanical engineering, although the book itself is intended and organized to be a reference source for professional engineers. In several respects this book parallels the development of another reference[16] by my late colleague Professor Raymond J. Roark, which was also drawn from classroom experience. Both books rely on a tabulated format for presenting results.

Although this book is not primarily a text, I believe it can be used as a text successfully. There is adequate material in Chapters 1 to 7 for an introductory course, with a second course following from the last nine chapters. Class problems are included. Instructors stressing derivations can assign any of the tabulated equations for verification as an analytical exercise, or students can be required to derive extensions of the present tabulations. I have presented derivational methods in my own classes, but I have found that student understanding is best obtained by problem solving.

All of the tabulations that follow are original; only the numerical values of beam coefficients given in Chapter 12 were obtained from the literature. I thus bear full responsibility for authenticity. Although I have made an extensive effort to check for accuracy, some errors are inevitable with this quantity of data, and I will appreciate readers calling my attention to any occurring in this first compilation.

Finally, I must express my appreciation to those who have contributed to this effort and made the book possible. These include the many engineering students with whom I have interacted, whose questions have stimulated much of the insight required. Several graduate students have been especially helpful and supportive: Jason Fan, William McNown, Wayde Thomann, Glenn Thomas, and Mohammad Zand. I also wish to thank John Olsen and Lorraine Loken for their considerable and capable assistance with the illustrations and with the typing of the manuscript, respectively.

RALPH J. HARKER

Madison, Wisconsin
December 1982

CONTENTS

6. TORSIONAL VIBRATION OF CRANKSHAFTS 156

7. VIBRATION ABSORBERS AND DAMPERS 181

NOTATION

a	=	complex variable representing displacement (amplitude)
a^r	=	relative complex displacement (amplitude)
A	=	area of cross section
C	=	dashpot constant
	=	real term in complex denominator
D	=	denominator
	=	diameter
e	=	radial eccentricity
E	=	Young's modulus of elasticity
E_d	=	rate of energy dissipation
EE	=	elastic energy
f	=	frequency
F	=	static force
F_0	=	initial force pulse
g_c	=	acceleration of gravity
G	=	center of gravity
	=	shear modulus of elasticity
I	=	mass moment of inertia about a transverse axis
	=	area moment of inertia of a cross section
j	=	complex operator
J	=	polar mass moment of inertia
	=	polar area moment of inertia
K	=	spring constant
KE	=	kinetic energy
l	=	length
me	=	rotating unbalance

M = mass
MF = vibratory magnification factor
n = natural frequency or mode
= number of cycles
= number of gear teeth
N = order of excitation
= order of whirl
p = total number of masses
= unit pressure
P = complex force (amplitude)
PE = potential energy
q = individual mass or spring
r = radius of gyration about centroid
R = crankshaft radius
= outside radius of cylinder
= pitch radius of gear
= mean ring radius
S = stress
t = time
T = static couple
= period of simple harmonic motion
= complex couple (amplitude)
= string tension
W = weight
= work
x = horizontal displacement
x_0 = amplitude of sinusoidal displacement
$\dot{x}$ = first derivative of displacement with respect to time
$\ddot{x}$ = second derivative of displacement with respect to time
y = vertical displacement

α = translational elastic compliance factor
β = normal modal amplitude
= elastic cross-compliance factor
γ = angular elastic compliance factor
δ = small term
= gravity deflection
θ = angular displacement
= crank angle
$\dot{\theta}$ = angular velocity
$\ddot{\theta}$ = angular acceleration
= summation of $mg^2\beta$ terms
μ = absolute viscosity
= mass per unit length
ρ = mass density
Σ = summation
ϕ = complex phase angle
= normal modal angular amplitude
= central ring angle
ψ = normalized elastic compliance ratio
= phase angle in star diagram
ω = circular frequency
Ω = rotational angular velocity

NORMALIZED PARAMETERS

(Dimensionless)

f = tuning ratio

$= \left(\dfrac{\omega_0}{\omega_1}\right) = \dfrac{\text{natural frequency of auxiliary system, } K_0, M_0}{\text{natural frequency of main system, } K_1, M_1}$

g = forced frequency ratio

$= \left(\dfrac{\omega}{\omega_1}\right) = \dfrac{\text{forcing frequency}}{\text{reference natural frequency}}$

g_{ni} = natural frequency factor

$= \left(\dfrac{\omega_{ni}}{\omega_1}\right) = \sqrt{\dfrac{k_{ni}}{m_{ni}}} = \dfrac{\text{modal natural frequency}}{\text{reference natural frequency}}$

g_i = modal forced frequency ratio

$= \left(\dfrac{\omega}{\omega_{ni}}\right) = \left(\dfrac{\mathsf{g}}{\mathsf{g}_{ni}}\right) = \left(\dfrac{\mathsf{g}_{n1}}{\mathsf{g}_{ni}}\right)\mathsf{g}_1 = \dfrac{\text{forcing frequency}}{\text{modal natural frequency}}$

k_q = normalized spring constant

$= \left(\dfrac{K_q}{K_1}\right) = \dfrac{\text{spring constant at } q}{\text{spring constant at 1}}$

k_{ni} = normalized modal spring

$= \left(\dfrac{K_{ni}}{K_1}\right) = \dfrac{\text{modal spring}}{\text{spring constant at 1}}$

m_q = normalized mass

$= \left(\dfrac{M_q}{M_1}\right) = \dfrac{\text{mass at } q}{\text{mass at 1}}$

m_{ni} = normalized modal mass

$= \left(\dfrac{M_{ni}}{M_1}\right) = \dfrac{\text{modal mass}}{\text{mass at 1}}$

$\overline{m}_i$ = normalized modal participation factor

$= \left(\dfrac{\Sigma m_q \beta_{qi}}{m_{ni}}\right) = \dfrac{\text{inertia excitation factor}}{\text{normalized modal mass}}$

α = initial velocity factor

$= \left(\dfrac{\dot{x}_0}{\omega_1 x_0}\right) = \dfrac{\text{initial velocity}}{\text{reference harmonic velocity}}$

β = damped frequency factor

$= \sqrt{1-\zeta^2}$ or $\sqrt{\zeta^2 - 1}$

ζ = damping ratio

$= \left(\dfrac{C}{C_c}\right) = \dfrac{\text{viscous damping factor}}{\text{critical damping factor}}$

λ_q = displacement excitation factor

$= \left(\dfrac{\lambda_q}{\lambda_1}\right) = \dfrac{\text{displacement amplitude at } q}{\text{displacement amplitude at 1}}$

ρ = radius of gyration factor

$= \left[1 - \left(\dfrac{r_p}{r}\right)^2 \left(\dfrac{1}{N}\right)\right] = \left(\dfrac{r_e}{r}\right) = \dfrac{\text{equivalent radius of gyration}}{\text{actual radius of gyration}}$

GENERALIZED METHODS OF VIBRATION ANALYSIS

1 ELEMENTS OF VIBRATION ANALYSIS

Vibration relates to the motion of masses and is thus concerned exclusively with dynamic phenomena. Motions are usually periodic. The most basic type of motion, called *simple harmonic*, varies sinusoidally with time; both the cyclic frequency and amplitude are constant. A familiar example of simple harmonic motion is the oscillation of a clock pendulum, with the constant frequency feature utilized to measure elapsed time.

Harmonic motion will be discussed in successive chapters as satisfying certain basic differential equations; however, it is introduced here to develop the concept of time-varying quantities. These quantities in turn are represented in complex variable notation.

There are three characteristic physical quantities, components, or building blocks usually present in vibratory systems: *mass*, *spring*, and *damping* elements. These are considered separately in this chapter. Discussion of excitation, another physical ingredient usually present, is deferred to Chapter 2.

Angular and translational effects are presented concurrently because they are similar fundamental vibratory forms, avoiding isolation of torsional vibration as a distinct subject. This interrelationship is emphasized by demonstrating equivalence and methods for converting from rectangular to polar coordinates, and vice versa.

1.1. SIMPLE HARMONIC MOTION

Sinusoidal displacement characterizes most vibratory theory and physical problems. It describes the free oscillatory motion of all undamped spring–mass combinations in which the spring is linear and the mass is constant. This important motion can be expressed mathematically in several forms, including the trigonometric expressions

$$x = x_0 \sin \omega t \qquad x = x_0 \cos \omega t \tag{1.1}$$

where x = the instantaneous displacement at time t

x_0 = maximum displacement, or amplitude, from midposition

ω = constant angular velocity related to the motion generation, or the circular frequency (rad/sec)

$= 2\pi f$

t = elapsed time from a starting condition (sec)

ωt = the function angle (rad)

f = vibratory frequency (cycles/sec, or hertz, Hz)

$T = 2\pi/\omega$ = the period, or time for one complete cycle (sec)

Using the cosine form in which the displacement is maximum initially corresponds to the displacement–time plot shown in Figure 1.1. The behavior is repetitive, with one cycle completed when $\omega t = 2\pi$. We also see that the construction of this function derives from the rotation of a radius vector of fixed length at the constant angular velocity ω with the radius representing maximum displacement, or amplitude. Instantaneous displacement is the vertical component of the vector, or its projection on the vertical axis, $x_0 \cos \omega t$.

If initial conditions do not correspond to zero, or maximum displacement with sine or cosine expressions, an initial time phase angle is involved and the equation becomes

$$x = A \sin \omega t + B \cos \omega t \tag{1.2a}$$

or

$$= x_0 \cos(\omega t - \phi) \tag{1.2b}$$

where $x_0 = \sqrt{A^2 + B^2}$

$\phi = \tan^{-1}(A/B)$

This situation is shown in Figure 1.2. We may further conclude that several

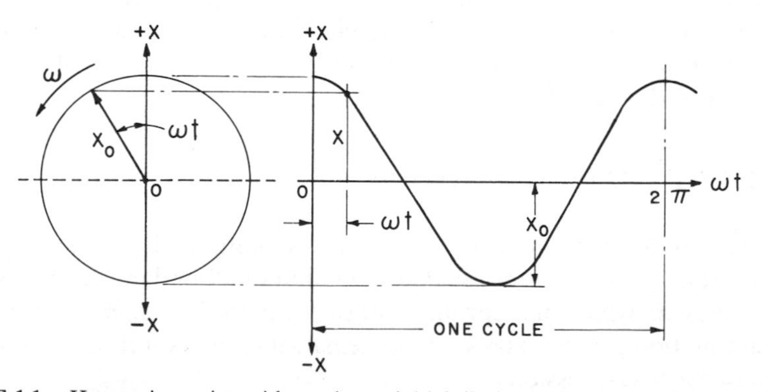

FIGURE 1.1. Harmonic motion with maximum initial displacement is defined by $x_0 \cos \omega t$, and generated by the projection of a vector rotating with constant velocity.

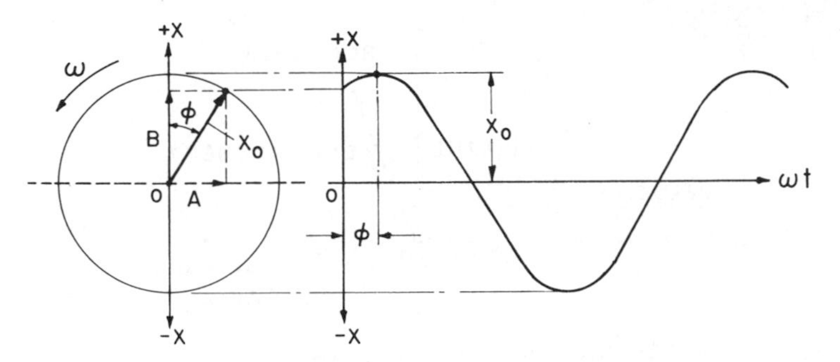

FIGURE 1.2. The displaced sinusoidal function has two components, $A \sin \omega t + B \cos \omega t$.

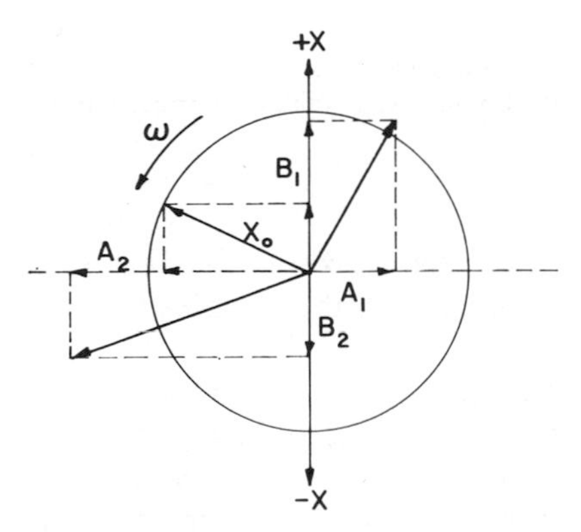

FIGURE 1.3. Harmonic variations at exactly the same frequency are synchronized vectorially, and combine to a single resultant rotating vector.

harmonic motions having exactly the same circular frequency can be combined into a single harmonic motion by vector addition (Figure 1.3).

$$x = (A_1 \sin \omega t + B_1 \cos \omega t) + (A_2 \sin \omega t + B_2 \cos \omega t)$$

$$= (A_1 + A_2) \sin \omega t + (B_1 + B_2) \cos \omega t \tag{1.3}$$

1.2. THE COMPLEX VARIABLE

From Figures 1.1, 1.2, and 1.3, it is seen that sinusoidal functions are conveniently related to a rotating vector, suggesting representation of harmonically varying quantities by means of a complex variable. For instance, a rotating vector a (Figure 1.4) has angular velocity ω. Projection on the *real* axis is the instantaneous displacement. Projection on the *imaginary* axis has no physical significance.

Use of the complex variable has a number of advantages in the analysis of sinusoidally varying quantities. The single variable a replaces the complete expressions shown in Equations 1.2a and 1.2b. As a radius its absolute value

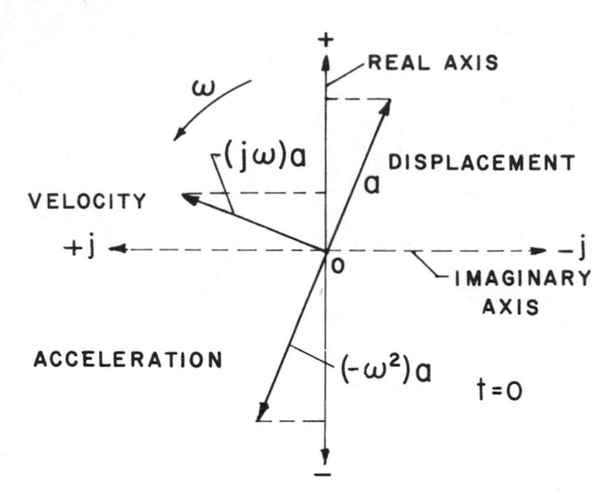

FIGURE 1.4. As represented in the complex plane, the displacement, velocity, and acceleration of a point correspond to mutually perpendicular vectors. Instantaneous values are projections on the real axis.

represents amplitude, occurring each time the vector crosses the real axis. Circular frequency ω is understood to be associated with the complex variable but specified separately.

As will be shown, operators on **a** conveniently alter its magnitude and/or direction. When used in equations it may be treated as an ordinary algebraic variable. It is important, however, always to visualize the complex variable as a rotating vector relative to real and imaginary axes. Once we achieve this concept, the time-varying quantities in vibratory systems are readily interpreted, and simply and conveniently expressed.

1.3. VELOCITY AND ACCELERATION

First and second derivatives of displacement with respect to time are important terms in vibratory analysis. With simple harmonic displacement both velocity and acceleration are also cyclic and harmonic. Using the general algebraic form shown in Equation 1.2a,

$$\text{Displacemen} = x = A \sin \omega t + B \cos \omega t \tag{1.4a}$$

$$\text{Velocity} \quad = \frac{dx}{dt} = \dot{x} = A\omega \cos \omega t - B\omega \sin \omega t \tag{1.4b}$$

$$\text{Acceleration} = \frac{d^2x}{dt^2} = \ddot{x} = -A\omega^2 \sin \omega t - B\omega^2 \cos \omega t \tag{1.4c}$$

In complex notation a single differentiation with respect to time consists of

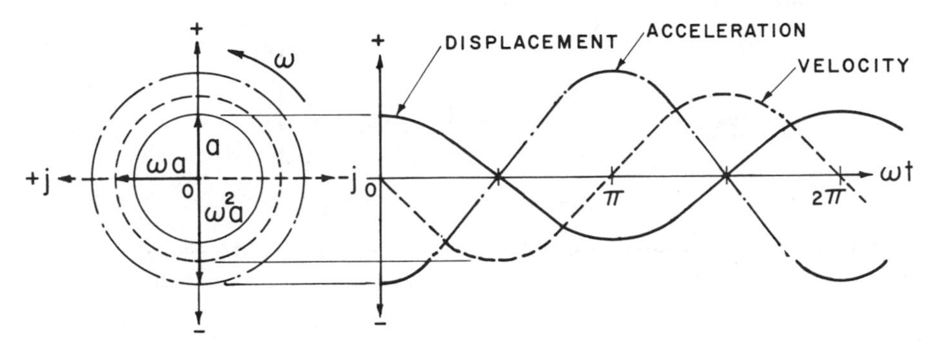

FIGURE 1.5. Complex variables representing displacement, velocity, and acceleration as vectors in the complex plane correspond to harmonically varying quantities in rectangular coordinates.

applying a $(j\omega)$ operator, or factor, to the displacement a. Then we have simply

$$\text{Displacement} = \mathsf{a} \tag{1.5a}$$

$$\text{Velocity} = (j\omega)\mathsf{a} \tag{1.5b}$$

$$\text{Acceleration} = (j\omega)^2\mathsf{a} = (-\omega^2)\mathsf{a} \tag{1.5c}$$

As seen in Figure 1.4, the displacement vector is converted to the velocity vector by rotating it forward 90° in the complex plane due to the $+j$ operator, and modifying its length by ω. Similarly, the velocity vector converts to acceleration by a further forward rotation of 90° and a further length multiplication by ω. The final phase position of the acceleration vector is therefore exactly opposite the displacement vector. The quadrature relationship of these vectors is a fundamental characteristic, demonstrated in Figures 1.4 and 1.5.

1.4. ORIENTATION OF COORDINATES

The previous discussion has related to harmonic motion in a translational sense with the x direction assumed vertical in Figures 1.1–1.5. This is convenient for plotting the sinusoidal curves horizontally as in the typical cases just given. In a physical situation in which the harmonic motion applies to the translation of a mass, the translation could be horizontal, or at any angular position in space. We must then visualize the real axis as coincident with the real direction, and the imaginary axis as perpendicular to the real to form the complex axes. Intersection of these axes must correspond to the midpoint of the motion, and the center about which the complex vector rotates in the complex plane.

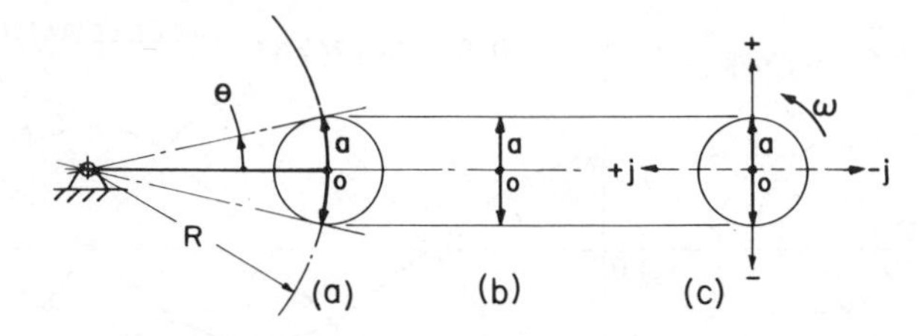

FIGURE 1.6. Angular harmonic motion (a) can be visualized as approximately equivalent to a chordal translational motion (b) generated by a rotating vector in the complex plane (c).

1.5. ROTATIONAL COORDINATES

If harmonic motion occurs in rotation or torsion (Figure 1.6) rather than in translation, the sinusoidal representations developed are equally applicable, with the dimensional units altered. Angles replace translational quantities, and with normal variables,

$$\theta = \theta_0 \cos \omega t \tag{1.6a}$$

$$\dot{\theta} = (-\omega\theta_0)\sin \omega t \tag{1.6b}$$

$$\ddot{\theta} = \left(-\omega^2\theta_0\right)\cos \omega t \tag{1.6c}$$

We treat θ as the complex variable a in radians, and Equations 1.5a, 1.5b, and 1.5c apply unchanged. Figure 1.5 shows the relationship of the angular quantities. As angular amplitudes are usually small, there is an approximate equivalent linear motion on the chord as shown.

1.6. VIBRATORY TRANSLATION OF THE RIGID MASS

A mass having translational acceleration develops an inertia load, or force, equal to the product of the mass and the acceleration. If this inertia effect is treated as a reversed effective load, or D'Alembert force, the direction of this load is exactly opposite to that of the acceleration.

In Figure 1.7 we see a sinusoidal motion at constant frequency of the mass. As shown in a positively displaced position, the acceleration is negative and the inertia force is $+(M\omega^2)\mathsf{a}$. When the mass has maximum displacement to the right, acceleration is maximum to the left, and the inertia force is maximum and to the right. Thus in steady state the inertia force is always in phase with the displacement, but it is greater than the displacement by the factor $(M\omega^2)$.

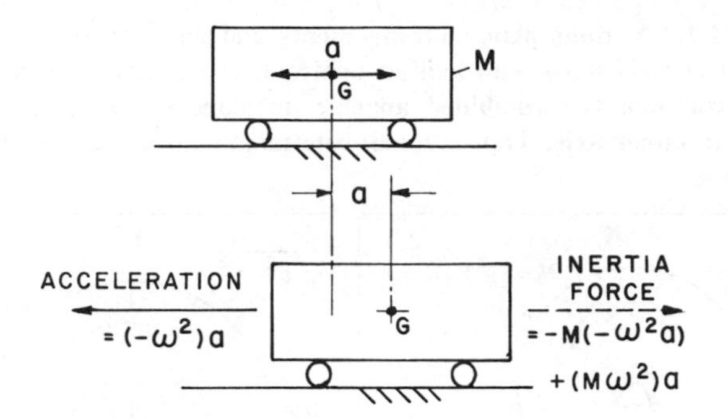

FIGURE 1.7. The sinusoidal inertia load associated with the oscillation of a mass is in phase with and proportional to the sinusoidal displacement, a.

1.7. VIBRATORY ROTATION OF THE RIGID MASS

Table 1.1*a* shows the simplest arrangement, rotation about a centroid coincident with the pivot point. As in translation, a harmonically varying angular displacement corresponds to a sinusoidal inertia torque having the same sense (Figure 1.8).

More generally the centroid is some distance l from the pivot (Table 1.1*b*). Combining the mass inertia effect about the centroid with the radial distance to the centroid is possible by the fundamental transfer of axes theorem.

$$J_0 = M(r^2 + l^2) = Ml_2^2 \tag{1.7}$$

where r = radius of gyration about the mass centroid

l = rotation radius to the centroid

l_2 = radius of gyration about pivot

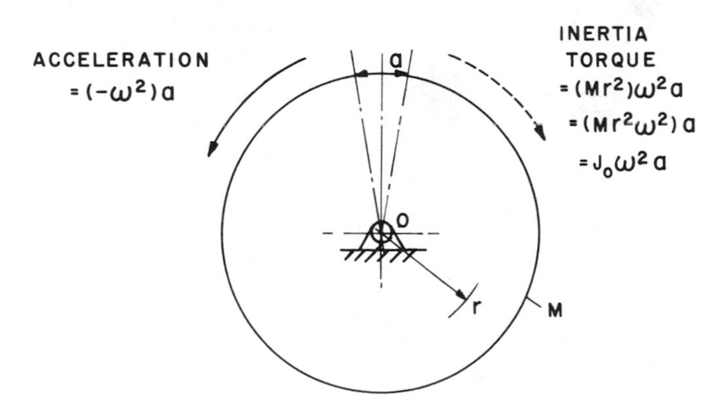

FIGURE 1.8. Sinusoidally varying angular displacement develops a sinusoidal inertia torque in phase with the angular displacement.

TABLE 1.1. Various planar arrangements and equivalents for rotational masses: (*a*) rigid mass with axis at centroid; (*b*) centroid displaced from axis of rotation; (*c*) combined angular unbalances; (*d*) geared inertias referred to either axis. Translational–rotational equivalence assumes small amplitudes.

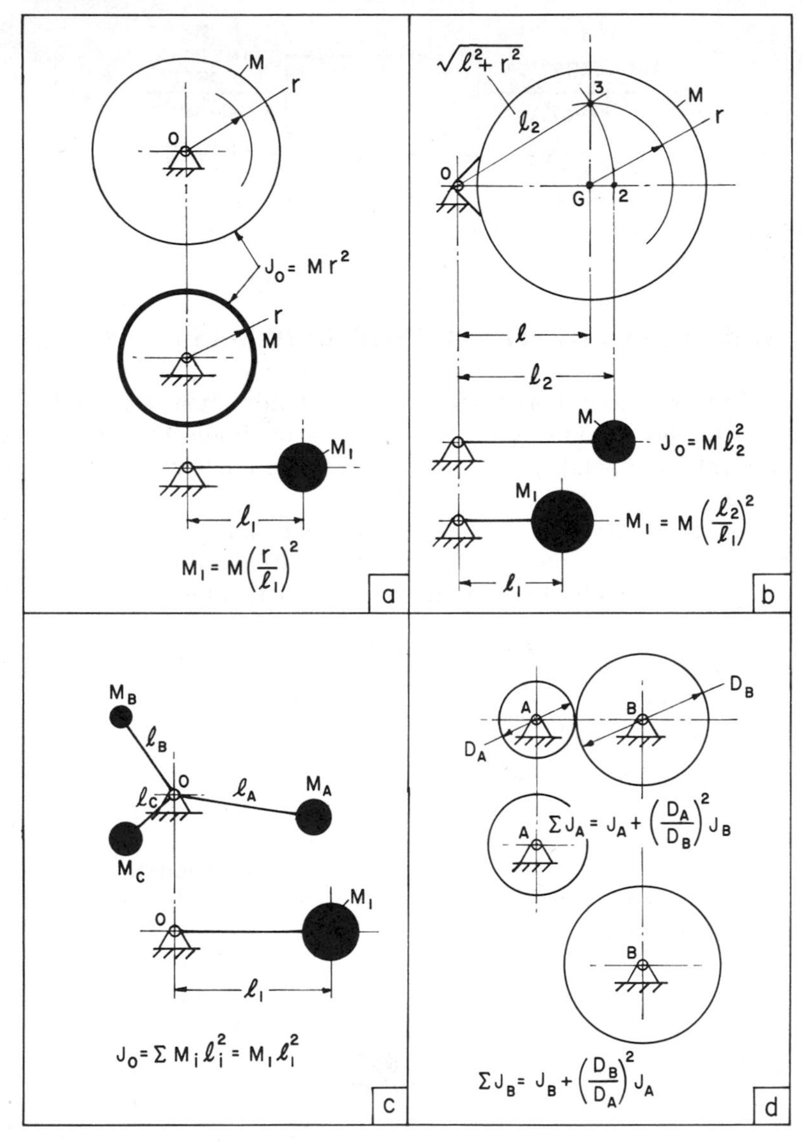

This condition is seen in Table 1.1*b*, with l_2 obtainable geometrically as the hypotenuse of a right triangle, $\sqrt{r^2 + l^2}$. Several equivalent combinations are also shown in the table; in these the same moment of inertia is maintained about the pivot, and l_1 is any selected distance.

We note the interchange possible from rotation to translational coordinates when relatively small vibratory motions are involved, and the x coordinates denote a tangential displacement. We also note that rotational mass equivalence depends on radial distance squared.

$$J_0 = Ml_2^2 = M_1 l_1^2$$

A further discussion of this subject will be found in Chapter 8, where the equivalent two-mass link is discussed. This refinement is necessary if the mass reaction at the pivot is to be considered.

1.8. VIBRATORY KINETIC ENERGY

A mass element translating with harmonic motion (Figure 1.7) has kinetic energy varying as the square of the sinusoidal velocity. With maximum initial displacement,

$$x = x_0 \cos \omega t$$

$$\dot{x} = (-\omega x_0) \sin \omega t$$

$$\mathrm{KE} = \tfrac{1}{2} M (\dot{x})^2 = \tfrac{1}{2} M (\omega x_0)^2 \sin^2 \omega t$$

$$= \left(\frac{M \omega^2 x_0^2}{4} \right) (1 - \cos 2\omega t) \tag{1.8}$$

Thus the kinetic energy variation cycles at double the vibratory frequency (Figure 1.9*b*). This energy is zero at extreme displacement, $t = 0$, and maximum at the midposition, $\omega t = \pi/2$, and equal to $(M\omega^2 x_0^2/2)$. It is proportional to the square of the frequency and the square of the amplitude.

Angular kinetic energy developed in vibratory torsional inertias becomes

$$\mathrm{KE} = \left(\frac{M r^2 \omega^2 \theta_0^2}{4} \right) (1 - \cos 2\omega t) \tag{1.9}$$

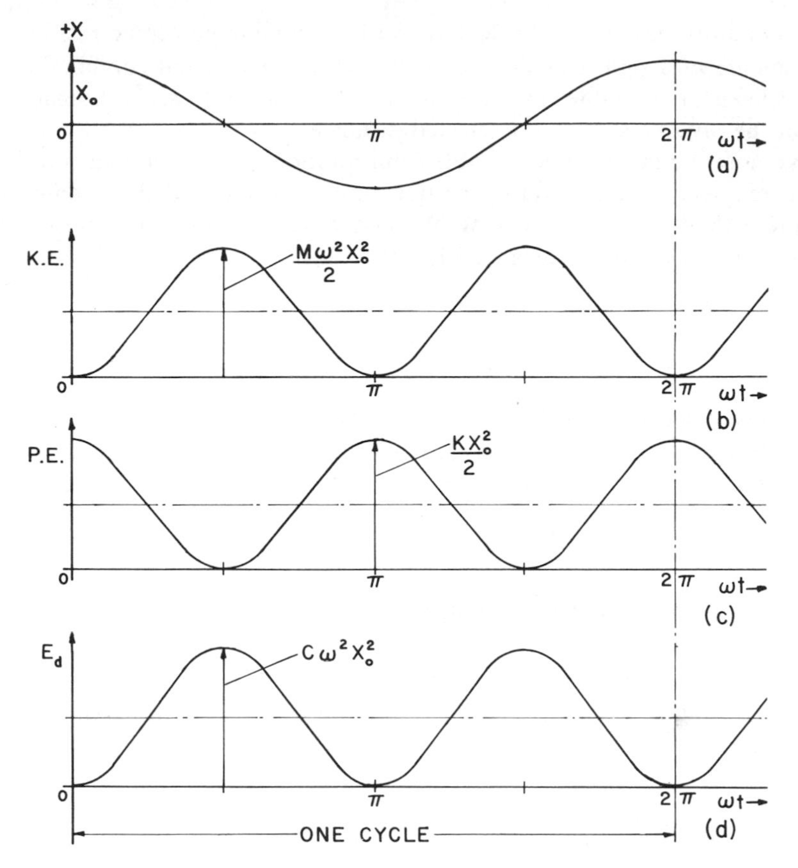

FIGURE 1.9. Absolute energy levels fluctuate harmonically at double frequency (*b*) and (*c*), kinetic and potential respectively. Rate of energy dissipation by a viscous dashpot also varies harmonically at double frequency (*d*).

1.9. THE VIBRATORY TRANSLATIONAL SPRING

Stiffness of an elastic element with respect to the deflection caused by a load in a given direction is often defined as *spring rate*, or the force required to produce unit deflection. This factor is more correctly defined as the slope of the load-deflection curve, although typically it is a constant to the proportional limit of the material under stress (Figure 1.10).

Determination of deflection characteristics is based on strength of materials methods and formulas, although experimental data are sometimes available. The analytical techniques can cover a wide range of mechanical systems including the ordinary helical spring, beams, and more involved elastic structures.

Whereas vibratory inertia forces relate to accelerations, spring forces relate to displacements. There is the feature of opposition, however, in both cases. In

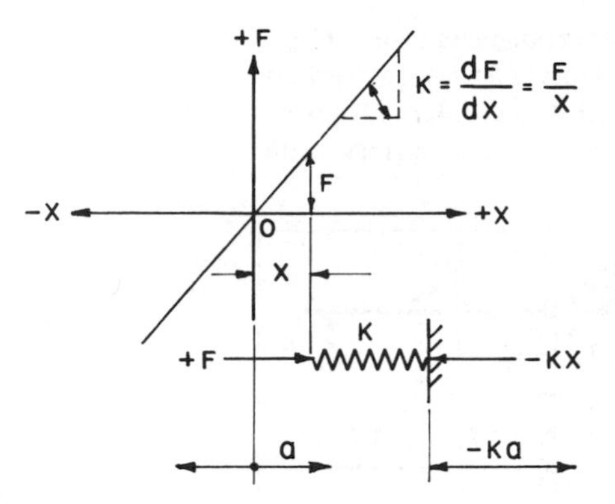

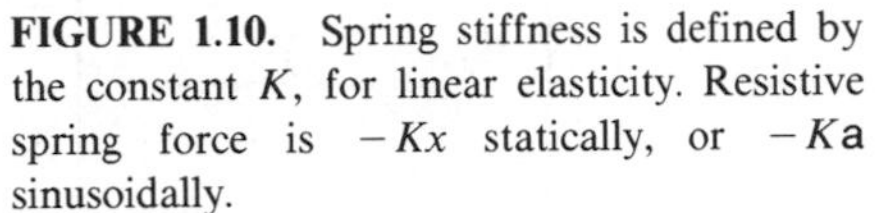

FIGURE 1.10. Spring stiffness is defined by the constant K, for linear elasticity. Resistive spring force is $-Kx$ statically, or $-K\mathrm{a}$ sinusoidally.

Figure 1.10 a force to the right produces a deformation to the right, but the spring resists the deformation, developing a force in static loading of $-Kx$. Similarly with harmonic motion the sinusoidal force developed in the spring is $-K\mathrm{a}$, where a is the complex variable representing the displacement.

1.10. THE VIBRATORY TORSIONAL SPRING

In angular coordinates the most fundamental example of the elastic element is a cylindrical shaft twisted about its longitudinal axis (Figure 1.11). As with the inertia elements, there is a convenient and important correspondence between translational and rotational elasticity (Table 1.2).

The basic conversion in Table 1.2*a* results from considering the torque about the pivot as equivalent to a force of (T_0/l_1) at radius l_1, and a tangential displacement of $(l_1\theta)$. Then we have the equivalent translational tangential

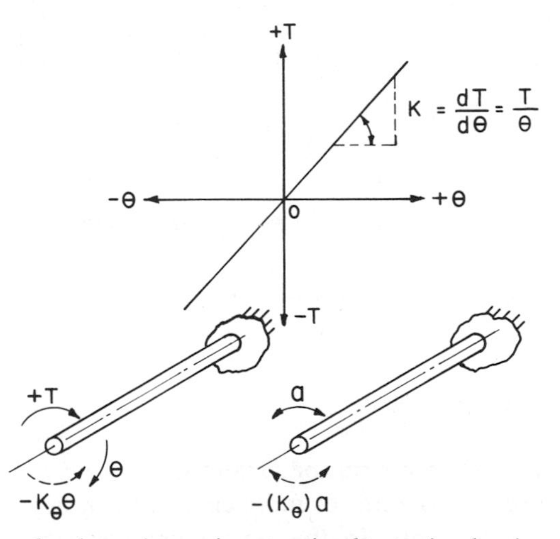

FIGURE 1.11. Torsional spring resistance is proportional to torsional spring rate and to angular displacement, static or vibratory.

TABLE 1.2. Various planar arrangements of springs: (*a*) simple torsional equivalents; (*b*) several springs on rigid, straight link, pivoted; (*c*) angularly disposed tangential springs; (*d*) two torsional springs with interposed gears.

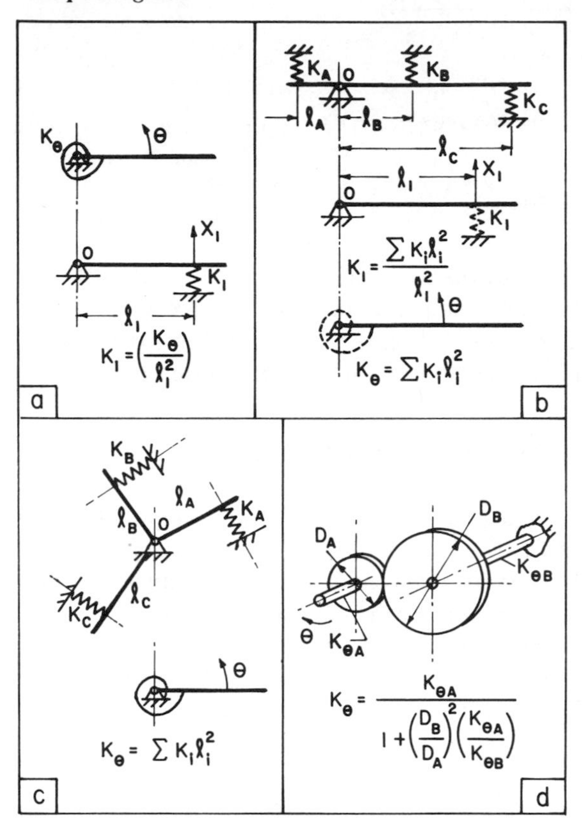

spring,

$$K_1 = \frac{F_1}{x_1} = \frac{T_0/l_1}{l_1\theta} = \frac{T_0/\theta}{l_1^2} = \frac{K_\theta}{l_1^2} \tag{1.10}$$

Other combinations of multiple springs are also indicated in Table 1.2.

1.11. COMBINED SPRINGS

Two or more spring elements can be arranged in *parallel* (Figure 1.12*a*) or in *series* (Figure 1.12*b*). Also, three or more elements can be combined in various series–parallel arrangements. In the simple parallel case all springs have a

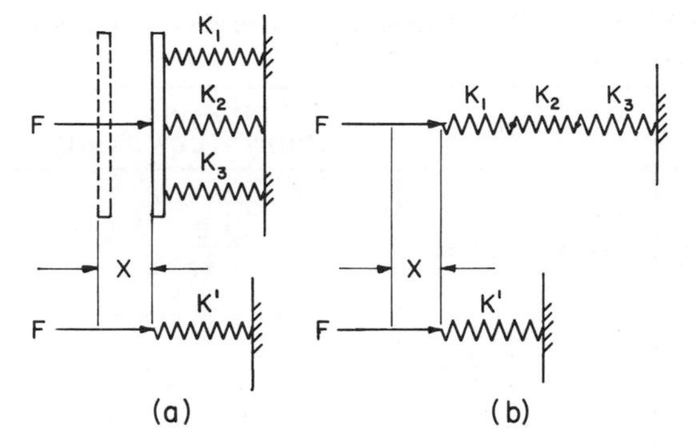

FIGURE 1.12. Springs in parallel (a) and in series (b) can be combined into a single equivalent spring.

TABLE 1.3. Translational springs in various combinations, including simple elastic beams.

	CASE	SPRING CONSTANT
a		$\frac{AE}{\ell}$
b	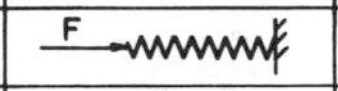	$\frac{Gd^4}{8nD^3}$ 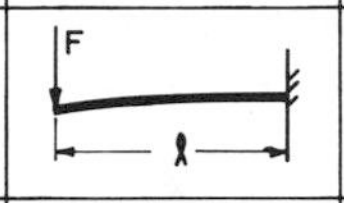
c	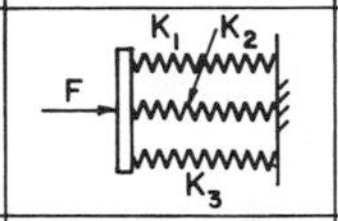	$\sum K_i$ (PARALLEL) 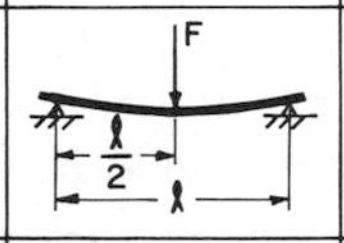
d	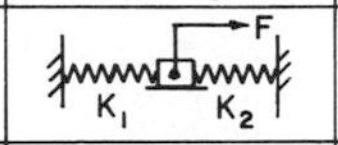	$K_1 + K_2$ (INDEPENDENT OF PRELOAD) 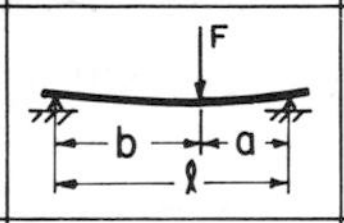
e	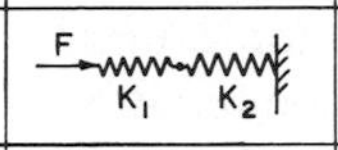	$\frac{1}{\sum \frac{1}{K_i}}$ (SERIES) 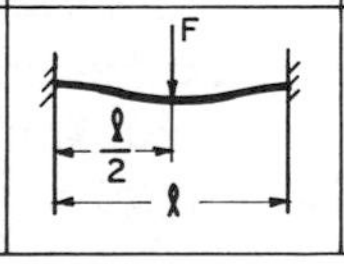
f	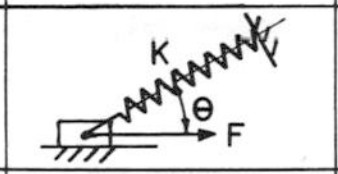	$K \cos^2\theta$
g	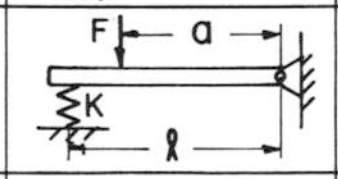	$K\left(\frac{\ell}{a}\right)^2$
h	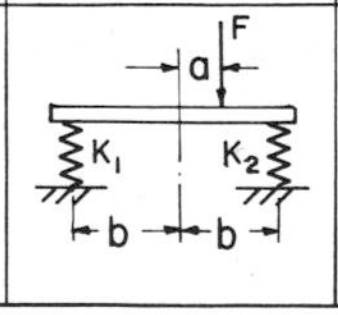	$\frac{4K_1K_2}{K_1\left(1+\frac{a}{b}\right)^2 + K_2\left(1\ \frac{a}{b}\right)^2}$
i		$\frac{3K}{1+\frac{3}{2}\left(\frac{a}{b}\right)^2}$
j		$\frac{3EI}{\ell^3}$
k		$\frac{48EI}{\ell^3}$
l		$\frac{3EI}{\ell^3}\left(\frac{\ell}{a}\right)^2\left(\frac{\ell}{b}\right)^2$
m		$\frac{192EI}{\ell^3}$

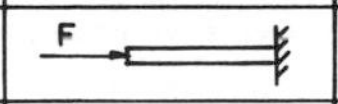

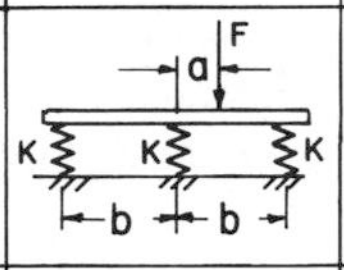

G = SHEAR MODULUS　n = NUMBER OF ACTIVE COILS
d = WIRE DIAMETER　E = YOUNG'S MODULUS OF ELASTICITY
D = COIL DIAMETER　I = AREA MOMENT OF INERTIA OF BEAM

TABLE 1.4. Torsional spring arrangements for several basic cases.

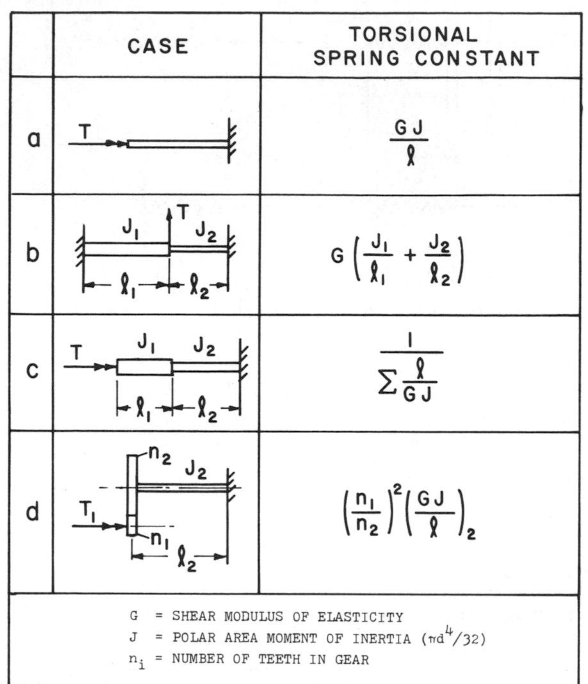

	CASE	TORSIONAL SPRING CONSTANT
a		$\frac{GJ}{\ell}$
b		$G\left(\frac{J_1}{\ell_1} + \frac{J_2}{\ell_2}\right)$
c		$\frac{1}{\sum \frac{\ell}{GJ}}$
d		$\left(\frac{n_1}{n_2}\right)^2 \left(\frac{GJ}{\ell}\right)_2$

G = SHEAR MODULUS OF ELASTICITY
J = POLAR AREA MOMENT OF INERTIA ($\pi d^4/32$)
n_i = NUMBER OF TEETH IN GEAR

common deflection and share the load, and spring constants are additive. In series springs have a *common load* with individual deflections additive.

$$x = x_1 + x_2 + \cdots = \frac{F}{K_1} + \frac{F}{K_2} + \cdots = F\left(\frac{1}{K_1} + \frac{1}{K_2} + \cdots\right)$$

$$\frac{1}{K'} = \frac{x}{F} = \left(\frac{1}{K_1} + \frac{1}{K_2} + \cdots\right) = \sum \frac{1}{K_i} \tag{1.11}$$

Tables 1.3 and 1.4 provide equations for spring constants for a number of fundamental geometries including several series and parallel combinations.

1.12. VIBRATORY POTENTIAL ENERGY

The spring elements just discussed, assumed weightless, involve no kinetic energy; however, the elastic deformations correspond to *potential* energy in the form of *strain* energy. As with kinetic, harmonic potential energy fluctuates at double frequency (Figure 1.9*c*).

$$\begin{aligned} \text{PE} &= \tfrac{1}{2}Kx^2 = \tfrac{1}{2}K(x_0 \cos \omega t)^2 \\ &= \left(\frac{Kx_0^2}{4}\right)(1 + \cos 2\omega t) \end{aligned} \tag{1.12a}$$

In torsion the energy becomes, similarly,

$$\mathrm{PE} = \left(\frac{K_\theta \theta_0^2}{4}\right)(1 + \cos 2\omega t) \tag{1.12b}$$

We see that the vibratory potential energy is a maximum at maximum displacement and zero when the vibratory spring displacement is zero, or at midposition.

1.13. THE VIBRATORY DASHPOT

An important feature of all vibratory systems is the capacity to dissipate energy to some degree. This characteristic is more difficult to identify and quantify than mass or spring elements. It usually results from distributed system damping, often including hysteretic material effects and fluid damping wherever induced by vibratory frictional motion. Although most types of physical damping are complex and nonlinear, they often behave approximately as simple equivalent viscous dashpots in vibratory systems.

The viscous dashpot element (Figure 1.13) resembles the automotive shock absorber. It represents a condensation of all sources of system damping into a single unit with assumed Newtonian viscous shear and a shear force of

$$\begin{aligned} P &= -\left(\mu \frac{A}{h}\right)\dot{x} = -C\dot{x} = (C\omega x_0)\sin \omega t \\ &= -C(j\omega a) = (-jC\omega)a \end{aligned} \tag{1.13}$$

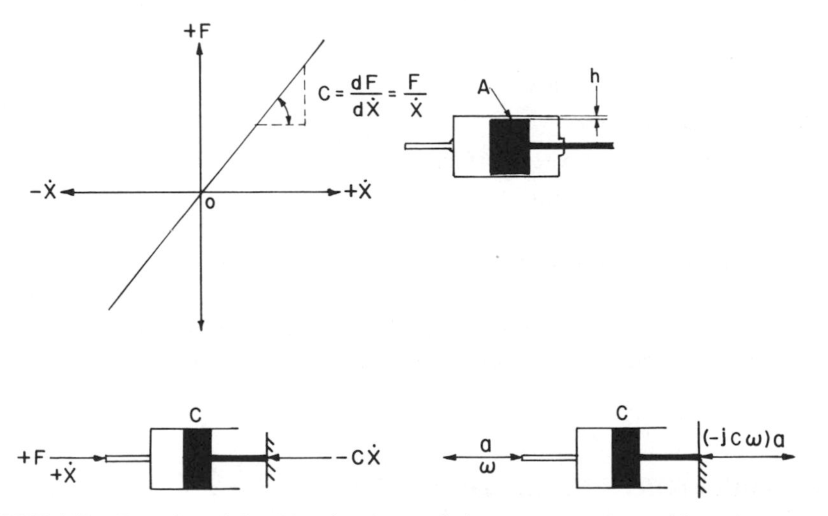

FIGURE 1.13. Damping, defined by the viscous dashpot constant C, provides a force proportional to the radial clearance.

TABLE 1.5. Equations of translational–rotational equivalence for dashpot elements, and the parallel condition.

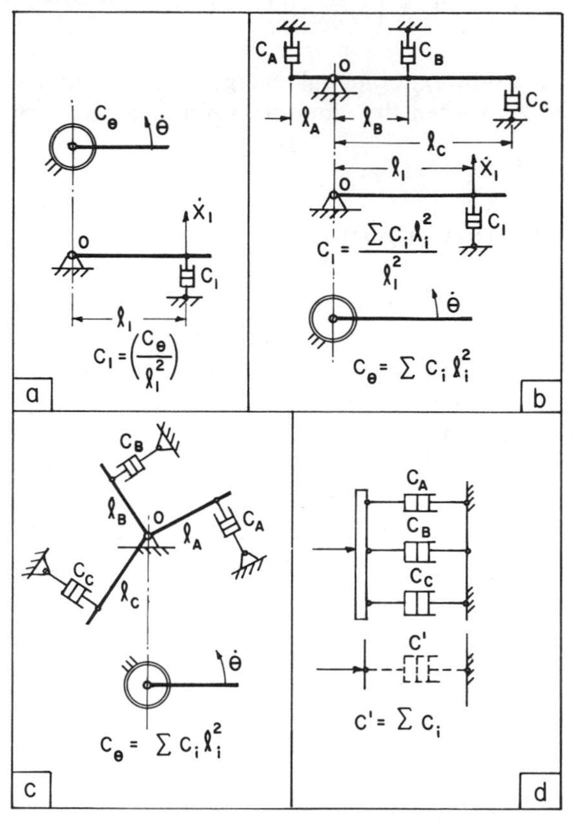

where μ = absolute viscosity
A = area in shear
h = clearance between surfaces
C = dashpot constant

We see from the complex form and from Figure 1.13 that the resistive damping force on the ungrounded end of the dashpot is out of phase with the vibratory velocity. As shown in Table 1.5, the viscous dashpot can be applied in torsion, and in various combinations. Series arrangements, however, are not meaningful.

1.14. VIBRATORY ENERGY DISSIPATION

The dashpot element with sinusoidal motion does not have instantaneous absolute energy level, as with a mass or spring. Rather, it provides energy dissipation by continuously converting energy to heat. The modulating rate at which this occurs (E_d) is obtained from the product of the vibratory velocity and the variable viscous force.

$$E_d = (C\dot{x})\dot{x} = C(\omega x_0)^2 \sin^2\omega t$$
$$= \left(\frac{C\omega^2 x_0^2}{2}\right)(1 - \cos 2\omega t) \tag{1.14}$$

This function is shown in Figure 1.9*d* with the maximum rate occurring at midposition or at maximum force and velocity, also at double frequency. The average rate is the mean ordinate ($C\omega^2 x_0^2/2$), which is the equivalent rate of continuous or constant energy dissipation, representing the rate of power loss. On a per-cycle basis the work done by the dashpot (W_d) becomes

$$W_d = \left(\frac{C\omega^2 x_0^2}{2}\right)\left(\frac{2\pi}{\omega}\right) = \pi C\omega x_0^2 \tag{1.15}$$

The energy rate is typically extremely small, but it is an index of the heat generation and of the capacity of a system to resist excessive vibratory response.

1.15. VIBRATORY FORCE SUMMARY

The three types of forces associated with classical mass, spring, and damping components have been discussed. Each is quantified by a factor M, K, or C, respectively. With sinusoidal motion, forces induced in these elements vary sinusoidally, and they can be expressed as complex variables:

$$\text{Inertia force} = M\omega^2 \mathsf{a}$$
$$\text{Damping force} = -jC\omega \mathsf{a}$$
$$\text{Spring force} = -K\mathsf{a}$$

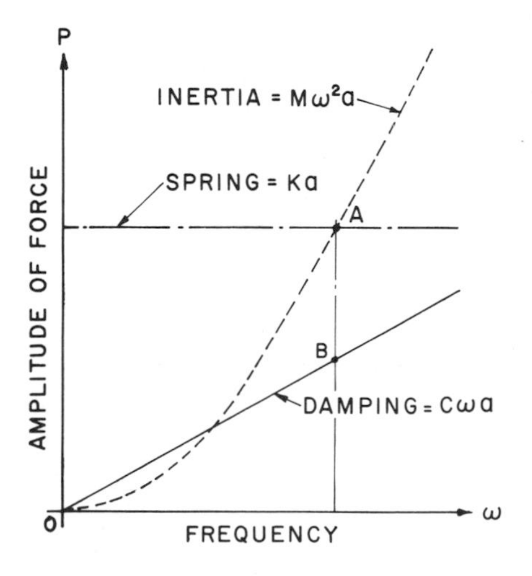

FIGURE 1.14. For a given harmonic amplitude a, maximum mass and damping related forces depend on the harmonic frequency. All three are directly proportional to amplitude.

TABLE 1.6. Suggested symbols and dimensional units, with SI in parentheses. When symbol duplication exists, proper identification results from specified units.

<table>
<tr><th></th><th>QUANTITY</th><th colspan="2">TRANSLATION</th><th colspan="2">ROTATION</th></tr>
<tr><td rowspan="3">SCALAR ELEMENTS</td><td>MASS</td><td>M</td><td>lb,
lb. sec^2/in.
(kg)</td><td>J,
Mr^2</td><td>lb.in^2
lb. in. sec^2
($kg \cdot m^2$)</td></tr>
<tr><td>SPRING</td><td>K</td><td>lb/in.
(N/m)</td><td>K_θ,
K</td><td>in. lb/rad
($\frac{N \cdot m}{r}$)</td></tr>
<tr><td>DAMPING</td><td>C</td><td>lb. sec/in.
($\frac{N \cdot s}{m}$)</td><td>C_θ,
C</td><td>lb. in. sec
($N \cdot m \cdot s$)</td></tr>
<tr><td rowspan="7">COMPLEX VARIABLES (SINUSOIDAL)</td><td>DISPLACEMENT</td><td>a</td><td>in.
(m)</td><td>a</td><td>rad
(r)</td></tr>
<tr><td>VELOCITY</td><td>$j\omega a$</td><td>in./sec
(m/s)</td><td>$j\omega a$</td><td>rad/sec
(r/s)</td></tr>
<tr><td>ACCELERATION</td><td>$-\omega^2 a$</td><td>in./sec^2
(m/s^2)</td><td>$-\omega^2 a$</td><td>rad/sec^2
(r/s^2)</td></tr>
<tr><td>INERTIA FORCE</td><td>$M\omega^2 a$</td><td rowspan="4">lb
(N)</td><td>$J\omega^2 a$,
$Mr^2\omega^2 a$</td><td rowspan="4">in. lb
(N.m)</td></tr>
<tr><td>SPRING FORCE</td><td>-Ka</td><td>$-K_\theta a$,
$-K\ell^2 a$</td></tr>
<tr><td>DASHPOT FORCE</td><td>$-jC\omega a$</td><td>$-jC_\theta \omega a$
$-jC\ell^2\omega a$</td></tr>
<tr><td>APPLIED FORCE</td><td>P</td><td>T,
$P\ell$</td></tr>
</table>

ω = Circular Frequency = $2\pi f$, rad/sec. r = Radius of Gyration
f = Sinusoidal Frequency, cps, Hz (r) = Radians
j = Complex Operator ℓ = Reference Radius

TABLE 1.7. Conversion factors for relating SI to traditional engineering units for parameters commonly used in vibration analysis. Also see Table 2.2.

QUANTITY	SI		BRITISH	
LENGTH	0.0254 *	m	39.37 **	in.
AREA	$645(10)^{-6}$	m^2	1550	$in.^2$
VOLUME, SECTION MODULUS V I/c, J/r	$16.39(10)^{-6}$	m^3	61,024	$in.^3$
AREA MOMENT OF INERTIA I, J	$0.4162(10)^{-6}$	m^4	$2.403(10)^6$	$in.^4$
VELOCITY	0.0254	m/s	39.37	in./sec
ACCELERATION	0.0254	m/s^2	39.37	$in./sec^2$
FORCE	4.4482	N	0.2248	lb
TORQUE, MOMENT, TORSIONAL SPRING	0.1130	N·m	8.851	lb in.
BEAM, SHAFT STIFFNESS EI, GJ	$2870(10)^{-6}$	$N{\cdot}m^2$	348.5	lb $in.^2$
LINEAR SPRING	175.1	N/m	$5710(10)^{-6}$	lb/in.
PRESSURE, MODULUS p E, G	6895	N/m^2	$145(10)^{-6}$	$lb/in.^2$
TORSIONAL DASHPOT	0.1130	N·m·s	8.851	lb in. sec
LINEAR DASHPOT	175.1	N·s/m	$5710(10)^{-6}$	lb sec/in.
MASS	0.4536	kg	2.2046	lb
UNBALANCE	0.01152	kg·m	86.79	lb in.
MASS MOMENT OF INERTIA	$292.6(10)^{-6}$	$kg{\cdot}m^2$	3417	lb $in.^2$
DENSITY	27,678	kg/m^3	$36.13(10)^{-6}$	$lb/in.^3$
WORK, ENERGY	0.1130	J	8.851	in. lb
POWER	0.7463	kw	1.34	H.P.

* Multiply British values by these factors to obtain SI equivalent.

** Multiply SI values by these factors to obtain British equivalent.

Obviously all forces are directly proportional to displacement and to the parameters M, C, and K; however, the frequency effect is quite different. As seen in Figure 1.14, the behavior with frequency involves exponents of ω of 2, 1, and 0, respectively. Point A indicates resonance with the ordinate B the corresponding amplitude of the damping force.

Phase relationships of these three quantities are always in quadrature, deriving from the vectors of Figure 1.5.

1.16. DIMENSIONAL DATA

With two major systems of units in use, it is desirable to accommodate both conventional and SI units in numerical calculations. Tables 1.6 and 1.7 indicate dimensional quantities applicable to most vibration problems; however, variations are occasionally necessary. Examples are worked using both systems.

Fortunately, much analysis can be developed using dimensionless ratios, as will be done in the following chapters. In any ratio it is only necessary that numerator and denominator have consistent units. This dimensionless approach minimizes the problem of units and increases the generality of the results.

Although there are considerable variations in material constants depending on the particular alloy involved, Table 1.8 provides approximate values. SI units are selected in combinations most applicable to vibratory calculations.

TABLE 1.8. Approximate material constants for common metals. Although unusual dimensionally, SI units are compatible with volume and area moment of inertia terms in centimeters.

MATERIAL	DENSITY		MODULUS OF ELASTICITY			
			$E(10)^{-6}$		$G(10)^{-6}$	
	kg/cm^3	lb/in.^{3}	N/cm^2	lb/in.^{2}	N/cm^2	lb/in.^{2}
ALUMINUM ALLOYS	0.0028	0.10	6 - 7	9 - 10	3 - 4	4 - 6
BRASSES	0.0083	0.30	9 - 10	13 - 15	3.5 - 4	5 - 6
BRONZES	0.0083	0.30	9 - 10	13 - 15	3.5 - 4	5 - 6
CAST IRON	0.0077	0.28	10 - 12	14 - 18	3.5 - 5	5 - 7
MAGNESIUM ALLOYS	0.0018	0.065	4.5	6.5	1.7	2.5
STEELS	0.0080	0.29	20	30	8	12

EXAMPLES

EXAMPLE 1.1. A sinusoidal motion occurs at a displacement amplitude of 2 in. and at a frequency of 5 Hz. Initial displacement is maximum positive. Find:

(a) The instantaneous displacement at an elapsed time of 0.08 sec.

(b) The corresponding velocity.

(c) The corresponding acceleration.

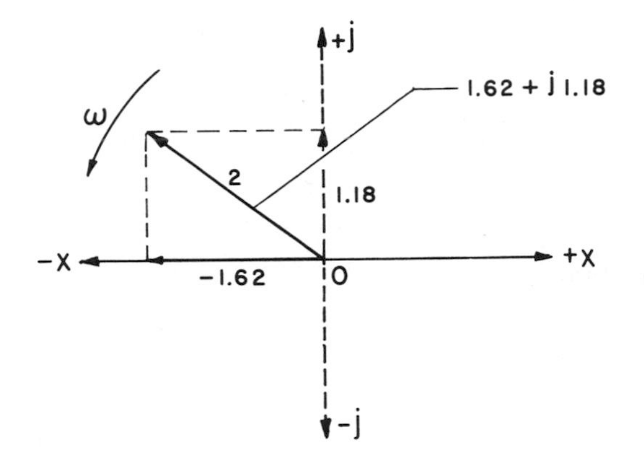

Solution

(a)
$$\omega = 2\pi f = 2\pi 5 = 31.4 \text{ rad/sec}$$

$$\omega t = 31.4(0.08) = 2.51 \text{ rad} = 144°$$

$$x = x_0 \cos \omega t = 2(-0.809) = -1.62 \text{ in.}$$

(b)
$$dx/dt = \dot{x} = -x_0 \omega \sin \omega t$$

$$= -2(31.4) \sin 144° = -36.9 \text{ in./sec}$$

(c)
$$d^2x/dt^2 = \ddot{x} = -x_0 \omega^2 \cos \omega t$$

$$= -2(31.4)^2 \cos 144° = +1595 \text{ in./sec}^2$$

EXAMPLE 1.2. A second harmonic displacement is added to that of Example 1.1, with an amplitude of 3 in. at exactly the same frequency; however, initial displacement is zero, with displacement then increasing positively. Determine:

(a) The amplitude of the resulting displacement.

(b) The elapsed time to maximum positive displacement.

(c) The elapsed time to the first midposition condition.

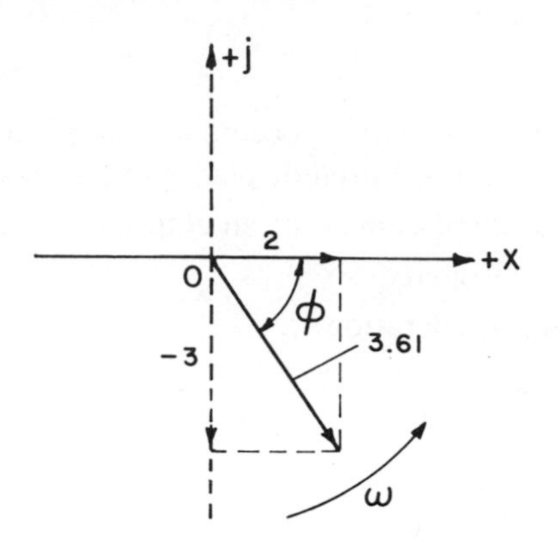

Solution

(a) From Equation 1.3,

$$x = 2\cos\omega t + 3\sin\omega t$$

$$x_0 = \sqrt{2^2 + 3^2} = 3.61 \text{ in.}$$

(b) The rotating vector first crosses the positive real axis when

$$\omega t = \phi = \tan^{-1}\left(\frac{3}{2}\right) = 0.983 \text{ rad} = 56.3°$$

$$t = \frac{\phi}{\omega} = \frac{0.983}{31.4} = 0.0313 \text{ sec}$$

(c) It first crosses the j axis when

$$\omega t = \phi + 90° = 0.983 + \frac{\pi}{2} = 2.55 \text{ rad}$$

$$t = \frac{2.55}{31.4} = 0.0813 \text{ sec}$$

EXAMPLE 1.3. Express the displacement, velocity, and acceleration of the results of Example 1.2 as complex variables.
21-12

Solution

$$\text{Displacement} = \mathbf{a}_1 + \mathbf{a}_2 = 2 - j3 \text{ in.}$$

$$\text{Velocity} = j\omega(2 - j3) = 31.4(3 + j2) \text{ in./sec}$$

$$\text{Acceleration} = -\omega^2(2 - j3) = (31.4)^2(-2 + j3)$$

$$= -1972 + j2958 \text{ in./sec}^2$$

EXAMPLE 1.4. A disk oscillates harmonically with an amplitude of 1.7° at a frequency of 22 Hz. Find the complex representation for the following:

(a) Displacement.
(b) Velocity.
(c) Acceleration.

Solution

(a) $\mathsf{a} = 1.7/57.3 = 0.0297$ rad
Understood to be maximum positive at $t = 0$, and at a circular frequency of $22(2\pi) = 138.2$ rad/sec.

(b) $j\omega \mathsf{a} = j[22(2\pi)]\mathsf{a} = j(138.2)\mathsf{a}$ r/s

(c) $-\omega^2 \mathsf{a} = -(138.2)^2 \mathsf{a}$ rad/sec^2 $= -567$ rad/sec^2

EXAMPLE 1.5. A reciprocating mass of 4.8 kg (Figure 1.7) has a vibratory frequency of 12 Hz. Amplitude is 0.3 cm. Find:

(a) The amplitude of the inertia force.
(b) The amplitude of the kinetic energy.

Solution

(a) $P = M\omega^2 x_0 = 4.8[2\pi(12)]^2(0.003) = 81.9$ N

(b) From Equation 1.8,

$$\text{KE} = \frac{4.8(24\pi)^2(0.003)^2\,2}{4} = 0.123 \text{ J}$$

EXAMPLE 1.6. A round steel bar in torsion (Figure 1.11) is 3.2 cm in diameter and 50 cm long. The free end experiences a harmonic displacement of 2° at 40 Hz. Find:

(a) The amplitude of the applied torque.
(b) The amplitude of the strain (potential) energy developed in the bar.

Solution

(a) Using Table 1.4*a*, and with $G = 8(10)^6$ N/cm^2,

$$K_\theta = \frac{GJ}{l} = \frac{8(10)^6(10.3)}{50} = 1.65(10)^6 \text{ N} \cdot \text{cm/r}$$

$$\text{T} = \text{K}_\theta \theta = 1.65(10)^6\left(\frac{2}{57.3}\right) = 58{,}000\ N \cdot cm = 580\ N \cdot m$$

(b) From Equation 1.12b,

$$\text{PE} = \frac{K_\theta \theta^2}{4}(2) = \frac{16{,}500(2/57.3)^2}{2} = 10.1 \text{ J}$$

EXAMPLE 1.7. A dashpot with a viscous coefficient of 3.5 lb sec/in. is located on a rigid pivoted link as shown. The amplitude at 1 is 0.60 in. at 20 Hz. Calculate:

(a) The amplitude of the equivalent damping force at 1.
(b) The equivalent torsional dashpot constant at 1.
(c) The average rate of energy dissipation.
(d) The horsepower (HP) of the energy dissipation.
(e) The energy dissipated per cycle.

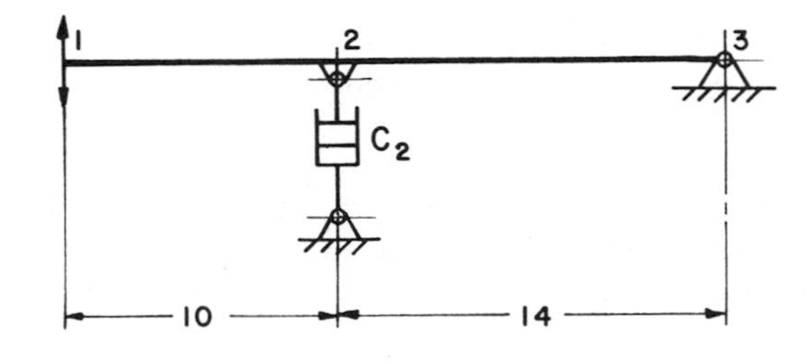

Solution. From Table 1.5*b*,

(a) $$C_1 = C_2\left(\frac{l_2}{l_1}\right)^2 = 3.5\left(\frac{14}{24}\right)^2 = 1.19 \text{ lb sec/in.}$$

$$\omega = 2\pi f = 2\pi(20) = 125.6 \text{ rad/sec}$$

$$P_1 = C_1\omega a_1 = 1.19(125.6)(0.60) = 89.7 \text{ lb}$$

(b) $$C_\theta = C_1 l_1^2 = C_2 l_2^2 = 1.19(24)^2 = 3.5(14)^2 = 685 \text{ lb in. sec}$$

(c) From Equation 1.14,

$$\frac{C_1\omega^2 x_1^2}{2} = \frac{1.19(125.6)^2(0.6)^2}{2} = 3380 \text{ in. lb/sec}$$

(d) $\text{HP} = \dfrac{3380}{12(550)} = 0.51$

(e) Work per cycle $= 3380T = 3380(2\pi/\omega) = 169$ in. lb/cycle

2 STEADY-STATE RESPONSE OF SINGLE-DEGREE SYSTEMS

Vibration problems may be broadly categorized as *free* or *forced*. There is some logic in presenting the topic of free, or transient, response initially, and then developing forced response; however, a sinusoidal forcing function on an elementary system is in some respects simpler conceptually and analytically than the free. Also, steady-state phenomena at constant machine speeds are predominant engineering-type vibration difficulties and are therefore introduced first. The elements considered independently in the previous chapter are now assembled and excited in various combinations.

2.1. DEGREES OF FREEDOM

Vibratory spring–mass systems are characterized by the number of discrete masses and by the corresponding number of coordinates necessary to specify the position of all masses at any instant. This number represents the *degrees of freedom*. Typically the coordinates are linear in a translational system and angular in a torsional system, but sometimes both occur in the same system. Freedom to translate, or to rotate indefinitely in a given direction rigidly with no spring displacement and all masses having identical displacement, constitutes a degree of freedom not possible in systems connected to ground.

Once determined, the number of degrees of freedom usually corresponds numerically to the number of *natural frequencies*, and each frequency in turn will exhibit a characteristic *natural mode* or deflection distribution. Thus the same integer specifies the degrees of freedom, natural frequencies, and natural modes. In this chapter this integer is unity for the single-mass cases.

2.2. FREE VIBRATION WITHOUT DAMPING

Figure 2.1 shows the most basic of all single-degree systems with one mass and one grounded spring. Assuming release from the position of spring compres-

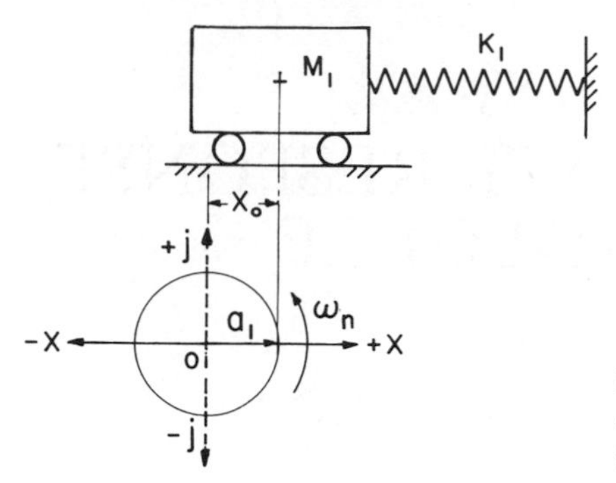

FIGURE 2.1. The simple spring–mass system oscillating in free vibration at its natural frequency is shown at the extreme positive displacement.

sion, the mass oscillates horizontally with simple harmonic motion $x_1 = x_0 \cos \omega t$, which satisfies the differential equation for equilibrium of horizontal forces.

$$\Sigma F_x = -M_1 \ddot{x}_1 - K_1 x_1 = 0$$

$$M_1 \omega^2 x_0 \cos \omega t - K_1 x_0 \cos \omega t = 0$$

$$\omega_n = \omega_1 = \sqrt{\frac{K_1}{M_1}} \tag{2.1}$$

where ω_1 is the unique natural frequency of the system. It is independent of amplitude and time. Or, using the complex notation and Figures 1.7 and 1.10, now that sinusoidal motion is established,

$$M_1 \omega_1^2 \mathbf{a}_1 - K_1 \mathbf{a}_1 = 0$$

$$\omega_n = \omega_1 = \sqrt{\frac{K_1}{M_1}}$$

These equations relate to the equilibrium of spring and inertia forces. On the basis of conservation of energy, Equations 1.8 and 1.12a can be used to equate the maximum potential energy at maximum displacement to the maximum kinetic energy at midposition for the same result.

$$\frac{K_1 x_0^2}{2} = \frac{M_1 \omega_1^2 x_0^2}{2}$$

$$\omega_1 = \sqrt{\frac{K_1}{M_1}}$$

The subscript 1 applied to the spring, mass, and frequency appears unnecessary and most common notation for natural frequency is ω_n; however, this subscript is used to be consistent with multiple systems to follow. For instance, in Table 2.1*c* the base excitation corresponds to 2, or $\mathbf{a}_2$. Similarly, with several

TABLE 2.1. **Responses of the basic spring–mass system to (*a*) sinusoidal force excitation on the mass, (*b*) centrifugally generated force excitation on the mass, (*c*) displacement excitation by movement of the base, and (*d*) positive sinusoidal displacement *e* between the ground and the mass, contraction positive. E_d is the average constant rate of energy dissipation.**

(a)		(b)		(c)		(d)	
$\frac{a_1}{P_1/K_1}$	$\frac{1}{C+jB}$	$\frac{a_1}{em/M_1}$	$\frac{g^2}{C+jB}$	$\frac{a_1}{a_2}$	$\frac{1+jB}{C+jB}$	$\frac{a_1}{e}$	1
$\frac{P_2}{P_1}$	$\frac{-(1+jB)}{C+jB}$	$\frac{P_2}{me\omega_1^2}$	$\frac{-g^2(1+jB)}{C+jB}$	$\frac{a_{12}}{a_2}$	$\frac{g^2}{C+jB}$	$\frac{P_{e1}}{K_1 e}$	$C+jB$
$\frac{E_d}{P_1^2\omega_1/K_1}$	$\frac{\frac{1}{2}Bg}{C^2+B^2}$	$\frac{E_d}{(me)^2\omega_1^5/K_1}$	$\frac{\frac{1}{2}Bg^5}{C^2+B^2}$	$\frac{P_2}{K_1 a_2}$	$\frac{-g^2(1+jB)}{C+jB}$	$\frac{P_2}{K_1 e}$	$-g^2$
				$\frac{E_d}{a_2^2 K_1\omega_1}$	$\frac{\frac{1}{2}Bg^5}{C^2+B^2}$	$\frac{E_d}{K_1\omega_1 e^2}$	Bg

$C = (1-g^2)$ $\quad B = 2\zeta_1 g$

$g = \left(\frac{\omega}{\omega_1}\right)$ $\quad \omega_1 = \sqrt{\frac{K_1}{M_1}}$ $\quad \zeta_1 = \left(\frac{C_1}{2M_1\omega_1}\right)$ $\quad a_{12} = (a_1 - a_2)$

TABLE 2.2. Coefficient modification required to adapt reference natural frequency calculation to different dimensional combinations of stiffness and mass.

TRANSLATION		TORSION		A	
K_1	M_1	K_1	M_1	rad/sec	Hz (cps)
lb/in.	lb sec^2/in.	in.lb/rad	lb in.sec^2	1	0.159
	lb		lb in.^{2}	19.64	3.127
N/m	kg	N.m/r	kg.m^2	1	0.159
N/cm	kg	N.cm/r	kg.cm^2	10	1.592
kN/m	kg	kN.m/r	kg.m^2	31.62	5.033
N/cm	gm	N.cm/r	gm.cm^2	316.2	50.33

$$\omega_1 = A\sqrt{\frac{K_1}{M_1}}$$

masses or springs subscripts will proceed from left to right and a reference natural frequency for the first spring–mass pair of $\omega_1 = \sqrt{K_1/M_1}$ will be used (Table 4.1).

Fundamental units of the circular frequency ω_1 are radians per second, although cycles per second or hertz (Hz) is usually a more meaningful designation of frequency. Also in Equation 2.1 dimensions in the ratio K_1/M_1 must be consistent, translational or torsional, SI or British, and M_1 must properly incorporate the gravitational constant. Table 2.2 is provided to facilitate the calculation of natural frequency by specifying the factor A which is applied to the various combinations of K_1/M_1 to yield the correct natural frequency in radians per second or Hz.

2.3. UNDAMPED FORCED RESPONSE

Given a sinusoidal force P_1 at forcing frequency ω, understood, on M_1 (Figure 2.2) but neglecting the dashpot force, equilibrium is

$$\Sigma F_x = M_1\omega^2 \mathrm{a}_1 - K_1\mathrm{a}_1 + P_1 = 0 \tag{2.2a}$$

Dividing by K_1,

$$\mathrm{a}_1\left(\frac{\omega}{\omega_1}\right)^2 - \mathrm{a}_1 = -\left(\frac{P_1}{K_1}\right)$$

$$\mathrm{a}_1(1 - \mathrm{g}^2) = \left(\frac{P_1}{K_1}\right) \tag{2.2b}$$

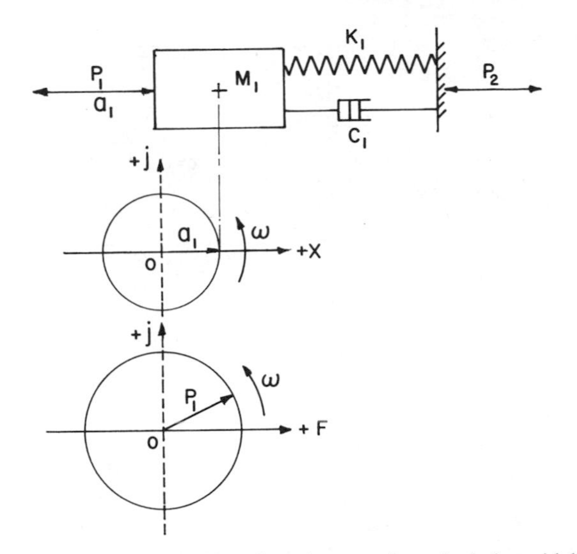

FIGURE 2.2. A forced vibration results when the mass is excited sinusoidally by P_1. Viscous damping, C_1, and ground reaction, P_2, are also shown.

where

$$g = (\omega/\omega_1) = \text{ratio of forcing to natural frequency}$$

$$P_1/K_1 = \text{equilibrium amplitude}$$

Equilibrium amplitude can be interpreted as follows:

1. The deflection of the spring statically with P_1 applied as a constant force.
2. The sinusoidal amplitude of the mass as ω approaches zero, and the inertia effect becomes negligible.
3. The sinusoidal amplitude of the spring at 1 if the mass is removed. Without the inertia effect the sinusoidal amplitude is independent of the frequency of excitation.

However we visualize it, (P_1/K_1) is an important quantity. Although K_1 is a scalar term, P_1 is a complex variable that can be displayed as a rotating vector. As an absolute ratio (P_1/K_1) is the reference displacement with which the vibratory amplitude is compared.

$$\frac{a_1}{(P_1/K_1)} = \frac{1}{1 - \left(\dfrac{\omega}{\omega_1}\right)^2} = \left(\frac{1}{1 - g^2}\right) \tag{2.3}$$

The ratio in Equation 2.3 is often termed the *vibratory magnification factor*, indicating the response of the mass as a function of frequency or of the *forced frequency ratio* g. For low frequencies the factor is nearly one and the static

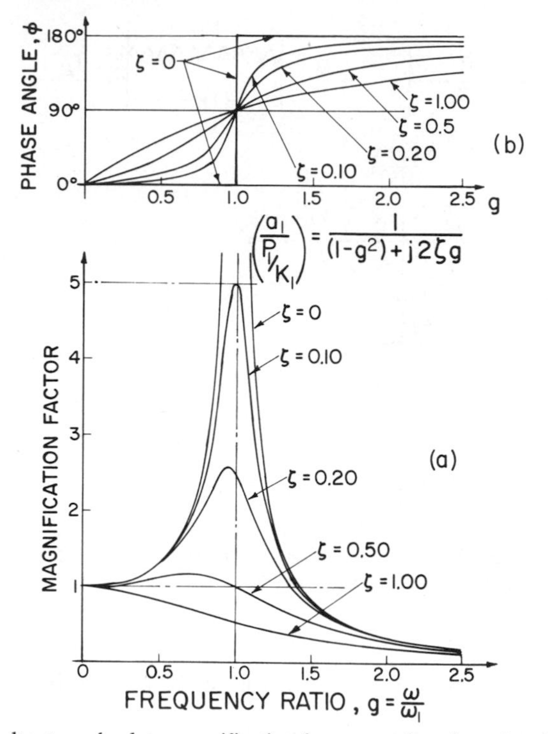

FIGURE 2.3. Resultant or absolute magnification factor as a function of exciting frequency for constant damping (a). Corresponding variations in phase angle are shown in (b).

case is approached. For resonance, $\omega = \omega_1$, and $\mathsf{g} = 1$, and the factor is infinite. At high frequency ratios the factor approaches zero as a limit.

Whenever operation is above resonance the magnification factor is negative, indicating that a_1 and P_1 are out of phase. This requires a reversal of P_1 relative to a_1 at resonance.

Characteristic behavior of Equation 2.3 is shown in Figure 2.3a for $\zeta_1 = 0$, but with the negative sign disregarded above $\mathsf{g} = 1$.

2.4. DAMPED FORCED RESPONSE

The dashpot in Figure 2.2 significantly modifies the previous undamped analysis. It is a source of energy dissipation, limits resonant amplitudes, and creates phase relationships. The damping force on the mass introduces the term $(-jC_1\omega\mathsf{a}_1)$ in Equation 2.2, with the negative sign indicating a force in opposition to the velocity of the mass.

$$\Sigma F_x = M_1\omega^2\mathsf{a}_1 - K_1\mathsf{a}_1 - jC_1\omega\mathsf{a}_1 + P_1 = 0 \tag{2.4}$$

Normalizing by K_1, the terms become displacements dimensionally. Letting

$$\zeta_1 = \frac{C_1\omega}{K_1} = \frac{C_1}{C_c} = \text{damping ratio} \quad \text{(see equation 3.3)}$$

$$C_c = 2M_1\omega_1 = \text{critical damping coefficient}$$

we have

$$a_1(1 - g^2 + j2\zeta_1 g) = \left(\frac{P_1}{K_1}\right) \tag{2.5a}$$

$$\frac{a_1}{(P_1/K_1)} = \frac{1}{(1 - g^2) + j2\zeta_1 g} = \frac{1}{C + jB} \tag{2.5b}$$

Although phase relations are available from Equation 2.5b, the absolute or resultant ratio represents the actual magnitude ratio found from the hypotenuse solution.

$$\left|\frac{a_1}{(P_1/K_1)}\right| = \frac{1}{\sqrt{C^2 + B^2}} \tag{2.5c}$$

Unless otherwise specified, absolute values of this ratio are understood and shown in Figure 2.3*a* for several damping ratios. We see that the peaks at higher damping ratios occur at less than $g = 1$; however, defining resonance as $\omega = \omega_1$, the response at $g = 1$ is

$$\frac{a_1}{(P_1/K_1)} = \frac{1}{j2\zeta_1} \tag{2.6a}$$

$$\left|\frac{a_1}{(P_1/K_1)}\right| = \frac{1}{2\zeta_1} \tag{2.6b}$$

From Equation 2.6a, the j operator indicates a 90° phase relation between a_1 and P_1 at resonance. In Equation 2.6b the magnification factor is inversely proportional to ζ_1.

2.5. PHASE RELATIONSHIPS

Equation 2.5b contains two complex variables, a_1 and P_1, each represented by a rotating vector in the complex plane. Also, (P_1/K_1) is a complex vector in displacement units and both sides of the equation are dimensionless. The phase relation between the response and the forcing function is obtained by constructing these vectors (Figure 2.4). The numerator is positive, unity, and real. The denominator has a real component, $(1 - g^2)$, and a positive j component, $2\zeta_1 g$. As these vectors correspond phasewise to the numerator and denomina-

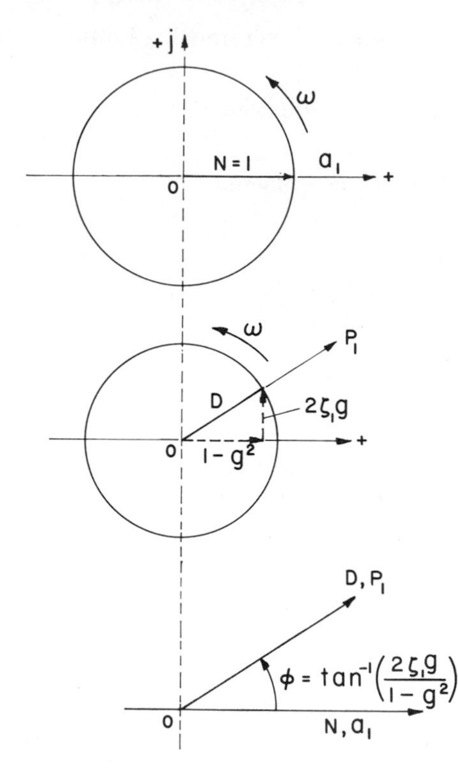

FIGURE 2.4. Response Equation 2.5b is interpreted in the complex plane. Exciting force leads displacement by the same phase angle by which denominator D leads numerator N.

tor on the left of the equation, the angle ϕ is determined, with P_1 leading a_1.

$$\phi = \tan^{-1}\left(\frac{2\zeta_1 \mathsf{g}}{1 - \mathsf{g}^2}\right) \tag{2.7}$$

The angle ϕ can vary from 0 to 180°, and is greater than 90° when $(1 - \mathsf{g}^2)$ is negative, or above resonance. At resonance the real component is zero, and the force leads the displacement by exactly 90°, independent of the damping ratio, as seen in Figure 2.3*b*. The transition at resonance is gradual with large damping, but abrupt at low damping.

2.6. EQUILIBRIUM OF VECTOR FORCES

Returning to Equation 2.4, the complex forces summed to zero are inertia, spring, dashpot, and excitation, respectively. Division of Equation 2.5a by a_1 provides a normalized interpretation of the vectorial equilibrium.

$$\mathsf{g}^2 - 1 - j2\zeta_1\mathsf{g} + \frac{P_1}{K_1\mathsf{a}_1} = 0 \tag{2.8}$$

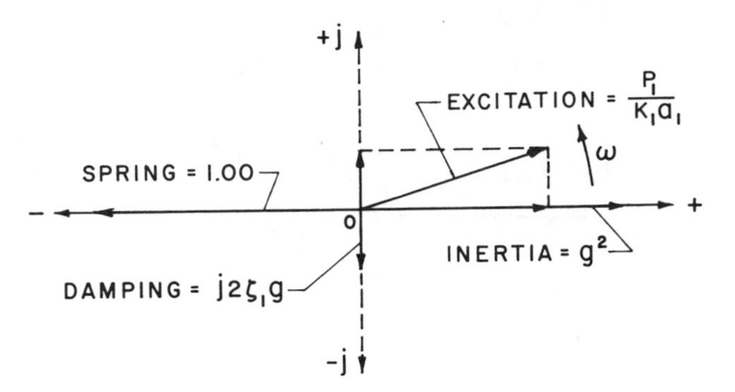

FIGURE 2.5. Equilibrium of normalized force terms (Equation 2.8) results in a resultant vector of zero.

The vectors in Figure 2.5 are dimensionless but proportional to and directionally identical to the several sinusoidal force amplitudes that must sum to zero for equilibrium. On the real axis the real component of $(P_1/K_1\mathsf{a}_1)$ adds to the inertia term g^2 to equal $+1$. The j component of $(P_1/K_1\mathsf{a}_1)$ balances the damping-related term, $j2\zeta_1\mathsf{g}$. The former is in phase and the latter out of phase with the velocity vector, corresponding to energy added by the excitation and dissipated by damping. The mutual opposition of these vectors occurs regardless of the frequency ratio. Above resonance, however, the excitation vector shifts to the second quadrant, with its real component out of phase with the displacement, as the inertia term g^2 exceeds $+1$. At resonance the spring and inertia terms are exactly equal and opposite with the excitation oriented to counter only the dashpot term.

2.7. REACTION TO GROUND

In Figure 2.2 the vibratory system transmits a sinusoidal force P_2 to the base, determined by equilibrium of the two external forces and the inertia load of the mass. Note that the spring and dashpot forces are internal and do not enter the summation.

$$\Sigma F_x = M_1\omega^2\mathsf{a}_1 + P_1 + P_2 = 0 \tag{2.9a}$$

$$\frac{P_2}{P_1} = -\left(\frac{M_1\omega^2}{P_1}\right)\mathsf{a}_1 - 1 = \frac{-(1 + j2\zeta_1\mathsf{g})}{1 - \mathsf{g}^2 + j2\zeta_1\mathsf{g}} = \frac{-(1 + jB)}{C + jB} \tag{2.9b}$$

Figure 2.6 shows equilibrium of the total forces in below-resonant conditions. Phase angles are located from normalized Equation 2.9b. For instance, the P_2 directional follows from the coordinates of the numerator, -1 and $-j2\zeta_1\mathsf{g}$.

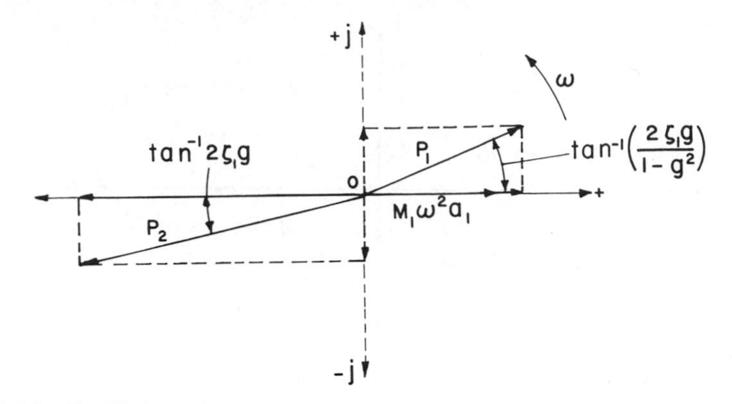

FIGURE 2.6. Equilibrium of actual forces on complete system (Figure 2.2) requires vector balance of P_1, P_2, and the mass inertia effect.

Thus P_1 is effectively altered in magnitude and phase by the interposition of the spring, mass, dashpot combination before reacting as P_2 (Figure 2.6).

2.8. TRANSMISSIBILITY

It is usually desirable to minimize the effect of a given vibratory excitation upon the mounting structure. This is equivalent to minimizing the P_2/P_1 ratio in Figure 2.2 using Equation 2.9b. Disregarding phase angle, the transmissibil-

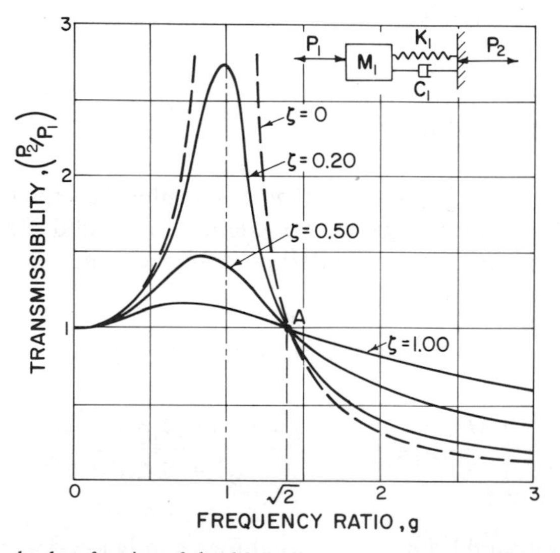

FIGURE 2.7. The absolute fraction of the driving force carried by the mounting approaches zero at high values of g for zero damping. Reduction can only be achieved if g is greater than $\sqrt{2}$, or beyond the independent point, A.

ity on the basis of the force amplitudes is indicated in Figure 2.7. The greatest reductions are obtained at high values of g, or low system natural frequency relative to the exciting frequency, and for minimum damping above A, or above $g = \sqrt{2}$. Complete isolation can theoretically be achieved at $g = \infty$, with zero damping.

As seen in Figure 2.7, if we must operate below the frequency ratio at A, damping is desirable to reduce the force ratio, but above A damping increases the transmissibility ratio. This characteristic arises from the dashpot coupling effects that persist at the higher frequencies.

2.9. ENERGY DISSIPATION

It is necessary for the excitation P_1 to supply the energy dissipated cyclically by the dashpot in steady state (Figure 2.2). In Equation 1.14 the average or constant damping energy rate was shown to be $\frac{1}{2}(C\omega^2 x_0^2)$, and the amplitude has now been defined in Equations 2.5 for the simple system in terms of forcing frequency and damping ratio. Then using the abbreviated notation of Table 2.1 for the square of the resultant amplitude,

$$E_d = \left(\frac{C_1\omega^2}{2}\right)\left(\frac{P_1}{K_1}\right)^2\left(\frac{1}{C^2 + B^2}\right)$$

$$\frac{E_d}{\left(P_1^2\omega_1/K_1\right)} = \frac{\frac{1}{2}Bg}{C^2 + B^2} \tag{2.10}$$

Behavior of the normalized average rate of energy dissipation is seen in Figure 2.8 for $\zeta_1 = 0.20$. There is a peak condition at resonance, with zero approached at low and high frequency ratios. At resonance, $g = 1$, and Equation 2.10 reduces to

$$\frac{E_d}{\left(P_1^2\omega_1/K_1\right)} = \frac{1}{4\zeta_1} \tag{2.11}$$

For a given system and excitation, the resonant energy rate is inversely proportional to the damping ratio. Thus a lightly damped system with high resonant amplitudes dissipates at a greater rate than a highly damped system with smaller amplitudes. This can be attributed to the square of the amplitude factor. In Equation 2.10, as in all response formulas given in the tables to follow, the dependent ratio on the left is dimensionless. The right term, or independent variable, is the factor determined by system frequency ratio and damping ratio only. In a literal sense all such terms are vibratory magnification or multiplication factors in which the principal parameter is the frequency factor.

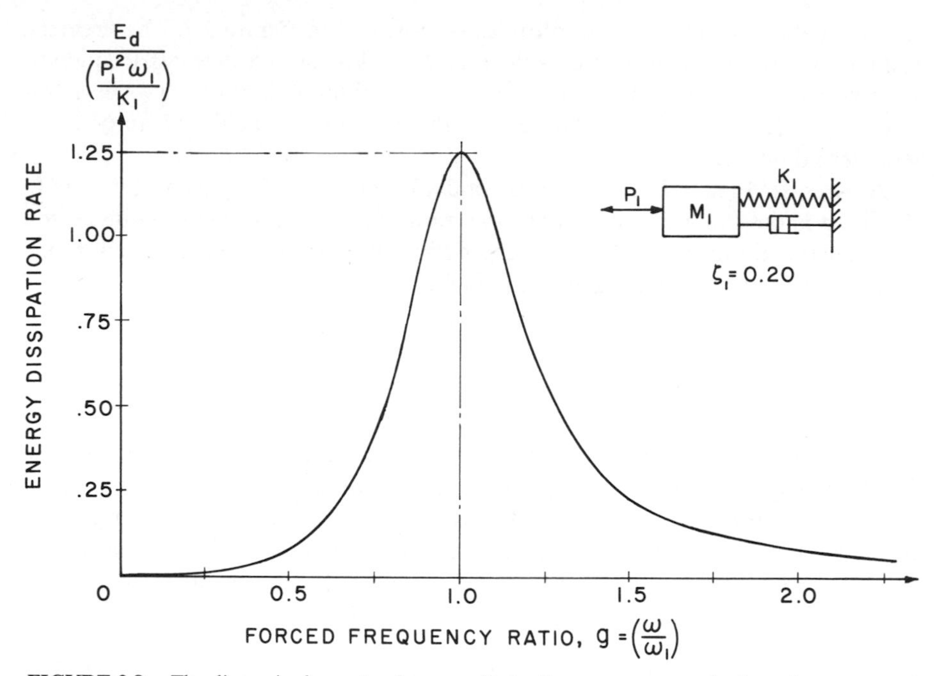

FIGURE 2.8. The dimensionless rate of energy dissipation on an average basis peaks near $g = 1$ for the simple system, and approaches zero at low and high frequency ratios. Values are from Table 2.1*a*.

2.10. CENTRIFUGAL EXCITATION

Rotating unbalance is a common source of sinusoidal excitation, with rotational frequency corresponding to vibratory circular frequency. Although the centrifugal force vector due to the unbalance, (me), rotates in space (Table 2.1*b*), the vertical component is assumed constrained at the mass so only horizontal effects are present. The forcing is sinusoidal with the amplitude of the force increasing parabolically from zero with increasing speed.

By substituting $(me\omega^2)$ for P_1 (Table 2.1*a*), and transferring the g^2 term to the right, Table 2.1*b* is obtained. The unbalance mass m is included in M_1 for mass ratio and natural frequency purposes, as it participates in the translation. This response factor approaches zero at low speeds and unity at high speeds, as in Figure 2.9. Also, the denominator $(me\omega_1^2)$ in the expression for P_2 (Table 2.1*b*) is a constant reference force, which would occur with rotation at the natural frequency.

2.11. DISPLACEMENT EXCITATION

The simple system can be subjected to sinusoidal forcing if the base or ground point is moved sinusoidally (see a_2 in Table 2.1*c*). Although base motion is specified as a displacement, a simultaneous force P_2 is also required. In the same sense, in Table 2.1*a* the excitation on the mass requires an exciting displacement as well as a force, but the force is defined. Summing forces on the mass,

$$\Sigma F_x = M_1\omega^2 \mathsf{a}_1 + K_1(\mathsf{a}_2 - \mathsf{a}_1) + jC_1\omega(\mathsf{a}_2 - \mathsf{a}_1) = 0$$

Dividing by K_1 and rearranging,

$$\frac{\mathsf{a}_1}{\mathsf{a}_2} = \frac{1 + jB}{C + jB} \tag{2.12}$$

Another variable of interest is the relative displacement seen by the spring and dashpot. This complex variable is

$$\mathsf{a}_{12} = (\mathsf{a}_1 - \mathsf{a}_2)$$

$$\frac{\mathsf{a}_{12}}{\mathsf{a}_2} = \left(\frac{\mathsf{a}_1}{\mathsf{a}_2}\right) - 1 \tag{2.13}$$

with the final expression in Table 2.1*c*.

At low frequency the absolute motion of M_1 approaches a_2, but the relative motion approaches zero (Figure 2.9). At high frequency the mass tends to remain stationary or seismic, with the relative motion approximately equal to

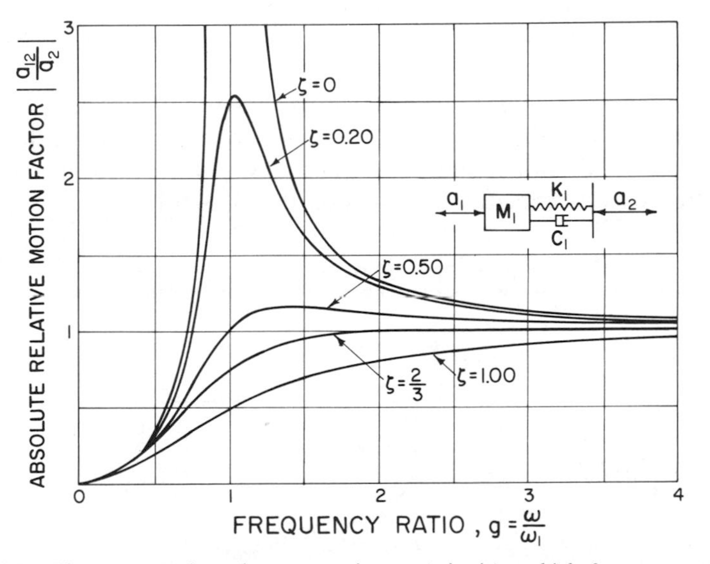

FIGURE 2.9. The mass tends to become stationary (seismic) at high frequency ratios, with relative motion approaching the input motion on an absolute basis. The curve is essentially constant at 1.00 above $\mathsf{g} = 2$, if $\zeta = \frac{2}{3}$.

the input motion. Vibratory load on the spring is determined from the product ($K_1\mathbf{a}_{12}$) and is maximum timewise as the $\mathbf{a}_{12}$ vector crosses the real axis in the complex plane.

2.12. SEISMIC BEHAVIOR

Steady-state displacements are often measured utilizing the feature of the relative motion $\mathbf{a}_{12}$ as a function of the $\mathbf{a}_2$ motion to be measured. Ideally the relative motion $\mathbf{a}_{12}$ will closely approximate sensed motion $\mathbf{a}_2$. A transducer such as a coil and magnet located between the mass and the input will then generate a voltage signal proportional to $\mathbf{a}_{12}$ and to the impressed frequency.

In Figure 2.9, we would tend to favor zero damping if a phase shift is to be minimized, although some damping can provide a degree of stability by suppressing transients. If phase shift is not a consideration, damping of $\zeta_1 = \frac{2}{3}$ provides the closest approximation to $\mathbf{a}_{12} = \mathbf{a}_2$. As we seek high frequency ratios for seismic purposes in instruments the problem is analogous to that of the vibration isolation problem (Section 2.8). The objective is to design for low natural frequencies by using large mass with respect to the spring stiffness.

2.13. RELATIVE DISPLACEMENT EXCITATION

Another basic form of sinusoidal excitation results from a positive displacement provided mechanically between the mass and the ground (Table 2.1*d*). For instance, a rotating cam with radial eccentricity *e* might drive a mass (Figure 2.10).

If the eccentric is removed, this system has a natural frequency determined by K_1 and M_1. With the eccentric in place, the mass must have the constant amplitude of the eccentric, and cannot peak, even if operated at the nominal natural frequency. Summation of forces on the mass becomes

$$\Sigma F_x = M_1\omega^2\mathbf{a}_1 - K_1\mathbf{a}_1 - jC_1\omega\mathbf{a}_1 + P_{el} = 0 \quad (2.14a)$$

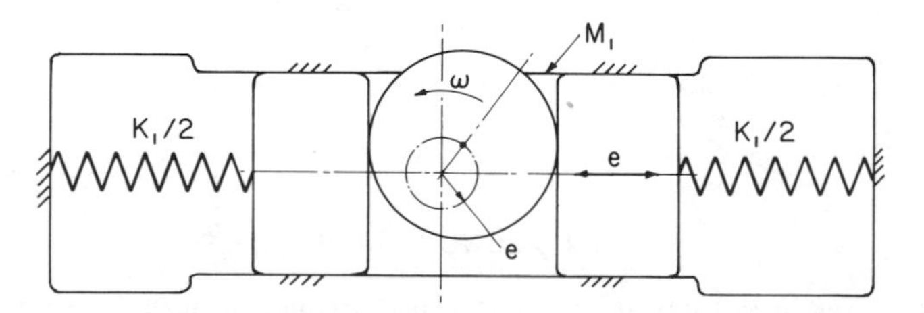

FIGURE 2.10. Cyclic loads on the driving eccentric are reduced by the addition of the auxiliary spring, which tends to counter the inertia effects of the mass (Table 2.1*d*).

TABLE 2.3. Positive relative displacement induced by mechanical means in various branches of the single-degree system.

	a	b	c
	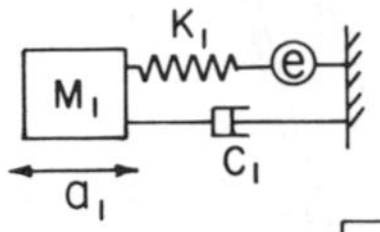	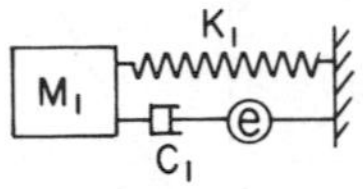	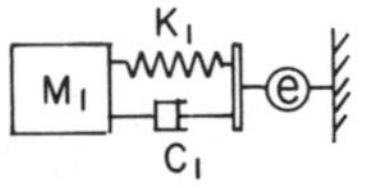
$\frac{a_1}{e}$	$\frac{1}{C + jB}$	$\frac{jB}{C + jB}$	$\frac{1 + jB}{C + jB}$
$\frac{P_{K1}}{K_1 e}$	$\frac{-g^2 + jB}{C + jB}$	$\frac{-jB}{C + jB}$	$\frac{-g^2}{C + jB}$
$\frac{P_{C1}}{K_1 e}$	$\frac{-jB}{C + jB}$	$\frac{jB(1 - g^2)}{C + jB}$	$\frac{-jg^2 B}{C + jB}$
	$C = (1 - g^2)$ $B = 2\zeta_1 g$	$g = \frac{\omega}{\omega_1}$ $\zeta_1 = \frac{C_1}{2M_1\omega_1}$	$\omega_1 = \sqrt{\frac{K_1}{M_1}}$

where P_{e1} = the contact force between the eccentric and the mass
$a_1 = e$ = radial eccentricity

The equation normalizes to

$$\left(\frac{P_{e1}}{K_1 e}\right) = (1 - g^2 + j2\zeta_1 g) \tag{2.14b}$$

The phase relation indicated is between the contact force vector and the impressed vector direction of e (Table 2.1d). If a positive relative displacement excitation is provided in various branches of a system, we again have resonant behavior. Solutions for these combinations are shown in Table 2.3.

2.14. MECHANICAL IMPEDANCE

Behavior of the required driving force P_{e1} with frequency (Table 2.1d) is interesting in the sense that we have the exact reciprocal of the conventional magnification factor. The driving force is minimized at resonance or $g = 1$. Practically, this suggests an auxiliary spring in this type of mechanism to reduce contact loads at a particular frequency. If the spring force exactly balances the inertia force at $g = 1$, only damping resistance remains. Figure

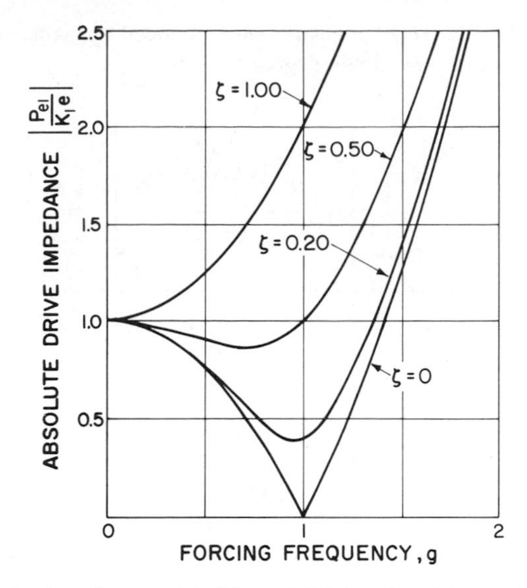

FIGURE 2.11. The load on the eccentric (Figure 2.10) is reduced to zero if forced at the natural frequency, $g = 1$. With damping, there is a value of $g < 1$ for which the contact load is minimized.

2.11 illustrates this behavior. With damping present, we achieve a minimum slightly below the resonant frequency. For zero damping the curve is algebraic, continuing down negatively beyond $g = 1$, corresponding to the shift from spring to mass loading on the driver.

As implied, the term *impedance* refers to the resistance of the spring–mass system to being reciprocated. This can be altered and minimized by the addition of auxiliary springs as shown.

2.15. CHARACTERISTIC RESPONSES

Systems in Tables 2.1*a–c* are related in many respects. Excepting energy functions, all denominators are identical, and there are only four possible numerators: $N = 1$, g^2, $(1 + jB)$, and $g^2(1 + jB)$. There four responses, divided by the common denominator, are shown in Figure 2.12 for an assumed damping ratio of 0.10.

Although peaks for all functions occur near (1,5), only the $N = 1$ and $N = g^2$ curves pass through this exact point when enlarging the peaks in this region. Also, no peak reaches a maximum exactly at $g = 1$; however, differences are negligible for this damping for practical purposes.

Another unexpected characteristic occurs for $N = g^2(1 + jB)$, which does not level at high frequencies but continues to rise after reaching a minimum at about $g = 2.5$.

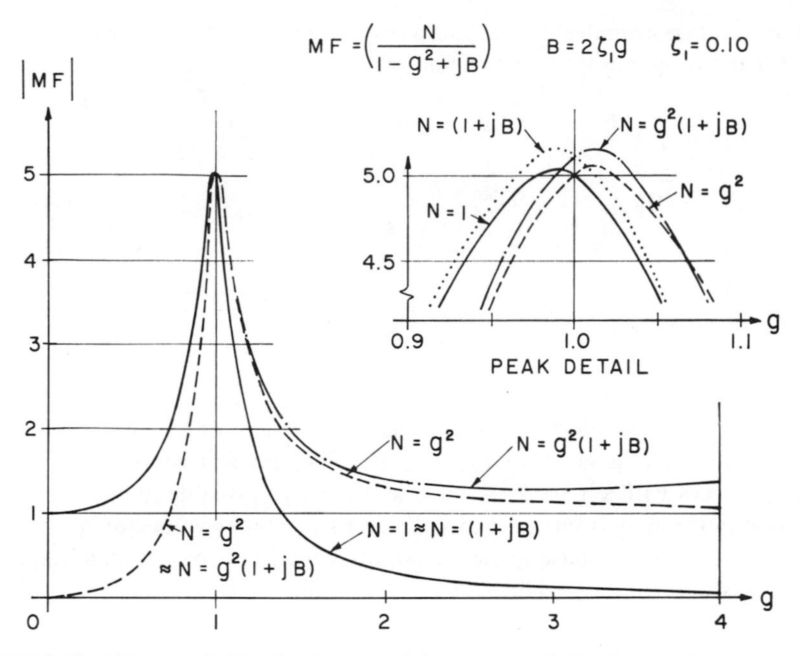

FIGURE 2.12. When studied in detail, none of the responses in Table 2.1*a*, *b*, and *c* actually peak at the nominal peak coordinates (1, 5).

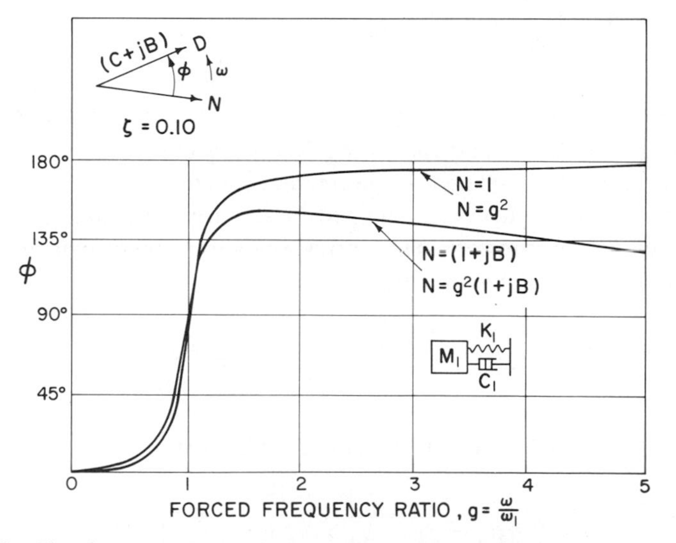

FIGURE 2.13. The phase angle by which the forcing function leads the response for parameters in Table 2.1*a*, *b*, and *c* has two possible variations with frequency.

The exact coordinates of maximum response can be determined analytically, and for the basic numerators become

$$
\begin{array}{lll}
 & N = 1 & N = \mathsf{g}^2 \\
\mathsf{g}^2_{\max} = & (1 - 2\zeta_1^2) & \left(\dfrac{1}{1 - 2\zeta_1^2}\right) \quad (2.15a) \\
(MF)_{\max} = & \left(\dfrac{1}{2\zeta_1\sqrt{1 - \zeta_1^2}}\right) & \left(\dfrac{1}{2\zeta_1\sqrt{1 - \zeta_1^2}}\right) \quad (2.15b)
\end{array}
$$

The behavior of the phase angle between the numerator and the denominator of these functions is given in Figure 2.13. The classical curve for $N = 1$ and $N = \mathsf{g}^2$ agrees with Figure 2.3*b*, but the other two possible numerators have a common curve that is considerably different at higher values of g. There are thus two alternative phase angle variations with frequency, depending on the type of system and the variables or vectors connected by ϕ.

2.16. ALTERNATIVE COMBINATIONS

The single mass can be restrained by several springs and dashpots. If these are in simple series or parallel arrangements, methods shown in Chapter 1 can be used for reducing these to single equivalent elements for which Table 2.1 can be applied. There are, however, other basic cases that can be analyzed by the procedures indicated (Tables 2.4–2.6).

With several springs or dashpots, we introduce the concept of system normalization. In Table 2.4, $k_0 = K_0/K_1$, treating K_1 as the reference spring. Similarly, the reference frequency is $\omega_1 = \sqrt{K_1/M_1}$, and the reference simple system is the first mass and the spring following the first mass, left to right. Dashpots are similarly related to the reference system.

System natural frequency is obtained in all cases by equating the real term C in the denominator to zero. This corresponds to resonance of the undamped system, and yields the *natural frequency factor* g_n.

$$C_n = 0 \qquad \mathsf{g}_n = \frac{\omega_n}{\omega_1} = \frac{\omega_n}{\sqrt{K_1/M_1}} \tag{2.16}$$

Note that g_n is only a ratio, as in Table 2.4 we have

$$\mathsf{g}_n = \sqrt{1 + k_0} = \sqrt{1 + \left(\frac{K_0}{K_1}\right)}$$

TABLE 2.4. Single mass coupled by two springs and two dashpots and subjected to (*a*) sinusoidal force on the mass, (*b*) sinusoidal displacement at the ungrounded spring–dashpot, and (*c*) simultaneous displacement at both ends.

(a)		(b)		(c)	
$\frac{a_1}{P_1/K_1}$	$\frac{1}{C+jD}$	$\frac{a_1}{a_2}$	$\frac{1+jB_1}{C+jD}$	$\frac{a_1}{a_2}$	$\frac{A+jD}{C+jD}$
$\frac{P_2}{P_1}$	$\frac{-(1+jB_1)}{C+jD}$	$\frac{a_{12}}{a_2}$	$\frac{g^2-k_0-jB_0}{C+jD}$	$\frac{a_{12}}{a_2}$	$\frac{g^2}{C+jD}$
$\frac{P_0}{P_1}$	$\frac{-(k_0+jB_0)}{C+jD}$	$\frac{P_2}{K_1 a_2}$	$\frac{A_1+jB_2}{C+jD}$	$\frac{P_2}{K_1 a_2}$	$\frac{-g^2(A+jD)}{C+jD}$
		$\frac{P_0}{K_1 a_2}$	$\frac{A_2-jB_3}{C+jD}$		

$A = (k_0+1)$ $\quad g = \frac{\omega}{\omega_1}$ $\quad A_1 = k_0 - g^2 - B_0B_1$

$C = (A - g^2)$ $\quad A_2 = g^2 - A_1$

$B_1 = 2\zeta_1 g$ $\quad D = (B_1 + B_0)$ $\quad \zeta_1 = \frac{C_1}{2M_1\omega_1}$ $\quad B_2 = B_1(k_0 - g^2) + B_0$

$B_0 = 2\zeta_0 g$ $\quad k_0 = \frac{K_0}{K_1}$ $\quad \zeta_0 = \frac{C_0}{2M_1\omega_1}$ $\quad B_3 = (k_0B_1 + B_0)$

Any system of this type has the natural frequency increased by the addition of K_0, and this is a relative increase. The factor g_n is independent of the absolute system parameters, applying equally to translational, torsional, large, and small systems, without regard to dimensional units if they are consistent. To convert to absolute frequency, however, we must calculate ω_1 using actual K_1 and M_1 values.

Another important observation is that the system natural frequency is independent of the applied excitation. In Table 2.1*c* the displacement excitation moves the spring end, but this has no bearing on the result when $C_n = 0$. The frequency is identical with that of a system with grounded springs.

2.17. SPRING CONSIDERATIONS

In the foregoing we have considered a concentrated mass carried by a massless, horizontal, helical spring, which is an idealized condition. There are many types of elastic suspensions, some of which are indicated in Tables 1.3 and 1.4, and include beam and torsional types. Practically it is impossible to have a

TABLE 2.5. Displacement-defined excitation applied to the single mass with various coupling arrangements.

a		b		c		d	
$\frac{a_1}{a_2}$	$\frac{jB_1}{C + jB_1}$	$\frac{a_1}{a_2}$	$\frac{jB_1}{C + jD}$	$\frac{a_1}{a_2}$	$\frac{1}{C + jB_0}$	$\frac{a_1}{a_2}$	$\frac{1 + jB_1}{C + jD}$
$\frac{a_{12}}{a_2}$	$\frac{-C}{C + jB_1}$	$\frac{a_{12}}{a_2}$	$\frac{-(C + jB_0)}{C + jD}$	$\frac{a_{12}}{a_2}$	$\frac{g^2 - jB_0}{C + jB_0}$	$\frac{a_{12}}{a_2}$	$\frac{g^2 - jB_0}{C + jD}$
$\frac{P_2}{K_0 a_2}$	$\frac{jCB_1}{C + jB_1}$	$\frac{P_2}{K_0 a_2}$	$\frac{B(B_0 - j)}{C + jD}$	$\frac{P_2}{K_1 a_2}$	$\frac{-g^2 + jB_0}{C + jB_0}$	$\frac{P_2}{K_1 a_2}$	$\frac{-(A_1 + jA_2)}{C + jD}$
$\frac{P_0}{K_0 a_2}$	$\frac{-jB_1}{C + jB_1}$	$\frac{P_0}{K_0 a_2}$	$\frac{B_1(B_0 + j)}{C + jD}$	$\frac{P_0}{K_1 a_2}$	$\frac{-jB_0}{C + jB_0}$	$\frac{P_0}{K_1 a_2}$	$\frac{B_0(B_1 - j)}{C + jD}$
$D = (B_0 + B_1)$		$g = \omega/\omega_0$	$\omega_0 = \sqrt{K_0/M_1}$	$g = \omega/\omega_1$	$\omega_1 = \sqrt{K_1/M_1}$	$A_1 = (g^2 + B_0 B_1)$	
$B_0 = 2\zeta_0 g$		$\zeta_0 = \frac{C_0}{2M_1\omega_0}$	$B_1 = 2\zeta_1 g$	$\zeta_1 = \frac{C_1}{2M_1\omega_1}$	$C = (1 - g^2)$	$A_2 = (B_1 g^2 - B_0)$	

TABLE 2.6. **The single mass with series spring–dashpot coupling subjected to (*a*), (*b*) force excitation, and (*c*), (*d*) displacement excitation.**

	a	b		c	d
$\frac{a_1}{P_1/K_1}$	$\frac{1+jB}{-g^2+jD}$		$\frac{a_1}{a_3}$	$\frac{jB}{-g^2+jD}$	
$\frac{a_2}{P_1/K_1}$	$\frac{1}{-g^2+jD}$	$\frac{jB}{-g^2+jD}$	$\frac{a_2}{a_3}$	$\frac{jD}{-g^2+jD}$	$\frac{-g^2+jB}{-g^2+jD}$
$\frac{a_{12}}{P_1/K_1}$	$\frac{+jB}{-g^2+jD}$	$\frac{+1}{-g^2+jD}$	$\frac{a_{12}}{a_3}$	$\frac{+jBg^2}{-g^2+jD}$	$\frac{+g^2}{-g^2+jD}$
$\frac{P_3}{P_1}$	$\frac{-jB}{-g^2+jD}$		$\frac{a_{23}}{a_3}$	$\frac{+g^2}{-g^2+jD}$	$\frac{+jBg^2}{-g^2+jD}$
			$\frac{P_3}{K_1 a_3}$	$\frac{-jBg^2}{-g^2+jD}$	

$B = 2\zeta_1 g$ $\quad D = B(1-g^2)$ $\quad g = \frac{\omega}{\omega_1}$

$\zeta_1 = \frac{C_1}{2M_1\omega_1}$ $\quad a_{12} = (a_1 - a_2)$ $\quad a_{23} = (a_2 - a_3)$ $\quad \omega_1 = \sqrt{\frac{K_1}{M_1}}$

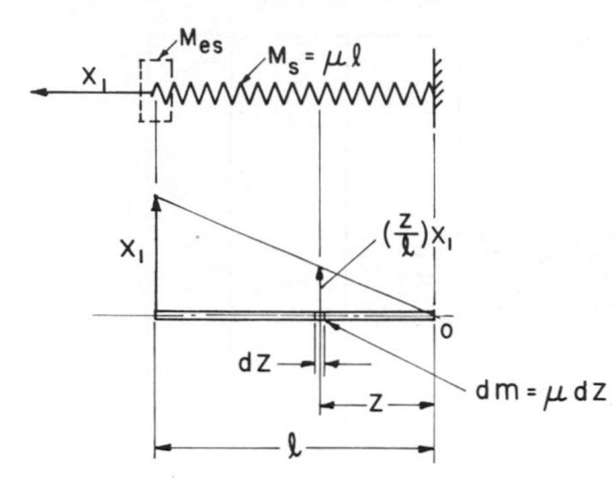

FIGURE 2.14. The effect of a spring element with uniformly distributed mass can be included in the concentrated end mass by integration of the related energy.

massless spring, although the mass can be negligible relative to the concentrated mass M_1. The effect of spring mass is to lower the natural frequency to some degree (Figure 2.14), for the simplest case of an elastic element in which the displacement of any point is proportional to the distance of the point from the fixed end. This includes the helical spring, the constant bar, and the simple torsional shaft.

For the triangular displacement pattern we sum the elemental masses in terms of the elemental kinetic energy and equate this to the kinetic energy of the equivalent spring mass at the free end.

$$\frac{1}{2} M_{es} (x_1 \omega)^2 = \int_0^l \frac{1}{2} \left(\frac{z}{l} x_1 \omega \right)^2 \mu \, dz$$

$$M_{es} = \mu \int_0^l \left(\frac{z}{l} \right)^2 dz = \frac{\mu}{l^2} \left(\frac{l^3}{3} \right) = \frac{\mu l}{3}$$

$$= \frac{M_s}{3} \tag{2.17}$$

where M_{es} = equivalent spring mass
M_s = actual spring mass
μ = mass per unit length = M_s/l

Thus one-third of the spring mass will be added to M_1 if it is to be considered.

Another possible result of spring mass is the effect of vibratory frequencies within the spring. This phenomenon is discussed in Chapter 12, relating to uniform bars. Spring modes, or surging, can exist within a spring coupled to a mass or in a plain spring. Depending on exciting frequencies and relative masses, combined interactions can occur, but usually do not.

2.18. SPRINGLESS SINGLE-DEGREE SYSTEMS

The mass–dashpot arrangement (Table 2.7), with only dashpot coupling to ground, can be analyzed by methods already indicated for steady-state re-

TABLE 2.7. Springless systems excited by (*a*) force on the mass, (*b*) centrifugal force on the mass, (*c*) base displacement, and (*d*) positive relative displacement.

a		b		c		d	
$\frac{a_1}{P_1/K_1}$	$\frac{-(j/g)}{1+jg}$	$\frac{a_1}{e(m/M_1)}$	$\frac{-jg}{1+jg}$	$\frac{a_1}{a_2}$	$\frac{1}{1+jg}$	$\frac{a_1}{e}$	1
$\frac{P_2}{P_1}$	$\frac{-1}{1+jg}$	$\frac{P_2}{me\omega_1^2}$	$\frac{g^2}{1+jg}$	$\frac{a_{12}}{a_2}$	$\frac{-jg}{1+jg}$	$\frac{P_{e1}}{K_1 e}$	$-g^2+jg$
				$\frac{P_2}{K_1 a_2}$	$\frac{-g^2}{1+jg}$	$\frac{P_2}{K_1 e}$	$-g^2$

$\omega_1 = \frac{C_1}{M_1} = \frac{K_1}{C_1}$ = REF. NATURAL FREQUENCY

$K_1 = \frac{C_1^2}{M_1}$ = REFERENCE SPRING

$g = \frac{\omega}{\omega_1} = \frac{\omega}{C_1/M_1}$ = REF. FORCED FREQ. RATIO

sponse to sinusoidal forcing. There is neither natural frequency nor critical damping factor, but the equilibrium equation that governs the behavior is

$$\Sigma F_x = M_1\omega^2 \mathsf{a}_1 - jC_1\omega \mathsf{a}_1 + P_1 = 0 \tag{2.18}$$

where ω is the forcing frequency. Normalized form is obtained by division by (C_1^2/M_1).

$$\frac{\mathsf{a}_1}{(P_1/K_1)} = \frac{-1}{\mathsf{g}^2 - j\mathsf{g}} \tag{2.19}$$

where $\mathsf{g} = \omega/\omega_1$ = forced frequency ratio
$\omega_1 = C_1/M_1 = K_1/C_1$ = reference natural frequency
$K_1 = C_1^2/M_1$ = reference spring

These reference parameters have no particular physical significance but correspond dimensionally to their respective counterparts and provide a means for generalized analysis.

Equation 2.19 can be multiplied by (j/g):

$$\frac{\mathsf{a}_1}{(P_1/K_1)} = \left[\frac{-j(1/\mathsf{g})}{1 + j\mathsf{g}}\right] \tag{2.20}$$

Although the numerator is in unusual form, we see in Table 2.7 that $(1 + j\mathsf{g})$ is the common and most simple denominator for the springless systems. Other relations are tabulated, analogous to Table 2.1 from which magnitudes and phase relations can be obtained.

2.19. RESPONSE BEHAVIOR OF SPRINGLESS SYSTEMS

Table 2.6 contains identical denominators and four types of numerators. These response factors are similar to magnification factors for conventional systems, although they do not exhibit a response peak. As seen in Figure 2.15, there are again four characteristic curves for the several numerators. There is, furthermore, a common factor of 0.707 at $\mathsf{g} = 1.00$. The one continuously rising factor is related to P_2, or $N = \mathsf{g}^2$.

2.20. GENERAL COMMENTS

In this, as in following chapters, we have consistently used horizontal models, with horizontal coordinates related to x. There are several reasons for this convention. The horizontally supported mass causes no vertical deflection or gravitational sag due to weight. Horizontal is also the normal direction of the real axis in the complex plane.

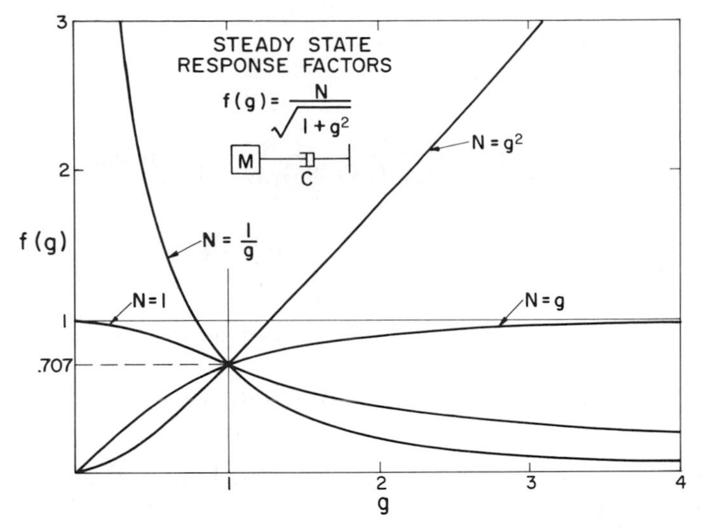

FIGURE 2.15. Absolute values of the four possible response rations for the springless forced systems in Table 2.5*a*, *b*, and *c* intersect at $g = 1$.

There is no implication, however, that the analyses are limited in this respect. Natural frequencies and responses are identical with vertical springs and motions, except for the gravity effects that lower the datum position. Angular mounting is similarly applicable.

As shown in Chapter 1, torsional systems have angular coordinates but are otherwise identical to the translational cases, and should not be treated as separate entities. Mass, stiffness, and damping parameters are simply adapted, as in Table 1.6. Complex variables $\mathbf{a}_1$ and $\mathbf{a}_2$ are in radians dimensionally, excitation P_1 becomes a couple, and so on.

EXAMPLES

EXAMPLE 2.1. The basic system (Table 2.1*a*) has the following values:

$$K_1 = 160 \text{ lb/in.} \qquad C_1 = 1.3 \text{ lb sec/in.}$$
$$M_1 = 25 \text{ lb} \qquad P_1 = 14 \text{ lb at } 15 \text{ Hz}$$

Find:

(a) The natural frequency.

(b) The amplitude of M_1.

Solution. Equating $C = 0$ and $\omega_n = \omega_1$,

(a)
$$\omega_1 = \sqrt{\frac{K_1}{M_1}} = \sqrt{\frac{(160)(386)}{25}} = 49.7 \text{ rad/sec}$$

$$= 49.7/2\pi = 7.91 \text{ Hz}$$

(b)

$$\mathsf{g} = \frac{\omega}{\omega_1} = \frac{15}{7.91} = 1.896$$

$$\mathsf{g}^2 = 3.596$$

$$\zeta_1 = \frac{C_1}{2M_1\omega_1} = \frac{1.3}{(2)(25/386)(49.70)} = 0.202$$

$$\frac{\mathsf{a}_1}{(P_1/K_1)} = \frac{1}{(1 - 3.596) + j(2)(0.202)(1.896)}$$

$$= \frac{1}{-2.596 + j(0.766)} = \frac{1}{\sqrt{(2.596)^2 + (0.766)^2}}$$

$$\mathsf{a}_1 = \left(\frac{14}{160}\right)(0.369) = (0.0875)(0.369) = 0.0323 \text{ in.}$$

EXAMPLE 2.2. A system (Table 2.1*b*) is excited by a rotating unbalance. Given the following,

$$K_1 = 22{,}000 \text{ N/m} \qquad me = 0.027 \text{ kg} \cdot \text{m}$$
$$M_1 = 6.0 \text{ kg} \qquad \omega = 1{,}800 \text{ RPM}$$
$$C_1 = 0$$

Find:

(a) The vibratory exciting force.

(b) The transmitted force.

(c) The rate of energy dissipation.

Solution

(a)

$$P_1 = me\omega^2 = (0.027)(188.5)^2 = 959\ N$$

$$\omega = (1800/60)2\pi = 188.5 \text{ rad/sec}$$

(b)

$$\omega_1 = \sqrt{K_1/M_1} = \sqrt{22{,}000/6} = 60.55 \text{ rad/sec}$$

$$= 60.55/2\pi = 9.64 \text{ Hz}$$

$$\mathsf{g} = \frac{\omega}{\omega_1} = \frac{1800/60}{9.64} = 3.13$$

$$\mathsf{g}^2 = 9.690$$

$$me\omega_1^2 = (0.027)(60.55)^2 = 99 \text{ N}$$

$$\frac{P_2}{me\omega_1^2} = \frac{-9.69}{(1 - 9.69)} = +1.115$$

$$P_2 = 1.115(99) = 110.4 \text{ N}$$

(c) $$E_d = 0 \quad \text{(there is no damping)}$$

EXAMPLE 2.3. A displacement excited system, as shown in Table 2.1*c*, has the following specifications:

$$K_1 = 5000 \text{ lb/in.} \qquad \zeta_1 = 0.12$$
$$M_1 = 310 \text{ lb} \qquad \mathsf{a}_2 = 0.15 \text{ in. at } 14 \text{ Hz}$$

Find:

(a) The amplitude of the spring force.
(b) The rate of energy dissipation.

Solution

(a) $$\omega_1 = \sqrt{\frac{K_1}{M_1}} = \sqrt{\frac{(5000)(386)}{310}} = 78.9 \text{ rad/sec}$$

$$= 78.9/2\pi = 12.56 \text{ Hz}$$

$$\mathsf{g} = \omega/\omega_1 = 14/12.56 = 1.115$$

$$\mathsf{g}^2 = 1.243$$

$$B = 2(0.12)(1.115) = 0.268$$

$$\frac{\mathsf{a}_{12}}{\mathsf{a}_2} = \frac{1.243}{(1 - 1.243) + j(0.268)} = 3.436$$

$$(K_1)\mathsf{a}_{12} = 5000(0.15)(3.436) = 2577 \text{ lb}$$

(b) $$\frac{E_d}{\mathsf{a}_2^2 K_1 \omega_1} = \frac{(0.5)(0.268)(1.115)^5}{(0.243)^2 + (0.268)^2} = 1.76$$

$$E_d = (0.15)^2(5000)(78.9)(1.76) = 15{,}600 \text{ in. lb/sec}$$

$$H.P. = \frac{15{,}600}{(550)(12)} = 2.37$$

EXAMPLE 2.4. A torsional system is excited simultaneously from both ends, with equal, in-phase displacements. The shaft is steel, and its mass effect is to be neglected. Find:

(a) The system natural frequency in torsion.
(b) Amplitude of the mass.
(c) Amplitude of the exciting torque.

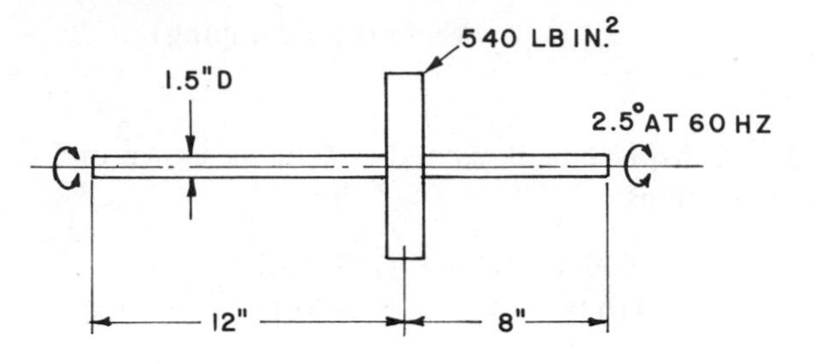

Solution

(a) $K_0 = GJ/l_0 = 0.497(10)^6$ in. lb/rad $\quad J = \pi d^4/32 = 0.497$ in.^{4}

$K_1 = GJ/l_1 = 0.746(10)^6$ in. lb/rad $\quad G = 12(10)^6$ lb/in^2.

Applying Table 2.4*c*,

$$\omega_1 = \sqrt{\frac{K_1}{M_1}} = \sqrt{\frac{(0.746)(10)^6 386}{540}} = 730 \text{ rad/sec} = 116.2 \text{ Hz}$$

$$C = A - \mathsf{g}_n^2 = 0$$

$$\mathsf{g}_n^2 = A = 1.667$$

$$\mathsf{g}_n = 1.291$$

$$A = k_0 + 1 = 497/746 + 1 = 8/12 + 1 = 1.667$$

$$\omega_n = \mathsf{g}_n \omega_1 = 1.291(116.2) = 150.0 \text{ Hz}$$

(b)

$$\mathsf{g} = 60/116.2 = 0.516$$

$$\mathsf{g}^2 = 0.267$$

$$\frac{\mathsf{a}_1}{\mathsf{a}_2} = \frac{1.667}{1.667 - 0.267} = +1.19$$

$$\mathsf{a}_1 = 2.98°$$

(c)

$$\frac{P_2}{K_1 \mathsf{a}_2} = \frac{-(0.267)(1.667)}{1.40}$$

$$P_2 = (0.746)(10)^6(2.5/57.3)(0.318) = 10{,}350 \text{ in. lb}$$

EXAMPLE 2.5. A system (Table 2.3*a*) has the following:

$K_1 = 75{,}000$ N/m $\quad C_1 = 60$ N · m · s
$M_1 = 2.40$ kg $\quad e = 1.8$ cm at 30 Hz

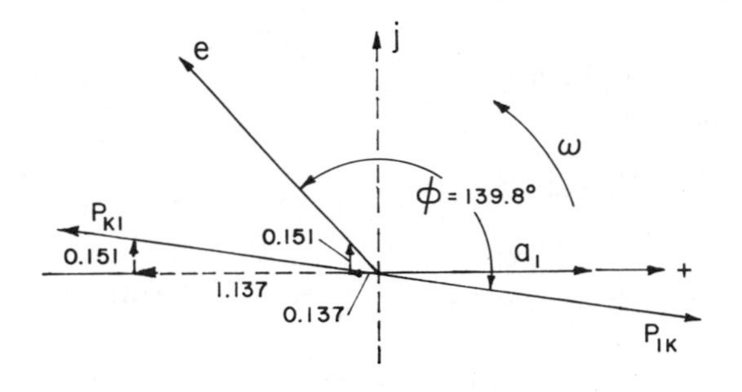

Find:

(a) The motion of the mass.

(b) The required exciting force amplitude.

(c) The phase angle between the force on the exciter from K_1 and the eccentricity vector.

Solution

(a)
$$\omega_1 = \sqrt{K_1/M_1} = \sqrt{75{,}000/2.40} = 176.8 \text{ rad/sec} = 28.13 \text{ Hz}$$

$$\mathsf{g} = \omega/\omega_1 = 30/28.13 = 1.066$$

$$\mathsf{g}^2 = 1.137$$

$$2\zeta_1\mathsf{g} = B - 2(1.066)\left[\frac{60}{2(2.4)(176.8)}\right] - 0.151$$

$$\frac{\mathsf{a}_1}{e} = \frac{1}{(1 - 1.137) + j(0.151)} = 4.905$$

$$\mathsf{a}_1 = 4.905(1.8) = 8.83 \text{ cm}$$

(b)
$$\frac{P_{K1}}{K_1 e} = \frac{-1.137 + j(0.151)}{-0.137 + j(0.151)} = 5.626$$

$$P_{K1} = P_e = (75{,}000)(0.018)(5.626) = 7595 \text{ N}$$

Here the spring force equals the driving force.

(c) From part b the vector relation is known, and can be drawn in the complex plane

Note that in the tabulation P_{K1} indicates the force exerted by the spring on the mass; therefore, the force on the exciter is reversed.

EXAMPLE 2.6. A mass–dashpot system (Table 2.7*a*) is excited at the mass, and

$$C_1 = 1.0 \text{ lb sec/in.} \qquad P_1 = 20 \text{ lb at 10 Hz}$$
$$M_1 = 15 \text{ lb}$$

Find:

(a) The amplitude of the mass.

(b) The reaction force.

Solution

(a)
$$K_1 = \frac{C_1^2}{M_1} = \frac{1}{15/386} = 25.73 \text{ lb/in.}$$

$$\omega_1 = C_1/M_1 = 25.73 \text{ rad/sec} = 4.096 \text{ Hz}$$

$$\mathsf{g} = \omega/\omega_1 = 10/4.096 = 2.442$$

$$\frac{\mathsf{a}_1}{(P_1/K_1)} = \frac{-j(1/\mathsf{g})}{(1 + j\mathsf{g})} = \frac{1/2.442}{\sqrt{1 + 2.442^2}} = 0.155$$

$$\mathsf{a}_1 = 0.155(20/25.73) = 0.12 \text{ in.}$$

(b)
$$\frac{P_2}{P_1} = \frac{-1}{(1 + j\mathsf{g})} = \frac{1}{\sqrt{1 + 2.442^2}} = 0.379$$

$$P_2 = 0.379(20) = 7.58 \text{ lb}$$

3 FREE VIBRATION OF THE DAMPED SYSTEM

A free vibration is one in which no external forcing is applied to a system after an initial disturbance. The general transient case can involve various types of input functions, such as forces or displacements that are not periodic, and the response varies dynamically but not sinusoidally. Although the broader transient analysis has important aspects, this discussion concerns mainly free motion, or simple decay behavior.

The free response is especially important to steady-state analysis, providing a philosophical basis for natural frequency and damping ratio. Also, the free vibration is useful in experimental or laboratory techniques relating to the decay phenomenon, as equations developed facilitate the calculation of theoretical responses for the interpretation of experimentally obtained decay data.

3.1. EQUATIONS OF MOTION

A spring–mass–dashpot system (Figure 3.1) is disturbed from its equilibrium position and allowed to oscillate freely. Various initial conditions are possible, including initial displacement, initial velocity, or both, at the start of the free cycle. Equilibrium of forces on the mass at any instant requires that

$$-M\ddot{x} - C\dot{x} - Kx = 0 \tag{3.1}$$

Taking as a solution

$$x = Ae^{at} \qquad \dot{x} = aAe^{at} \qquad \ddot{x} = a^2Ae^{at}$$

$$(Ma^2 + Ca + K)Ae^{at} = 0 \tag{3.2}$$

where A and a are constants, the latter corresponding dimensionally to circular frequency. The quadratic expression within parentheses is converted to dimen-

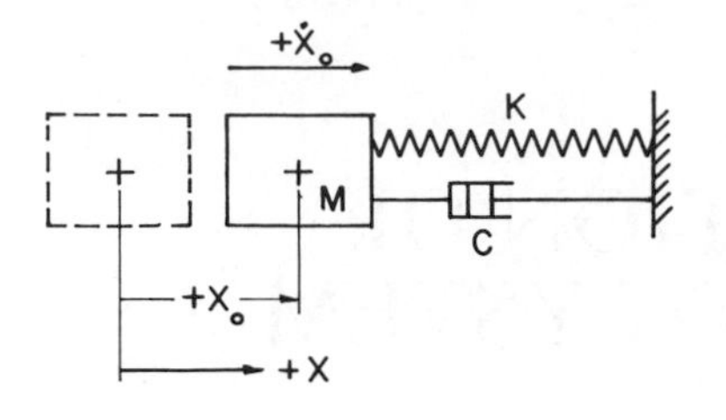

FIGURE 3.1. Initial conditions for the single-degree system can include both displacement and velocity at $t = 0$.

sionless form by dividing by K.

$$\left(\frac{a}{\omega_1}\right)^2 + 2\zeta\left(\frac{a}{\omega_1}\right) + 1 = 0 \tag{3.3}$$

$$\text{where } \omega_1 = \sqrt{K/M} \qquad \zeta = (C/C_c) = C/2M\omega_1$$

and the solution is

$$\left(\frac{a}{\omega_1}\right)_{1,2} = -\zeta \pm \sqrt{\zeta^2 - 1} \tag{3.4}$$

A significant value of the damping ratio occurs when the radical becomes zero.

$$\zeta = \frac{C}{C_c} = 1 \qquad C = C_c = 2M\omega_1$$

This damping ratio indicates the damping coefficient to be the *critical damping coefficient*, C_c. Any other coefficient is conveniently related to C_c as the ratio ζ.

Returning to the displacement relation and introducing two arbitrary constants depending on initial conditions,

$$x = C_1 e^{(-\zeta + \sqrt{\zeta^2 - 1})\omega_1 t} + C_2 e^{(-\zeta - \sqrt{\zeta^2 - 1})\omega_1 t} \tag{3.5}$$

We note the importance of ζ as a system parameter, which can vary from zero to infinity, with four distinct response characteristics possible at specific damping ratios. These basic categories are each considered in some detail.

3.2. INITIAL DISPLACEMENT

Typical free responses of the damped vibratory system are shown in Figure 3.2. The simplest case involves the displacement of a mass and its subsequent

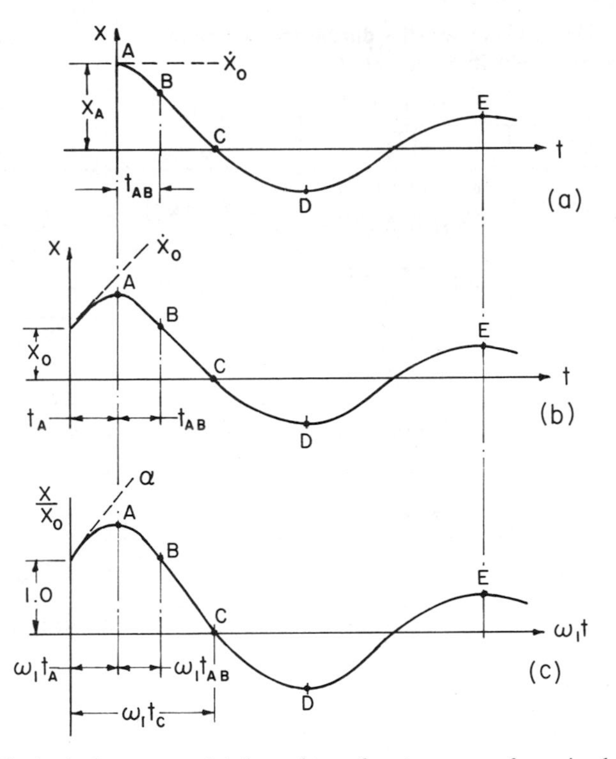

FIGURE 3.2. The basic decay curve (*a*) for a damped system ensues from simple release at an initial displacement. Response with initial velocity, (*b*) and (*c*), constitutes a preliminary phase, leading to *A*. Absolute and dimensionless coordinates are shown.

release (*a*). Damping continuously reduces the displacement with respect to the cosine function which would represent zero damping. Several significant coordinate points are identified in Table 3.1:

A = maximum initial positive displacement
B = maximum recovery velocity
C = first instance of zero spring force
D = first maximum reversed amplitude
E = second maximum positive displacement

Initial velocity (Figure 3.2*b*), related to the inertia or kinetic energy of the mass at $t = 0$, causes overshoot to the maximum point A. With further time the system decays exactly as in (*a*). Complete response then can be obtained by finding the coordinates of A and combining the initial and the subsequent phase.

TABLE 3.1. Decay characteristics during free vibration with initial displacement only. Maximum recovery velocity occurs at *B*.

	$\zeta = 0$	$0 < \zeta < 1$	$\zeta = 1$	$\zeta > 1$
β	1.00	$\sqrt{1-\zeta^2}$	0	$\sqrt{\zeta^2-1}$
$\dfrac{x}{x_A}$	$\cos \omega_1 t$	$e^{-\zeta\omega_1 t}\left[\cos \beta\omega_1 t + \left(\zeta/\beta\right)\sin \beta\omega_1 t\right]$	$e^{-\omega_1 t}\left[1+\omega_1 t\right]$	$e^{-\zeta\omega_1 t}\left[\cosh \beta\omega_1 t + \left(\zeta/\beta\right)\sinh \beta\omega_1 t\right]$
$\dfrac{\dot{x}}{\omega_1 x_A}$	$-\sin \omega_1 t$	$-e^{-\zeta\omega_1 t}\left[\dfrac{\sin \beta\omega_1 t}{\beta}\right]$	$-e^{-\omega_1 t}\left[\omega_1 t\right]$	$-e^{-\zeta\omega_1 t}\left[\dfrac{\sinh \beta\omega_1 t}{\beta}\right]$
$\dfrac{\ddot{x}}{\omega_1^2 x_A}$	$-\cos \omega_1 t$	$e^{-\zeta\omega_1 t}\left[-\cos \beta\omega_1 t + \left(\zeta/\beta\right)\sin \beta\omega_1 t\right]$	$e^{-\omega_1 t}\left[\omega_1 t - 1\right]$	$e^{-\zeta\omega_1 t}\left[-\cosh \beta\omega_1 t + \left(\zeta/\beta\right)\sinh \beta\omega_1 t\right]$
$\omega_1 t_B$	$\dfrac{\pi}{2}$	$\dfrac{1}{\beta}\tan^{-1}\left(\dfrac{\beta}{\zeta}\right)$	1.00	$\dfrac{1}{\beta}\tanh^{-1}\left(\dfrac{\beta}{\zeta}\right)$
$\dfrac{x_B}{x_A}$	0	$2\zeta e^{-\zeta\omega_1 t_B}$	$2e^{-1}$	$2\zeta e^{-\zeta\omega_1 t_B}$
$\dfrac{\dot{x}_B}{\omega_1 x_A}$	-1.00	$-e^{-\zeta\omega_1 t_B}$	$-e^{-1}$	$-e^{-\zeta\omega_1 t_B}$
$\omega_1 t_C$	$\dfrac{\pi}{2}$	$\dfrac{1}{\beta}\tan^{-1}\left(-\dfrac{\beta}{\zeta}\right)$	DECAY DURING FREE VIBRATION $x_0 = x_A$ $\dot{x}_0 = 0$ 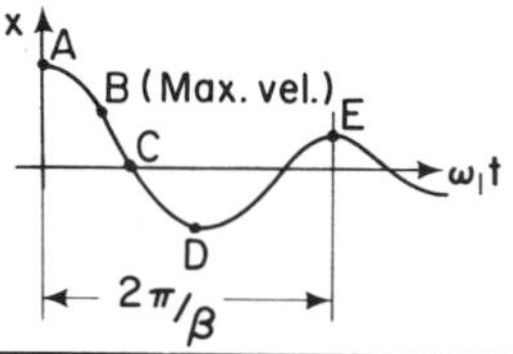	x A B (Max. vel.) C D E $\omega_1 t$ $2\pi/\beta$
$\dfrac{x_E}{x_A}$	1.00	$e^{-2\pi\left(\frac{\zeta}{\beta}\right)}$		

3.3. UNDAMPED INITIAL DISPLACEMENT

Equation 3.5 can be converted to trigonometric form to describe the simple harmonic motion of the undamped system.

$$x = A_1 \cos \omega_1 t + A_2 \sin \omega_1 t \tag{3.6}$$

and the velocity becomes

$$\dot{x} = -A_1 \omega_1 \sin \omega_1 t + A_2 \omega_1 \cos \omega_1 t \tag{3.7}$$

When $t = 0$, $x = x_0 = A_1$, and $\dot{x} = \dot{x}_0 = A_2 \omega_1$, and displacement of the free undamped system is then

$$x = x_0 \cos \omega t + \left(\frac{\dot{x}_0}{\omega_1}\right) \sin \omega t$$

$$\left(\frac{x}{x_0}\right) = \cos \omega t + \alpha \sin \omega t \tag{3.8}$$

where $\alpha = (\dot{x}_0/\omega_1 x_0)$, or the dimensionless initial velocity relating the actual initial velocity $\dot{x}_0$ to the conventional harmonic velocity $\omega_1 x_0$, (Table 3.2).

The nature of this vibratory response is shown in Figure 3.3 for positive initial displacement and velocity, with harmonic motion continuing indefinitely at the natural frequency. The maximum displacement at A corresponds to zero velocity, with the complete velocity expressed by

$$\dot{x} = -\omega_1 x_0 \sin \omega_1 t + \dot{x}_0 \cos \omega_1 t$$

$$\left(\frac{\dot{x}}{\omega_1 x_0}\right) = -\sin \omega_1 t + \alpha \cos \omega_1 t \tag{3.9}$$

Or, from Figure 3.3, the phase angle at which maximum displacement occurs is

$$\omega_1 t_A = \tan^{-1}\alpha \tag{3.10}$$

TABLE 3.2. Free vibratory response with both initial displacement and initial velocity.

	$\zeta = 0$	$0 < \zeta < 1$	$\zeta = 1$	$\zeta > 1$
β	1.00	$\sqrt{1-\zeta^2}$	0	$\sqrt{\zeta^2-1}$
$\dfrac{x}{x_0}$	$\cos \omega_1 t + \alpha \sin \omega_1 t$	$e^{-\zeta\omega_1 t}\left[\cos \beta\omega_1 t + \left(\frac{\alpha+\zeta}{\beta}\right) \sin \beta\omega_1 t\right]$	$e^{-\omega_1 t}[1+\omega_1 t(1+\alpha)]$	$e^{-\zeta\omega_1 t}\left[\cosh \beta\omega_1 t + \left(\frac{\alpha+\zeta}{\beta}\right) \sinh \beta\omega_1 t\right]$
$\dfrac{\dot{x}}{\omega_1 x_0}$	$\alpha \cos \omega_1 t - \sin \omega_1 t$	$e^{-\zeta\omega_1 t}\left[\alpha \cos \beta\omega_1 t - \left(\frac{1+\alpha\zeta}{\beta}\right) \sin \beta\omega_1 t\right]$	$e^{-\omega_1 t}[\alpha - \omega_1 t(1+\alpha)]$	$e^{-\zeta\omega_1 t}\left[\alpha \cosh \beta\omega_1 t - \left(\frac{1+\alpha\zeta}{\beta}\right) \sinh \beta\omega_1 t\right]$
$\dfrac{\ddot{x}}{\omega_1^2 x_0}$	$-\cos \omega_1 t - \alpha \sin \omega_1 t$	$e^{-\zeta\omega_1 t}\left[-(1+2\alpha\zeta)\cos\beta\omega_1 t + \frac{\alpha(2\zeta^2-1)+\zeta}{\beta} \sin\beta\omega_1 t\right]$	$e^{-\omega_1 t}[(1+\alpha)\omega_1 t - (1+2\alpha)]$	$e^{-\zeta\omega_1 t}\left[-(1+2\alpha\zeta)\cosh\beta\omega_1 t + \frac{\alpha(2\zeta^2-1)+\zeta}{\beta} \sinh \beta\omega_1 t\right]$
$\omega_1 t_A$	$\tan^{-1}\alpha$	$\frac{1}{\beta}\tan^{-1}\left(\frac{\alpha\beta}{1+\alpha\zeta}\right)$	$\left(\frac{\alpha}{1+\alpha}\right)$	$\frac{1}{\beta}\tanh^{-1}\left(\frac{\alpha\beta}{1+\alpha\zeta}\right)$
$\dfrac{x_A}{x_0}$	$\sqrt{1+\alpha^2}$	$e^{-\zeta\omega_1 t_A}\sqrt{1+2\alpha\zeta+\alpha^2}$	$e^{-\frac{\alpha}{1+\alpha}}[1+\alpha]$	$e^{-\zeta\omega_1 t_A}\sqrt{1+2\alpha\zeta+\alpha^2}$
$\dfrac{x_B}{x_0}$	0	$e^{-\zeta(\omega_1 t_A+\omega_1 t_B)}[2(1+\alpha)]$	$e^{-\left(\frac{1+2\alpha}{1+\alpha}\right)}[2(1+\alpha)]$	$e^{-\zeta(\omega_1 t_A+\omega t_B)}[2(1+\alpha)]$
$\dfrac{\dot{x}_B}{\omega_1 x_0}$	$-\sqrt{1+\alpha^2}$	$-e^{-\zeta(\omega_1 t_A+\omega_1 t_B)}\sqrt{1+2\alpha\zeta+\alpha^2}$	$-e^{-\left(\frac{1+2\alpha}{1+\alpha}\right)}[1+\alpha]$	$-e^{-\zeta(\omega_1 t_A+\omega_1 t_B)}\sqrt{1+2\alpha\zeta+\alpha^2}$
$\omega_1 t_C$	$\tan^{-1}\left(-\frac{1}{\alpha}\right)$	$\frac{1}{\beta}\tan^{-1}\left(\frac{-\beta}{\alpha+\zeta}\right)$	$-\left(\frac{1}{1+\alpha}\right)$ (REQUIRES NEGATIVE α)	$\frac{1}{\beta}\tanh^{-1}\left(\frac{-\beta}{\alpha+\zeta}\right)$

$\alpha = \dfrac{\dot{x}_0}{\omega_1 x_0}$ $\quad \omega_1 = \sqrt{\dfrac{K}{M}}$ $\quad \zeta = \dfrac{C}{C_c}$

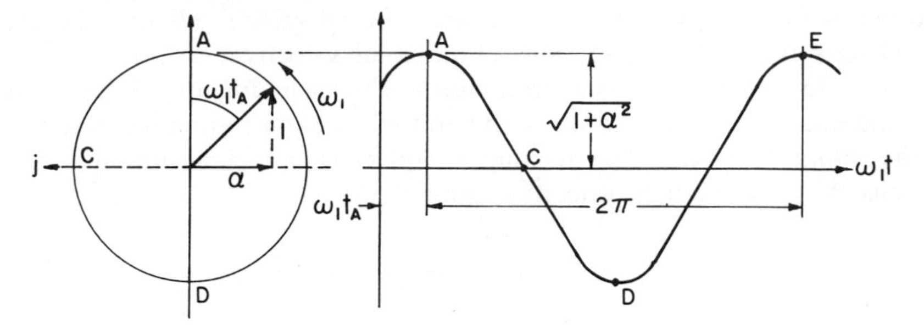

FIGURE 3.3. Undamped system, $\zeta = 0$, reaches maximum excursion at A, with amplitude determined by the radius in the complex plane ($\alpha = \dot{x}_0/\omega_1 x_0$).

and the corresponding ordinate is

$$\frac{x_A}{x_0} = \sqrt{1 + \alpha^2} \tag{3.11}$$

3.4. UNDERDAMPED INITIAL DISPLACEMENT

From the original equation of motion (Equation 3.5), it is seen that intermediate damping ratios between zero and unity result in an expression with complex exponents.

$$\begin{aligned} x &= C_1 e^{(-\zeta + j\sqrt{1-\zeta^2})\omega_1 t} + C_2 e^{(-\zeta - j\sqrt{1-\zeta^2})\omega_1 t} \\ &= e^{-\zeta\omega_1 t}\left(C_1 e^{j\beta\omega_1 t} + C_2 e^{-j\beta\omega_1 t}\right) \end{aligned} \tag{3.12}$$

where $\beta = \sqrt{1 - \zeta^2}$.

Transforming to trigonometric functions,

$$x = e^{-\zeta\omega_1 t}(A_1 \cos\beta\omega_1 t + A_2 \sin\beta\omega_1 t) \tag{3.13}$$

Equation 3.13 is similar to Equation 3.6, with two exceptions, both due to damping. First, the harmonic motion indicated within the brackets has a frequency reduced by the factor β, but this effect is small at usual damping. Second, the amplitude is reduced by the decreasing exponential factor which multiplies the oscillatory terms. Circular frequency of the exponent, $\zeta\omega_1$, is also less than ω_1.

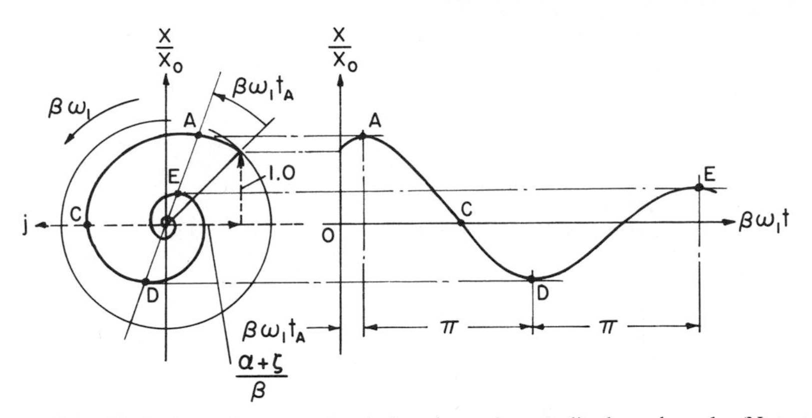

FIGURE 3.4. Underdamped system, $\zeta < 1$, has decreasing amplitude each cycle. (Note $t_{AC} > t_{CD}$.)

The resulting displacement variation is therefore vibratory, the amplitude decreasing with time, shown normalized in Figure 3.4. The greater the damping, the faster the decay.

It is interesting to observe the representation of Equation 3.13 in the complex plane (Figure 3.5*a*). Generation of the harmonic motion by the single rotating vector is normal representation, consisting of a constant resultant amplitude C, an initial phase angle ϕ, and a real axis projection. Similarly, Equation 3.12 is represented in Figure 3.5*b* by two counterrotating vectors which produce the same harmonic variation on the real axis.

Velocity is obtained by differentiating Equation 3.13, and by introducing initial conditions.

$$A_1 = x_0 \qquad A_2 = \left[\frac{(\dot{x}_0/\omega_1) + \zeta x_0}{\beta}\right]$$

from which the dimensionless displacement is

$$\frac{x}{x_0} = e^{-\zeta\omega_1 t}\left[\cos\beta\omega_1 t + \left(\frac{\alpha + \zeta}{\beta}\right)\sin\beta\omega_1 t\right] \tag{3.14}$$

The sine term coefficient depends on the initial displacement and velocity, the natural frequency, and the damping ratio.

The dimensionless velocity becomes

$$\left(\frac{\dot{x}}{\omega_1 x_0}\right) = e^{-\zeta\omega_1 t}\left[\alpha\cos\beta\omega_1 t - \left(\frac{1 + \alpha\zeta}{\beta}\right)\sin\beta\omega_1 t\right] \tag{3.15}$$

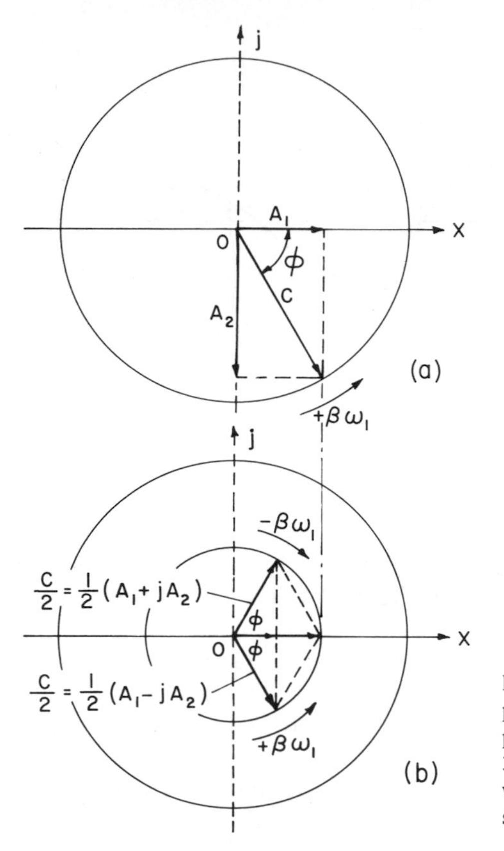

FIGURE 3.5. The trigonometric term in parentheses in Equation 3.13 and the exponential term in parentheses in Equation 3.12 generate equivalent harmonic motions when projected on the real axis. Vectors are shown in the complex plane for $t = 0$.

Maximum response at A (Figure 3.4) is obtained by equating Equation 3.15 to zero.

$$\tan \beta\omega_1 t_A = \left(\frac{\alpha\beta}{1 + \alpha\zeta}\right) \tag{3.16}$$

As shown in Figure 3.4, the generating vector in the complex plane starts from an initial angle and is reduced as it rotates by the factor $e^{-\zeta\omega_1 t}$, developing a spiral curve for projection. It can be seen from the spiral geometry that the time interval from A to C is greater than that from C to D, but that the AD and DE intervals are both π.

We note that in $\beta\omega_1 t_A$ time cannot be negative, and the angle must be measured in a positive sense for negative values of the tangent.

3.5. CRITICALLY DAMPED INITIAL DISPLACEMENT

If the viscous damping coefficient is exactly equal to $2M\omega_1$, critical damping exists and the exponents in Equation 3.5 are no longer complex. It is necessary to modify the equation as follows:

$$x = e^{-\omega_1 t}(C_1 + C_2 t) \tag{3.17}$$

$$\dot{x} = e^{-\omega_1 t}[C_2 - \omega_1(C_1 + C_2 t)] \tag{3.18}$$

where $C_1 = x_0$
$C_2 = (\dot{x}_0 + \omega_1 x_0)$

The dimensionless displacement becomes

$$\left(\frac{x}{x_0}\right) = e^{-\omega_1 t}[1 + \omega_1 t(1 + \alpha)] \tag{3.19}$$

Generalized transient response curves are shown in Figure 3.6. As shown, it is necessary for α to be greater than 1.0 in a negative sense in order for the displacement to cross the axis. In any case, there is no reversal of sign after a maximum peak A has been reached, and critical damping represents the threshold between periodic and this nonperiodic vibration. Also, the concept of damped circular frequency $\beta\omega_1$ does not now exist, although the nominal or undamped natural frequency is still an important parameter.

Additional expressions applying to critically damped systems are given in Tables 3.1 and 3.2, including both velocity and acceleration. It can be noted

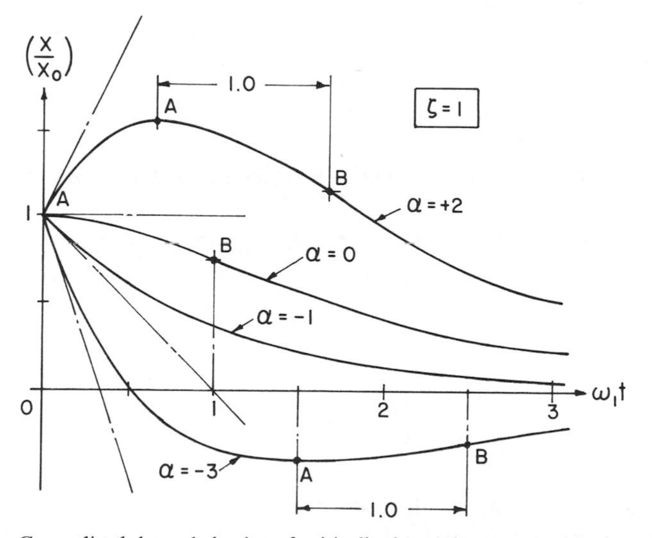

FIGURE 3.6. Generalized decay behavior of critically damped system is a function of the initial velocity factor, $\alpha = \dot{x}_0/\omega_1 x_0$.

that the time phase angle from maximum displacement to maximum recovery velocity, A to B, is exactly 1 rad.

3.6. OVERDAMPED INITIAL DISPLACEMENT

If damping exceeds critical, Equation 3.5 is directly applicable but with real exponents.

$$x = e^{-\zeta\omega_1 t}\left(C_1 e^{\beta\omega_1 t} + C_2 e^{-\beta\omega_1 t}\right) \tag{3.20}$$

β now represents $\sqrt{\zeta^2 - 1}$ rather than $\sqrt{1 - \zeta^2}$, as in the underdamped case. The constants become

$$C_1 = \left[\frac{x_0(\beta + \zeta) + (\dot{x}_0/\omega_1)}{2\beta}\right]$$

$$C_2 = \left[\frac{x_0(\beta - \zeta) - (\dot{x}_0/\omega_1)}{2\beta}\right]$$

And the equation for displacement in reduced form is

$$\frac{x}{x_0} = e^{-\zeta\omega_1 t}\left[\cosh\beta\omega_1 t + \left(\frac{\zeta + \alpha}{\beta}\right)\sinh\beta\omega_1 t\right] \tag{3.21}$$

As seen in Tables 3.1 and 3.2, expressions for the overdamped system are identical in all respects to the underdamped except that the trigonometric functions become hyperbolic.

3.7. EXPERIMENTAL DETERMINATION OF DAMPING

By taking two successive peak amplitudes and letting $\beta\omega_1 t = 0$ and 2π in Equation 3.14, we find this to be a constant ratio of $e^{-2\pi\zeta/\beta}$. Considering more than one cycle, peak amplitudes continue to decrease by multiples of this constant (Figure 3.7).

$$\frac{x_n}{x_A} = \left(e^{-2\pi\zeta/\beta}\right)^n \tag{3.22}$$

Successive amplitude ratios also decay in a similar fashion if we consider intermediate or half-cycle ordinates. The dashed lines that form the envelope of the peaks are accordingly symmetrical with respect to the time axis.

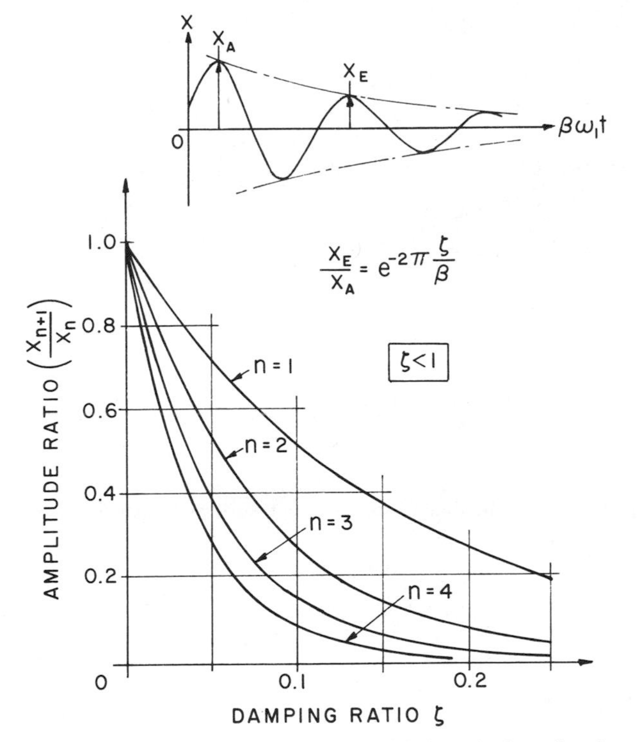

FIGURE 3.7. Progressive or exponential decay in free vibration results in successive amplitudes that are constantly reduced by a fixed factor.

It is usually convenient to plot decay data on semilog graph paper, as cycles versus peak values will become a straight line for viscous damping because of the exponential nature of Equation 3.22.

Damping can also be quantified by observing a portion of the initial decay curve (Figure 3.8). The displacement, say at $\pi/2$, varies from zero to one, depending on damping. Thus if natural frequency is known, we can study damping ratios above critical for which cyclic data are not possible.

3.8. ENERGY DISSIPATION

Referring to the Figure 3.7 decay curves, it is apparent that the level of total energy in a system decreases with both time and displacement, and that the dashpot is the source of this loss. In the underdamped case then we can compare the decrease in elastic energy in the spring from A to E when no

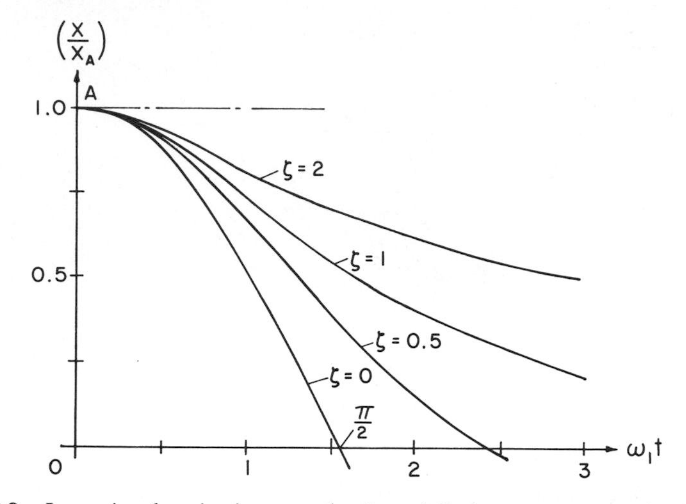

FIGURE 3.8. Increasing damping increases the time of displacement recovery in simple decay.

kinetic energy exists.

$$(\mathrm{EE})_A - (\mathrm{EE})_E = \tfrac{1}{2}K\left(x_A^2 - x_E^2\right)$$

$$\left(\frac{E_{A+1}}{E_A}\right) = \left(1 - e^{-4\pi\zeta/\beta}\right) \tag{3.23}$$

This dimensionless factor represents the energy dissipated per cycle relative to the total elastic energy at the beginning of the cycle under consideration.

3.9. INITIAL VELOCITY

Although a system with only initial velocity and no displacement at $t = 0$ seems similar to previous cases, it must be treated somewhat differently analytically. If $x_0 = 0$ the preceding reference terms become zero; therefore, it is necessary to use a different reference basis for the generalized variables. These involve the initial velocity $\dot{x}_0$, now the only initial condition. As shown in Figure 3.9, the dimensionless displacement passes through the origin with the slope of the tangent proportional to $\dot{x}_0$. Again the response can be considered as having two component phases. One is the initial from 0 to A, leading to the coordinates of A. The other is then a simple decay from A with initial displacement of x_A, and zero initial velocity.

3.10. UNDAMPED INITIAL VELOCITY

With $x_0 = 0$, Equation 3.6 provides the following simple relations:

$$\left(\frac{x}{\dot{x}_0/\omega_1}\right) = \sin \omega_1 t \tag{3.24}$$

$$\left(\frac{\dot{x}}{\dot{x}_0}\right) = \cos \omega_1 t \tag{3.25}$$

and the reference displacement $(\dot{x}_0/\omega_1)$ corresponds to the amplitude of a steady-state simple harmonic motion in which the maximum velocity at midposition is $\dot{x}_0$. Other variables for zero damping and the ensuing sinusoidal response are given in Table 3.3.

TABLE 3.3. Free vibratory response with initial velocity but with zero initial displacement.

	$\zeta = 0$	$0 < \zeta < 1$	$\zeta = 1$	$\zeta > 1$
β	1.00	$\sqrt{1-\zeta^2}$	0	$\sqrt{\zeta^2 - 1}$
$\dfrac{x}{(\dot{x}_0/\omega_1)}$	$\sin \omega_1 t$	$e^{-\zeta\omega_1 t}\left[\dfrac{\sin \beta\omega_1 t}{\beta}\right]$	$e^{-\omega_1 t}[\omega_1 t]$	$e^{-\zeta\omega_1 t}\left[\dfrac{\sinh \beta\omega_1 t}{\beta}\right]$
$\dfrac{\dot{x}}{\dot{x}_0}$	$\cos \omega_1 t$	$e^{-\zeta\omega_1 t}[\cos \beta\omega_1 t - \zeta/\beta \sin \beta\omega_1 t]$	$e^{-\omega_1 t}[1 - \omega_1 t]$	$e^{-\zeta\omega_1 t}[\cosh \beta\omega_1 t - \zeta/\beta \sinh \beta\omega_1 t]$
$\dfrac{\ddot{x}}{\omega_1 \dot{x}_0}$	$-\sin \omega_1 t$	$e^{-\zeta\omega_1 t}\left[-2\zeta \cos \beta\omega_1 t + \left(\dfrac{2\zeta^2-1}{\beta}\right) \sin \beta\omega_1 t\right]$	$e^{-\omega_1 t}[\omega_1 t - 2]$	$e^{-\zeta\omega_1 t}\left[-2\zeta \cosh \beta\omega_1 t + \left(\dfrac{2\zeta^2-1}{\beta}\right) \sinh \beta\omega_1 t\right]$
$\omega_1 t_A = \omega_1 t_{AB}$	$\dfrac{\pi}{2}$	$\dfrac{1}{\beta} \tan^{-1}\left(\dfrac{\beta}{\zeta}\right)$	1.00	$\dfrac{1}{\beta} \tanh^{-1}\left(\dfrac{\beta}{\zeta}\right)$
$\dfrac{x_A}{(\dot{x}_0/\omega_1)}$	1.00	$e^{-\zeta\omega_1 t_A}$	e^{-1}	$e^{-\zeta\omega_1 t_A}$
$\dfrac{x_B}{(\dot{x}_0/\omega_1)}$	0	$2\zeta e^{-2\zeta\omega_1 t_A}$	$2e^{-2}$	$2\zeta e^{-2\zeta\omega_1 t_A}$
$\dfrac{\dot{x}_B}{\dot{x}_0}$	-1.00	$-e^{-2\zeta\omega_1 t_A}$	$-e^{-2}$	$-e^{-2\zeta\omega_1 t_A}$
$\omega_1 t_C$	π	$\dfrac{\pi}{\beta}$	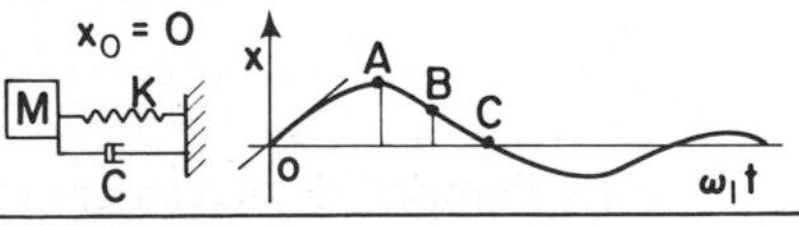 $x_0 = 0$ M K C	x A B C O $\omega_1 t$

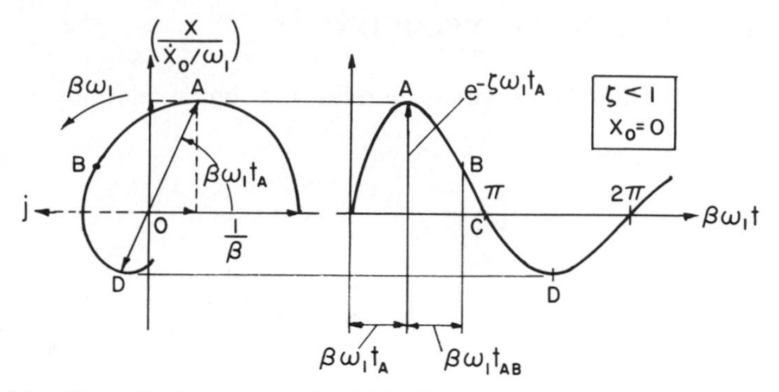

FIGURE 3.9. Generalized response with initial velocity only is shown generated in the complex plane.

3.11. UNDERDAMPED INITIAL VELOCITY

For moderate damping, Equation 3.13 yields the following coefficients:

$$A_1 = 0 \qquad A_2 = \frac{1}{\beta}\left(\frac{\dot{x}_0}{\omega_1}\right)$$

from which

$$\left(\frac{x}{\dot{x}_0/\omega_1}\right) = e^{-\zeta\omega_1 t}\left(\frac{\sin \beta\omega_1 t}{\beta}\right) \tag{3.26}$$

The nature of the complex vector relationships is shown in Figure 3.9, with the displacement vector on the imaginary axis at $t = 0$ and equal to $1/\beta$. Characteristics of the response can be determined from Table 3.3; several are indicated in Figure 3.9.

3.12. CRITICALLY DAMPED INITIAL VELOCITY

If $\zeta = 1$, the Equation 3.17 coefficients become

$$C_1 = 0 \qquad C_2 = \dot{x}_0$$

The dimensionless displacement is then

$$\left(\frac{x}{\dot{x}_0/\omega_1}\right) = e^{-\omega_1 t}(\omega_1 t) \tag{3.27}$$

This critically damped displacement (Figure 3.10) indicates that the maximum

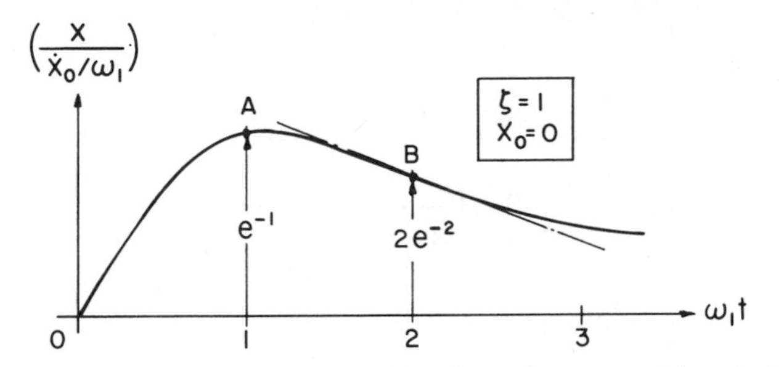

FIGURE 3.10. Generalized response of critically damped system with only initial velocity achieves maximum displacement at $\omega_1 t = 1$.

is reached at exactly $\omega_1 t = 1$, with a normalized displacement at A of e^{-1}. Thus the peak motion is only 37% of the undamped or sinusoidal amplitude associated with the same initial velocity. Maximum recovery velocity occurs at $\omega_1 t = 2$, with displacement eventually approaching zero asymptotically.

3.13. OVERDAMPED INITIAL VELOCITY

Applying initial conditions to Equation 3.20 with $\zeta > 1$,

$$C_1 = \frac{1}{2\beta}\left(\frac{\dot{x}_0}{\omega_1}\right) \qquad C_2 = -\frac{1}{2\beta}\left(\frac{\dot{x}_0}{\omega_1}\right)$$

This results in a generalized displacement expression:

$$\left(\frac{x}{\dot{x}_0/\omega_1}\right) = e^{-\zeta\omega_1 t}\left(\frac{\sinh \beta\omega_1 t}{\beta}\right) \tag{3.28}$$

Other relations follow in Table 3.3. It is interesting to note that the time interval to maximum displacement is exactly one-half the total time to maximum recovery velocity, $\omega_1 t_A = \omega_1 t_{AB}$, *for any value of damping ratio.*

3.14. SUMMARY OF INITIAL PHASE COORDINATES

Figure 3.11 indicates the effects of damping ratio on both the time to reach peak displacement and the magnitude of that displacement for all ratios from 0 to 3. The overshoot time interval is reduced with increasing damping from a maximum of $\pi/2$ at zero damping or sinusoidal motion (upper curve). Unit damping ratio corresponds to unit angle.

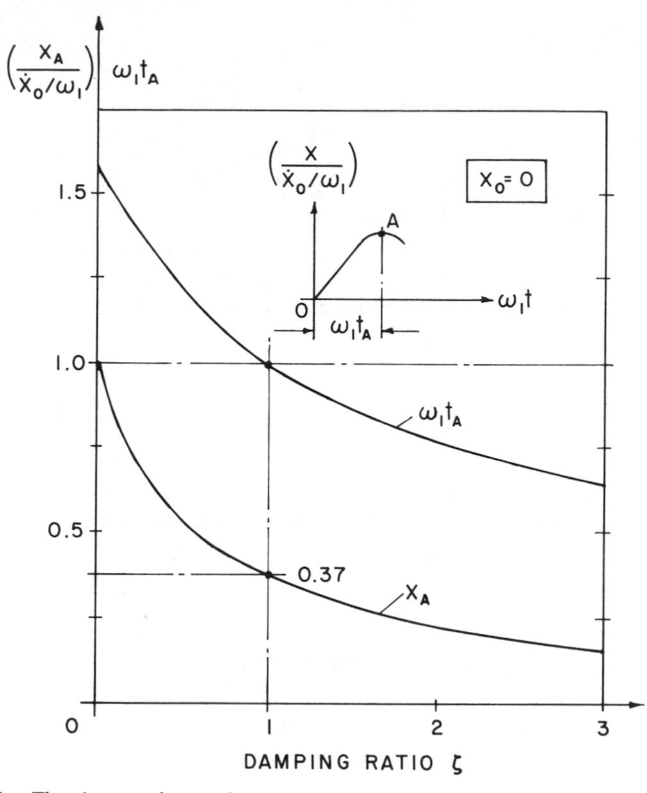

FIGURE 3.11. The time to the maximum point, A, decreases with damping ratio, as does the normalized displacement at the resulting peak.

Normalized displacement drops from unity (lower curve) on a continuous basis from zero to hypercritical damping.

3.15. THE FREE VERTICAL SYSTEM

In most respects the behavior of a vibratory system is independent of its spatial orientation. If vertical, gravity effects cancel in the force equation, resulting only in a lowered operational position; however, weight terms should technically be included in Equation 2.1. We note that in all systems taken horizontally, as in Figure 2.1, there is no weight contribution and the spring at rest is unstressed. With vertical position (Figure 3.12), free vibration occurs relative to the deflected position and the equilibrium equation becomes

$$\Sigma Fy = -M\ddot{y} - W - K(y - b) = 0$$

$$= M\ddot{y} - Ky = 0 \tag{3.29}$$

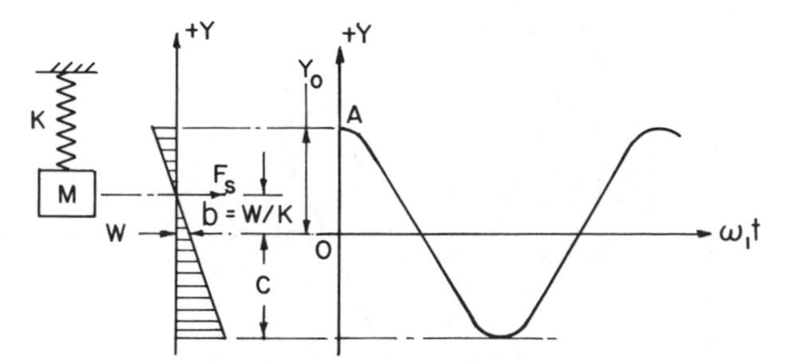

FIGURE 3.12. The vertical system with initial positive displacement oscillates about the position of static sag.

Thus a sinusoidal oscillation ensues with displacements defined as shown. Force in the spring is the sum of the constant weight load and the vibratory forces due to disturbance. If a dashpot is present, its force becomes zero as the system decays to the static position having no influence upon the terminal state $(y = 0)$.

The subsequent force equation is then for the vertical system Equation 3.1, with x replaced by y, and the relations given in Tables 3.1–3.3 are generally applicable.

If initial conditions should correspond to release with a vertical displacement of $+y_0$ above the deflected position (Figure 3.12), initial energy is the elastic energy in the spring and the potential energy at this position.

$$E_1 = (\mathrm{EE})_1 + (\mathrm{PE})_1 = \tfrac{1}{2}K(y_0 - b)^2 + W(c + y_0)$$

After release, the mass reaches a maximum displacement and zero velocity at the potential energy datum with only elastic energy stored in the spring.

$$E_2 = (\mathrm{EE})_2 = \tfrac{1}{2}K(c + b)^2$$

From conservation of energy, $E_1 = E_2$, and

$$(\mathrm{EE})_1 + (\mathrm{PE})_1 = (\mathrm{EE})_2$$

which reduces to

$$c = y_0$$

We conclude that the excursion of the mass is y_0 equally above and below the position of static deflection, with the mass recovering completely to the upper position (1). Motion is harmonic at the natural frequency.

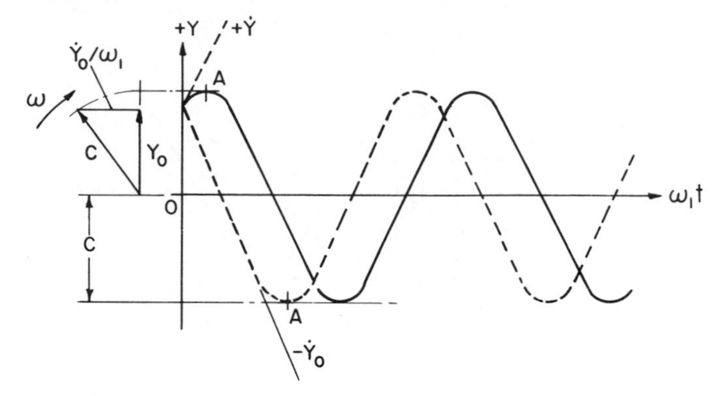

FIGURE 3.13. With initial displacement and positive or negative velocity, oscillation similarly occurs about the position of static equilibrium from which y is measured.

Similarly, with initial velocity superimposed at the upper position, kinetic energy is introduced (Figure 3.13).

$$E_1 = (\text{EE})_1 + (\text{PE})_1 + (\text{KE})_1 = \tfrac{1}{2}(y_0 - b)^2 + W(c + y_0) + \tfrac{1}{2}M\dot{y}_0^2$$

$$E_2 = (\text{EE})_2 = \tfrac{1}{2}K(c + b)^2$$

from which

$$c^2 = y_0^2 + \left(\frac{\dot{y}_0}{\omega_1}\right)^2 \quad \text{or} \quad \frac{c}{y_0} = \frac{y_A}{y_0} = \sqrt{1 + \alpha^2} \tag{3.30}$$

where $\omega_1 = \sqrt{K/M}$
$\alpha = (\dot{y}_0/\omega_1 y_0)$

Note that the resultant peak amplitude is represented by the hypotenuse of a right triangle in Figure 3.13. Also, since the initial velocity term is squared, the same vibratory amplitude is induced regardless of the direction of $\dot{y}_0$.

3.16. LOADING BY CONTINUOUS FORCE

The vertical system of Section 3.15 is illustrative of a spring–mass system subjected to a sustained constant force, the weight of the mass element. We have seen that the mass oscillates about the deflected position of static equilibrium, regardless of the initial conditions. With a horizontal system the behavior is similar if a sustained external horizontal force is exerted on the

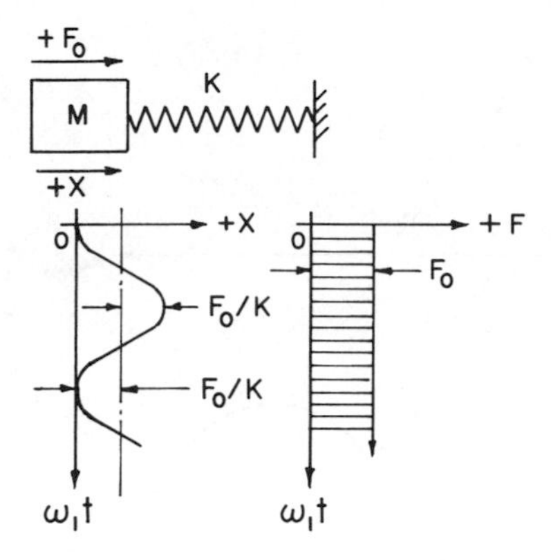

FIGURE 3.14. The mass subjected to a sustained constant force oscillates at the natural frequency about a midposition corresponding to a bias of F_0/K.

mass (Figure 3.14).

$$\Sigma F_x = -M\ddot{x} - Kx + F_0 = 0$$

$$\left(\frac{\ddot{x}}{\omega_1^2}\right) + x = \left(\frac{F_0}{K}\right) \tag{3.31}$$

If the constant force F_0 is applied to the right with zero initial spring displacement and velocity (see Figure 3.14), the maximum displacement is $2F_0/K$ and the mean steady-state position is at $+F_0/K$, with a solution

$$x = \left(\frac{F_0}{K}\right)(1 - \cos \omega_1 t) \tag{3.32a}$$

$$\dot{x} = \left(\frac{F_0}{K}\right)\omega_1 \sin \omega_1 t \tag{3.32b}$$

$$\ddot{x} = \left(\frac{F_0}{K}\right)\omega_1^2 \cos \omega_1 t \tag{3.32c}$$

where $\omega_1 = \sqrt{K/M}$, the natural frequency.

3.17. LOADING BY STEP FORCE FUNCTION

If F_0 is removed after an elapsed time t_0 (Figure 3.15), a free vibrational phase follows the initial forced phase, harmonic at the natural frequency. This secondary phase has initial values x_0 and $\dot{x}_0$ corresponding to the displacement and velocity existing at the termination of F_0; therefore, Equations 3.32 can be

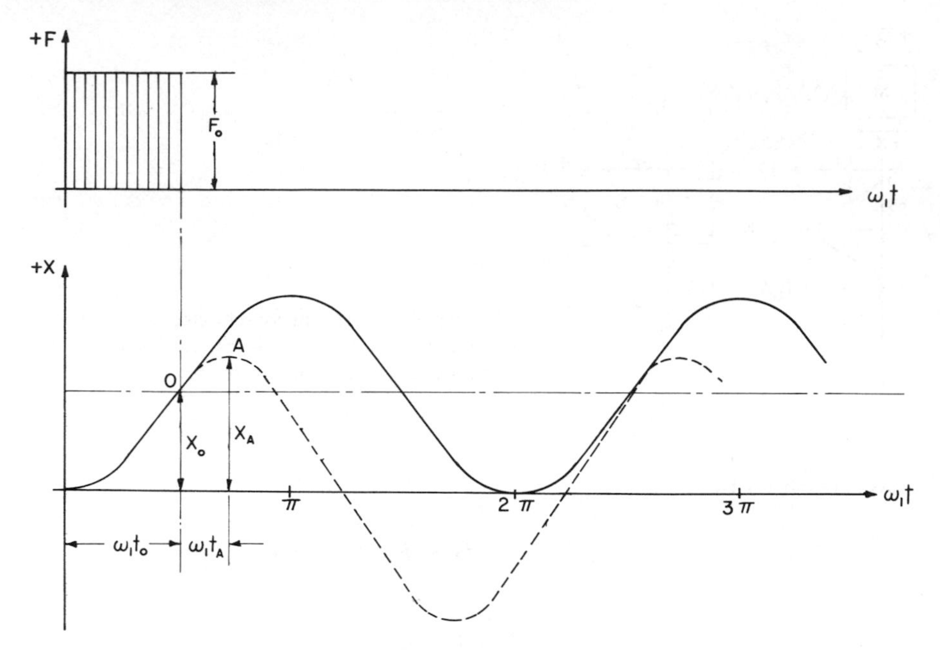

FIGURE 3.15. A continuous force loading produces the sinusoidal response (solid curve) but with no negative displacement. Free vibration following termination of the force (dashed curve) oscillates about the unloaded position of the spring.

used to relate the two phases, and

$$\alpha = \left(\frac{\dot{x}_0}{\omega_1 x_0}\right) = \left(\frac{\sin \omega_1 t_0}{1 - \cos \omega_1 t_0}\right) = \left[\frac{\sin \omega_1 t_0}{2 \sin^2(\omega_1 t_0/2)}\right] \tag{3.33}$$

Then use of Table 3.2 for $\zeta = 0$ permits complete determination of the secondary phase. At the first peak,

$$\frac{x_A}{x_0} = \sqrt{1 + \alpha^2} \tag{3.34}$$

$$\omega_1 t_A = \tan^{-1}\alpha \tag{3.35}$$

Equations 3.34 and 3.35 are conveniently visualized vectorially[9] in Figure 3.16, drawn with radius F_0/K and a rotational time angle of $\omega_1 t_0$ to determine $\overline{OP}$. From the geometry,

$$\overline{OP} = \left(\frac{F_0}{K}\right) 2 \sin\left(\frac{\omega_1 t_0}{2}\right) = x_A$$

Similarly the required time to arrive at the subsequent peak response is

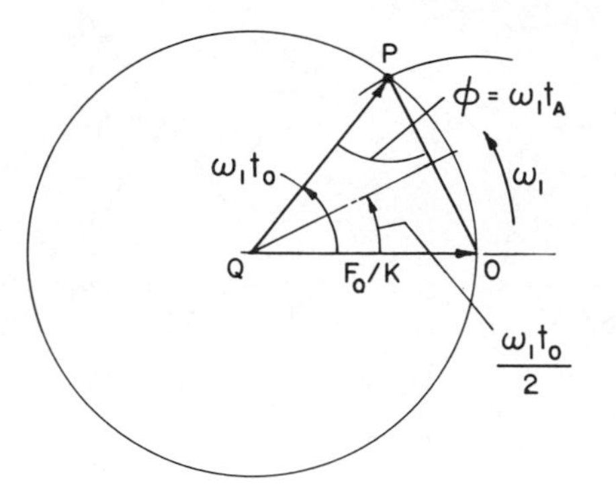

FIGURE 3.16. Vector solution for coordinates of peak, A, of secondary free vibration, with $OP = x_A$.

displayed as ϕ, since

$$\tan\phi = \frac{(F_0/K)\cos(\omega_1 t_0/2)}{(F_0/K)\sin(\omega_1 t_0/2)} = \cot\left(\frac{\omega_1 t_0}{2}\right)$$

and

$$\tan\omega_1 t_A = \alpha = \left(\frac{\sin\omega_1 t_0}{1 - \cos\omega_1 t_0}\right)$$

from which

$$\omega_1 t_A = \phi \tag{3.36}$$

Thus the peak response x_A is given by the distance $\overline{OP}$ and the phase angle $\omega_1 t_A$ by the angle ϕ in the vector diagram shown. F_0 can be applied for more than one system period, in which case the generating point P exceeds one complete revolution. With a continuous force the point and vector continue to rotate.

3.18. LOADING BY GENERAL FORCE–TIME FUNCTION

Forcing of the mass obviously need not be restricted to the simple step function. For instance if $F(t)$ is an arbitrary function or variation, it can be approximated as a series of rectangular pulses acting successively, each equivalent to a Figure 3.16 diagram. This can involve several possibilities, as seen in Figure 3.17:

1. The magnitude of F_0 can increase, with the center of the next arc, Q_2, farther away from P_1 (Figure 3.17*a*).
2. The magnitude of the force can decrease, reducing the radius to P_1Q_2 (Figure 3.17*b*).

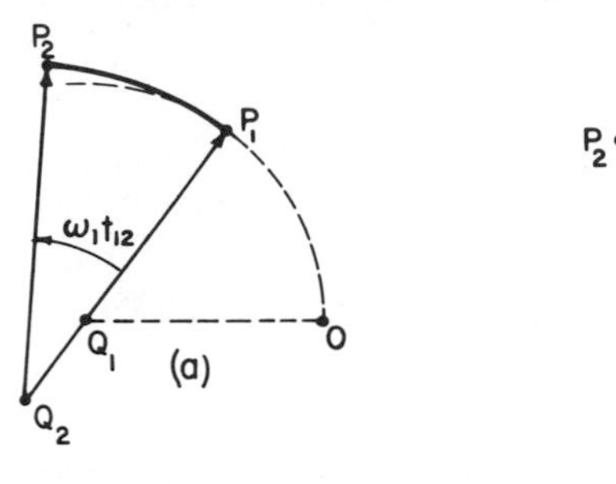

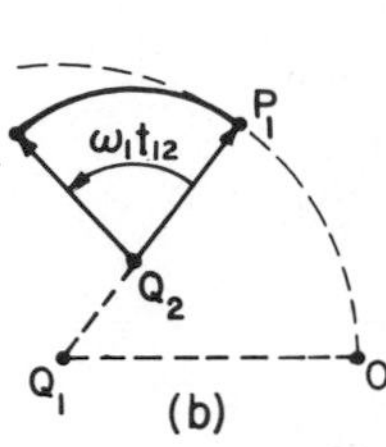

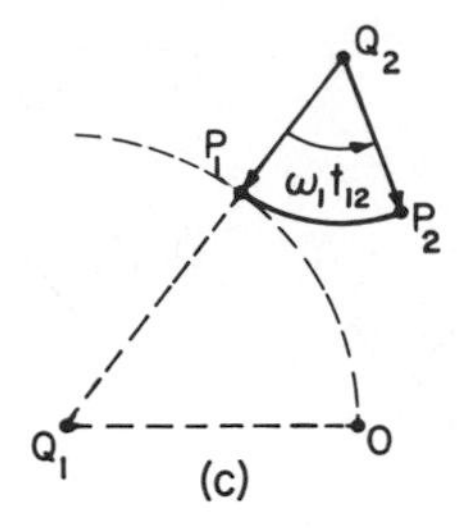

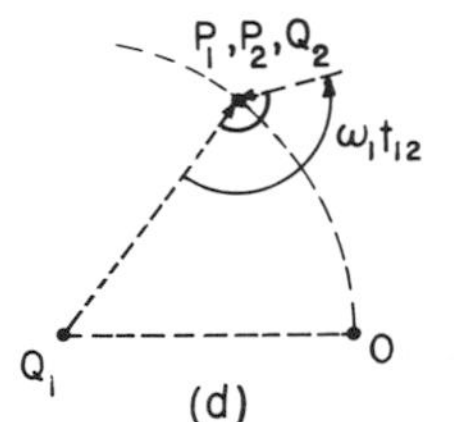

FIGURE 3.17. Transition from one pulse to the next can involve modification of the length of the generating vector (a) or (b); reversal of sense (c); or a zero (null) vector (d). At transition the new center must lie on the directional of the previous terminal vector.

3. F_0 can become negative, requiring a reversal of Q_2 to a position outside of the previous circle (Figure 3.17c).
4. F_0 can become zero for some time interval, resulting in the rotation of a null vector (Figure 3.17d).

In all cases we note the adjacent pulse arcs have a common tangent at the transitional instant, and the vectors all rotate in the counterclockwise direction. As suggested by Figures 3.15 and 3.16, the radius $\overline{OP}$ represents a maximum displacement x_A at time t_A, usually after cessation of a force application; however, from Figure 3.14, a maximum can also occur during the force application. The condition or test for a maximum response point is that the rotational vector must be coincident with a line drawn from the instantaneous position of P to the origin, O.

For example, in Figure 3.16 the vector position is QP as F_0 is removed. As time elapses with a zero vector, a null vector continues to rotate at ω_1, and when this null vector aligns with $\overline{OP}$, after an angle of $\omega_1 t_A$, maximum displacement x_A is achieved instantaneously. Alternatively there is then a common tangent at P between the circle of radius $\overline{OP}$ and the radius of the null circle under generation.

To illustrate this analysis Figure 3.18 indicates a conventional sinusoidal excitation for an undamped resonant situation, $\omega = \omega_1$. The system is unstable and cannot reach steady state. By using approximate reversing rectangular steps and applying the vectorial solution, we see the amplitude increasing each half-cycle. The factors are 2, 4, 6, 8,..., or a $4n$ series, where n is the number

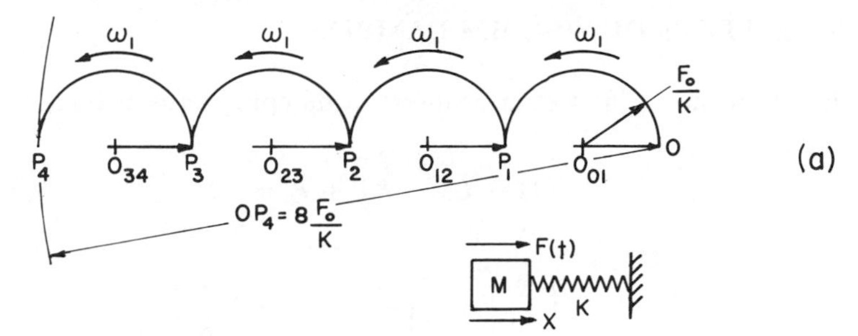

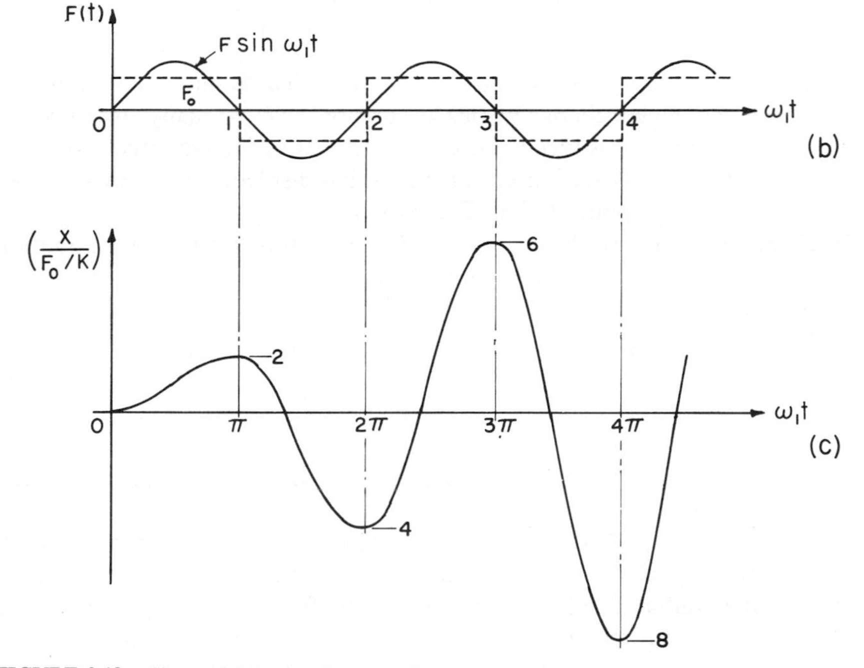

FIGURE 3.18. Sinusoidal forcing is approximated by reversing pulses (*b*), leading to a vector solution (*a*). The corresponding divergent resonant amplitude is shown in (*c*).

of completely reversed cycles. Maxima correspond to the alignment of the vectors with the horizontal at the instant of reversal. As expected, the successive peaks build toward infinity. We also observe in Figure 3.18*c* that the sense of F_0 is at all times in the direction of the vibratory velocity, so net external energy is added to the system during each pulse. The incremental energy increases as both the velocity and displacement levels increase progressively.

This then illustrates the nature of the divergent amplitude of a resonant undamped system and the mechanism of this divergence on an approximately quantitative basis, and indicates the time effects in the development of resonant amplitudes.

3.19. EFFECTS OF VISCOUS DAMPING

Inclusion of the dashpot element inserts a damping term in Equation 3.31.

$$-M\ddot{x} - C\dot{x} - Kx + F_0 = 0$$

$$\left(\frac{\ddot{x}}{\omega_1^2}\right) + 2\zeta\left(\frac{\dot{x}}{\omega_1}\right) + x = \left(\frac{F_0}{K}\right) \tag{3.37}$$

In Figure 3.19 we note that the addition of subcritical damping modifies and reduces the undamped characteristics in Figure 3.14. Actually, referring to Figure 3.2a, the motion is seen to be closely related to the recovery pattern of the initially displaced system; that is, if the initial displacement is $x_A = F_0/K$, the coordinate z corresponds to x in Figure 3.2a.

In Figure 3.19 then we have $x = (F_0/K) - z$. Substituting from Equation 3.14 with $\zeta < 1$, $x_0 = x_A = F_0/K$, and $\alpha = 0$,

$$x = \left(\frac{F_0}{K}\right)\left[1 - e^{-\zeta\omega_1 t}\left(\cos\beta\omega_1 t + \frac{\zeta}{\beta}\sin\beta\omega_1 t\right)\right] \tag{3.38}$$

Thus the continuously applied force represents the superposition of a fixed displacement and a free vibration, with various degrees of damping easily introduced (see Figure 3.20 and Table 3.4). Similarly, if the force is removed, displacement and velocity at the instant of termination can be established which become initial conditions for the truly free damped phase readily defined from previous relations. Several are indicated in Figure 3.20.

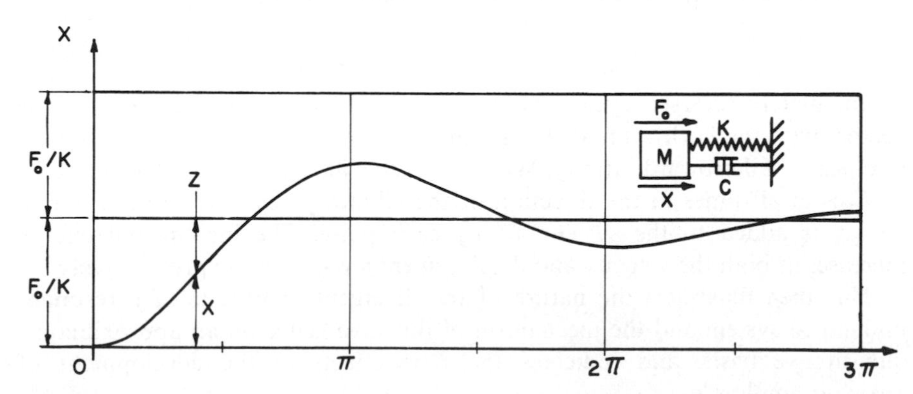

FIGURE 3.19. Response due to external force is equivalent to a damped free release, Z, as shown for $\zeta = 0.25$.

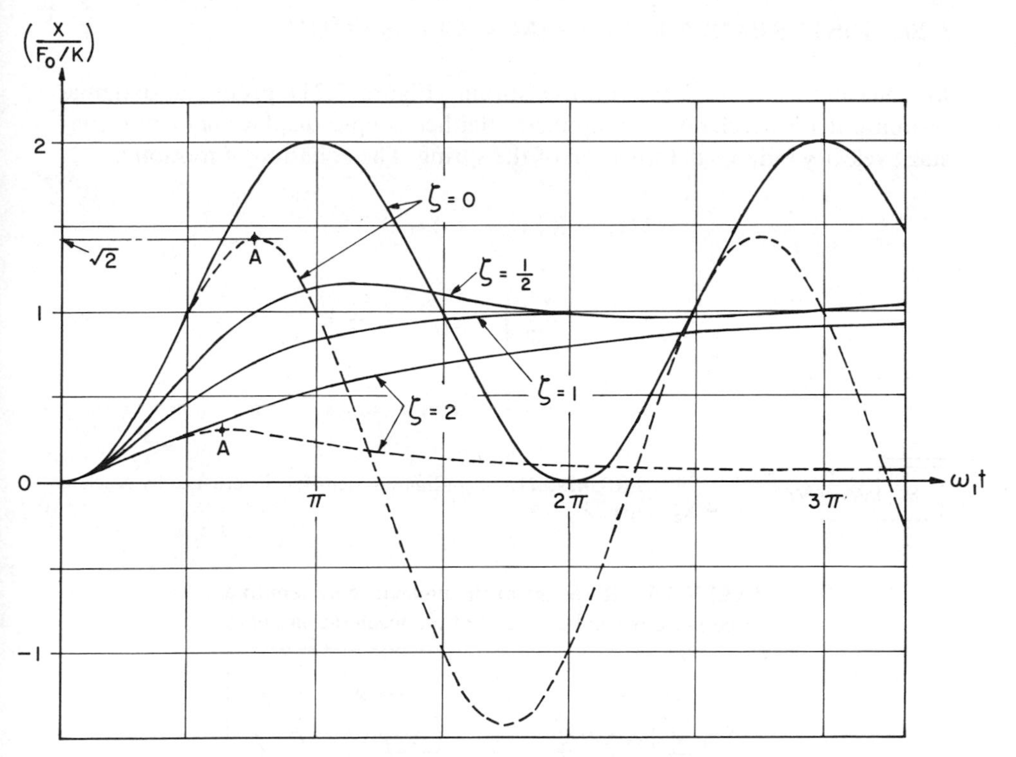

FIGURE 3.20. Response characteristics with various damping for continuous F_0 (solid curves), indicating final asymptotic displacement as F_0/K. After F_0 is removed (dashed curves), the asymptote is zero.

TABLE 3.4. Equations for determination of displacement, velocity, and acceleration for various damping when disturbed by a constant force pulse F_0.

	$\zeta = 0$	$0 < \zeta < 1$	$\zeta = 1$	$\zeta > 1$
β	1.00	$\sqrt{1-\zeta^2}$	0	$\sqrt{\zeta^2-1}$
$\frac{X}{F_0/K}$	$1-\cos\omega_1 t$	$1-e^{-\zeta\omega_1 t}\left[\cos\beta\omega_1 t + \frac{\zeta}{\beta}\sin\beta\omega_1 t\right]$	$1-e^{\omega_1 t}[1+\omega_1 t]$	$1-e^{\zeta\omega_1 t}\left[\cosh\beta\omega_1 t + \frac{\zeta}{\beta}\sinh\beta\omega_1 t\right]$
$\frac{\dot{X}}{\omega_1(F_0/K)}$	$\sin\omega_1 t$	$\frac{e^{-\zeta\omega_1 t}}{\beta}\left[\sin\beta\omega_1 t\right]$	$e^{-\omega_1 t}(\omega_1 t)$	$\frac{e^{-\zeta\omega_1 t}}{\beta}\left[\sinh\beta\omega_1 t\right]$
$\frac{\ddot{X}}{\omega_1^2(F_0/K)}$	$\cos\omega_1 t$	$e^{-\zeta\omega_1 t}\left[\cos\beta\omega_1 t - \frac{\zeta}{\beta}\sin\beta\omega_1 t\right]$	$e^{-\omega_1 t}\left[1-\omega_1 t\right]$	$e^{-\zeta\omega_1 t}\left[\cosh\beta\omega_1 t - \frac{\zeta}{\beta}\sinh\beta\omega_1 t\right]$
α	$\frac{\sin\omega_1 t}{2\sin^2\frac{\omega_1 t}{2}}$	FROM ABOVE	$\frac{\omega_1 t}{e^{\omega t}-(1+\omega_1 t)}$	FROM ABOVE

3.20. DISTURBANCE BY SYSTEM ACCELERATION

Movement of the free end of the spring (Figure 3.21) produces dynamic response if an acceleration is applied. Neither simple displacement nor constant velocity induces deformation of the spring. The equation of motion is

$$-M\ddot{x}_1 + K(x_2 - x_1) = 0$$

$$\left(\frac{\ddot{x}_{12}}{\omega_1^2}\right) + x_{12} = \left(\frac{\ddot{x}_2}{\omega_1^2}\right) \tag{3.39}$$

FIGURE 3.21. Coordinate system for disturbance by acceleration, $\ddot{x}_2$.

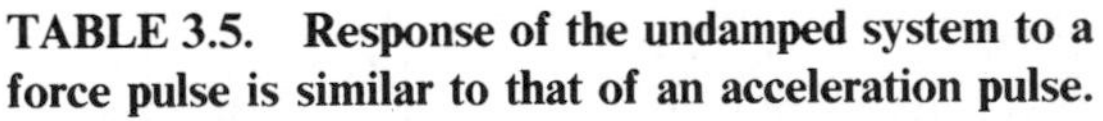

TABLE 3.5. Response of the undamped system to a force pulse is similar to that of an acceleration pulse.

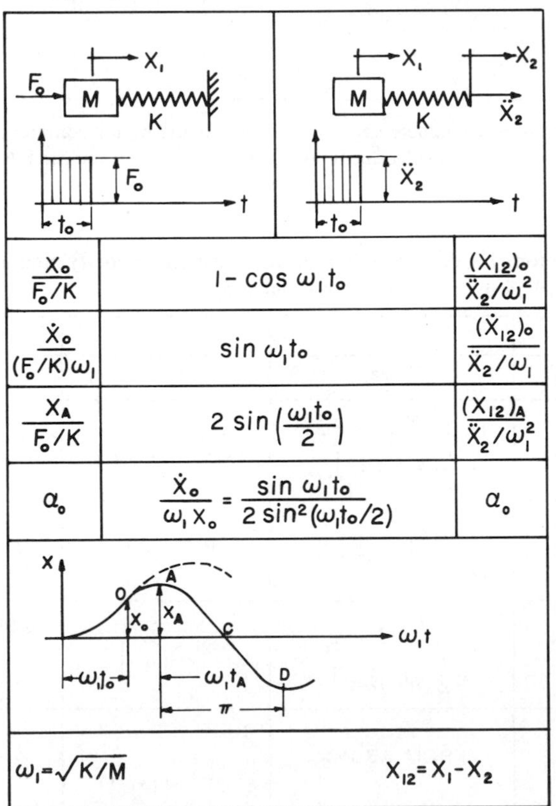

Force pulse		Acceleration pulse
$\frac{X_o}{F_o/K}$	$1 - \cos \omega_1 t_o$	$\frac{(X_{12})_o}{\ddot{X}_2/\omega_1^2}$
$\frac{\dot{X}_o}{(F_o/K)\omega_1}$	$\sin \omega_1 t_o$	$\frac{(\dot{X}_{12})_o}{\ddot{X}_2/\omega_1}$
$\frac{X_A}{F_o/K}$	$2 \sin\left(\frac{\omega_1 t_o}{2}\right)$	$\frac{(X_{12})_A}{\ddot{X}_2/\omega_1^2}$
α_o	$\frac{\dot{X}_o}{\omega_1 X_o} = \frac{\sin \omega_1 t_o}{2 \sin^2(\omega_1 t_o/2)}$	α_o
$\omega_1 = \sqrt{K/M}$		$X_{12} = X_1 - X_2$

where $\ddot{x}_{12}=(\ddot{x}_1-\ddot{x}_2)=$ relative acceleration
$x_{12}=(x_1-x_2)=$ relative displacement
$\ddot{x}_2=$ instantaneous acceleration of 2

Equation 3.39 is similar to Equation 3.31, but with $F_0/K=\ddot{x}_2/\omega_1^2$.

This analogy permits us to apply all previous relations developed for both the continuous and pulse-type force loading, as indicated in Table 3.5 for zero damping. Viscous damping can also be included and patterned after the F_0 solutions. Note that for $\ddot{x}_2$ disturbance, however, both displacements and velocities are *relative* (see Table 3.5 and Equation 3.39).

3.21. THE SPRINGLESS SYSTEM

If a mass reacts on a viscous dashpot in the absence of a spring (Figure 3.22), there is no natural frequency and only system parameters C and M can be used in the analysis. Velocity is the only possible initial condition, as the dashpot is neutral with respect to the horizontal position of the mass. For equilibrium,

$$-M\ddot{x}-C\dot{x}=0 \tag{3.40}$$

Letting $x=Ae^{at}$,

$$(Ma^2+Ca)Ae^{at}=0 \tag{3.41}$$

$$a=-\frac{C}{M}=-\omega_1$$

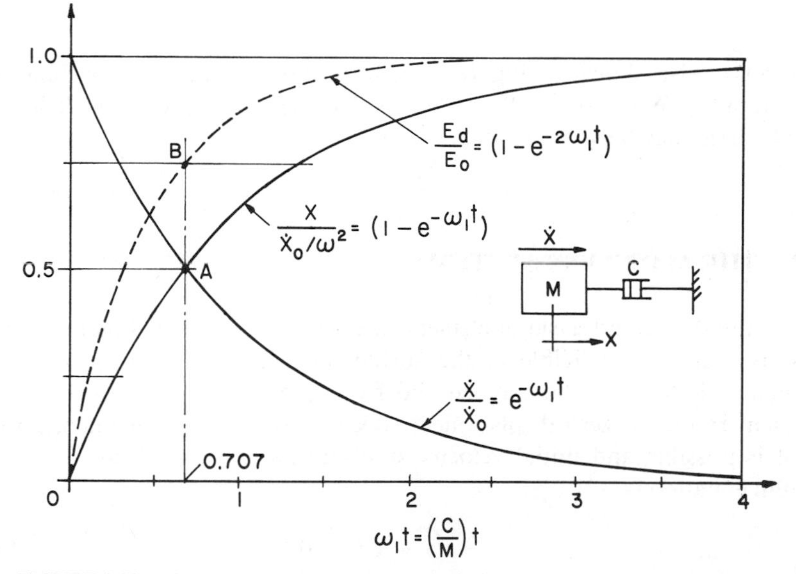

FIGURE 3.22. Generalized behavior of mass–dashpot system with initial velocity.

where ω_1 is an *equivalent* frequency, and since the initial velocity must exist at $t = 0$,

$$\left(\frac{\dot{x}}{\dot{x}_0}\right) = e^{-\omega_1 t} \tag{3.42}$$

Integrating, the generalized displacement becomes

$$\left(\frac{x}{\dot{x}_0/\omega_1}\right) = (1 - e^{-\omega_1 t}) \tag{3.43}$$

As shown, the displacement increases in the direction of initial velocity, approaching asymptotically a limiting displacement of $\dot{x}_0/\omega_1$. The velocity decrease is also shown, representing the variation in dashpot force with time. These two curves intersect at an ordinate of 0.50, indicating that when the velocity drops to half its original value half of the terminal displacement has been reached.

One of the main functions of a dashpot is the dissipation of energy. As the transient cycle progresses, the total energy of the mass decreases from the initial kinetic energy, $\frac{1}{2}M\dot{x}_0^2$. The remaining energy is $\frac{1}{2}M\dot{x}^2$, and the energy dissipated is given by

$$\frac{E_d}{E_0} = \left(1 - \frac{(\mathrm{KE})}{(\mathrm{KE})_0}\right) = 1 - \left(\frac{\dot{x}}{\dot{x}_0}\right)^2 = (1 - e^{-2\omega_1 t}) \tag{3.44}$$

This curve, also shown in Figure 3.22, indicates the ratio approaching unity quite rapidly. When one-half of the displacement has occurred, 75% of the initial energy has been eliminated.

3.22. THE MASSLESS SYSTEM

If only the spring and dashpot elements are coupled together (Figure 3.23), and mass is assumed negligible in the spring and dashpot, there is no natural frequency. Release of the coupled end from a displaced position results in a transient return to zero displacement. Because of the lack of mass, no overshoot is possible and initial velocity similarly has no significance. The equilibrium equation is

$$-Kx - C\dot{x} = 0 \tag{3.45}$$

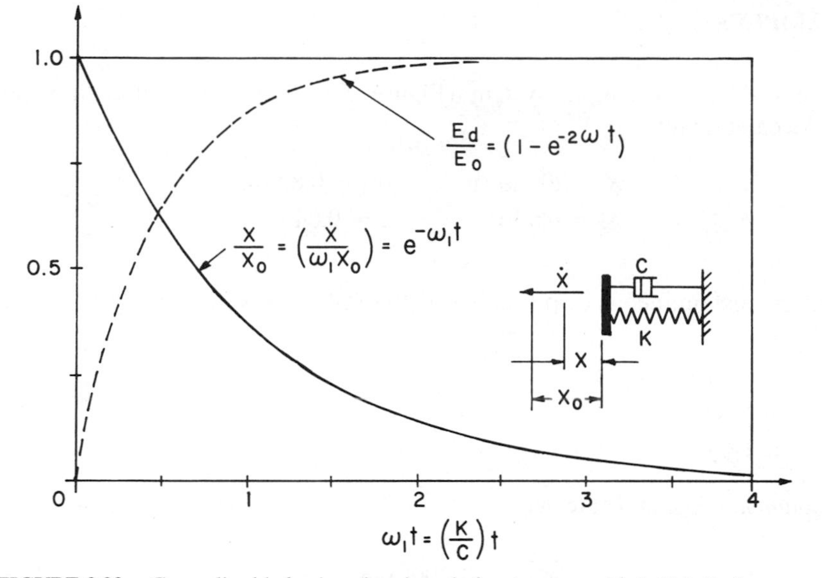

FIGURE 3.23. Generalized behavior of spring–dashpot system with initial displacement.

With $x = Ae^{at}$, and $a = -(K/C) = -\omega_1$, the displacement is

$$\left(\frac{x}{x_0}\right) = e^{-\omega_1 t} \tag{3.46}$$

and the velocity is

$$\left(\frac{\dot{x}}{\omega_1 x_0}\right) = -e^{-\omega_1 t} \tag{3.47}$$

Again, ω_1 is an *equivalent* natural frequency defining the system for purposes of generalized analysis, having dimensions of 1/sec. The exponential return to equilibrium is shown in Figure 3.23. Both the displacement and the velocity have the same pattern as normalized.

Initial system energy is the strain energy in the spring $\frac{1}{2}Kx_0^2$. As the system relaxes, residual strain energy is proportional to x^2, with the difference dissipated.

$$\frac{(\text{EE})}{(\text{EE})_0} = 1 - \left(\frac{x}{x_0}\right)^2 = (1 - e^{-2\omega_1 t}) \tag{3.48}$$

The energy level decreases rapidly with only about 10% remaining at $\omega_1 t =$ 1 rad.

EXAMPLES

EXAMPLE 3.1. A simple system (Figure 3.1) is released after a positive displacement, and

$$K = 200 \text{ lb/in.} \qquad x_0 = 0.80 \text{ in.}$$
$$M = 6.5 \text{ lb} \qquad t = 0.04 \text{ sec}$$

Find the instantaneous displacement if the following values apply:
(a) $\zeta = 0$
(b) $\zeta = 0.30$
(c) $\zeta = 1.00$
(d) $\zeta = 1.90$

Solution. Using Table 3.1,

$$\omega_1 = \sqrt{\frac{K}{M}} = \sqrt{\frac{200(386)}{6.5}} = 108.9 \text{ rad/sec}$$

$$\omega_1 t = 4.36 \text{ rad}$$

(a)
$$x/x_A = \cos \omega_1 t = \cos 4.36 = -0.346$$

$$x = (0.80)(-0.349) = -0.277 \text{ in.}$$

(b)
$$\beta = \sqrt{1 - \zeta^2} = 0.954$$

$$\beta\omega_1 t = (0.954)(4.36) = 4.16$$

$$\zeta\omega_1 t = (0.30)(4.36) = 1.31$$

$$x/x_A = e^{-1.31}[\cos 4.16 + (0.30/0.954)\sin 4.16] = -0.216$$

$$x = (0.80)(-0.216) = -0.173 \text{ in.}$$

(c)
$$x/x_A = e^{-4.36}(1 + 4.36) = 0.069$$

$$x = (0.80)(0.069) = +0.055 \text{ in.}$$

(d) $$\beta = \sqrt{(1.9)^2 - 1} = 1.616$$

$$x/x_A = e^{-8.28}[\cosh 7.04 + (1.9/1.616)\sinh 7.04] = 0.316$$

$$x = (0.80)(0.316) = +0.253 \text{ in.}$$

EXAMPLE 3.2. The basic system of Figure 3.1 has both initial positive displacement and velocity. Data are as follows:

$$K = 7 \text{ kN/m} \qquad x_0 = +0.40 \text{ cm}$$
$$M = 20 \text{ kg} \qquad \dot{x}_0 = +60 \text{ cm/sec}$$
$$C = 150 \text{ N} \cdot \text{s/m}$$

Find:

(a) Time required to reach maximum displacement.

(b) The maximum displacement.

Solution. Using Table 3.2,

$$\omega_1 = \sqrt{K/M} = \sqrt{7000/20} = 18.7 \text{ rad/sec}$$

$$\zeta = \frac{C}{2M\omega_1} = \frac{175}{2(20)(18.7)} = 0.234$$

$$\beta = \sqrt{1 - \zeta^2} = 0.972$$

$$\alpha = \frac{60}{18.7(0.40)} = 8.02$$

(a) $$\omega_1 t_A = \frac{1}{0.972}\tan^{-1}\left[\frac{(8.02)(0.972)}{1 + (8.02)(0.234)}\right]$$

$$t_A = 1.252/18.7 = 0.067 \text{ sec}$$

(b) $$x_A/x_0 = e^{-(0.234)(1.252)}\sqrt{1 + 2(8.02)(0.234) + (8.02)^2}$$

$$= (0.746)(8.31) = 6.20$$

$$x_A = (0.40)(6.20) = +2.48 \text{ cm}$$

EXAMPLE 3.3. The single-degree system has negative initial displacement and positive initial velocity as shown. Determine the maximum force experienced by the dashpot.

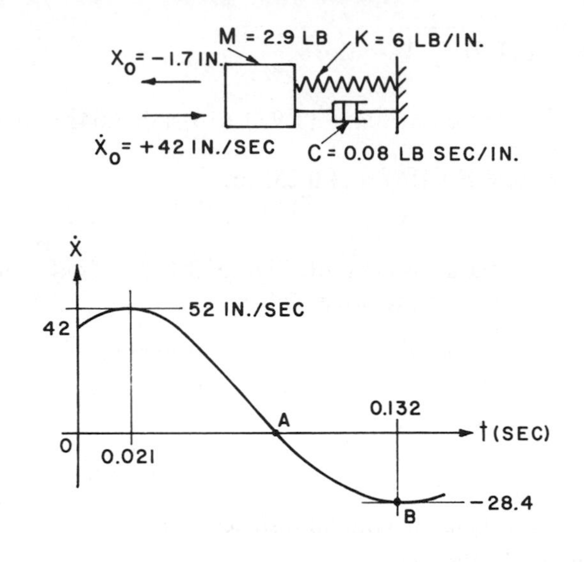

Solution. Using Table 3.2, and equating the acceleration to zero,

$$\omega_1 = \sqrt{6(386)/2.9} = 28.26 \text{ rad/sec}$$

$$\zeta = \frac{0.08}{2(2.90/386)(28.26)} = 0.188$$

$$\beta = \sqrt{1 - 0.188^2} = 0.982$$

$$\alpha = 42/(28.3)(-1.7) = -0.874$$

$$0 = e^{-0.188(\omega_1 t)}\Bigg(-\left[1 - 2(0.874)(0.188)\right]\cos 0.982\omega_1 t$$

$$+\left\{\frac{-0.874\left[2(0.188)^2 - 1\right] + 0.188}{0.982}\right\}\sin 0.982\,\omega_1 t\Bigg).$$

$$-0.671\cos 0.982\omega_1 t + 1.019\sin 0.982\omega_1 t = 0$$

$$\tan 0.982\omega_1 t = 0.671/1.019 = 0.659$$

$$\omega_1 t = 0.583/0.982 = 0.593$$

$$t = 0.021 \text{ sec}$$

Substituting in the velocity expression, the corresponding velocity is 52 in./sec. Maximum recovery velocity, $\dot{x}_B$, is computed as -28.4 in./sec. Therefore the maximum velocity occurs at the initial phase of the transient motion, and the maximum dashpot force is

$$F = C\dot{x} = (0.08)(52) = 4.16 \text{ lb}$$

EXAMPLE 3.4. A translational system with an initial horizontal velocity to the left of 10 ft/sec impacts a rigid wall as shown. Find:

(a) The maximum force in the spring.
(b) The maximum spring force if the dashpot is removed.
(c) The energy dissipated by the dashpot to the instant of zero velocity.

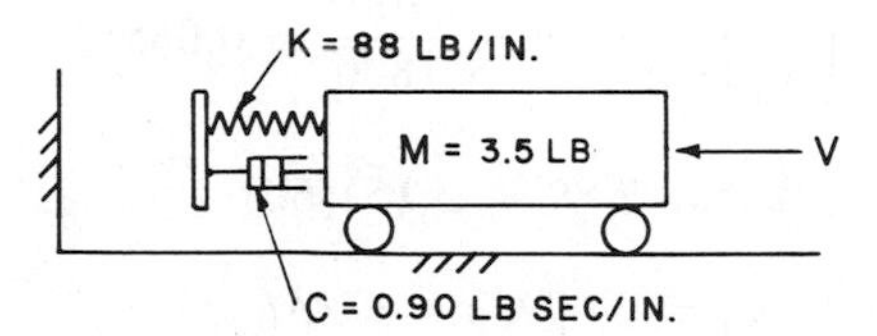

Solution. Using Table 3.3 for initial velocity only,

$$\omega_1 = \sqrt{\frac{88(386)}{3.5}} = 98.51 \text{ rad/sec}$$

$$\zeta = \frac{0.90}{2(3.5/386)98.51} = 0.504$$

$$\beta = \sqrt{1 - (0.504)^2} = 0.864$$

(a)

$$\omega_1 t_A = (1/0.864)\tan^{-1}(0.864/0.504) = 1.207$$

$$\frac{x_A}{\dot{x}_0/\omega_1} = e^{-(0.504)(1.207)} = 0.544$$

$$x_A = \frac{(120)(0.544)}{98.51} = 0.663 \text{ in.}$$

$$Kx_A = 58.3 \text{ lb}$$

(b)

$$x_A = 120/98.51 = 1.22 \text{ in.}$$

$$Kx_A = 107.2 \text{ lb}$$

(c) Initial KE $= \frac{1}{2}MV^2 = \frac{1}{2}(3.5/386)(120)^2 = 65.3$ in. lb
Final elastic energy $= \frac{1}{2}Kx^2 = \frac{1}{2}(88)(0.663)^2 = 19.3$ in. lb
Energy dissipated $= 65.3 - 19.3 = 46.0$ in. lb

EXAMPLE 3.5. A flywheel of 26 kg · m^2 coasts from a speed of 1800 RPM to 100 RPM in 5 min, subjected only to simple viscous bearing drag. Find:

(a) The torsional damping coefficient.
(b) The equivalent translational (tangential) damping coefficient if the bearing diameter is 9 cm.
(c) The total number of revolutions that occurred.
(d) The energy dissipated in the bearing.

Solution

(a) Applying Equation 3.42 in angular coordinates,

$$\left(\frac{\dot{x}}{\dot{x}_0}\right) = e^{-\omega_1 t} = \frac{100}{1800} = 0.0555$$

$$\omega_1 t = 2.89 = \omega_1(5)(60)$$

$$\omega_1 = 0.00963 = C/M$$

$$C_\theta = (0.00963)(26) = 0.25 \text{ N} \cdot \text{m} \cdot \text{s}$$

(b) From Table 1.5, converting,

$$C_1 = \left[\frac{0.25}{(0.045)^2}\right] = 123.4 \text{ N} \cdot \text{s/m}$$

(c) Applying Equation 3.43,

$$\left(\frac{x}{\dot{x}_0/\omega_1}\right) = (1 - e^{-2.89}) = 0.944$$

$$x = 0.944\left(\frac{60\pi}{0.00963}\right) = 18{,}490 \text{ rad} = 2940 \text{ revolutions}$$

(d) From Equation 3.44,

$$\frac{E_d}{E_0} = 1 - e^{-2(2.89)} = 0.9969$$

$$E_0 = \tfrac{1}{2}M\Omega^2 = \tfrac{1}{2}(26)(60\pi)^2 = 462{,}000 \text{ N} \cdot \text{m}$$

$$E_d = (462{,}000)(0.9969) = 460{,}500 \text{ N} \cdot \text{m}$$

EXAMPLE 3.6. The free end of a spring–mass system is forced by a constant positive acceleration for a finite time, as shown. Then 2 becomes stationary, or grounded. Determine:

(a) The final position of the end of the spring.

(b) The relative vibratory amplitude after termination of the disturbance.

(c) The final relative amplitude using a vectorial solution.

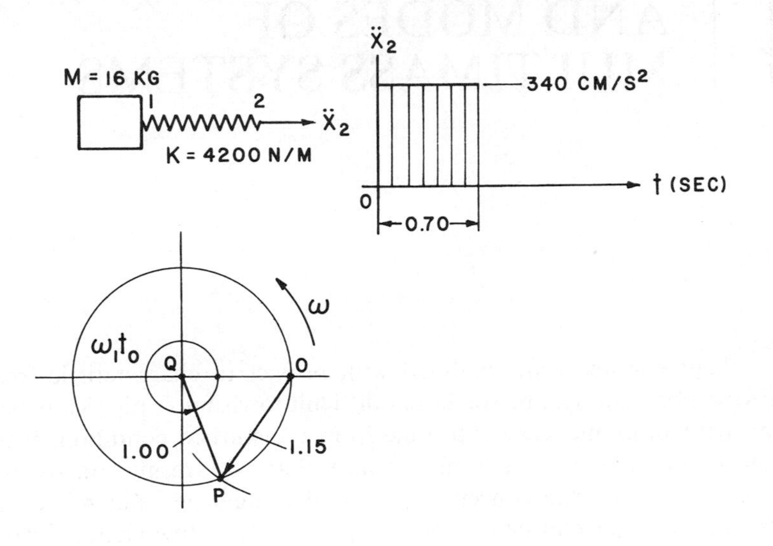

Solution

(a) $x_2 = \frac{1}{2}\ddot{x}_2 t^2 = \frac{1}{2}(340)(0.7)^2 = 83.3$ cm

(b) Using Table 3.5,

$$\omega_1 = \sqrt{\frac{4200}{16}} = 16.2 \text{ rad/sec}$$

$$\frac{(x_{12})_A}{\ddot{x}_2/\omega_1^2} = 2\sin\frac{(16.2)(0.70)}{2} = -1.15$$

$$(x_{12})_A = \left(\frac{340}{263}\right)(-1.15) = 1.49 \text{ cm}$$

(c) As in Figure 3.16, with

$$\omega_1 t_0 = 16.2(0.70) = 11.34 \text{ rad} = 649.8°$$

$$\overline{OP} = 1.15$$

$$(x_{12})_A = \left(\frac{340}{263}\right)(1.15) = 1.49 \text{ cm}$$

4 FREQUENCIES AND MODES OF MULTIMASS SYSTEMS

In this chapter systems are analyzed with respect to characteristic free sustained vibration. This can be somewhat difficult to visualize physically because all real free vibrations decay with time from a disturbed condition. It is also difficult to provide an initial disturbance that will result exactly in one particular mode, but the concepts of natural frequencies and modes are of paramount importance in determining response. These free characteristics are obtainable by analytical methods now indicated for the undamped, unforced vibration of discrete linear systems. Systems with many springs and masses are studied by the use of iterative solutions.

4.1. NATURAL FREQUENCIES

Chapter 2 introduced the natural frequency for the single-mass system as

$$\omega_n = \omega_1 = \sqrt{\frac{K_1}{M_1}}$$

Or, with several springs and a single mass,

$$\omega_n = \mathsf{g}_n\sqrt{\frac{K_1}{M_1}} = \mathsf{g}_n\omega_1$$

where ω_1 is the reference natural frequency.

We now expand the concept of natural frequency factor to several degrees of freedom using g_{ni} to denote this ratio, where i is any mode. The different factors relate system frequencies to the single reference frequency.

With two masses coupled to ground (Figure 4.1*a*), two simultaneous equilibrium equations can be written:

$$\begin{cases} \Sigma F_1 = M_1\omega^2 a_1 + K_1(a_2 - a_1) = 0 & (4.1a) \\ \Sigma F_{12} = M_1\omega^2 a_1 + M_2\omega^2 a_2 - K_2 a_2 = 0 & (4.1b) \end{cases}$$

Normalizing with respect to the reference system (K_1, M_1) by dividing by K_1, we have forces on M_1 and on the complete system,

$$\begin{cases} a_1(g^2 - 1) + a_2 = 0 & (4.2a) \\ a_1 g^2 + a_2(m_2 g^2 - k_2) = 0 & (4.2b) \end{cases}$$

where $k_2 = K_2/K_1$ and $m_2 = M_2/M_1$

Although these equations involve the complex unknowns a_1 and a_2, they can be treated algebraically and conveniently solved by the determinant format. We have, as previously, taken the sinusoidal steady-state conditions as satisfying the differential equations of motion.

$$a_1 = \left| \frac{\begin{matrix} 0 & 1 \\ 0 & (m_2 g^2 - k_2) \end{matrix}}{\begin{matrix} (g^2 - 1) & 1 \\ g^2 & (m_2 g^2 - k_2) \end{matrix}} \right| = \frac{0}{m_2 g^4 - (m_2 + k_2 + 1)g^2 + k_2} \quad (4.3)$$

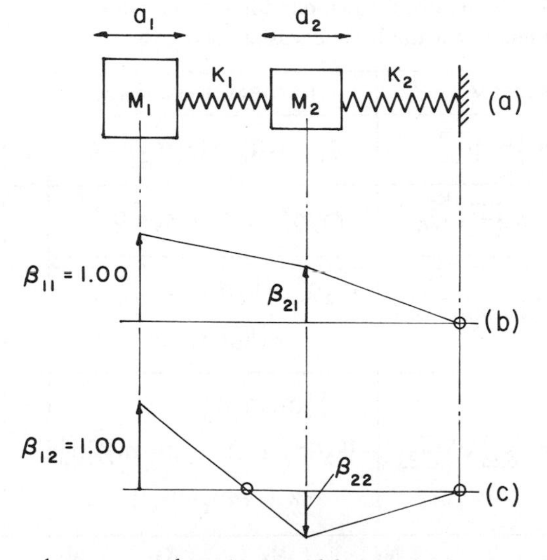

FIGURE 4.1. The two-degree system has two natural frequencies. Typical natural modes at the first and second natural frequencies are shown in (*b*) and (*c*).

Since a_1 is not zero the only interpretation is that the denominator must be zero and the ratio then an indeterminate quantity. It is logical that a solution not exist for amplitude because the forcing has not been indicated. Thus equating the denominator to zero provides the normalized natural frequency quadratic.

$$m_2 g_n^4 - (m_2 + k_2 + 1) g_n^2 + k_2 = 0 \tag{4.4}$$

where g_n is the particular value of the frequency ratio for which the denominator vanishes. Equation 4.4 is a quadratic in g_n^2 from which two positive roots are obtained, then used to calculate the two absolute system natural frequencies. It is a general relation with coefficients determined by ratios only. For instance, with two equal masses and two equal springs, $m_2 = 1$ and $k_2 = 1$. The equation becomes.

$$g_n^4 - 3g_n^2 + 1 = 0 \tag{4.5}$$

$$g_n^2 = \frac{3 \pm \sqrt{5}}{2} = 0.382, 2.618$$

$$g_n = 0.618, 1.618$$

The two equal spring–mass systems in series combine to form a coupled system in which the lower natural frequency is less than the single by a factor of 0.618. The higher system natural frequency is greater than the single system by the factor 1.618.

TABLE 4.1. Generalized equations for system natural frequency factors and normal modes for the basic discrete mass cases.

	SYSTEM	NATURAL FREQUENCY	β_n	MODE
a	M_1 –K_1– M_2	$g_n^2[m_2 g_n^2 - (1 + m_2)] = 0$	2	$(1 - g_n^2)$
b	M_1 –K_1– M_2 –K_2– wall	$m_2 g_n^4 - A g_n^2 + k_2 = 0$	2	$(1 - g_n^2)$
c	wall –K_0– M_1 –K_1– M_2 –K_2– wall	$m_2 g_n^4 - [k_0 m_2 + A] g_n^2 + [k_0 + k_0 k_2 + k_2] = 0$	2	$(k_0 + 1 - g_n^2)$
d	M_1 –K_1– M_2 –K_2– M_3	$g_n^2\{m_2 m_3 g_n^4 - [k_2(m_2+m_3) + m_3(1+m_2)] g_n^2 + k_2(1 + m_2 + m_3)\} = 0$	2	$(1 - g_n^2)$
			3	$\dfrac{k_2(1 - g_n^2)}{k_2 - m_3 g_n^2}$

$A = (1 + m_2 + k_2)$, $m_2 = \dfrac{M_2}{M_1}$, $m_3 = \dfrac{M_3}{M_1}$, $k_2 = \dfrac{K_2}{K_1}$, $g_n = \dfrac{\omega_n}{\omega_1}$, $\omega_1 = \sqrt{\dfrac{K_1}{M_1}}$

TABLE 4.2. Similar to Table 4.1, but for any number of equal springs and masses.

	System ($\beta_{qi} = \beta_{1i}$, β_{2i}, $\beta_{pi} = 1$; $q = 1, 2, \ldots, p$)	MODE i	ϕ_i	$\left(\frac{\beta_q}{\beta_p}\right)_{ni}$
a	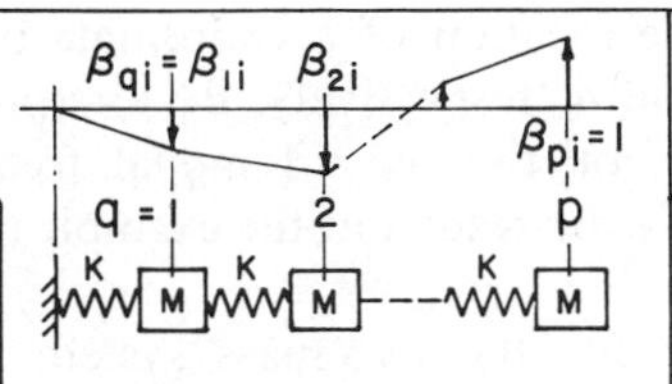 K M K M --- K M	1, 2, ...	$\pi\left(\frac{i-\frac{1}{2}}{p+\frac{1}{2}}\right)$	$\left(\frac{\sin q\phi_i}{\sin p\phi_i}\right)$
b	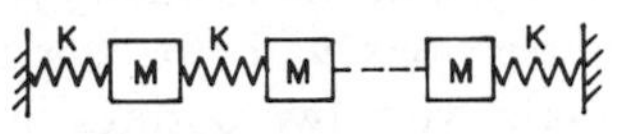 K M K M --- M K	1, 2, ...	$\pi\left(\frac{i}{p+1}\right)$	$\left(\frac{\sin q\phi_i}{\sin p\phi_i}\right)$
c	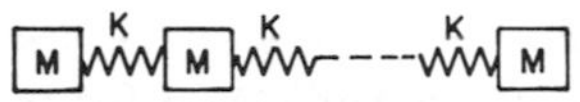 M K M K --- K M	0, 1, 2, ..	$\pi\left(\frac{i}{p}\right)$	$\left(\frac{\cos(q-\frac{1}{2})\phi_i}{\cos(p-\frac{1}{2})\phi_i}\right)$
q = NUMBER OF MASS p = TOTAL NUMBER	$g_{ni} = \left(\frac{\omega_{ni}}{\omega_1}\right) = 2\sin\left(\frac{\phi_i}{2}\right)$, $\omega_1 = \sqrt{\frac{K}{M}}$			

It is further noted that the coefficients of Equation 4.4 are independent of the relation of k_2 to m_2, and only dependent on the separate K_2/K_1 and M_2/M_1 ratios. We can change the springs and/or the masses in this example as long as each ratio remains unity without altering the factors 0.618 and 1.618. The equation applies to both translational and torsional arrangements and the k_2 and m_2 ratios are in any consistent units, since they define only *relative* sizes.

Actual system natural frequencies must involve the reference frequency and therefore K_1 and M_1, also in consistent units. The mass term must incorporate the gravitational constant.

Several spring–mass combinations are shown in Table 4.1, with the normalized frequency equations shown. These are derived by the method indicated, and limited to the cases given because of the increasing complexity as the system involves more masses. Table 4.2 applies to systems with equal springs and equal masses with various end conditions.[15] Formulas are given for the frequency factors g_{ni}, and these can be applied with no limit on the number of elements.

4.2. BEHAVIOR OF g_n FOR BASIC SYSTEMS

In Figure 4.1*a* a fundamental two-mass system is shown and coefficients in Equation 4.4 can be used to provide a generalized study of the system frequency variation with k_2 and m_2 (Figures 4.2 and 4.3). Curves of constant g_n are straight lines of positive slope with k_2 the abscissa and m_2 the ordinate.

Lower modal roots are shown by solid lines and upper modal roots by dashed lines. Figure 4.2 is detailed for small values of m_2 and k_2.

Use of these figures requires only the location of a coordinate point at a particular value of m_2 and k_2, say 8 and 4, respectively. By interpolation in Figure 4.3, $g_{n1} = 0.64$ and $g_{n2} = 1.10$, or the second modal frequency is $1.10/0.64 = 1.72$ times the first. Similarly the results in the example (Equation 4.5) correspond to A in Figure 4.2.

The vertical axes provide g_n values for the two-mass system with one connecting spring free to translate (Table 4.1a), and the vibratory frequency is greater than the reference frequency. Intersections with the horizontal axis provide system frequency data for one mass with two springs in series, or $m_2 = 0$. For this case g_n values are less than one, as the natural frequency is reduced from ω_1 by the additional spring K_2.

These figures not only facilitate frequency determination for the two-mass system but allow visualization of the effects of changes in the spring and mass ratios in altering system frequencies. They also allow comparison of first and second modal frequencies when this information is of interest.

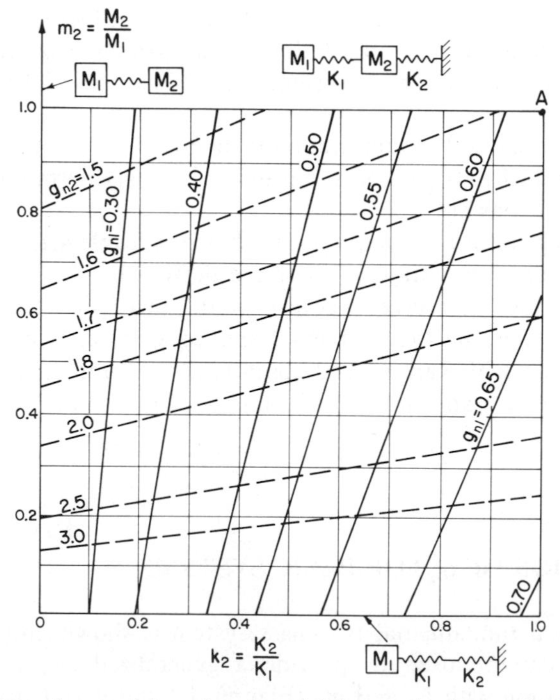

FIGURE 4.2. Effect of m_2 and k_2 ratios on modal frequencies for two-degree system. Higher modal factors are shown dashed.

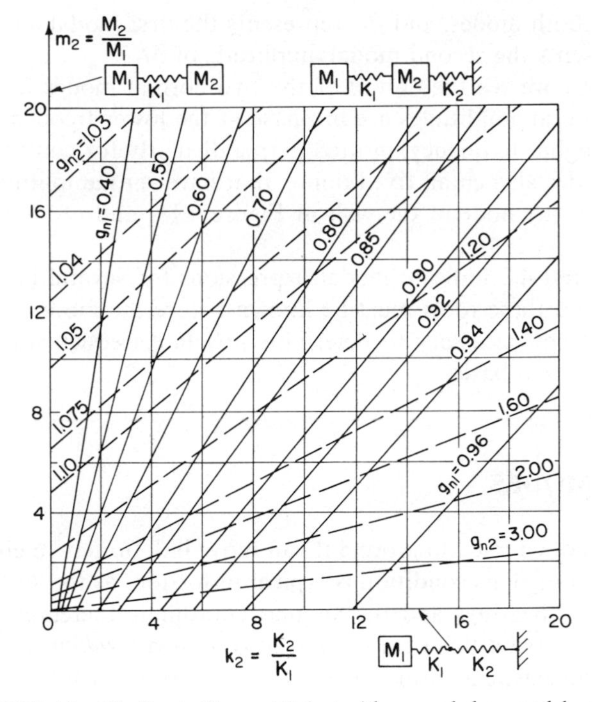

FIGURE 4.3. Similar to Figure 4.2, but with expanded m_2 and k_2 scales.

4.3. NORMAL MODES

If a system vibrates freely at a natural frequency, it does so with characteristic vibratory amplitudes, or, more correctly, amplitude ratios, termed *natural modes*. If a reference amplitude of unity is used, the corresponding normalized amplitudes define a *normal mode*, with a particular normal mode associated with each natural frequency.

Equation 4.3 is of no help in calculating the normal mode, as the a_1 expression degenerates to 0/0 at $\mathsf{g} = \mathsf{g}_n$; however, the original equation, 4.2a, is applicable.

$$\left(\frac{\mathsf{a}_2}{\mathsf{a}_1}\right)_n = \left(\frac{\beta_2}{\beta_1}\right)_n = \left(1 - \mathsf{g}_n^2\right) \tag{4.6}$$

Two values of the g_n natural frequency factors will yield the two modal amplitude ratios, which in special recognition of their importance are designated β_n or β_{ni}, with the n specifying *normal* and the i specifying the modal number. In the previous example (Equation 4.5), $\beta_{21} = 1 - 0.382 = +0.618$ and $\beta_{22} = 1 - 2.618 = -1.618$, where $\beta_{11} = \beta_{12} = 1$ is the reference amplitude of M_1 in both modes, and β_{21} represents the first modal amplitude of M_2 and β_{22} represents the second modal amplitude of M_2.

In Figure 4.1 we see the nature of the two normal modes for the classical system. Horizontal amplitudes are in phase at the lower frequency and out of phase at the higher frequency, plotted perpendicularly for convenience.

Normal modes also entail fixed points that have zero amplitude, designated *nodes*. There is one node at the wall in Figure 4.1*b*, and two nodes in Figure 4.1*c*.

Tables 4.1 and 4.2 indicate modal expressions for several basic systems in terms of g_n^2, and these roots must be known before the modes are calculated. However, it is not necessary to determine absolute frequencies. All relations are on a normalized basis.

4.4. RIGID MODES

If a system is not coupled to ground it can move indefinitely in either direction without constraint. This condition is typical in torsion, as the system is usually free to rotate in bearings and the angular coordinate increases without limit rotationally. In torsion this *rigid mode* is also termed a *rolling mode*. Motion is involved, but no elastic deformation or stress. As seen in Tables 4.1*a*, 4.1*d*, and 4.2*c*, these *free–free* systems have roots of $g_n = 0$. This is a zero natural frequency corresponding to infinitely slow cycling. The normal mode for this condition is simple unison with all amplitudes $+1$. Inertia forces are also zero in this mode.

4.5. ORTHOGONALITY OF NORMAL MODES

An interesting property of normal modes is that the summation of the products of the amplitudes in any two modes and the mass at each coordinate equals zero. In normalized form,

$$\Sigma M_q \beta_{qi} \beta_{qj} = \Sigma m_q \beta_{qi} \beta_{qj} = 0 \tag{4.7}$$

In the previous example, with $k_2 = m_2 = 1$, using the numerical modal amplitudes,

$$m_1 \beta_{11} \beta_{12} + m_2 \beta_{21} \beta_{22} = 0 \tag{4.8}$$

$$(1)(1)(1) + (1)(0.618)(-1.618) = 0$$

With normalized values the first term is always unity. This principle is often useful in verifying modal results numerically, and the zero mode, if present, is applicable.

4.6. EQUIVALENT MODAL SYSTEMS

The multimass system has characteristic natural frequencies and normal modes. Once determined, these parameters enable us to represent each mode by an equivalent modal single spring–mass system at a particular point or coordinate of the multimass system, usually at M_1. The equivalent system simulates the steady-state dynamic characteristics of a given mode with respect to excitation at the chosen mass. The concept is important in the superposition of modal responses, particularly in systems with low damping. These relations are discussed now as inherent properties related to the modes and frequencies of systems in the free condition.

Consider the case in Figure 4.4 with known first modal amplitudes and natural frequency. An equivalent modal spring and mass at β_1 must involve the same kinetic energy at the midposition of the harmonic motion as the original multimasses in free vibration at the natural frequency.

$$\frac{1}{2}M_{n1}(\beta_{11}\omega_{n1})^2 = \frac{1}{2}\Sigma M_q(\beta_{q1}\omega_{n1})^2$$

$$M_{n1} = \Sigma M_q \beta_{q1}^2 \tag{4.9}$$

This equation directly determines the modal mass (Figure 4.4*c*).

In addition to the energy equivalence, the modal equivalent must have the same natural frequency, assumed in Equation 4.9.

$$\omega_{n1} = \sqrt{\frac{K_{n1}}{M_{n1}}}$$

$$K_{n1} = \omega_{n1}^2 M_{n1} = \omega_{n1}^2 \Sigma M_q \beta_{q1}^2 \tag{4.10a}$$

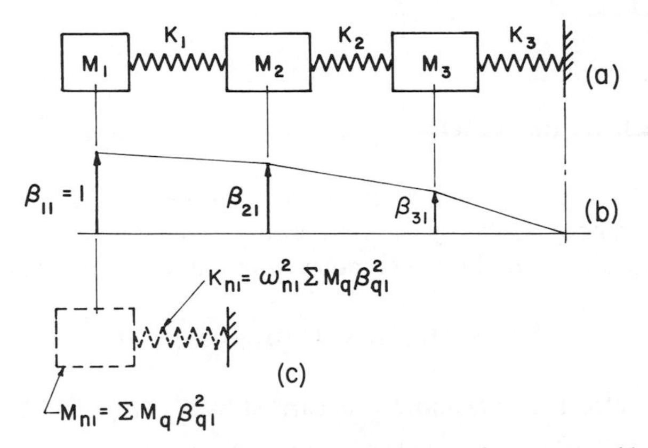

FIGURE 4.4. Each vibratory mode of a multi mass system can be represented by a single-degree analog system, which has identical modal vibratory characteristics.

The equivalent modal spring is of particular importance in the determination of responses, and there are three such characteristic stiffnesses K_{ni} for this three-mass system.

In the ungrounded system (Table 4.1*a*), Equations 4.9 and 4.10a are also applicable to the vibratory mode, and K_{ni} is to ground. The zero mode will be considered later as having a variable rather than a constant modal spring, also to ground.

Equation 4.10a can be normalized to

$$K_{ni} = K_1 \mathsf{g}_{ni}^2 \sum_{q=1}^{q=p} m_q \beta_{qi}^2 = K_1 \mathsf{g}_n^2 \Sigma m \beta^2 \tag{4.10b}$$

4.7. HOLZER METHOD FOR NATURAL FREQUENCIES

One of the most versatile methods for obtaining quantitative evaluation of vibratory characteristics of a system with discrete masses is the Holzer tabulation. It can be adapted to translational or torsional systems, and to various end conditions to include damping; it can also be used to obtain forced response as well as natural frequencies and normal modes. Additionally, Holzer-generated data can be used to determine load and stress distribution throughout a multimass system, and to calculate modal energy. The method involves trial frequencies and is not convergent; however, satisfactory accuracy is obtained quite rapidly by interpreting results of trial computations or by programming techniques.

For natural frequencies the method requires determining inertia loads and relative spring deflections that satisfy equilibrium working incrementally across a system. Normalized parameters are again used to facilitate the procedure, and numerical examples are given.

4.8. HOLZER DERIVATION

In Figure 4.5*a* a three-mass system is free to translate, having a rigid mode and two vibratory modes. Isolating the first spring and mass, assuming harmonic motion at frequency ω, and using complex notation for Figure 4.5*b*,

$$\Sigma F_1 = M_1 \omega^2 \mathsf{a}_1 + K_1(\mathsf{a}_2 - \mathsf{a}_1) = 0 \tag{4.11a}$$

Taking inertia effects of M_1 and M_2 as carried by K_2 (Figure 4.5*c*),

$$\Sigma F_{12} = M_1 \omega^2 \mathsf{a}_1 + M_2 \omega^2 \mathsf{a}_2 + K_2(\mathsf{a}_3 - \mathsf{a}_2) = 0 \tag{4.11b}$$

Finally, all inertia effects must be resisted by an external force, P_3 on M_3 (Figure 4.5*d*):

$$\Sigma F_{13} = M_1\omega^2 a_1 + M_2\omega^2 a_2 + M_3\omega^2 a_3 + P_3 = 0$$

$$P_3 = \sum M_q\omega^2 a_q \tag{4.11c}$$

Equation 4.11a involves the relative displacement of 1 with respect to 2; 4.11b, of 2 with respect to 3.

$$a_{12} = \frac{M_1\omega^2 a_1}{K_1} \qquad a_{23} = \frac{M_1\omega^2 a_1 + M_2\omega^2 a_2}{K_2} \tag{4.12}$$

Equation 4.12 determines the forced mode corresponding to an assumed a_1 and ω (Figure 4.6). Thus we are able to subtract the difference a_{12} from a_1 to find a_2, and a_{23} from a_2 for a_3. Knowing the actual amplitudes we can evaluate P_3 from Equation 4.11*c*. Physically this implies an external excitation at M_3 of P_3 with amplitude a_3 at forcing frequency ω.

But in our system a free vibration at a natural frequency is self-contained and *the external forcing must be zero*. This therefore provides the Holzer test.

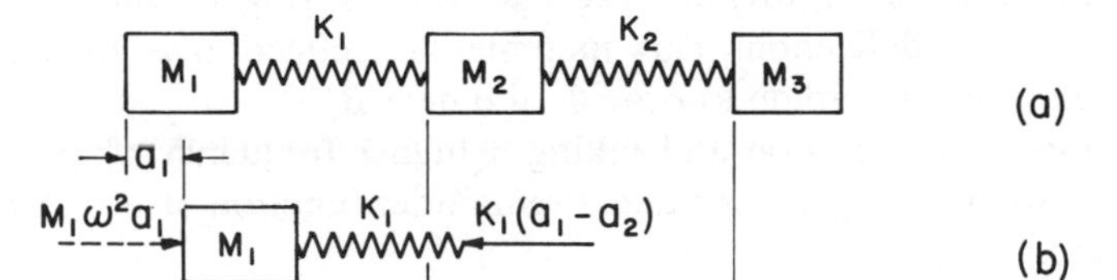
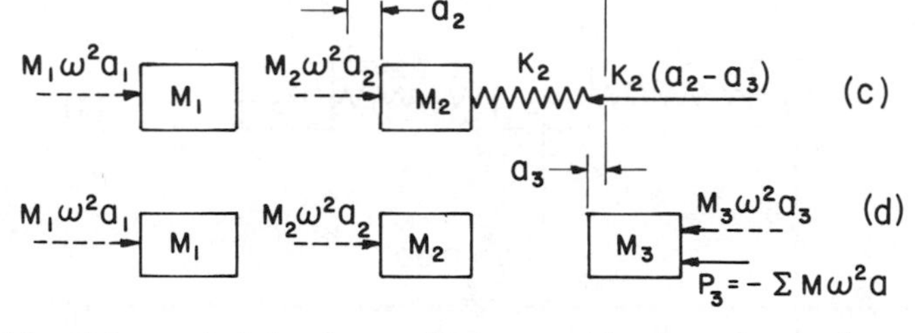

FIGURE 4.5. Holzer analysis involves equilibrium considerations of successive components.

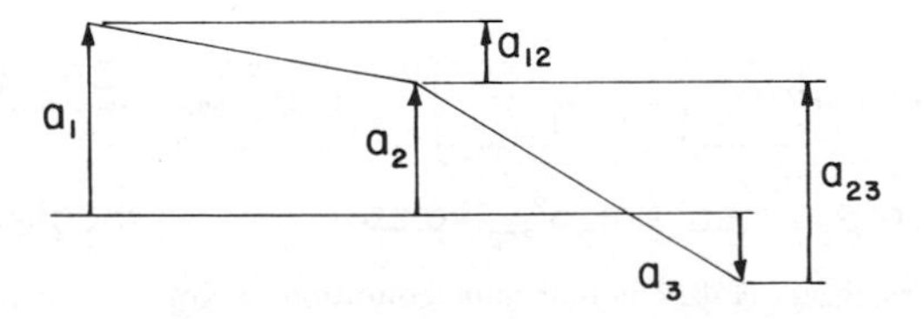

FIGURE 4.6. Successive amplitudes are determined by differences, or by relative amplitudes caused by spring deflections.

Only as P_3 approaches zero does the assumed frequency ω approach a system natural frequency ω_{ni}. Similarly, $\Sigma M_q \omega^2 \mathsf{a}_q$ approaches zero as the inertia forces balance internally. We have then also approximated the natural free mode in terms of a_1, a_2, and a_3. If $\mathsf{a}_1 = 1.00 = \beta_1$, the resulting normal amplitudes are β_2 and β_3.

4.9. HOLZER CALCULATIONS

As an elementary numerical example, assume that in Figure 4.7 there are three equal masses and two equal springs. By normalizing Equations 4.11 and 4.12, dividing by K_1, we have a format that does not require absolute values and produces generalized results. With m_2 and $m_3 = k_2 = 1$, we take $\mathsf{g} = 0.707$, or $\mathsf{g}^2 = 0.500 = \omega^2/(K_1/M_1)$. In other words the assumed frequency is 0.707 of the reference frequency. Then the Holzer results are as shown in Figure 4.8*a* with a remainder ΣF of 0.625, not zero.

Figure 4.7*b* shows the forced mode obtained in this calculation. Note the first amplitude difference is numerically equal to g^2 or 0.500. Figure 4.7*c* shows the equilibrium of inertia forces, requiring an external reaction factor of -0.625.

Briefly the tabular calculation involves proceeding horizontally as indicated by the column headings, dividing the load on a spring by the spring rate to obtain the spring deflection. This incremental deflection is then subtracted from the previous amplitude to provide the next β.

Returning to the solution and taking a higher frequency of $\mathsf{g} = 0.949$, or $\mathsf{g}^2 = 0.900$, we have Figure 4.8*b* and a remainder function of $+0.189$, reduced, but not zero.

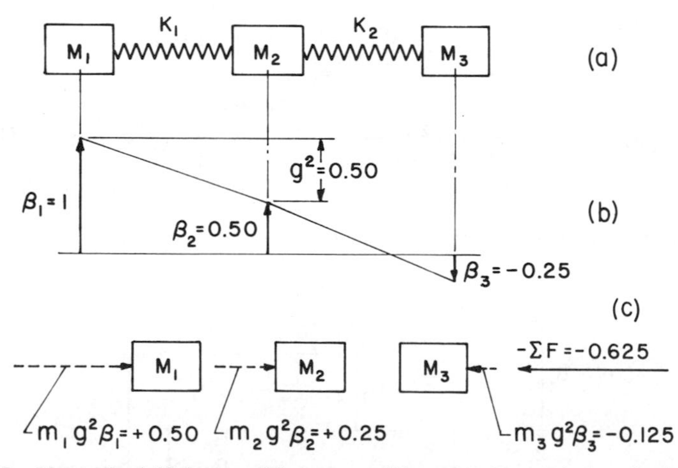

FIGURE 4.7. Normalized Holzer equilibrium conditions involve equivalent loads, or actual inertia loads divided by K_1. Although displacements dimensionally, they simulate and are proportional to the respective forces.

(a) $g^2 = 0.50$						
q	m	β	$mg^2\beta$	Σ	k	Σ/k
1	1	1.000	0.500	0.500	1	0.500
2	1	0.500	0.250	0.750	1	0.750
3	1	-0.250	-0.125	0.625		
(b) $g^2 = 0.90$						
q	m	β	$mg^2\beta$	Σ	k	Σ/k
1	1	1.000	0.900	0.900	1	0.900
2	1	0.100	0.090	0.990	1	0.990
3	1	-0.890	-0.801	0.189		
(c) $g^2 = 1.00$						
q	m	β	$mg^2\beta$	Σ	k	Σ/k
1	1	1.000	1.000	1.000	1	1.000
2	1	0	0	1.000	1	1.000
3	1	-1.000	-1.000	0		
(d) $g^2 = 3.00$						
q	m	β	$mg^2\beta$	Σ	k	Σ/k
1	1	1.000	3.000	3.000	1	3.000
2	1	-2.000	-6.000	-3.000	1	-3.000
3	1	1.000	3.000	0		

FIGURE 4.8. Sample normalized Holzer calculations for the Figure 4.7 system with equal masses and springs. First and second modes correspond to (*c*) and (*d*).

Taking a simpler value of $g^2 = 1$ proves more fortuitous (Figure 4.8*c*). The remainder is exactly zero with the mode having a node at M_2 and being antisymmetrical, as M_1 and M_3 move equally but in opposition. This system is obviously a special case with a system frequency equal to the reference frequency when M_2 is a node.

To find the higher modal frequency we must increase the assumed g^2 considerably above g_{n1}^2, in this case to $g^2 = 3$. This satisfies the zero remainder (Figure 4.8*d*), with a second mode of $+1$, -2, $+1$. Now M_2 moving in one direction balances M_1 and M_3 moving in the opposite direction at half the amplitude.

The nature of the remainder polynomial with g^2 is shown in Figure 4.9 for this calculation, with frequency roots at $g_n^2 = 0$, 1, and 3. Systems with any number of masses will have a similar behavior with alternate positive and negative loops crossing the axis at each frequency root.

Figure 4.7*b* shows the trial mode obtained in Figure 4.8*a*. Note that with normalization the first amplitude difference is equal to $g^2 = 0.50$. Figure 4.7*c* shows the equilibrium of the inertia forces requiring an external reaction factor of -0.625.

4.10. BEHAVIOR OF THE REMAINDER FORCE FUNCTION

In the simple case just discussed the cubic polynomial pattern derived from successive Holzer calculations can be compared with the closed solution (Table 4.1*d*).

$$f(g^2) = g^2(g^4 - 4g^2 + 3) = 0 \tag{4.13a}$$

Solution of the quadratic results in the same roots, and differentiation describes the slope of the tangent to the curve (Figure 4.9).

$$f(g^2) = (g^2)^3 - 4(g^2)^3 + 3(g^2)$$

$$\frac{df(g^2)}{dg^2} = 3g^4 - 8g^2 + 3 \tag{4.13b}$$

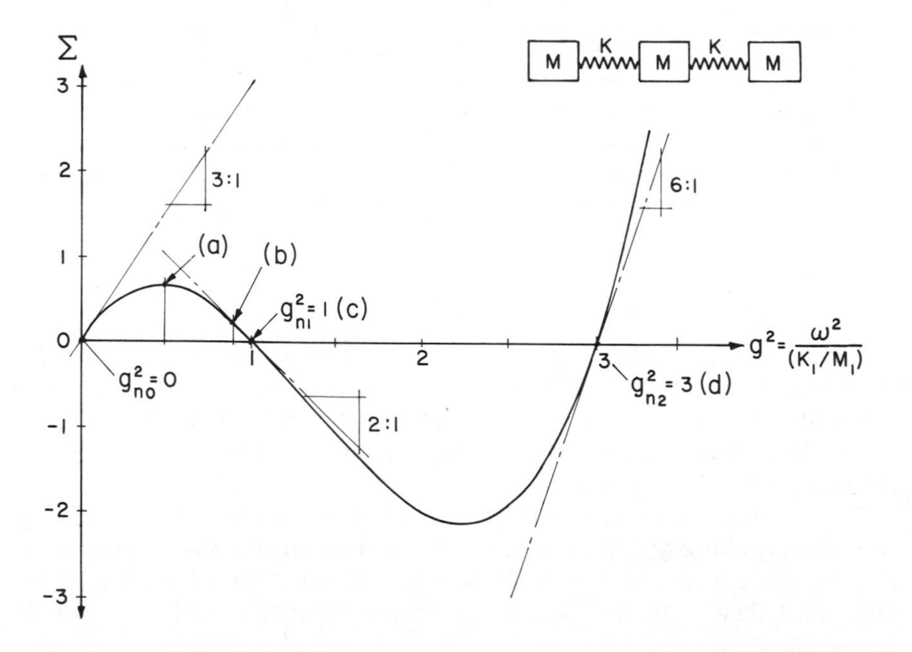

FIGURE 4.9. The remainder function from Figure 4.8 type computations plots as a polynomial, with frequency roots at 1 and 3. Tangent slopes at the roots are also indicated.

Slopes calculated at the roots, $g_n^2 = 0$, 1, and 3 are then respectively +3, −2, and +6. These values represent the normalized rate of change in the excitation required at M_3 with changes in g^2 and are indicative of the sensitivity of the computation at a root, typically increasing at the higher roots. The slope at $g^2 = 0$ is related to Σm; in this example the slope at the origin is $m_1 + m_2 + m_3 = 3$.

4.11. ACCURACY OF THE NATURAL FREQUENCIES

In Holzer, true values are usually only approximated, but can be refined to any required degree by successive interpolation. The most meaningful accuracy test is by bracketing two values of g^2 for which the remainder function changes sign. For instance, if g^2 values of 2.67 and 2.59 result in Σ values of opposite sign, the g_n^2 root lies between them. The maximum possible error at either trial is related to the square root of the difference, or $(1.6093 - 1.6031)/1.6031 = 0.00388$. Thus the error is less than 0.4%. The absolute value of Σ is not meaningful with respect to the degree of convergence achieved.

4.12. MODAL IDENTIFICATION

Reduction of the remainder forcing term to zero or near zero at the ungrounded right end of a system will locate a natural frequency. But the multimass system has several. The key to the number of the modes is found in the amplitude distribution. Typically one change in sign in the β column corresponds to one node, and the first natural frequency; two changes, to the second, and so on. Similarly, a next assumed g^2 value, when searching for a root, is related to both changes of sign of β and the known pattern of the remainder polynomial. By considering both effects the probable increase or decrease direction for the next trial frequency can be judged.

In this connection any single modal root can be found if desired, as these solutions are substantially independent.

4.13. MODES WITH FIXED ENDS

The previous Holzer examples have concerned the ungrounded vibratory system in Figure 4.5*a*. This is the conventional application of the Holzer method to rotational problems, for instance in crankshaft torsional analysis; however, the typical translational system is grounded, as are some torsional systems. In such cases the tabular calculation is somewhat modified.

With the ground at the left, M_1 is subjected to a spring force from K_0 to the left, numerically equal to K_0 for unit amplitude, $\beta_1 = 1$. This requires starting the table at $q = 0$, and taking $\beta_0 = 0$ and $\Sigma = -k_0$ in the normalized format.

Thus $\Sigma/k_0 = -1$, and $\beta_1 = +1$ when this difference is subtracted from β_0, and the computation proceeds from the unit amplitude so provided at M_1 or m_1. The final boundary condition is satisfied when $\Sigma = 0$ at the free end at the right (Figure 4.10).

Taking the identical previous system reversed, the table begins with $\Sigma = 0$ and $\beta_1 = 1$ as usual, but displacement must be zero at the right of the system, $\beta_3 = 0$. The ground point is capable of providing any required vibratory reaction, or Σ. Although the systems are the same, the reference frequency is now lower. We have the same reference spring K_1 but a larger reference mass;

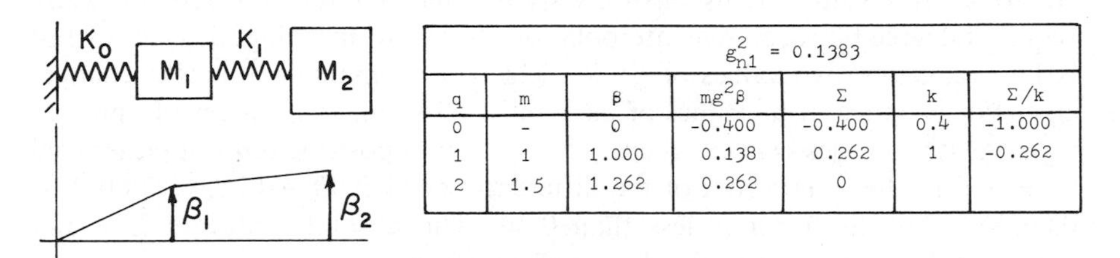

$g_{n1}^2 = 0.1383$						
q	m	β	$mg^2\beta$	Σ	k	Σ/k
0	-	0	-0.400	-0.400	0.4	-1.000
1	1	1.000	0.138	0.262	1	-0.262
2	1.5	1.262	0.262	0		

FIGURE 4.10. Holzer solution with first spring to ground and free end mass.

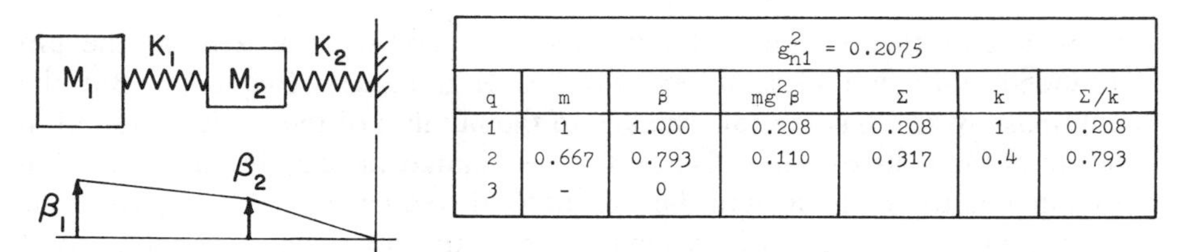

$g_{n1}^2 = 0.2075$						
q	m	β	$mg^2\beta$	Σ	k	Σ/k
1	1	1.000	0.208	0.208	1	0.208
2	0.667	0.793	0.110	0.317	0.4	0.793
3	-	0				

FIGURE 4.11. Holzer solution with free first mass and right spring grounded.

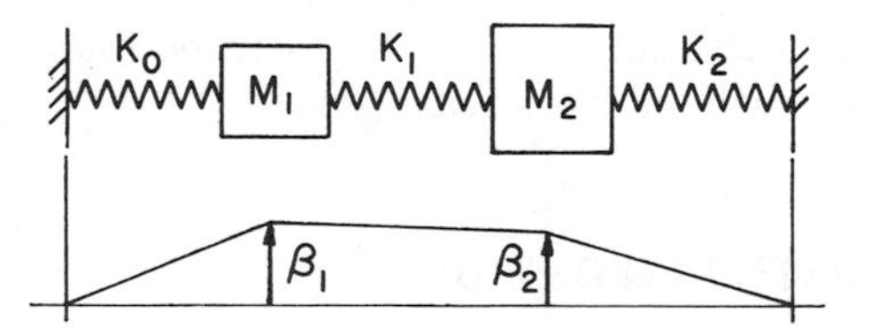

$g_{n1}^2 = 0.4394$						
q	m	β	$mg^2\beta$	Σ	k	Σ/k
0	-	0	-0.400	-0.400	0.4	-1.000
1	1	1.000	0.439	0.039	1	0.039
2	1.5	0.961	0.633	0.673	0.7	0.961
3		0				

FIGURE 4.12. Holzer solution with both end springs grounded.

therefore we must raise g_{n1}^2 to a factor of (1/1.5)(0.1383) = 0.2075 if we expect to arrive at the identical natural frequency, as shown in Figure 4.11.

Similarly, fixed–fixed conditions require $\beta_0 = 0$ and $\beta_3 = 0$, and the assumed g^2 varied to achieve this result, as shown in Figure 4.12.

4.14. THE SIMPLE GEARED BRANCH

If two single-degree torsional systems are connected by two gears in mesh (Figure 4.13*a*) and rotated at constant speed, the velocity and displacement ratios are governed by the number of teeth in each gear during rotational operation. In Figure 4.13*b*,

$$\left(\frac{\Omega_3}{\Omega_2}\right) = \left(\frac{\theta_3}{\theta_2}\right) = -\left(\frac{n_2}{n_3}\right) = -R \tag{4.14}$$

where n is the number of teeth on a gear, with the negative sign indicating the reversed sense of both resulting from the mesh. Torques carried by the respective shafts are

$$\left(\frac{T_2}{T_3}\right) = -R \tag{4.15}$$

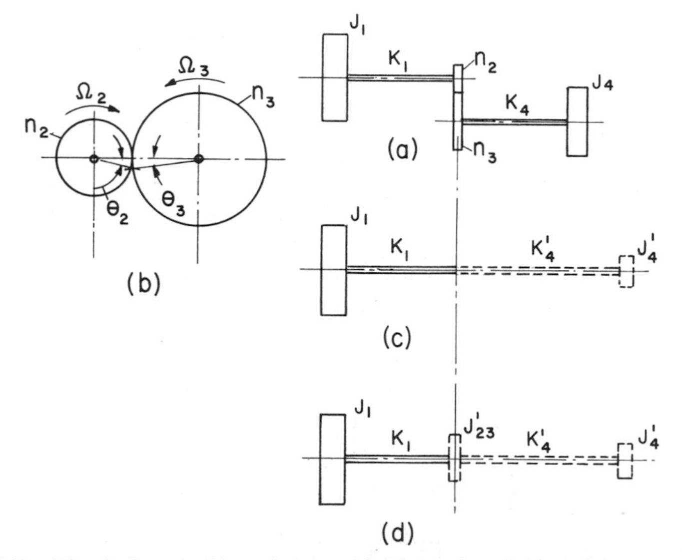

FIGURE 4.13. The single geared branch (*a*) can be reduced to the torsionally equivalent system (*c*) by modifying both the spring and the inertia. Equivalent gear inertias are included in (*d*).

Note that the sense of the torque transmitted from one branch to the other is reversed by the mesh; however, given J_1 as the driver and J_4 as the driven, the sense of the torque or twist in K_1 is the same with the gears or if K_4 is directly connected to K_1 (Figure 4.13*c*).

Assuming zero backlash in the gears, Equations 4.14 and 4.15 are equally applicable if a torsional vibration exists at zero speed. Similarly, the same relations apply if a vibratory condition is superimposed on a steady rotation, with respect to both vibratory and rotational components. We therefore consider the simpler concept of vibration without rotation.

4.15. THE EQUIVALENT COAXIAL SYSTEM

A conventional approach to finding the system natural frequency is to refer the branch to the main or reference axis, K_1. With negligible inertia at the gears

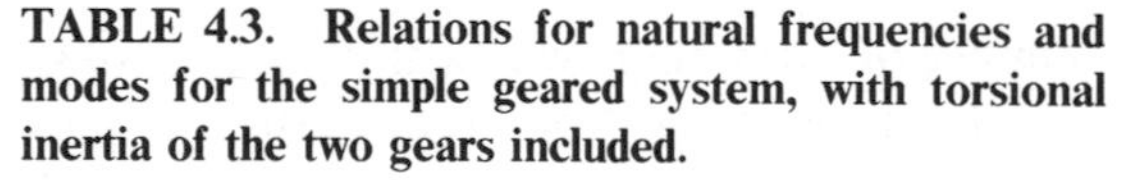

TABLE 4.3. Relations for natural frequencies and modes for the simple geared system, with torsional inertia of the two gears included.

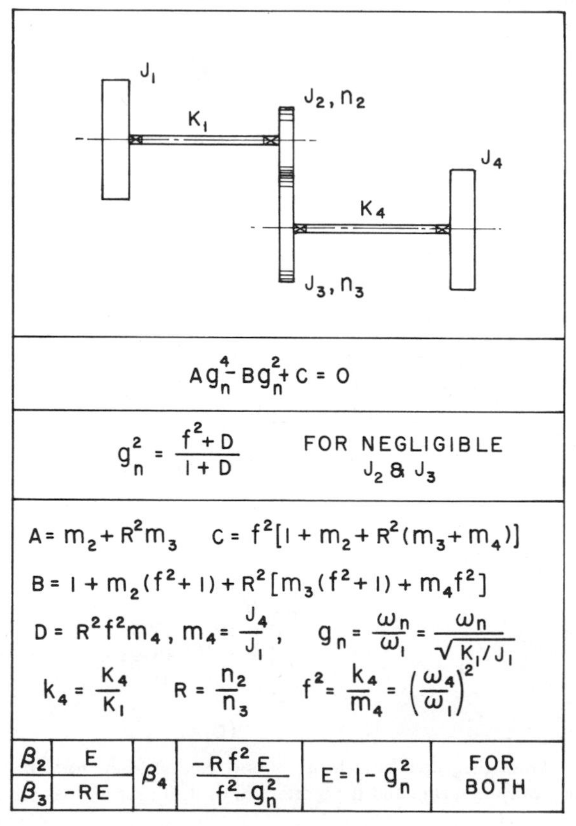

the equivalence becomes

$$K_4' = \left(\frac{n_2}{n_3}\right)^2 K_4 \qquad J_4' = \left(\frac{n_2}{n_3}\right)^2 J_4 \tag{4.16}$$

We must then combine the two springs, K_1 and K_4', in series, as shown in Table 1.3*e*, and finally use Table 4.1*a* for the modified system. These relations are incorporated in a single equation in Table 4.3.

To include polar inertias of the gears, the equivalent center disk becomes (Figure 4.13*d*)

$$J_{23}' = J_2 + \left(\frac{n_2}{n_3}\right)^2 J_3 \tag{4.17}$$

There are then two torsional modes that can be determined from Table 4.1*d*. The methods used can be extended to multiple successive branches.

There is additionally a rigid mode for which

$$\beta_1 = \beta_2 = 1 \qquad \beta_3 = \beta_4 = -R \tag{4.18}$$

where $R = n_2/n_3$.

4.16. SINGLE-DEGREE COMBINATIONS

By incorporating the gearing relations and recognizing that in Table 4.3 we have two single-degree systems, each with behavior determined by Table 2.1*c* the displacement excitations are related at the mesh by the ratio. Applying first the torque equation at the reference axis,

$$\Sigma T_1 = J_1\omega_n^2\beta_1 - \left(\frac{n_2}{n_3}\right) J_4\omega_n^2\beta_4 = 0$$

which normalizes to

$$1 - Rm_4\beta_4 = 0$$

The single-degree responses are

$$\left(\frac{\beta_1}{\beta_2}\right) = \frac{1}{1 - (\omega_n/\omega_1)^2} \quad \text{and} \quad \left(\frac{\beta_4}{\beta_3}\right) = \frac{1}{1 - (\omega_n/\omega_4)^2}$$

where

$$\omega_1^2 = \left(\frac{K_1}{J_1}\right) \quad \text{and} \quad \omega_4^2 = \left(\frac{K_4}{J_4}\right)$$

With g_n the system frequency factor ω_n/ω_1, we have from Table 2.1c

$$\beta_2 = \left(1 - \mathsf{g}_n^2\right) \quad \text{and} \quad \beta_3 = -R\beta_2$$

Further defining a ratio of the natural frequencies of the two branches as

$$\mathsf{f}^2 = \left(\frac{K_4/J_4}{K_1/J_1}\right) = \left(\frac{K_4/K_1}{J_4/J_1}\right) = \left(\frac{k_4}{m_4}\right)$$

The system natural frequency with gear inertia neglected becomes

$$\mathsf{g}_n^2 = \left(\frac{\mathsf{f}^2 + R^2\mathsf{f}^2 m_4}{1 + R^2\mathsf{f}^2 m_4}\right) \tag{4.19}$$

Extending this approach to include the gear inertias produces the general frequency quadratic in Table 4.3.

TABLE 4.4. Natural frequencies and modes for the simple double-mesh case, with gear inertias neglected.

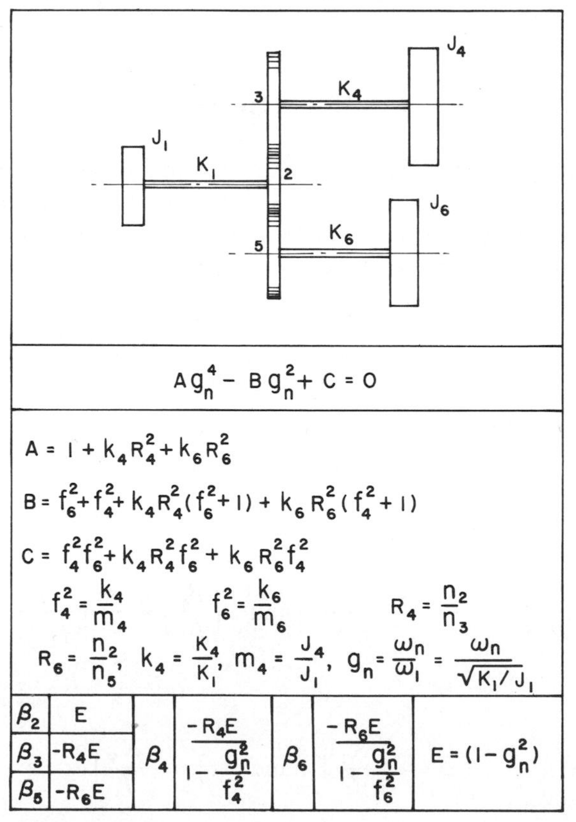

$$A g_n^4 - B g_n^2 + C = 0$$

$$A = 1 + k_4 R_4^2 + k_6 R_6^2$$

$$B = f_6^2 + f_4^2 + k_4 R_4^2 (f_6^2 + 1) + k_6 R_6^2 (f_4^2 + 1)$$

$$C = f_4^2 f_6^2 + k_4 R_4^2 f_6^2 + k_6 R_6^2 f_4^2$$

$$f_4^2 = \frac{k_4}{m_4} \qquad f_6^2 = \frac{k_6}{m_6} \qquad R_4 = \frac{n_2}{n_3}$$

$$R_6 = \frac{n_2}{n_5}, \quad k_4 = \frac{K_4}{K_1}, \quad m_4 = \frac{J_4}{J_1}, \quad g_n = \frac{\omega_n}{\omega_1} = \frac{\omega_n}{\sqrt{K_1/J_1}}$$

β_2	E					
β_3	$-R_4E$	β_4	$\dfrac{-R_4E}{1-\dfrac{g_n^2}{f_4^2}}$	β_6	$\dfrac{-R_6E}{1-\dfrac{g_n^2}{f_6^2}}$	$E = (1 - g_n^2)$
β_5	$-R_6E$					

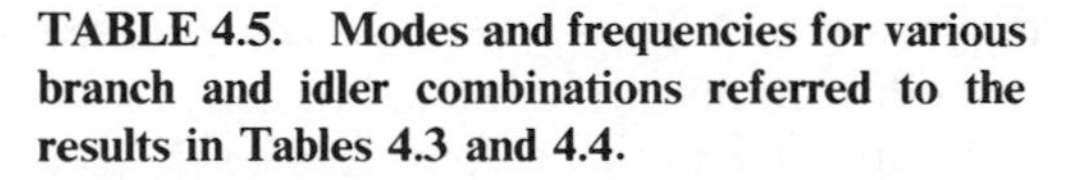

TABLE 4.5. Modes and frequencies for various branch and idler combinations referred to the results in Tables 4.3 and 4.4.

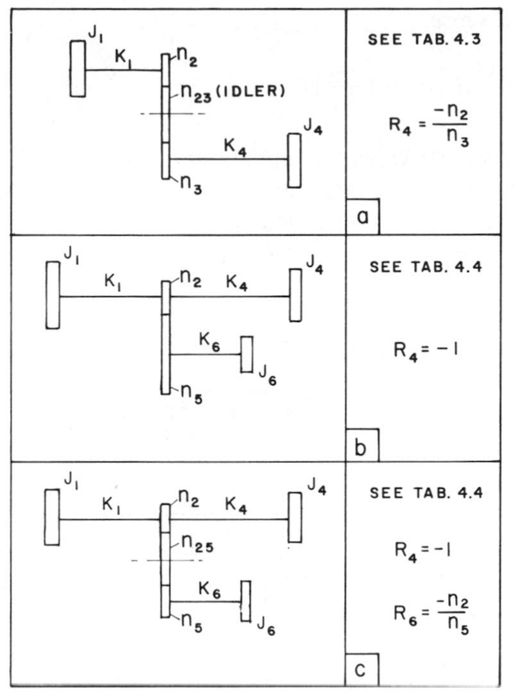

In Table 4.4 a second branch is included with the reference system at the common gear (2) and each of the three single-degree branches displacement excited at the gears. Neglecting gear inertias there are two frequencies and modes obtainable by a similar procedure. The normalized torque equation at the gears for equilibrium becomes

$$1 - R_4 m_4 \beta_4 - R_6 m_6 \beta_6 = 0 \tag{4.20}$$

with final relations as shown in the table.

Table 4.5 provides analytical methods for several more extensive cases including idler gears and lateral branches. As shown, these systems require negative R factors applied in Tables 4.3 and 4.4.

4.17. HOLZER FOR GEARED SYSTEMS

The previous analytical approaches become cumbersome if extended to more geared branches, or if branches have multiple masses. Holzer tabular calculations are adapted readily to any geared torsional system, expedited by normalization.

Taking the simple case of Figure 4.13*a*, we can proceed from J_1 to J_4 by introducing minor modifications at the mesh. Namely, as we move from 2 to 3 the amplitude and torque at 2 must be converted to the gear 3 using the gear ratio and reversed sense.

$$\beta_3 = -\left(\frac{n_2}{n_3}\right)\beta_2 \qquad \Sigma_3 = -\left(\frac{n_3}{n_2}\right)\Sigma_2 \tag{4.21}$$

Finally, for the proper g^2 value the remainder torque factor Σ_4 will be zero and $g^2 = g_n^2$. Note that K_4 and J_4 are actual, not equivalent, values. Also, normalized gear inertia can easily be included.

With specific values for a numerical example of

$$m_4 = 1.5$$

$$k_4 = 1$$

$$R = 0.833 = \frac{20}{24}$$

$g_n^2 = 0.8033$				
q	m	β	Σ	k
1	1	1.000	0.803	1
2	-	0.197	-	-
3	-	-0.164	-0.964	1
4	1.5	0.800	0	

(*a*)

q	m	β	Σ	k
1	1	1.000	0.803	1
2	-	0.197	-	

(*b*)

q	m	β	Σ	k
4	1.5	1.000	1.205	1
3	-	-0.205	-	

(*c*)

FIGURE 4.14. (*a*) Holzer solution proceeding continuously from J_1 to J_4. (*b*) Tabular results proceeding from J_1 to the mesh. (*c*) As in (*b*), proceeding from J_4 as a free end towards the mesh.

and using the converged frequency factor, in Figure 4.14*a*, where

$$\beta_3 = -\left(\frac{20}{24}\right)(0.197) = -0.164$$

$$\Sigma_3 = -\left(\frac{24}{20}\right)(0.803) = -0.964$$

These results are summarized in Figure 4.15, indicating the sense of the respective torques and displacements.

Another scheme, necessary when a gear meshes with two or more other gears, is a tabular procedure from each free end toward the common gear. As usual we assume $+1.00$ as the modal amplitude at each free end. Resultant amplitudes and torque conditions must then be tested for conformance with physical constraints at the gears. For the same system we tabulate for $g_n^2 = 0.803$ from 1 to 2 (Figure 4.14*b*). Similarly, from 4 to 3, we have Figure 4.14*c*.

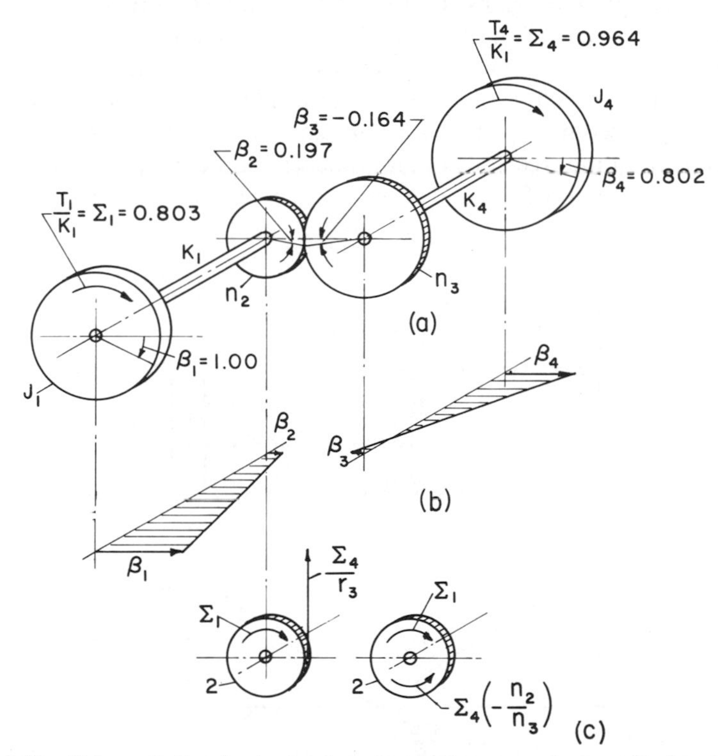

FIGURE 4.15. Holzer solution for torsional modes requires reversal of amplitudes at the gears (*a*), with the first mode shown in (*b*). The pinion, 2, is in equilibrium with equal and opposite torques (*c*).

A rectifying factor F must be applied to the latter tabulation and to all amplitudes and torques:

$$F = \frac{0.197(-20/24)}{-0.205} = +0.800$$

We note the agreement of the rectified values with those of Figure 4.14*a*.

Since we have forced the gear amplitudes to agree with the required rotations (Figure 4.16), it remains to test and to satisfy the torque situation. The equilibrium relation for the common or reference gear is

$$\Sigma_r - (R_j\Sigma_j + R_k\Sigma_k + \cdots) = 0 \tag{4.22}$$

In this example,

$$\Sigma_2 - R_3\Sigma_3 = 0.803 - (0.833)(1.205)(0.800) = 0$$

With trial values of g^2, Equation 4.22 becomes the test for modal frequency, with this value plotted against g^2 for interpolation (Figure 4.17). This procedure requires no sign changes in the branch tabulations, as we do not pass across a mesh in a continuous sense as in the previous method.

Although the branched case used as an example is simple for conciseness, the general procedure is applicable to complex torsional systems. Holzer calculations from the free ends of all branches can be integrated at a common gear with modification factors determined for each branch. The branches in

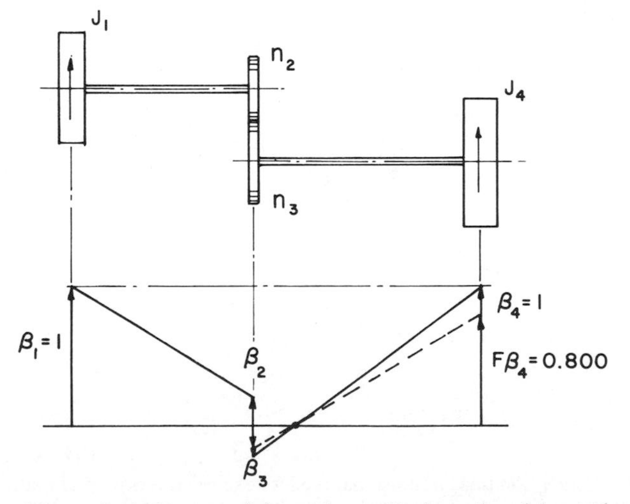

FIGURE 4.16. When calculating toward the mesh, amplitudes at 3 and 4 must be reduced by the factor F (dashed line), to satisfy the gear ratio, $R = 20/24$.

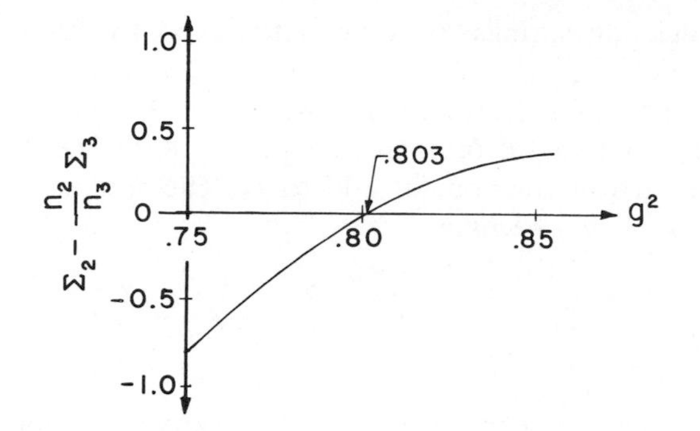

FIGURE 4.17. Agreement of torques with respect to the reference gear axis, 2, verifies modal frequency factor as the curve crosses the axis.

turn can consist of multiple discrete springs and masses, incorporated in the normalized Holzer computation.

4.18. THE STRING–MASS SYSTEM

Given a series of concentrated masses connected by a massless elastic string in tension (Figure 4.18a), grounded at the ends, we have an undamped multidegree system. With the string, however, the masses move transversely to the reference axis rather than along the axis. Masses are easily defined, but the

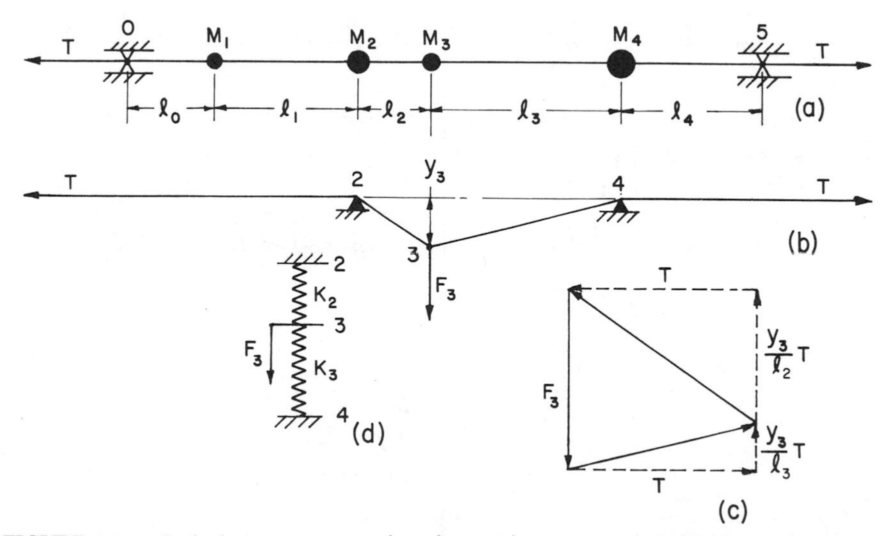

FIGURE 4.18. Equivalent transverse springs for a string system are derived by supporting two adjacent spans to ground.

equivalent connecting springs must be determined by static force analysis (Figure 4.18*b*, *c*).

We isolate a pair of adjacent spans, 2–3 and 3–4, and apply a transverse load F_3 resisted by two vector forces (Figure 4.18*c*). The horizontal forces cancel, and the vertical components add to F_3. The common deflection is y_3. Spring constants (F/y_3) become

$$K_2 = \frac{T}{l_2} \qquad K_3 = \frac{T}{l_3} \qquad K_q = \frac{T}{l_q} \tag{4.23}$$

It is true that in Figure 4.18*b* we have introduced auxiliary supports at 2 and 4, but if there were helical springs connecting the masses we would take them in parallel (*d*). Fixed points at 2 and 4 correspond to the auxiliary supports.

Normalizing, $k_2 = l_1/l_2$, $k_3 = l_1/l_3, \ldots$, and these stiffness ratios are easily determined. Frequencies and modes are then obtainable by any spring–mass method, including Holzer. With equal masses and spans, Table 4.2*b* can be applied. Results are given in Table 4.6 for several such cases.

As always, reference natural frequency for the normalized solution is based on M_1 and K_1, or now on M_1 and l_1, and $\omega_1 = \sqrt{T/M_1 l_1}$.

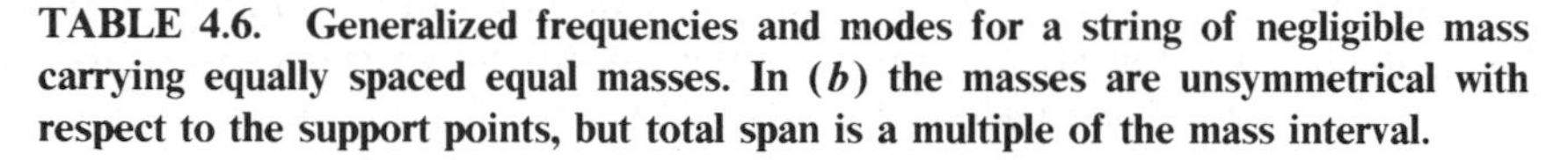

TABLE 4.6. Generalized frequencies and modes for a string of negligible mass carrying equally spaced equal masses. In (*b*) the masses are unsymmetrical with respect to the support points, but total span is a multiple of the mass interval.

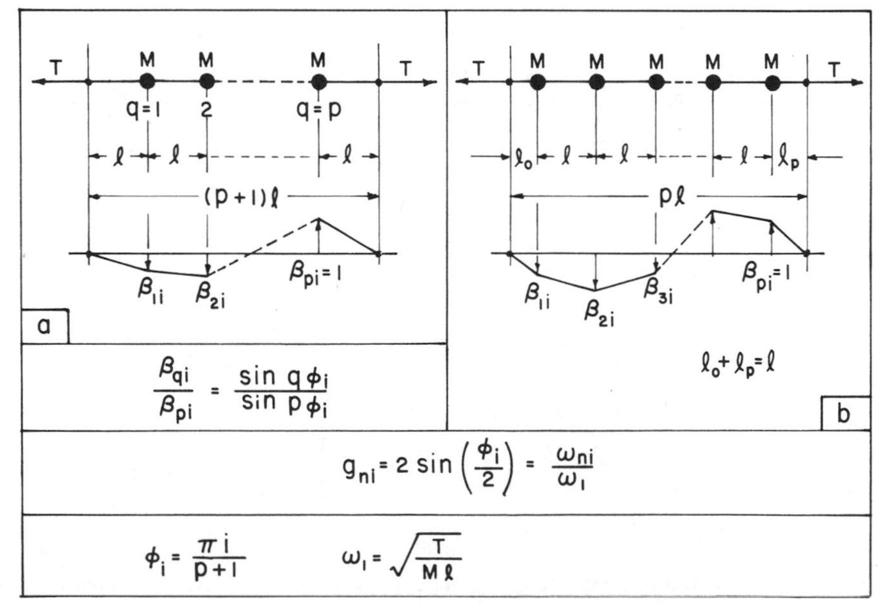

EXAMPLES

EXAMPLE 4.1. A two-mass system (Table 4.1*c*) has the following parameters:

$$K_0 = 24 \text{ N/m} \qquad M_1 = 8 \text{ kg}$$
$$K_1 = 30 \qquad M_2 = 3$$
$$K_2 = 17$$

Find:

(a) The two natural frequencies.

(b) The two normal modes.

(c) The two modal equivalent single-degree systems

Solution.

(a)
$$\omega_1 = \sqrt{K_1/M_1} = \sqrt{30/8} = 1.936 \text{ r/s} = 0.308 \text{ Hz}$$

$$m_2 g_n^4 - (k_0 m_2 + m_2 + k_2 + 1) g_n^2 + (k_0 + k_0 k_2 + k_2) = 0$$

$$k_0 = 24/30 = 0.800 \qquad m_2 = 3/8 = 0.375$$

$$k_2 = 17/30 = 0.567$$

$$0.375 g_n^4 - 2.242 g_n^2 + 1.820 = 0$$

$$g_{n1}^2 = 0.969 \qquad g_{n1} = 0.984$$

$$g_{n2}^2 = 5.010 \qquad g_{n2} = 2.238$$

$$\omega_{n1} = g_{n1}\omega_1 = 0.984(0.308) = 0.303 \text{ Hz}$$

$$\omega_{n2} = g_{n2}\omega_1 = 2.238(0.308) = 0.689 \text{ Hz}$$

(b)
$$\beta_{21} = k_0 + 1 - g_{n1}^2 = 1.80 - 0.969 = 0.831$$

$$\beta_{22} = 1.80 - g_{n2}^2 = 1.80 - 5.010 = -3.21$$

(c)
$$M_{n1} = M_1 \sum m\beta^2 = 8\left[(1)(1)^2 + (0.375)(0.831)^2\right]$$

$$= 8(1.259) = 10.07 \text{ kg}$$

$$M_{n2} = 8\left[(1)(1)^2 + (0.375)(-3.21)^2\right] = 8(4.864) = 38.9 \text{ kg}$$

$$K_{n1} = K_1 g_{n1}^2 \sum m\beta^2 = 30(0.969)(1.259) = 36.6 \text{ N/m}$$

$$K_{n2} = K_1 g_{n2}^2 \sum m\beta^2 = 30(5.010)(4.864) = 731 \text{ N/m}$$

EXAMPLE 4.2. Find the natural frequencies of two systems of the type shown in Table 4.1b, specified as follows:

A		*B*	
$M_1 = 50$ lb.	$K_1 = 400$ lb/in.	$M_1 = 35$ *Kg*	$K_1 = 8000$ N/m
$M_2 = 150$	$K_2 = 600$	$M_2 = 105$	$K_2 = 12000$ N/m

Solution. For both A and B, $m_2 = 3$ and $k_2 = 1.5$, and the generalized frequency quadratic is

$$m_2 g_n^4 - (m_2 + k_2 + 1) g_n^2 + k_2 = 0$$

$$3g_n^4 - 5.5g_n^2 + 1.5 = 0$$

$$g_{n1} = 0.577$$

$$g_{n2} = 1.224$$

For A,

$$\omega_1 = \sqrt{K_1/M_1} = \sqrt{400(386)/50} = 55.57 \text{ rad/sec} = 8.84 \text{ Hz}$$

$$\omega_{n1} = g_{n1}\omega_1 = 0.577(8.84) = 5.10 \text{ Hz}$$

$$\omega_{n2} = g_{n2}\omega_1 = 1.224(8.84) = 10.83 \text{ Hz}$$

For B,

$$\omega_1 = \sqrt{K_1/M_1} = \sqrt{8000/35} = 15.12 \text{ r/s} = 2.406 \text{ Hz}$$

$$\omega_{n1} = g_{n1}\omega_n = 0.577(2.406) = 1.39 \text{ Hz}$$

$$\omega_{n2} = g_{n2}\omega_1 = 1.224(2.406) = 2.95 \text{ Hz}$$

Normal modes for A and B will be identical since they are both calculated from the same g_n^2 values.

A and B are kindred systems with respect to the m_2 and k_2 ratios. They therefore share certain basic characteristics although the systems are of different absolute size and are specified in different systems of units.

One could further change the springs for A, say to 4000 and 6000 lb/in., without affecting k_2 or m_2, or in any way changing the generalized frequency or modal calculations. In fact, the only change will be an increase in the natural frequencies of A by a factor of $\sqrt{10}$.

EXAMPLE 4.3. A steel shaft is 5 cm in diameter and 80 cm long. Find the first natural torsional frequency of the distributed system using a six-mass approximation. One end of the shaft is rigidly clamped.

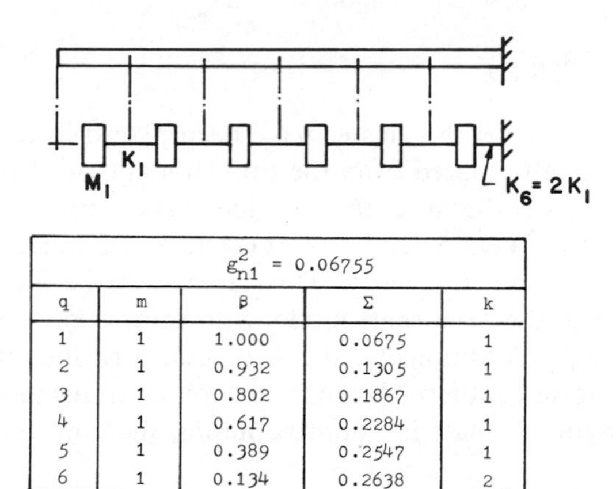

$g_{n1}^2 = 0.06755$				
q	m	β	Σ	k
1	1	1.000	0.0675	1
2	1	0.932	0.1305	1
3	1	0.802	0.1867	1
4	1	0.617	0.2284	1
5	1	0.389	0.2547	1
6	1	0.134	0.2638	2
		0.002		

Solution

$$M_1 = \frac{(\rho A l) r^2}{6} = \frac{(0.008)(19.63)(80)(1.77)^2}{6}$$

$$= \left(\frac{39.2}{6}\right) \text{kg} \cdot \text{cm}^2$$

$$K_1 = 6\left(\frac{GJ}{l}\right) = 6\left[\frac{8(10)^6 61.4}{80}\right] = 6(6.13)(10)^6 \text{ N} \cdot \text{cm/r}$$

$$\omega_1 = 10\sqrt{\frac{6(6.13)(10)^6}{(39.2/6)}} = 60\sqrt{\frac{6.13(10)^6}{39.2}}$$

$$= 23{,}7000 \text{ r/s} = 3780 \text{ Hz}$$

From Holzer,

$$\omega_{n1} = g_{n1}\omega_1 = \sqrt{0.06755}\,(3780) = 982 \text{ Hz}$$

The exact solution (Table 12.1*a*), is

$$\omega_{n1} = \frac{\pi}{2}\left(\frac{1}{0.80}\right)\sqrt{\frac{800(10)^8}{8000}} = \frac{\pi}{2}(2950) = 6206 \text{ r/s}$$

$$= 988 \text{ Hz}$$

In the discrete mass model the masses have been placed in the center of each span rather than equally spaced from the tip. Although not fitting Table 4.2*a*, this is a better model because the tip has maximum amplitude, and a concentrated mass at the tip tends to exaggerate the kinetic energy effect, resulting in a low natural frequency. This table, and the lumped model, are equally applicable to the axial shaft mode, with translational parameters used in the ω_1 calculation. Additionally, the β column provides the first normal mode for any uniform cantilever beam in torsion or translation, regardless of cross section, length, or material, approximating the sine function in Table 12.1*a*.

Although hand-held calculators are available for Holzer calculations with varying degrees of program capabilities, results are readily obtained on a basic model with only one memory. A suggested procedure in the latter case is as follows:

(a) Enter and store $\beta_1 = 1.000$.
(b) Multiply display by m_1 and $\mathsf{g}^2 = 0.06755$.
(c) Record $\Sigma_1 = 0.0675$.
(d) Divide display by $k_1 = 0.06755$ and reverse sign $= -0.06755$.
(e) Sum to memory and record $\beta_2 = 1 - 0.06755 = 0.932$.
(f) Multiply display by $m_2\mathsf{g}^2 = 0.0630$.
(g) Add to $\Sigma_1 = 0.1305$ and record as Σ_2.
(h) Divide Σ_2 by k_2, change sign, sum to memory, and continue.

The β values carried in memory will be relatively accurate. In the Σ terms we accept round-off error, and record only as many places as necessary. More places can be added as convergence is approached.

EXAMPLE 4.4. A system consists of three equal springs and three equal masses. The values are 56 lb/in. and 8.5 lb, respectively.

(a) Find the three natural frequencies.
(b) Determine the three normal modes.
(c) Check orthogonality.

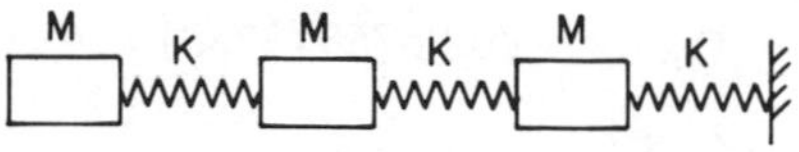

M K M K M K

Solution

(a)

$$\phi_1 = \pi\left(\frac{1 - \frac{1}{2}}{3 + \frac{1}{2}}\right) = 0.4488 \qquad (i = 1) \quad \text{(Table 4.2a)}$$

$$g_{n1} = 2\sin\left(\frac{0.4488}{2}\right) = 0.445$$

$$\omega_1 = \sqrt{\frac{K_1}{M_1}} = \sqrt{\frac{386(56)}{8.5}} = 50.43 \text{ or } 8.03 \text{ Hz}$$

$$\omega_{n1} = 0.445(8.03) = 3.57 \text{ Hz}$$

$$\phi_2 = \pi\left(\frac{2 - \frac{1}{2}}{3 + \frac{1}{2}}\right) = 1.3464 \qquad (i = 2)$$

$$\omega_{n2} = 10.01 \text{ Hz} \qquad \omega_{n3} = 14.47 \text{ Hz} \qquad (i = 3)$$

(b) First mode:

$$\frac{\beta_{11}}{\beta_{31}} = \frac{\sin 0.4488}{\sin 3(0.4488)} = 0.445 = g_{n1}$$

$$\frac{\beta_{21}}{\beta_{31}} = \frac{\sin 2(0.4488)}{\sin 3(0.4488)} = 0.802$$

$$\beta_{31} = 1.000$$

Second mode:

$$\beta_{12} = -1.247 \qquad \beta_{22} = -0.555 \qquad \beta_{32} = 1.000$$

Third mode:

$$\beta_{13} = 1.802 \qquad \beta_{23} = -2.247 \qquad \beta_{33} = 1.000$$

(c) From Equation 4.7,

$$i = 1,\ j = 2 \qquad (0.445)(-1.247) + (0.802)(-0.555) + 1$$

$$= -(0.555) - (0.445) + 1 = 0$$

$$i = 1,\ j = 3 \qquad (0.445)(1.802) + (0.802)(-2.247)$$

$$= 0.802 - 1.802 + 1 = 0$$

$$i = 2,\ j = 3 \qquad (-1.247)(1.802) + (-0.555)(-2.247) + 1$$

$$= -2.247 + 1.247 + 1 = 0$$

5 FORCED RESPONSE OF MULTIMASS SYSTEMS

The multimass system has been shown to have characteristic natural frequencies and modes in free vibration calculated from all the spring and mass parameters. As with the single-mass case (Chapter 2), resonance occurs if a sinusoidal forcing function acts upon the more complex system at any of the several natural frequencies. Excitation can be defined in terms of force or displacement amplitudes; however, natural frequencies and modes are unaffected by the type of forcing.

Total steady-state response is obtained most obviously by simultaneous solution of differential equations of motion and the introduction of the forcing function. With many masses, the Holzer tabular approach is adapted to determine forced response in a single calculation.

We can also obtain response by combining the effective contributions of each forced normal mode. This approach is important conceptually, providing a powerful analytical tool. It in fact reduces response of the most complex systems to a series of single-degree responses that can be superimposed. Thus we can focus on any particular resonance of interest as a single-degree phenomenon.

Damping effects have been introduced in Chapters 1, 2, and 3. The principal results of damping in steady-state cases are the limiting of resonant amplitudes, the introduction of phase relations between the excitation and the several amplitudes, and the dissipation of energy. Modal superposition methods apply, however, only approximately if appreciable damping is present. Holzer methods are again applied for total forced response, including any degree of damping and phase effects.

Damping in most mechanical and structural systems is relatively low. Often the modal damping ratio is less than 5%, and in some cases it is as low as 1%. Nevertheless the effects are significant, particularly near resonance, and methods of analysis are now presented for discrete systems.

5.1. EQUATIONS OF MOTION

The classical forced undamped two-mass system is shown in Figure 5.1*a*. Equilibrium equations are similar to Equations 4.1, with the addition of the external force on M_1, but whereas the previous equations assumed natural frequencies, we now understand all terms to vary at the *forcing frequency*. Force equilibrium in (*b*) and (*c*) of Figure 5.1 becomes, respectively,

$$\begin{cases} \Sigma F_1 = M_1\omega^2 a_1 + K_1(a_2 - a_1) + P_1 = 0 & (5.1a) \\ \Sigma F_{12} = M_1\omega^2 a_1 + M_2\omega^2 a_2 - K_2 a_2 + P_1 = 0 & (5.1b) \end{cases}$$

Normalizing with respect to the reference system (K_1, M_1),

$$\begin{cases} a_1(1 - g^2) - a_2 = \dfrac{P_1}{K_1} & (5.2a) \\ -a_1 g^2 + a_2(k_2 - m_2 g^2) = \dfrac{P_1}{K_1} & (5.2b) \end{cases}$$

Solving for a_1 and a_2,

$$\frac{a_1}{(P_1/K_1)} = \frac{\begin{vmatrix} 1 & -1 \\ 1 & (k_2 - m_2 g^2) \end{vmatrix}}{m_2 g^4 - (m_2 + k_2 + 1)g^2 + k_2} = \frac{1 + k_2 - m_2 g^2}{C} \tag{5.3a}$$

$$\frac{a_2}{(P_1/K_1)} = \frac{\begin{vmatrix} 1 - g^2 & 1 \\ -g^2 & 1 \end{vmatrix}}{C} = \frac{1}{C} \tag{5.3b}$$

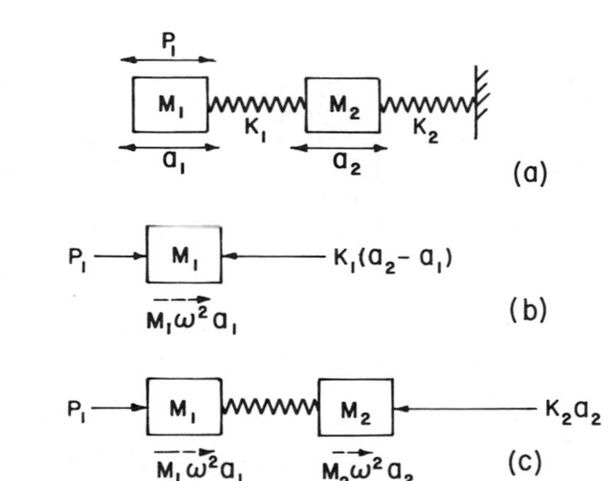

FIGURE 5.1. The two-mass grounded system is excited at M_1, with equilibrium conditions for M_1 (*b*), and for both M_1 and M_2 (*c*).

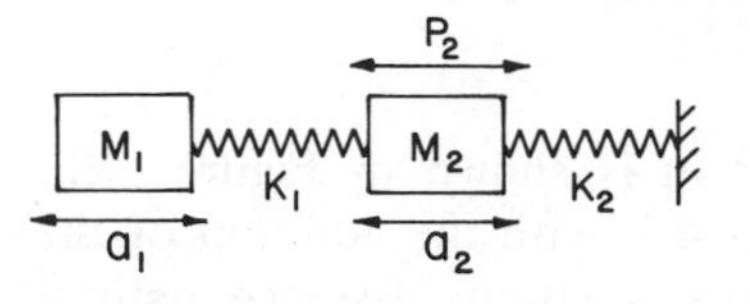

FIGURE 5.2. Similar to Figure 5.1, but with the sinusoidal exciting force applied to M_2.

Transferring the excitation to M_2 (Figure 5.2),

$$\begin{cases} \Sigma F_1 = M_1\omega^2 \mathrm{a}_1 + K_1(\mathrm{a}_2 - \mathrm{a}_1) = 0 & (5.4a) \\ \Sigma F_{12} = M_1\omega^2 \mathrm{a}_1 + M_2\omega^2 \mathrm{a}_2 - K_2 \mathrm{a}_2 + P_2 = 0 & (5.4b) \end{cases}$$

These reduce to the results in Table 5.1*a*,

$$\frac{\mathrm{a}_1}{(P_2/K_1)} = \frac{1}{C} \qquad \frac{\mathrm{a}_2}{P_2/K_1} = \left(\frac{1 - \mathrm{g}^2}{C}\right) \tag{5.5}$$

a_{12} is the relative amplitude of 1 with respect to 2.

TABLE 5.1. The classical undamped two-degree system, force excited on either mass, with (*a*) one ground, and (*b*) three springs, both ends grounded.

a	P_1	P_2	b	P_1	P_2
$\frac{a_1}{P/K_1}$	$\frac{A}{C}$	$\frac{1}{C}$	$\frac{a_1}{P/K_1}$	$\frac{A}{C}$	$\frac{1}{C}$
$\frac{a_{12}}{P/K_1}$	$\frac{k_2 - m_2 g^2}{C}$	$\frac{g^2}{C}$	$\frac{a_2}{P/K_1}$	$\frac{1}{C}$	$\frac{B}{C}$
$\frac{a_2}{P/K_1}$	$\frac{1}{C}$	$\frac{1-g^2}{C}$	$\frac{P_0}{P}$	$\frac{-k_0 A}{C}$	$\frac{-k_0}{C}$
$\frac{P_3}{P}$	$\frac{-k_2}{C}$	$\frac{-k_2(1-g^2)}{C}$	$\frac{P_3}{P}$	$\frac{-k_2}{C}$	$\frac{-k_2 B}{C}$
C	$m_2 g^4 - A_1 g^2 + k_2$		C	$m_2 g^4 - (A_1 + m_2 k_0) g^2 + A_2$	

$A = (1 + k_2 - m_2 g^2)$ $\quad A_2 = (k_0 + k_0 k_2 + k_2)$ $\quad g = \omega/\omega_1$

$A_1 = (1 + k_2 + m_2)$ $\quad B = (1 - g^2 + k_0)$ $\quad m_2 = \frac{M_2}{M_1}$ $\quad k_2 = \frac{K_2}{K_1}$

5.2. FREQUENCY RESPONSE BEHAVIOR

The normalized equations can now be used to observe the effect of frequency on amplitudes. In Figure 5.3 we assume forcing at M_1, and system parameters of $m_2 = 0.7$ and $k_2 = 1.2$, and calculate the natural frequency factors $\mathsf{g}_{n1} = 0.683$ and $\mathsf{g}_{n2} = 1.917$. Both a_1 and a_2 become infinite at resonance when the denominator is zero, $\mathsf{g} = \mathsf{g}_n$, or

$$m_2\mathsf{g}_n^4 - (m_2 + k_2 + 1)\mathsf{g}_n^2 + k_2 = 0 \tag{5.6}$$

which is also Equation 4.4.

Also, in Figure 5.3*a*, a_1 becomes zero if the numerator in Equation 5.3a is zero. This nodal condition occurs at $\mathsf{g} = 1.77$. There is no such condition for M_2.

At zero frequency, static deflections are, at $\mathsf{g} = 0$,

$$\frac{\mathsf{a}_1}{(P_1/K_1)} = \left(\frac{1 + k_2}{k_2}\right) = 1.833 \qquad \frac{\mathsf{a}_2}{P_1/K_1} = \frac{1}{k_2} = 0.833 \tag{5.7}$$

With increasing frequency a_1 exceeds a_2, both in phase with P_1, until the first resonance develops, at which $\mathsf{a}_2/\mathsf{a}_1$ approaches β_2/β_1 (Table 4.1*b*), and phase shifts of 180° occur. Between resonances a_1 shifts to in phase, with a_2 reaching a minimum out of phase. After the second resonance, both amplitudes approach zero, with M_1 and M_2 moving in opposition to each other.

Figure 5.4*a* displays representative total forced modes for the sample system due to P_1. Resonances are shown by dashed lines, indicating normal modes of undefined amplitude at the corresponding phase shift. In (*b*) responses are shown for P_2, with M_2 nodal at $\mathsf{g} = 1$.

The examples of behavior outlined are characteristic of all similar two-degree systems which will differ only in the effect of k_2 and m_2 on the several coefficients.

5.3. EXTERNAL REACTION

Table 5.1*a* contains expressions for P_3. Taking the previous system, we observe the effect of increasing forcing frequency on the force transmitted to ground (Figure 5.3*b*). Obviously this quantity approaches infinity with the amplitudes at the two resonances with a minimum between the two. Statically P_3 reacts P_1, as there are no inertia effects interposed at M_1 or M_2.

In the normalized coordinates the P_3/P_1 ratio is $+1.20$, or k_2, at two frequencies. One is at $\mathsf{g} = 1$. The other is at $\mathsf{g} = 1.77$, when M_2 is a node.

Transmissibility is never zero, but becomes small above $\mathsf{g} = 2$ for this system.

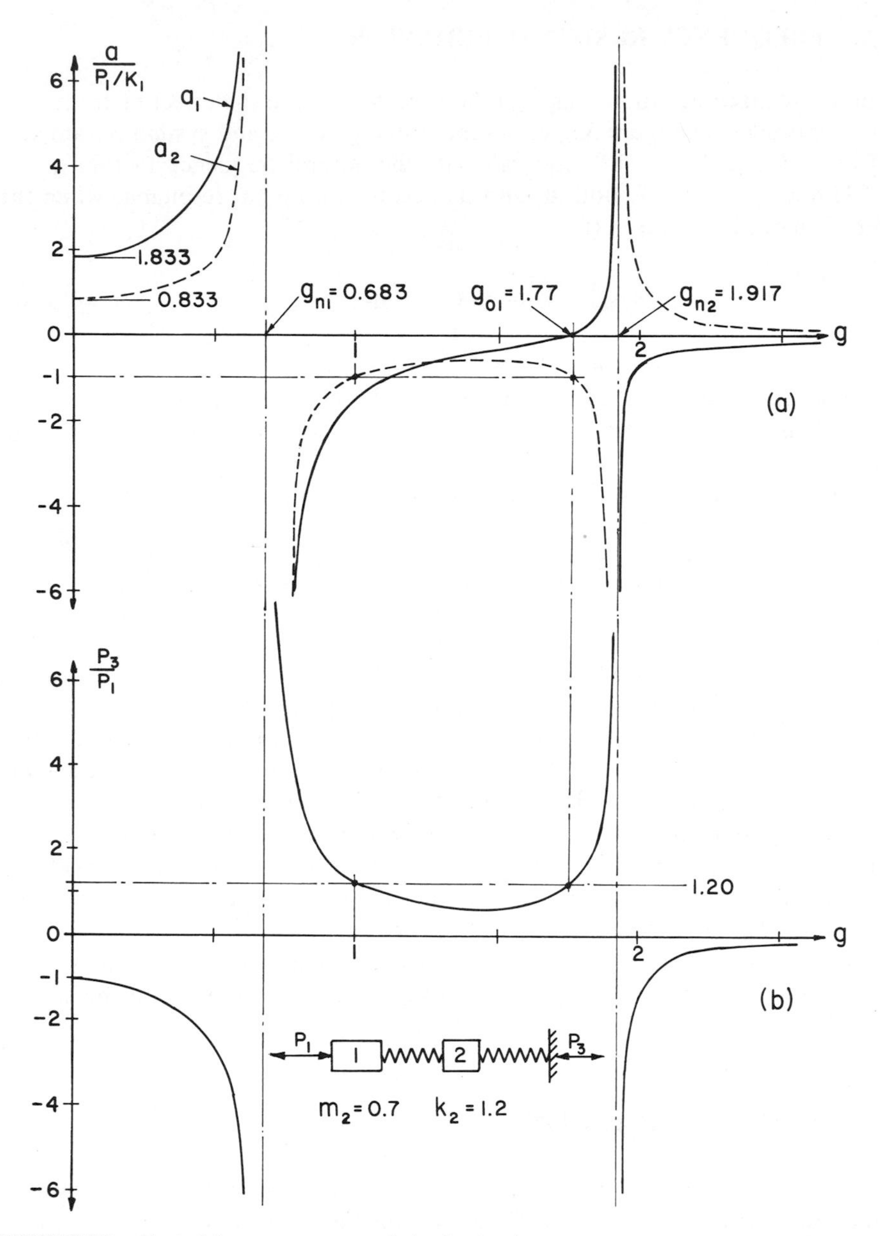

FIGURE 5.3. Typical frequency response behavior for the system of Table 5.1*a*, including several calculated coordinate values. Transmissibility is shown in (*b*).

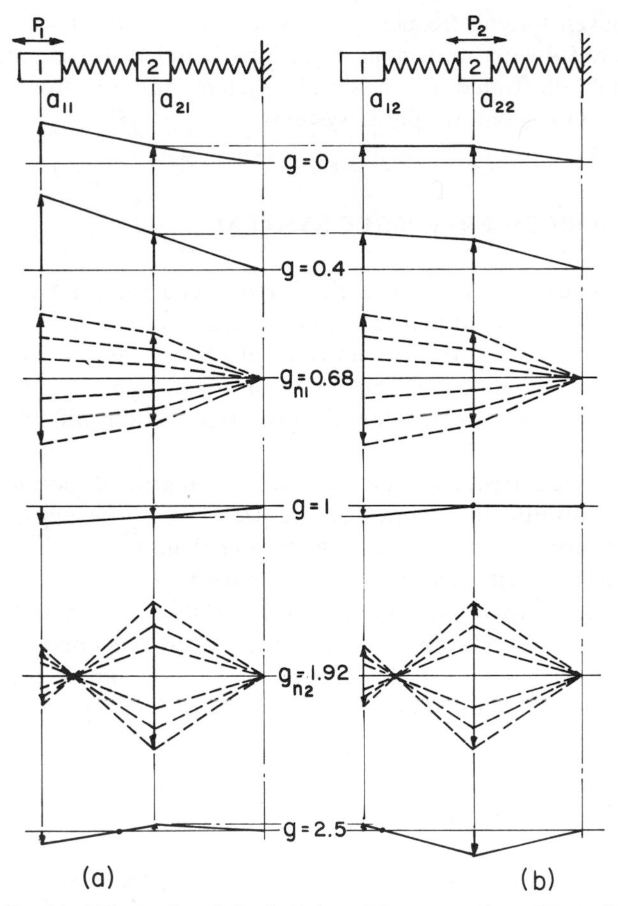

FIGURE 5.4. Graphical illustration of the forced modal responses from Figure 5.3 with increasing frequency of excitation. In (a) forcing is at M_1; in (b), at M_2. Dynamic reciprocity is shown by the horizontal projections.

5.4. DYNAMIC RECIPROCITY

Maxwell's theorem of reciprocal displacements in static elastic systems states that for any two points in an elastic structure, if a unit force applied at 1 produces a deflection at 2 in a given direction, a unit force at 2 in this direction will produce an identical deflection at 1, and in the direction of the load at 1; or $y_{12} = y_{21}$.

We now note a similar relationship in steady-state vibration for the two-degree case (Table 5.1):

$$\frac{a_{21}}{(P_1/K_1)} = \frac{a_{12}}{(P_1/K_1)} = \frac{1}{C} \tag{5.8}$$

Thus for a given forcing frequency, $a_{21} = a_{12}$ if $P_1 = P_2$. This is seen graphically in Figure 5.4 with horizontal projections from left to right, or from (*a*) to (*b*). Although illustrated for a simple system, the reciprocal relationship applies to the most complex linear systems.

5.5. THE FORCED FREE–FREE SYSTEM

Equilibrium equations applied to the ungrounded system lead to the Table 5.2*a* results. The only normalized system parameter is m_2. Taking $m_2 = 0.6$, and the excitation P_2 at M_2, we can calculate the amplitudes shown in Figure 5.5*a*. There is now only one normal vibratory mode at $g_{n1} = 1.63$, and a nodal condition for M_2 at $g = 1$. This absorber-type action will be explained in Chapter 7.

At low forcing frequency the absence of a ground permits amplitudes approaching infinity. In Figure 5.5*a* we see these to be negative, or out of phase with P_2 for both masses, with a_1 greater than a_2.

Relative motion between the masses (Figure 5.5*b*) defines the amplitude of the force in the spring K_1, the product of the relative motion and K_1. Resonant behavior is indicated. The apparent intercept at $g = 0$ is impossible physically, as P_2 must approach zero in this condition. Thus in simple translation there can be no relative motion, and the spring force is then zero.

5.6. DISPLACEMENT EXCITATION

Given sinusoidal displacement a_3 of the two-degree system at the spring (Figure 5.6), we have the following equations:

$$\begin{cases} \Sigma F_1 = M_1\omega^2 a_1 + K_1(a_2 - a_1) = 0 & (5.9a) \\ \Sigma F_{12} = M_1\omega^2 a_1 + M_2\omega^2 a_2 + K_2(a_3 - a_2) = 0 & (5.9b) \end{cases}$$

which normalize to

$$\begin{cases} a_1(1 - g^2) - a_2 = 0 & (5.10a) \\ -a_1 g^2 + a_2(k_2 - m_2 g^2) = k_2 a_3 & (5.10b) \end{cases}$$

Solutions are given in Table 5.3*a*.

Taking the spring–mass system of Figure 5.3, we obtain characteristic amplitude responses (Figure 5.6*a*) referenced to a_3. The required exciting force varies as in Figure 5.6*b*, approaching infinity to react the masses at the resonances. There is an interesting condition at $g = 1.56$, which is the resonant

TABLE 5.2. Free–free or ungrounded systems, having rigid translational modes, excited at any mass.

(a) M_1 —K_1— M_2; excitations P_1, P_2; responses a_1, a_2	P_1	P_2	(b) M_1 —K_1— M_2 —K_2— M_3; excitations P_1, P_2, P_3; responses a_1, a_3	P_1	P_2	P_3
$\frac{a_1}{P/K_1}$	$\frac{1-m_2g^2}{C}$	$\frac{1}{C}$	$\frac{a_1}{P/K_1}$	$\frac{-m_2m_3g^4+A_1g^2-k_2}{C}$	$\frac{m_3g^2-k_2}{C}$	$\frac{-k_2}{C}$
$\frac{a_2}{P/K_1}$	$\frac{1}{C}$	$\frac{1-g^2}{C}$	$\frac{a_2}{P/K_1}$	$\frac{m_3g^2-k_2}{C}$	$\frac{-m_3g^4+A_2g^2-k_2}{C}$	$\frac{k_2(g^2-1)}{C}$
$\frac{a_{12}}{P/K_1}$	$\frac{m_2g^2}{C}$	$\frac{g^2}{C}$	$\frac{a_3}{P/K_1}$	$\frac{-k_2}{C}$	$\frac{k_2(g^2-1)}{C}$	$\frac{m_2g^4+A_3g^2-k_2}{C}$
C	$g^2\left[m_2g^2-(1+m_2)\right]$		C	$g^2\left[m_2m_3g^4-(A_1+m_2m_3)g^2+k_2(1+m_2+m_3)\right]$		

$$g=\frac{\omega}{\sqrt{K_1/M_1}}\qquad m_2=\frac{M_2}{M_1}\qquad k_2=\frac{K_2}{K_1}$$

$$A_1=k_2(m_2+m_3)+m_3\,,\quad A_2=(k_2+m_3)\,,\quad A_3=(k_2+1+m_2)$$

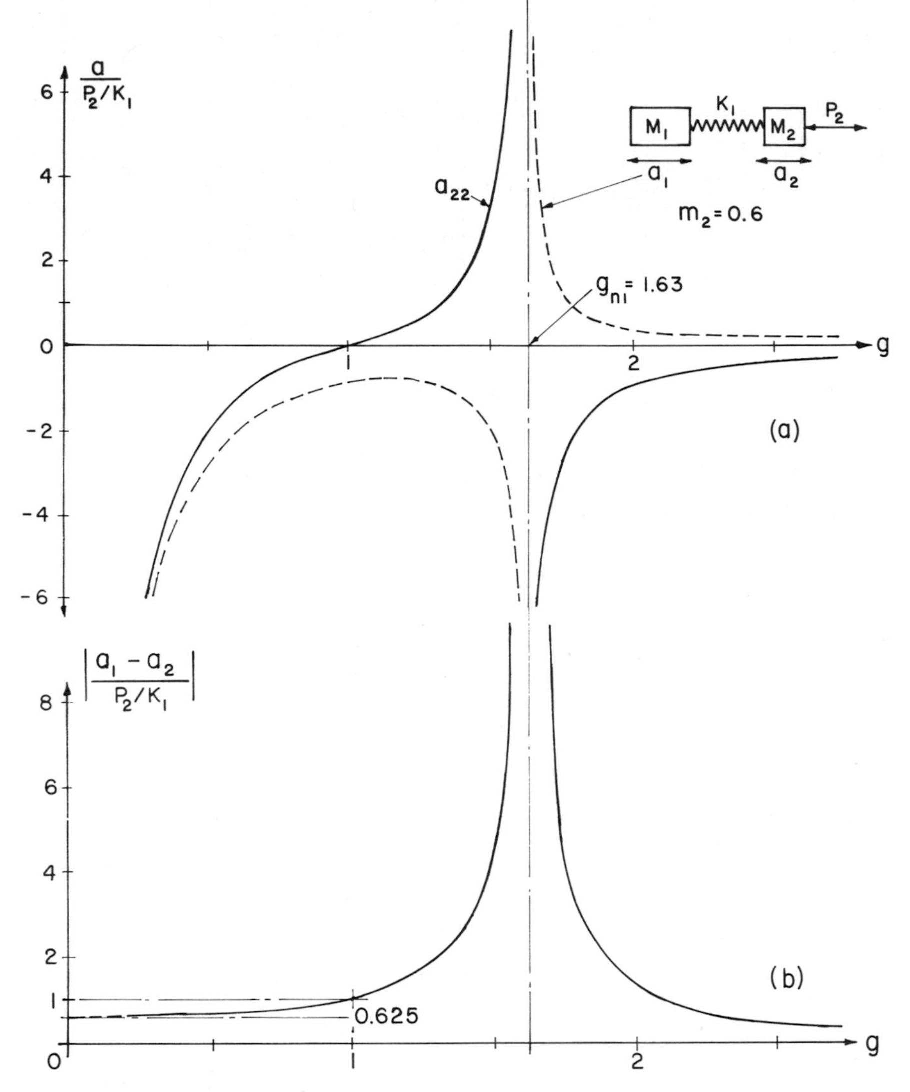

FIGURE 5.5. Example of the response variation from Table 5.2a. In (b) the ordinate represents the ratio of the amplitude of the spring force to the exciting force.

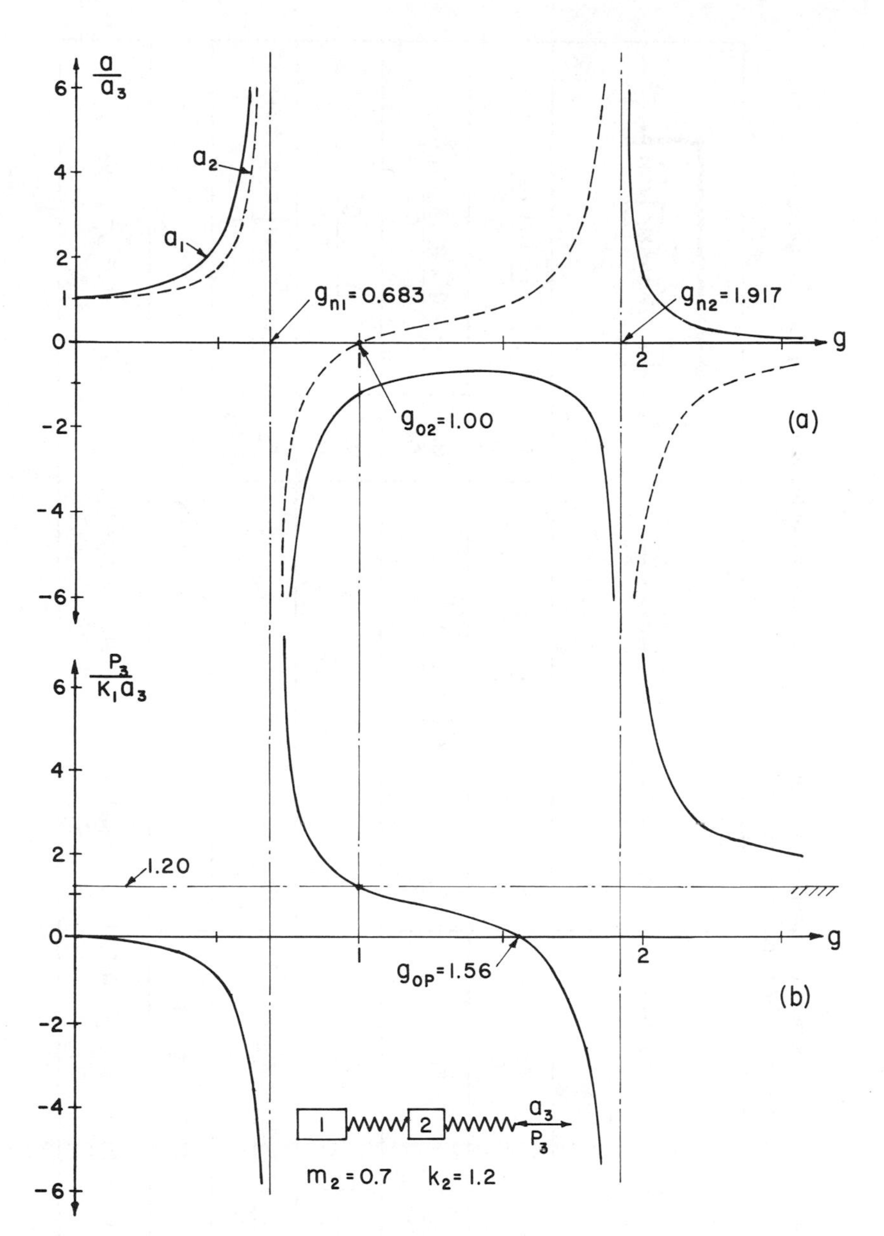

FIGURE 5.6. Illustration of the Table 5.3*a* response. Variation of the exciting force for a constant amplitude a_3 is shown in (*b*).

TABLE 5.3. **Displacement excitation of undamped two-mass systems excited at (*a*), (*b*) the free spring end, and (*c*) both ends simultaneously by in-phase excitations.**

a		b		c	
M_1, K_1, M_2, K_2, a_3, P_3, a_1, a_2		P_0, K_0, M_1, K_1, M_2, K_2, a_3, P_3		K_0, M_1, K_1, M_2, K_2, a_3, P_3	
$\frac{a_1}{a_3}$	$\frac{k_2}{C}$	$\frac{a_1}{a_3}$	$\frac{k_2}{C}$	$\frac{a_1}{a_3}$	$\frac{A_2 - k_0 m_2 g^2}{C}$
$\frac{a_{12}}{a_3}$	$\frac{k_2 g^2}{C}$	$\frac{a_2}{a_3}$	$\frac{k_2(1 + k_0 - g^2)}{C}$	$\frac{a_{12}}{a_3}$	$\frac{g^2(k_2 - k_0 m_2)}{C}$
$\frac{a_2}{a_3}$	$\frac{k_2(1 - g^2)}{C}$	$\frac{P_3}{K_2 a_3}$	$\frac{m_2 g^4 - A_1 g^2 + k_0}{C}$	$\frac{a_2}{a_3}$	$\frac{A_2 - k_2 g^2}{C}$
$\frac{P_3}{K_1 a_3}$	$\frac{-k_2 g^2[1 + m_2(1 - g^2)]}{C}$	$\frac{P_0}{K_2 a_3}$	$\frac{-k_0}{C}$	$\frac{P_3}{K_1 a_3}$	$\frac{-g^2(A_2[m_2 + 1] - m_2 g^2[k_0 + k_2])}{C}$
C	$m_2 g^4 - A g^2 + k_2$	C	$m_2 g^4 - (A_1 + k_2) g^2 + A_2$		

$A = (1 + m_2 + k_2)$, $A_1 = 1 + m_2(1 + k_0)$, $A_2 = (k_0 + k_0 k_2 + k_2)$, $m_2 \frac{M_2}{M_1}$, $k_0 = \frac{K_0}{K_1}$, $k_2 = \frac{K_2}{K_1}$

frequency of the two-mass system connected by K_1. Inertia forces of the masses then balance in this normal mode, requiring zero exciting force from K_2, or P_3.

5.7. RESPONSE TO FORCING BY HOLZER

The Holzer tabulation (Sections 4.7–4.13) has been shown to be a means for determining natural frequencies and modes, and this is its principal product. Any such calculation, however, assumes a forcing function and frequency, and a given calculation generates a specific forced mode. In Figures 4.7 and 4.8 this exciting force acts on M_3. The forced mode is determined, as well as the remainder function Σ, being minimized. The latter represents a driving function that satisfies the tabulated conditions.

In normalized form, system variables are divided by K_1. With this division in the equations of motion, P_1 becomes P_1/K_1, dimensionally a displacement but symbolizing a force. Figure 5.7 illustrates the forced solution of a two-mass system for $\omega = \omega_1$, or $\mathsf{g} = 1$. In the tabulation the amplitude a_1 is the unknown or variable, and the forcing term $28/300 = 0.0933$ introduced at Σ effectively adds to the inertia term, $m_1\mathsf{g}^2\mathsf{a}_1$, at $q = 1$. Finally, equating the amplitude at the fixed end to zero, we find the forced amplitudes shown. Negative signs indicate out of phase with P_1. Spring forces, denormalizing by multiplication by K_1, are respectively $300\ \Sigma_1 = -14$ and $300\ \Sigma_2 = -33.6$ lb, with negative signs indicating corresponding maximum tension in each.

Taking a similar system in torsion with displacement excitation and in SI units at $\mathsf{g} = 1$, we have Figure 5.8. An amplitude ratio of -1.2 is found for $\mathsf{a}_1/\mathsf{a}_3$, with $\mathsf{a}_2 = 0$. Spring force amplitudes are

$$K_1\Sigma_1 = 850(-1.2)(0.0524) = -53.4\ \text{N}\cdot\text{m}$$

$$K_1\Sigma_2 = 850(-1.2)(0.0524) = -53.4\ \text{N}\cdot\text{m}$$

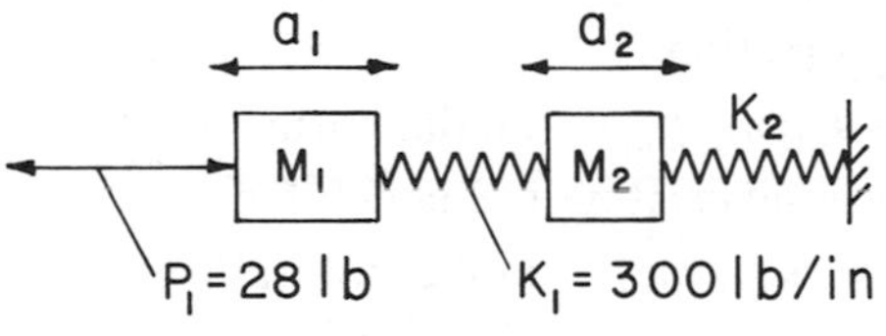

q	m	β	mg²β	Σ	k	Σ/k
1	1	a_1	a_1	$a_1 + 0.0933$	1	$a_1 + 0.0933$
2	0.7	$0 - 0.0933$	$0 - 0.0653$	$a_1 + 0.0280$	1.2	$0.833a_1 + 0.0233$
3	-	$-0.833a_1 - 0.1167 = 0$				

$a_1 = -0.140$, $a_2 = -0.0933$

FIGURE 5.7. Steady-state response obtained from a normalized Holzer calculation at $\mathsf{g} = 1.00$, with an applied exciting force.

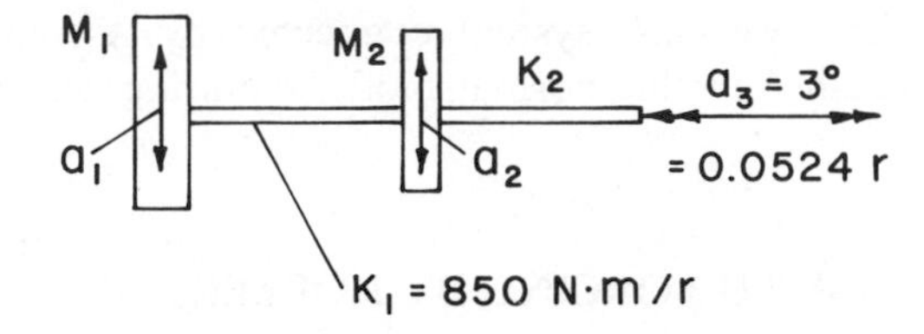

q	m	β	$mg^2\beta$	Σ	k	Σ/k
1	1	a_1	a_1	a_1	1	a_1
2	0.7	0	0	a_1	1.2	$0.833\,a_1$
3	-	$-0.833\,a_1 = a_3$				

$a_1/a_3 = -1.2, \quad a_1 = -3.6^\circ, \quad a_2 = 0$

FIGURE 5.8. Similar to Figure 5.7, but with displacement excitation of a torsional system.

5.8. FORCED MODAL COMPONENTS

The previous discussion has dealt exclusively with *total* steady-state vibratory responses. We will now investigate equivalent forced modes that retain the *normal modal distributions*. Since all modal components respond at the forcing frequency, they are additive algebraically, and the sum coincides with the total response.

As shown in Figure 4.4, any lumped system can be reduced to an equivalent modal mass and modal grounded spring, the latter of particular importance. By this analog we are able to calculate amplitude in any mode as a single-degree process (Figure 5.9).

$$\frac{a_{1i}}{(P_1/K_{ni})} = \left(\frac{1}{1 - g_i^2}\right) \tag{5.11}$$

where
a_{1i} = forced amplitude of M_1 in mode i
P_1 = exciting function
$K_{ni} = K_1 k_{ni} = K_1(m_{ni} g_{ni}^2)$ = modal spring
$g = \omega/\omega_1$ = basic exciting frequency factor
$g_{ni} = \omega_{ni}/\omega_1 = \sqrt{k_{ni}/m_{ni}}$ = modal natural frequency factor
$g_i = \omega/\omega_{ni} = g/g_{ni}$ = modal forced frequency ratio
$m_{ni} = \sum_1^p m_q \beta_{qi}^2$ = normalized modal mass $= M_{ni}/M_1$

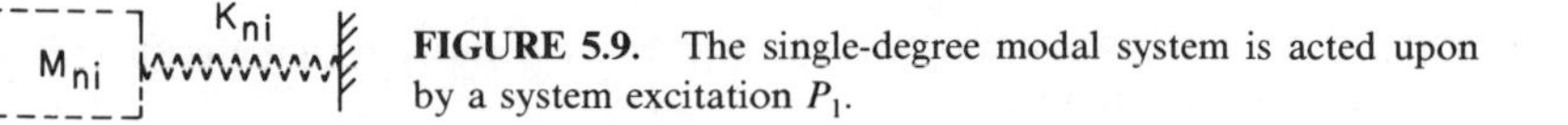

FIGURE 5.9. The single-degree modal system is acted upon by a system excitation P_1.

If a system is free to translate, there is also a rigid mode response to the exciting force. Summing forces in this mode (Figure 5.5),

$$\Sigma F_{12} = M_1\omega^2 a_0 + M_2\omega^2 a_0 + P_2 = 0$$

$$a_0 = \left[\frac{-P_2}{\omega^2(M_1 + M_2)}\right] \tag{5.12a}$$

Or, in the more general multimass case, with forcing at M_1, the common amplitude becomes

$$\frac{a_0}{(P_1/K_{n0})} = -1 \tag{5.12b}$$

where

$$K_{n0} = \omega^2(M_1 + M_2 + \cdots) = K_1 g^2 \Sigma m_q$$

$$\Sigma m_q = (1 + m_1 + m_2 + \cdots)$$

Thus the zero mode spring is not a constant, but varies with exciting frequency. Amplitude in the zero mode is inversely proportional to the total system mass

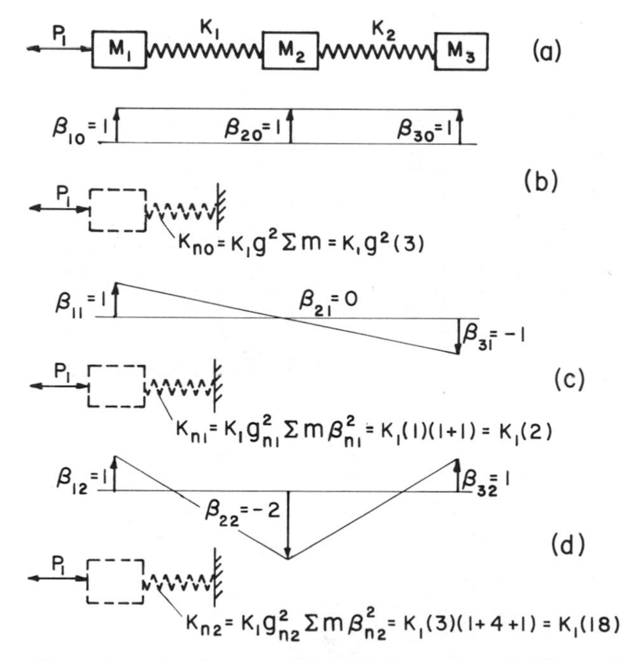

FIGURE 5.10. Illustration of a P_1 sinusoidal excitation acting simultaneously on the three normal modes, including the rigid mode (*b*). Masses and springs are equal.

and the square of the exciting frequency, but directly proportional to the exciting force.

As an example, including the translational mode, we take equal masses and springs (see Figure 5.10, which has normal modes and frequency factors as shown). Note that P_1 acts simultaneously, equally, and independently on all modes, and the second modal spring is nine times as stiff as the first.

Further taking an exciting frequency of $2\omega_1 = 2\sqrt{K_1/M_1}$, or $\mathsf{g} = 2$, modal amplitudes at the first mass are

$$\mathsf{a}_{10} = \frac{P_1}{3K_{14}}(-1) = -\frac{1}{12}\left(\frac{P_1}{K_1}\right)$$

$$\mathsf{a}_{11} = \frac{P_1}{2K_1}\left[\frac{1}{1-(4/1)}\right] = -\frac{1}{6}\left(\frac{P_1}{K_1}\right)$$

$$\mathsf{a}_{12} = \frac{P_1}{18K_1}\left[\frac{1}{1-(4/3)}\right] = -\frac{1}{6}\left(\frac{P_1}{K_1}\right)$$

$$\Sigma\mathsf{a}_1 = \mathsf{a}_0 + \mathsf{a}_{11} + \mathsf{a}_{12} = -\frac{5}{12}\left(\frac{P_1}{K_1}\right)$$

Normal mode distributions are seen in Figure 5.11*a*, and as plotted in (*b*) for the other masses.

MODE	β_1	β_2	β_3	$\frac{a_1}{P_1/K_1}$	$\frac{a_2}{P_1/K_1}$	$\frac{a_3}{P_1/K_1}$
0	1	1	1	$-\frac{1}{12}$	$-\frac{1}{12}$	$-\frac{1}{12}$
1	1	0	-1	$-\frac{1}{6}$	0	$+\frac{1}{6}$
2	1	-2	-1	$-\frac{1}{6}$	$+\frac{1}{3}$	$-\frac{1}{6}$
			Σ	$-\frac{5}{12}$	$+\frac{3}{12}$	$-\frac{1}{12}$

(a)

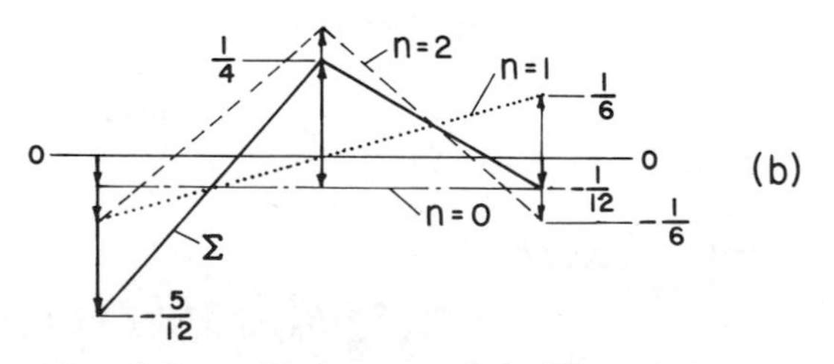

FIGURE 5.11. Summation of the modal responses of the Figure 5.10 system assuming $\mathsf{g} = 2$ provides the total forced response at the forcing frequency.

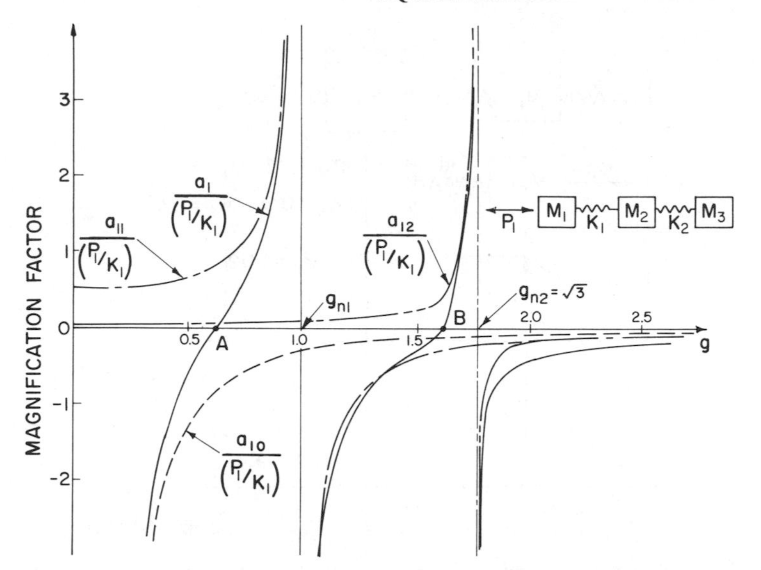

FIGURE 5.12. Frequency response characteristics of the Figure 5.10 case indicating behavior of the several modal components.

As exciting frequency varies for this system, there are two modal resonances (Figure 5.12). Each is characteristic of a single-degree case, but with different horizontal and vertical scales. Zero mode response approaches negative infinity at zero frequency and zero at infinite frequency.

To summarize,

$$\mathsf{a}_1 = \left(\frac{P_1}{K_{n0}}\right)(-1) + \Sigma\left(\frac{P_1}{K_{ni}}\right)\left(\frac{1}{1 - \mathsf{g}_i^2}\right) \tag{5.13}$$

With a grounded system, the first term is not present.

5.9. EQUIVALENT MODAL EXCITATION

If the exciting force is applied at other than M_1, we must refer it to the modal mass and spring at 1 using a modal factor, which is the normal mode amplitude. In Figure 5.13 we have the force P_2 applied to M_2.

Since this excitation is located at a mass that has slightly more amplitude in the first normal mode than M_1, we expect more system response from P_2 than from P_1. And this is so, with response directly proportional to the factor 1.02, or,

$$P_{ni} = P_q \beta_{qi} \tag{5.14a}$$

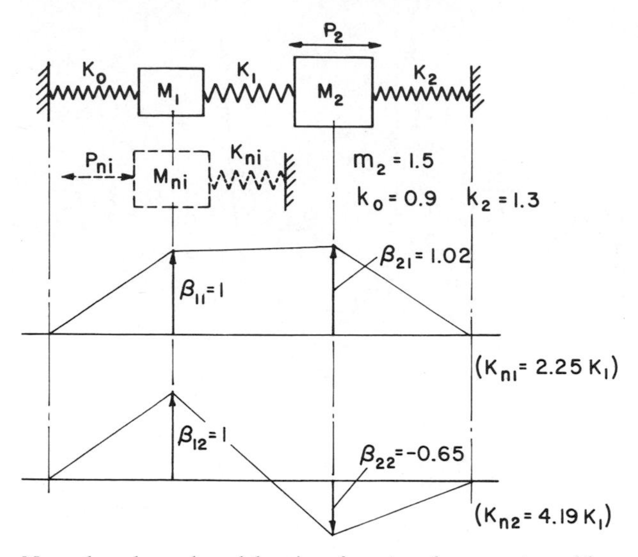

FIGURE 5.13. Normal modes and modal springs for a two-degree system with specific mass and spring ratios, with forcing at M_2.

In the second mode P_2 is 65% as effective in developing steady-state amplitudes, and there is a phase reversal indicated by the negative sign. If a mass is nodal, an excitation applied at this mass cannot excite the corresponding normal mode, as $\beta_{qi} = 0$. The amplitude factor is somewhat analogous to the leverage of the force on a mode.

If several forces act simultaneously on a system at *exactly* the same frequency, they can be combined to a single equivalent force on M_1. As multiple forces can have time phase relations, they can be combined vectorially in the complex plane.

$$P_{ni} = \Sigma P_q \beta_{qi} \tag{5.14b}$$

where β_{qi} is a scalar multiplier.

For simplicity all modes have been referenced to the first mass with amplitudes found at this point. Other mass amplitudes follow as simple normal modal distributions (Figure 5.11).

5.10. MODAL RESPONSE TO DISPLACEMENT EXCITATION

With displacement excitation (Figure 5.14), there are no external forces on any mass. Rather, a constant sinusoidal acceleration ($\omega^2 \mathbf{a}_3$) is impressed on all masses as if rigidly connected (Figure 5.14*b*). The masses have induced inertia forces on *all masses* of $M_q \omega^2 \mathbf{a}_3$, which simulate the source of excitation in *all modes* (Figure 5.14*c*).

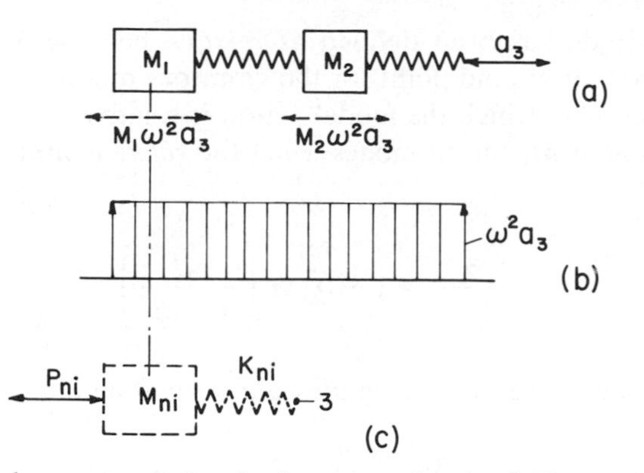

FIGURE 5.14. Displacement excitation develops an acceleration field (b), which induces force effects in the masses, combining to an effective force excitation (c).

As in Section 5.9 we combine these to an equivalent excitation on M_{n1} at M_1.

$$P_{n1} = \omega^2 \mathsf{a}_3 (M_1 \beta_{11} + M_2 \beta_{21})$$

$$= M_1 \omega^2 \mathsf{a}_3 (1 + m_2 \beta_{21}) \tag{5.15}$$

Then, applying Equation 5.11,

$$\mathsf{a}_{11}^r = \left(\frac{P_{n1}}{K_{n1}} \right) \left(\frac{1}{1 - \mathsf{g}_1^2} \right)$$

which reduces to the general result for any system and mode.

$$\frac{\mathsf{a}_{1i}^r}{\mathsf{a}_{p+1}} = \left(\frac{\Sigma m_q \beta_{qi}}{\Sigma m_q \beta_{qi}^2} \right) \left(\frac{\mathsf{g}_i^2}{1 - \mathsf{g}_i^2} \right)$$

$$= \overline{m}_i \left(\frac{\mathsf{g}_i^2}{1 - \mathsf{g}_i^2} \right) \tag{5.16}$$

where a_{1i}^r = amplitude of M_1 in mode i, relative to rigid mode

a_{p+1} = displacement excitation

$\Sigma m_q \beta_{qi} = 1 + m_2 \beta_{2i} + m_3 \beta_{3i} + \cdots$

$\Sigma m_q \beta_{qi}^2 = 1 + m_2 \beta_{2i}^2 + m_3 \beta_{3i}^2 + \cdots = m_{ni}$ = normalized modal mass

$\overline{m}_i = \left(\frac{\Sigma_{mq} \beta_{qi}}{m_{ni}} \right)$ = dimensionless modal participation factor

$\mathsf{g}_i = \left(\frac{\omega}{\omega_{ni}} \right)$ = modal forced frequency ratio.

Modal amplitude has been defined as *relative* because 3 in Figure 5.14*c* represents a node, or ground point for the vibratory mode. Actually all masses have the motion a_3 to which the modal action is relative.

Total response at M_1 for all modes using the *relative* approach is found by summation.

$$\frac{\mathsf{a}_1}{\mathsf{a}_{p+1}} = 1 + \sum_1^p \overline{m}_i \left(\frac{\mathsf{g}_i^2}{1 - \mathsf{g}_i^2} \right) \tag{5.17}$$

The corresponding expression using *absolute* normal modal components is

$$\frac{\mathsf{a}_1}{\mathsf{a}_{p+1}} = \sum_1^p \overline{m}_i \left(\frac{1}{1 - \mathsf{g}_i^2} \right) \tag{5.18}$$

Note the similarity of the magnification factors in Equations 5.17 and 5.18 to those of the simple system (Table 2.1*c*) for zero damping. Also, in Equation 5.18, zero exciting frequency causes all magnification factors to become unity; therefore, since $\mathsf{a}_1 = \mathsf{a}_{p+1}$, we conclude that

$$\sum_1^p \overline{m}_i = +1 \tag{5.19}$$

Taking the Figure 5.8 numerical example, in Figure 5.15 we calculate parameters for a two-degree system. As generalized we have five normalized factors that characterize the system defined by only two, $m_2 = 0.7$ and $k_2 = 1.2$.

	FIRST MODE	SECOND MODE		FIRST MODE	SECOND MODE
g_{ni}	0.683	1.917	β_{2i}	+0.535	-2.675
g_i	1.464	0.522	$\Sigma\, m\beta$	+1.375	-0.872
$\frac{1}{1 - g_i^2}$	-0.874	+1.374	m_{ni}	1.200	6.009
$\frac{g_i^2}{1 - g_i^2}$	-1.874	+0.374	$\overline{m}_i$	1.146	-0.146

FIGURE 5.15. Summary of modal parameters calculated for the Figure 5.6 system, and magnification factors if $\mathsf{g} = 1$. Note that $\Sigma(1/m_{ni}) = 1.00$ and $\Sigma\overline{m}_i = 1.00$.

	RELATIVE		ABSOLUTE	
MODE	$\frac{a_1^r}{a_3}$	$\frac{a_2^r}{a_3}$	$\frac{a_1}{a_3}$	$\frac{a_2}{a_3}$
1	-2.148	-1.148	-1.002	-0.537
2	-0.055	+0.148	-0.201	+0.537
Σ	-2.203	-1.000	-1.203	0
INPUT	+1	+1		
Σ	-1.203	0		

FIGURE 5.16. Calculated modal responses of displacement excited system (Figure 5.6) with factors from Figure 5.15.

Defining g we have three more terms that relate the forcing frequency to the normalized modal system, for a total of eight per mode.

Applying Equations 5.17 and 5.18 and using the tabulated quantities, we have the responses in Figure 5.16. There is agreement between the relative and absolute results. Additionally, total amplitudes coincide with those obtained by Holzer (Figure 5.8), since $-3.6/3.0 = -1.20$, and $0/3.0 = 0$.

5.11. EQUATIONS FOR TOTAL DAMPED RESPONSE

The two-mass system (Figure 5.17) is limited to one dashpot to simplify the analysis. For equilibrium in sinusoidal steady state,

$$\begin{cases} \Sigma F_1 = M_1\omega^2 a_1 + K_1(a_2 - a_1) + jC_1\omega(a_2 - a_1) + P_1 = 0 & (5.20a) \\ \Sigma F_{12} = M_1\omega^2 a_1 + M_2\omega^2 a_2 - K_2 a_2 + P_1 = 0 & (5.20b) \end{cases}$$

Normalizing,

$$\begin{cases} a_1(1 - g^2 + j2\zeta_1 g) - a_2(1 + j2\zeta_1 g) = \dfrac{P_1}{K_1} & (5.21a) \\ -a_1 g^2 + a_2(k_2 - m_2 g^2) = \dfrac{P_1}{K_1} & (5.21b) \end{cases}$$

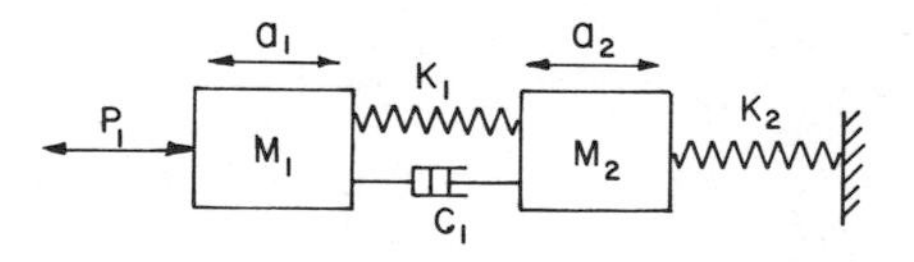

FIGURE 5.17. The two-degree damped system is forced at M_1.

TABLE 5.4. Steady-state response of two-degree viscously damped systems (*a*) free–free with internal coupling, (*b*) classical two-mass, forced at either mass, and (*c*) displacement excited.

(a)	P_1	P_2	(b)	P_1	P_2	(c)	
$\frac{a_1}{P/K_1}$	$\frac{A+jB}{C+jD}$	$\frac{1+jB}{C+jD}$	$\frac{a_1}{P/K_1}$	$\frac{A+jB}{C+jD}$	$\frac{1+j2\zeta_1 g}{C+jD}$	$\frac{a_1}{a_3}$	$\frac{A_1+jB_1}{C+jD}$
$\frac{a_{12}}{P/K_1}$	$\frac{-m_2 g^2}{C+jD}$	$\frac{g^2}{C+jD}$	$\frac{a_2}{P/K_2}$	$\frac{1+j2\zeta_1 g}{C+jD}$	$\frac{1-g^2+j2\zeta_1 g}{C+jD}$	$\frac{a_2}{a_3}$	$\frac{A_2+jB_2}{C+jD}$
$\frac{a_2}{P/K_1}$	$\frac{1+jB}{C+jD}$	$\frac{A_1+jB}{C+jD}$	$\frac{P_3}{P}$	$\frac{-(A_1+jB_1)}{C+jD}$	$\frac{-(A_2+jB_2)}{C+jD}$	$\frac{P_3}{K_1 a_3}$	$\frac{A_3-jB_3}{C+jD}$
C	$g^2[m_2 g^2-(1+m_2)]$		C	$m_2 g^4-(m_2+k_2+1+4\zeta_1\zeta_2)g^2+k_2$			
$A=(1-m_2g^2)$ $A_1=(1-g^2)$; $D=-Bg^2(1+m_2)$; $B=2\zeta_1 g$ $\zeta_1=\frac{C_1}{2M_1\omega_1}$			$A=(1+k_2-m_2g^2)$ $A_3=g^2[k_2m_2g^2-A_1(1+m_2)]$, $D=[B_1-2\zeta_1 g^3(1+m_2)]$; $A_1=(k_2-4\zeta_1\zeta_2 g^2)$, $B=2g(\zeta_1+\zeta_2)$, $B_1=2g(k_2\zeta_1+\zeta_2)$, $B_2=(B_1-2\zeta_2 g^3)$; $A_2=[k_2(1-g^2)-4\zeta_1\zeta_2 g^2]$, $B_3=2g^3\{\zeta_1 k_2(1-m_2)+\zeta_2[1+m_2(1-g^2)]\}$				

Solving by determinants for a_1,

$$\frac{a_1}{(P_1/K_1)} = \frac{\begin{vmatrix} 1 & -(1+j2\zeta_1 g) \\ 1 & (k_2 - m_2 g^2) \end{vmatrix}}{\begin{vmatrix} (1 - g^2 + j2\zeta_1 g) & -(1+j2\zeta_1 g) \\ -g^2 & (k_2 - m_2 g^2) \end{vmatrix}} = \frac{A + jB}{C + jD} \quad (5.22)$$

where the final coefficients are given in Table 5.4*b* including the ζ_2 terms. The ratio of the two complex vectors on the basis of the resultant length of the numerator to the denominator represents the absolute magnification factor. This is relative to the reference equilibrium amplitude P_1/K_1. Phase relationships also are obtainable from the complex fraction, as has been indicated in Chapter 2.

The equation for the second displacement a_2 is similarly arranged with the (P_1/K_1) or $+1$ terms transferred to the right column of the numerator.

If the excitation is moved to M_2, becoming P_2, the forcing term appears in Equation 5.21*b* only. All denominators for these simultaneous solutions are the same as in Equation 5.22. Displacements and reaction forces are given in Table 5.4*b*, with displacement excitation of the two-mass damped system presented in Table 5.4*c*.

5.12. FORCED HOLZER FOR RELATIVE DAMPING

Inclusion of damping elements in parallel with spring elements (Figure 5.18) requires both real and imaginary terms in a Holzer solution for the forced condition. Procedures are generally similar to those for the undamped case (Figure 5.8), with sinusoidal excitation applied at the right. Isolating a basic single-degree element of the chain, M_q, K_q, and C_q (Figure 5.19), the equi-

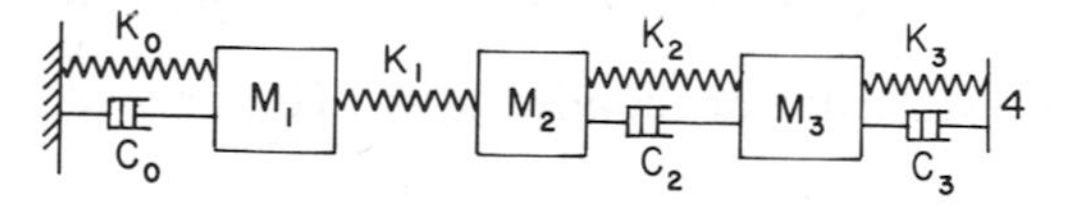

FIGURE 5.18. A multimass system coupled by springs and viscous dashpots, excited at the right coordinate.

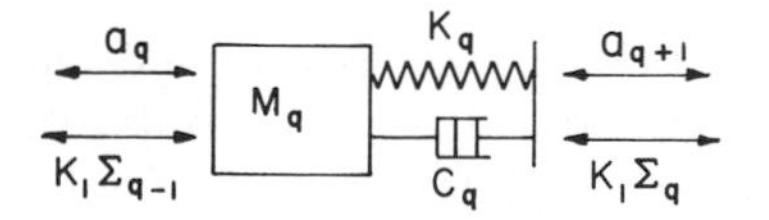

FIGURE 5.19. Any selected single-degree element is in equilibrium with respect to all the forces on the mass, spring, and dashpot.

librium equation is

$$\Sigma F_q = \left(K_1\Sigma_{q-1} + M_q\omega^2 \mathsf{a}_q\right) + K_q(\mathsf{a}_{q+1} - \mathsf{a}_q) + jC_q\omega(\mathsf{a}_{q+1} - \mathsf{a}_q) = 0 \tag{5.23}$$

where the first quantity in parentheses represents the accumulated load from the left, which must be carried by the spring and dashpot. Normalizing, with division by K_1,

$$\Sigma_{q-1} + m_q \mathsf{g}^2 \mathsf{a}_q = \Sigma_q = (k_q + j2\zeta_q \mathsf{g})\,\Delta_q \tag{5.24}$$

where $\Delta_q = (\mathsf{a}_q - \mathsf{a}_{q+1})$ = Holzer incremental displacement

$$\text{and} \qquad \zeta_q = \left(\frac{C_q}{2M_1\omega_1}\right) = \left(\frac{C_q}{C_1}\right)\zeta_1$$

Introducing complex components,

$$(\Delta\mathsf{a}_{rq} + j\Delta\mathsf{a}_{jq})(k_q + j2\zeta_q\mathsf{g}) = (\Sigma_{rq} + j\Sigma_{jq}) \tag{5.25}$$

Dividing by $(k_q + j2\zeta_z\mathsf{g})$, rationalizing the denominator, and separating real and imaginary terms,

$$\Delta\mathsf{a}_{rq} = \frac{k_q\Sigma_{rq} + (2\zeta_q\mathsf{g})\Sigma_{jq}}{k_q^2 + (2\zeta_q\mathsf{g})^2}$$

$$\Delta\mathsf{a}_{jq} = \frac{-(2\zeta_q\mathsf{g})\Sigma_{rq} + k_q\Sigma_{jq}}{k_q^2 + (2\zeta_q\mathsf{g})^2}$$

$$\Delta\mathsf{a}_{rq} = A_q\Sigma_{rq} + B_q\Sigma_{jq} \tag{5.26a}$$

$$\Delta\mathsf{a}_{jq} = -B_q\Sigma_{rq} + A_q\Sigma_{jq} \tag{5.26b}$$

where A and B are scalar factors applied to the Holzer load summation terms, calculated from

$$A = \frac{k_q}{k_q^2 + (2\zeta_q\mathsf{g})^2} \qquad B = \frac{(2\zeta_q\mathsf{g})}{k_q^2 + (2\zeta_q\mathsf{g})^2} \tag{5.27}$$

A sample normalized calculation is shown in Figure 5.20. Known system parameters are first entered, including m, k, ζ, the forced frequency ratio g, and A and B. Then we assume $\beta_1 = \beta_{r1} = 1.00$, proceeding across the figure,

$g^2 = 0.655$										
q	m	β_r	β_j	β	$mg^2\beta_r$	$mg^2\beta_j$	Σ_r	Σ_j	Σ	k
1	1.0	1.0000	0	1.0000	0.6550	0	0.6550	0	0.6550	1.0
2	0.5	0.3782	0.1438	0.4046	0.1239	0.0471	0.7790	0.0471	0.7800	3.6
3		0.1629	0.1530	0.2235						
q	ζ	A	B	$A\Sigma_r$	$B\Sigma_j$	$\Delta\beta_r$	$-B\Sigma_r$	$A\Sigma_j$	$\Delta\beta_j$	$\Delta\beta$
1	0.1428	0.9493	0.2194	0.6218	0	0.6218	-0.1438	0	-0.1438	0.6382
2	0.2298	0.2748	0.0284	0.2140	0.0013	0.2153	-0.0221	0.0129	-0.0092	0.2154

FIGURE 5.20. Displacement-forced Holzer calculation including relative viscous damping between masses requires real and complex components.

treating the r and j terms separately, obtaining the $\Delta\beta_r$ and $\Delta\beta_j$ displacements. These are then subtracted algebraically from the previous β_r and β_j terms, as in the conventional Holzer method. Resultant values, β, Σ, and $\Delta\beta$ are not part of the solution, and can be calculated after the sequential procedure.

5.13. QUANTIFYING THE HOLZER RESPONSE

Amplitudes in Figure 5.20 describe the steady-state behavior of the given two-mass system with a particular forcing frequency of $\omega = 0.81\sqrt{K_1/M_1}$. As with other Holzer calculations, we have taken $\beta_1 = 1$, which establishes the magnitudes of all other responses. The β notation is maintained, even though there is no direct relation to the normal modal amplitudes β_n; however, β now denotes a normalized forced mode with respect to $\beta_1 = 1$, as well as amplitudes referenced to the Holzer calculation.

Viewing the Figure 5.20 solution as causing a specific displacement excitation a_3, we can reduce the tabulated values to actual values, since in the linear system a direct proportionality exists. The ratios for absolute amplitudes are

$$\left(\frac{\beta_1}{\beta_3}\right) = \left(\frac{1.000}{0.2235}\right) = \left(\frac{a_1}{a_3}\right) = 4.47$$

$$\left(\frac{\beta_2}{\beta_3}\right) = \left(\frac{0.4046}{0.2235}\right) = \left(\frac{a_2}{a_3}\right) = 1.81$$

Similarly, the exciting force P_3 is obtained by denormalizing tabular loads:

$$P_3 = \left(\frac{a_3}{0.2235}\right)(K_1\Sigma_2)$$

$$\left(\frac{P_3}{K_1 a_3}\right) = 4.47(\Sigma_2) = 3.49$$

Or, as a magnification factor by reciprocation, with the exciting force given,

$$\frac{\mathsf{a}_3}{P_3/K_1} = \frac{1}{3.49} = 0.287$$

If an additional mass M_3 is present in the Figure 5.20 system, with P_3 excitation on M_3, the tabular solution can be adapted. Letting $m_3 = 1.3$, the final load terms become

$$\Sigma_{r3} = m_3 \mathsf{g}^2 \beta_{r3} + \Sigma_{r2}$$

$$= 1.3(0.655)(0.1629) + 0.779 = 0.918$$

$$\Sigma_{j3} = m_3 \mathsf{g}^2 \beta_{j3} + \Sigma_{j2}$$

$$= 1.3(0.655)(0.1539) + 0.0471 = 0.177$$

And,

$$\Sigma_3 = \Sigma_{r3} + \Sigma_{j3} = 0.935$$

The magnification factor with respect to a_3 becomes

$$\left(\frac{\mathsf{a}_3}{P_3/K_1}\right) = \left(\frac{\beta_3}{\Sigma_3}\right) = \left(\frac{0.2235}{0.935}\right) = 0.239$$

This result is for P_3 excitation on M_3. If P_1 acts on M_1, we can use the same procedure by reversing the system. It is advantageous to orient the system to locate the exciting force at the right termination, with the computation from left to right, as in Figure 5.20. The excitation terms then do not appear in the figure, and are considered with respect to scale effects as indicated.

5.14. PHASE RELATIONSHIPS

In addition to absolute displacements, Figure 5.20 contains real and imaginary components, providing components for display in the complex plane (Figure 5.21*b*). Plots of β_r and β_j define the angular positions of the several displacement vectors at the instant $t = 0$. As shown, the excitation β_3 leads β_2, and β_2 leads β_1, which is taken initially on the real axis.

Phase position of P_3 is determined by the Σ_{r2} and Σ_{j2} terms, locating a small positive angle, lagging β_3 by ϕ_P; however, since the Σ terms are positive to the right, representing inertia loads, equilibrium requires that the phase position of P_3 be reversed. The actual angle by which P_3 leads β_3 is $(\pi - \phi_P)$, (Figure 5.21*c*).

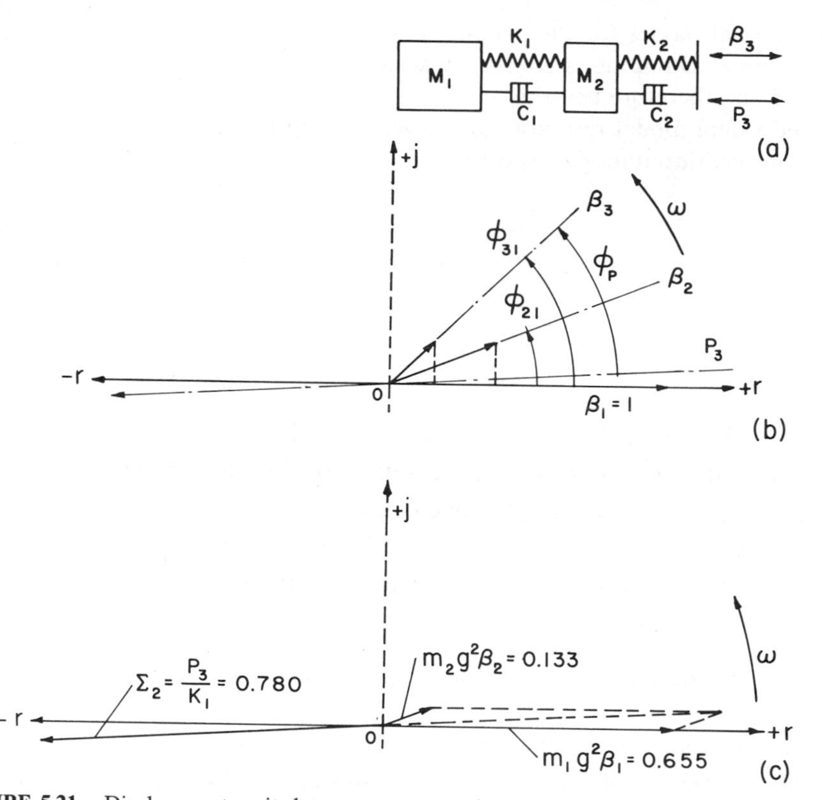

FIGURE 5.21. Displacement-excited two-mass system for tabular solution of Figure 5.20 has complex amplitudes, with phase angles indicated (*b*). In (*c*) equilibrium corresponds to zero resultant force in the complex plane.

Also, in Figure 5.21*c* we note the equilibrium of normalized force vectors in the complex plane. The two inertia loads from M_1 and M_2 are shown to the right, in phase with the respective displacements. The antiresultant of these two vectors is Σ_2; therefore the resultant rotating vector is zero and equilibrium is satisfied at any instant of time.

5.15. MODAL SUPERPOSITION

For typical system damping, which is relatively small, forced response can be viewed in terms of modal components (Sections 5.8–5.10), an approach desirable beyond a several-mass system. For large systems the closed solutions, as in Table 5.4, become unwieldy. Holzer solutions are possible, but they are somewhat complex when damping is included. The modal concept permits simplification conceptually, allowing us to focus on selected resonant peaks as single-degree phenomena. This does require calculation of natural frequencies and normal modes for the undamped system, but only for those of practical importance. Damping is an empirical quantity, usually estimated for various modes and therefore never exact.

Individual modal responses at a given forcing frequency are, for force and displacement excitations respectively,

$$\frac{\mathsf{a}_{1i}}{(P_1/K_{ni})} = \frac{1}{(1 - \mathsf{g}_i^2 + j2\zeta_i\mathsf{g}_i)} \tag{5.28}$$

$$\frac{\mathsf{a}_{1i}}{\mathsf{a}_{q+1}} = \frac{\overline{m}_i}{(1 - \mathsf{g}_i^2 + j2\zeta_i\mathsf{g}_i)} \tag{5.29}$$

where $K_{ni} = K_1\mathsf{g}_{ni}^2\Sigma(m\beta^2)_i$ = vibratory modal spring
$\mathsf{g}_i = (\omega/\omega_{ni}) = (\mathsf{g}/\mathsf{g}_{ni})$ = modal forced frequency ratio
$\zeta_{ni} = (C_{ni}/2M_{ni}\omega_{ni})$ = modal damping ratio
a_{q+1} = harmonic displacement excitation to right of M_q
$\overline{m}_i = [\Sigma(m\beta)_i/\Sigma(m\beta^2)_i]$ = modal excitation factor

5.16. CHARACTERISTIC MODAL RESPONSE

As an example, we take the simple three-mass forced case of Figure 5.10, and assume modal dampings of $\zeta_{n1} = 0.05$ and $\zeta_{n2} = 0.025$. The rigid mode by nature lacks internal damping. Equation 5.28 applies to the two vibratory modes, with rigid mode response as in Equation 5.12b.

$$\mathsf{a}_1 = \left(\frac{P_1}{K_{n0}}\right)(-1) + \left(\frac{P_1}{K_{n1}}\right)\left(\frac{1}{1 - \mathsf{g}_1^2 + j2\zeta_1\mathsf{g}_1}\right)$$

$$+ \left(\frac{P_1}{K_{n2}}\right)\left(\frac{1}{1 - \mathsf{g}_2^2 + j2\zeta_2\mathsf{g}_2}\right) \tag{5.30}$$

Respective first- and second-mode peaks are

$$(\mathsf{a}_1)_{n1} = \frac{P_1}{K_{n1}}\left(\frac{1}{2\zeta_1}\right) = \frac{P_1}{2K_1}\left[\frac{1}{2(0.05)}\right] = 5\left(\frac{P_1}{K_1}\right) \tag{5.31a}$$

$$(\mathsf{a}_1)_{n2} = \frac{P_1}{K_{n2}}\left(\frac{1}{2\zeta_2}\right) = \frac{P_1}{18K_1}\left[\frac{1}{2(0.025)}\right] = 1.11\left(\frac{P_1}{K_1}\right) \tag{5.31b}$$

Thus, though the higher mode has only half the damping of the lower, the higher mode peak magnification factor is only 22% of the lower, with both

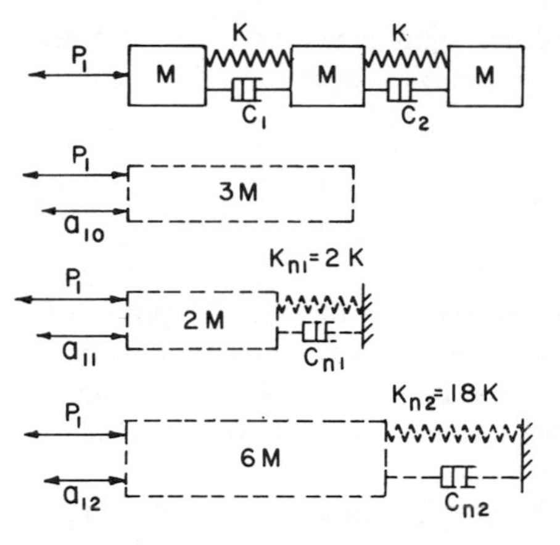

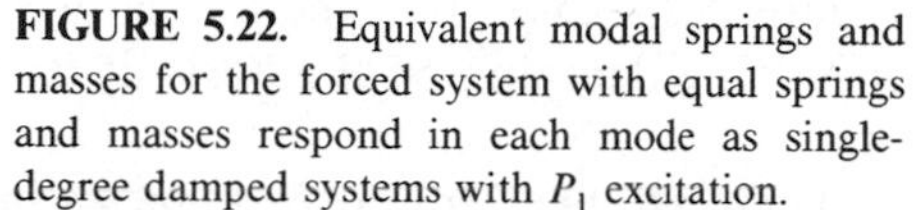
FIGURE 5.22. Equivalent modal springs and masses for the forced system with equal springs and masses respond in each mode as single-degree damped systems with P_1 excitation.

based on the reference K_1. We observe the second mode to be inhibited by a large dynamic stiffness, $K_{n2} = 18K_1$ (Figure 5.22).

The single-degree damped responses are shown in Figure 5.23 over the complete frequency range for this particular case of the ungrounded system with three equal masses and two equal springs, unsymmetrical only with respect to damping. Ordinates are resultant values without regard to phase. Amplitudes of masses other than M_1 would be assumed to agree with the normal mode distribution, with only slight distortion caused by the internal

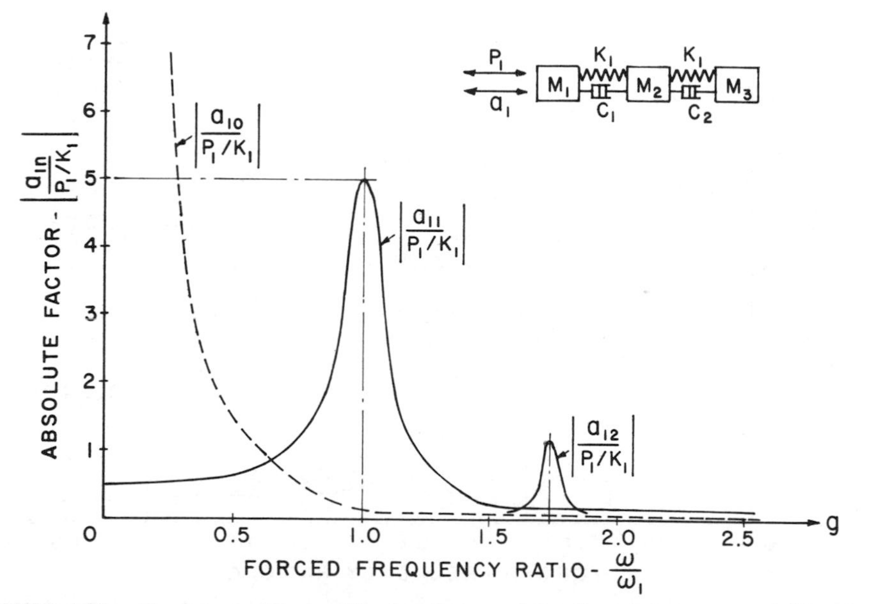

FIGURE 5.23. Absolute modal magnification factors of the three-degree system define two resonant peaks, with the translational mode shown dashed.

damping. In this figure we see the particular significance of the modal peak coordinates, and the comparatively brief analysis necessary for their determination.

5.17. TOTAL FORCED BEHAVIOR

As a comparison with the foregoing results, we plot the total response of a system in Figure 5.24 for the parameters shown. This is different from the Figure 5.23 case, but there are two vibratory modes. Note that we are specifying damping on the basis of dashpot values, rather than modal damping.

There are two natural frequencies in Figure 5.24 at $\mathsf{g}_{n1} = 0.53$ and $\mathsf{g}_{n2} = 1.42$, and total amplitudes are calculated for both masses from Table 5.4*b*. Although we normally expect two respectable peaks, we are disappointed. The lower peak has magnification factors of 7.6 and 5.2 for a_1 and a_2, respectively. At the higher mode, however, the peak is only a modest rise below the second resonance, and we have no second resonance for small dampings assumed.

At $\mathsf{g} = 0$, we have static spring deflections caused by P_1. The zero intercepts are 2.11 and 1.11 in Figure 5.24. These agree for the static condition, first with K_1 and K_2 in series and then with K_2 only.

Predominance of the first mode is a feature observed in both Figures 5.23 and 5.24 for different systems. It can be concluded that this is a typical, though

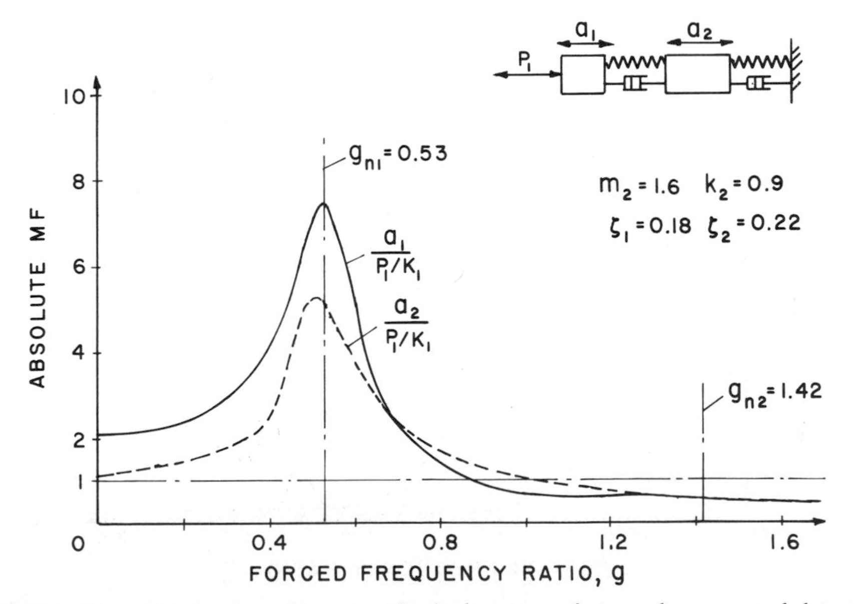

FIGURE 5.24. Calculated total response for both masses of a two-degree grounded system indicates a negligible peak associated with the second mode.

not completely general, conclusion. The second mode must certainly tend to infinite a_1 and a_2 for zero damping, as in Figures 5.3 and 5.6, and the higher mode, or modes, can develop for special combinations of spring, mass, and excitation. However, the more formidable nature of the first-mode resonance has been demonstrated.

5.18. THE FORCED DOUBLE PENDULUM

With two concentrated masses suspended in series by weightless links we have a vertical multimass system that can be excited sinusoidally in the horizontal direction. Pendular cases depend on gravity for forces tending to restore the masses to the central position, as will be further explained in Section 8.1. The effect is similar to that of an elastic spring resisting displacement.

Assuming small amplitudes, concentrated masses, no damping, forcing at M_1, and taking moments about the two fulcrums successively,

$$\begin{cases} \Sigma T_3 = M_1 a_1 \omega^2 (l_1 + l_2) + M_2 a_2 \omega^2 l_2 - W_1 a_1 - W_2 a_2 + P_1 (l_1 + l_2) = 0 & (5.32a) \\ \Sigma T_2 = M_1 a_1 \omega^2 l_1 - W_1 (a_1 - a_2) + P_1 l_1 = 0 & (5.32b) \end{cases}$$

These are generalized by normalization to

$$\begin{cases} a_1[1 - g^2(1 + \rho_2)] + a_2[m_2(1 - \rho_2 g^2)] = \dfrac{P_1 l_1}{W_1}(1 + \rho_2) & (5.33a) \\ a_1[1 - g^2] - a_2 = \dfrac{P_1 l_1}{W_1} & (5.33b) \end{cases}$$

with notation as indicated in Table 5.5. The reference natural frequency is that of the single pendulum (l_1, M_1),

$$\omega_1 = \sqrt{\frac{g_c}{l_1}}$$

where g_c = acceleration of gravity
l_1 = length of the reference pendulum.

Other relations for response based on these simultaneous equations are given in the table, as well as the case of forcing at M_2 (Table 5.5b) and displacement excitation of the entire system (Table 5.5c).

TABLE 5.5. The double pendulum with steady-state sinusoidal excitation (*a*) forced at M_1, (*b*) forced at M_2, and (*c*) displacement at the fulcrum. Note the reciprocal relations between (*a*), (*b*), and (*c*). Small amplitudes are assumed.

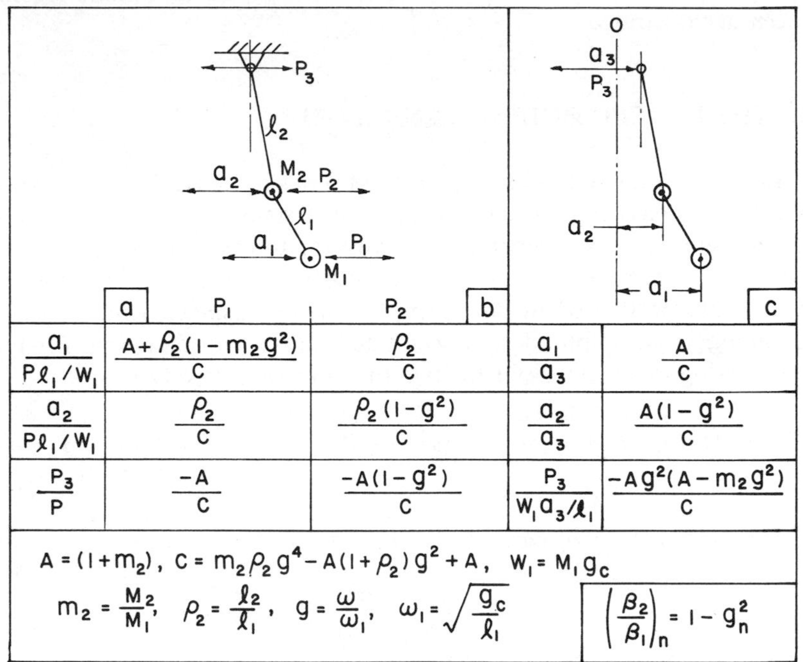

	a: P_1	b: P_2		c
$\dfrac{a_1}{P\ell_1/W_1}$	$\dfrac{A+\rho_2(1-m_2g^2)}{C}$	$\dfrac{\rho_2}{C}$	$\dfrac{a_1}{a_3}$	$\dfrac{A}{C}$
$\dfrac{a_2}{P\ell_1/W_1}$	$\dfrac{\rho_2}{C}$	$\dfrac{\rho_2(1-g^2)}{C}$	$\dfrac{a_2}{a_3}$	$\dfrac{A(1-g^2)}{C}$
$\dfrac{P_3}{P}$	$\dfrac{-A}{C}$	$\dfrac{-A(1-g^2)}{C}$	$\dfrac{P_3}{W_1a_3/\ell_1}$	$\dfrac{-Ag^2(A-m_2g^2)}{C}$

$A=(1+m_2)$, $C=m_2\rho_2g^4-A(1+\rho_2)g^2+A$, $W_1=M_1g_c$

$m_2=\dfrac{M_2}{M_1}$, $\rho_2=\dfrac{\ell_2}{\ell_1}$, $g=\dfrac{\omega}{\omega_1}$, $\omega_1=\sqrt{\dfrac{g_c}{\ell_1}}$ $\qquad \left(\dfrac{\beta_2}{\beta_1}\right)_n=1-g_n^2$

EXAMPLES

EXAMPLE 5.1. A torsional two-mass system (Table 5.2*a*) has the following parameters.

$$K_1 = 1400 \text{ kN} \cdot \text{m} \qquad M_1 = 53.5 \text{ kg} \cdot \text{m}^2$$
$$P_2 = 12.5 \text{ kN} \cdot \text{m at 36 Hz} \qquad M_2 = 34.2 \text{ kg} \cdot \text{m}^2$$

Find:

(a) The natural frequency.
(b) Vibratory torque in the spring.

Solution

$$\omega_1 = \sqrt{\frac{K_1}{M_1}} = \sqrt{\frac{1.4(10)^6}{53.5}} = 161.8 = 25.75 \text{ Hz}$$

$$g = \omega/\omega_1 = 36/25.75 = 1.40$$

$$\mathsf{g}^2 = 1.955$$

$$m_2 = 34.2/53.5 = 0.639$$

(a)
$$C_n = \mathsf{g}_n^2\left[m_2\mathsf{g}_n^2 - (1 + m_2)\right] = 0$$

$$\mathsf{g}_n^2 = (1 + m_2)/m_2 = 2.564$$

$$\mathsf{g}_n = 1.601$$

$$\omega_{n1} = (1.601)(25.75) = 41.23 \text{ Hz}$$

(b)
$$K_1\mathsf{a}_{12} = \frac{P_2}{m_2\mathsf{g}^2 - (1 + m_2)} = -\frac{12.5}{0.39}$$

$$= -32 \text{ kN} \cdot \text{m}$$

EXAMPLE 5.2. Verify the results of Example 5.1*b* using the forced Holzer method.

Solution. The normalized table, similar to Figure 5.8, is for $\mathsf{g}^2 = 1.96$:

q	m	β	Σ	k	Σ/k
1	1	a_1	$1.96\mathsf{a}_1$	1	$1.96\mathsf{a}_1$
2	0.64	$-0.96\mathsf{a}_1$	$0.76\mathsf{a}_1 + P_2/K_1 = 0$		

$$\mathsf{a}_1 = \left(\frac{1}{0.76}\right)\left(\frac{-12.5}{1400}\right) = -0.0117\, r$$

Actual torque or load is obtained from the real value of Σ_1,

$$K_1\mathsf{a}_{12} = K_1(1.96\mathsf{a}_1) = 32 \text{ kN} \cdot \text{m}$$

EXAMPLE 5.3. A system (Table 5.3*a*) is translational, with

$K_1 = 300$ lb/in. $\quad M_1 = 22$ lb $\quad \mathsf{a}_3 = 0.15$ in. at 20 Hz
$K_2 = 600 \quad M_2 = 22$ lb

Find:

(a) The total response of each mass.
(b) The modal response of each mass.

Solution

$$m_2 = 1$$

$$k_2 = 2$$

$$K_1\mathsf{a}_3 = (300)(0.15) = 45 \text{ lb}$$

(a)

$$\omega_1 = \sqrt{\frac{K_1}{M_1}} = \sqrt{\frac{300(386)}{22}} = 72.55 = 11.55 \text{ Hz}$$

$$\mathsf{g} = \frac{20}{11.55} = 1.732$$

$$\mathsf{g}^2 = 3.000$$

$$\left(\frac{\mathsf{a}_1}{\mathsf{a}_3}\right) = \frac{k_2}{C} = \left[\frac{2}{(1)(3)^2 - (4)(3) + 2}\right] = -2$$

$$\mathsf{a}_1 = (-2)(0.15) = -0.30 \text{ in.}$$

$$\left(\frac{\mathsf{a}_2}{\mathsf{a}_3}\right) = \left[\frac{2(1-3)}{-1}\right] = +4$$

$$\mathsf{a}_2 = 4(0.15) = 0.60 \text{ in.}$$

(b) From Table 4.1*b*,

$$C_n = \mathsf{g}_n^4 - 4\mathsf{g}_n^2 + 2 = 0 \qquad \mathsf{g}_{n1}^2 = 0.586$$

$$\mathsf{g}_{n2}^2 = 3.414$$

$$\beta_{21} = \left(1 - \mathsf{g}_{n1}^2\right) = 0.414$$

$$\beta_{22} = \left(1 - \mathsf{g}_{n2}^2\right) = -2.414$$

Using Equation 5.17,

$$\overline{m}_1 = \left[\frac{1 + 1(0.414)}{1 + 1(0.414)^2}\right] = 1.207$$

$$\overline{m}_2 = -0.207$$

$$(\Sigma\overline{m} = 1.000)$$

Forced modal frequency factors are

$$g_1 = (20/8.84) = 2.26$$

$$g_2 = (20/21.34) = 0.937$$

$$(a_1/a_3) = 1 + (1.207)(-1.242) + (-0.207)(7.216) = -2.00$$

$$a_1 = (0.15)(-2.00) = -0.30 \text{ in.}$$

$$(a_2/a_3) = 1 - 1.50(0.414) - 1.50(-2.414) = 1 - 0.621 + 3.621 = 4$$

$$a_2 = 0.15 - 0.093 + 0.543 = +0.60 \text{ in.}$$

EXAMPLE 5.4. A system (Table 5.1*b*) has unequal masses, $m_2 = 2$, and three equal springs. $P_1 = 42$ and $P_2 = 65$ lb are at the same forcing frequency, but are 90° out of phase, with P_1 leading P_2 ($P_2 = -jP_1$). Find:

(a) The equivalent excitation to be used for the first modal responses.

(b) The equivalent excitation for the second mode.

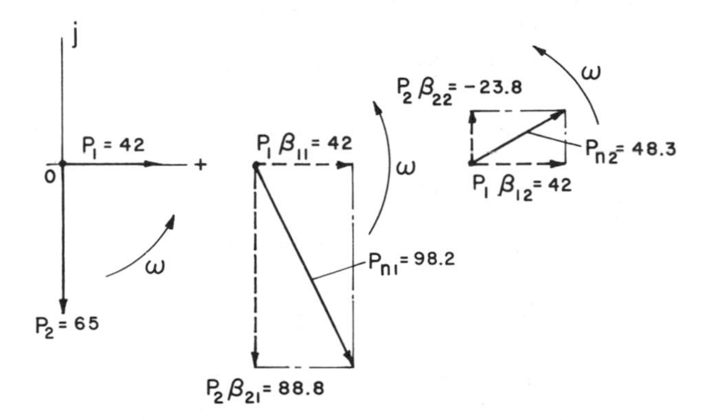

Solution. From Table 4.1*c*,

$$2g_n^4 - 6g_n^2 + 3 = 0 \qquad g_{n1}^2 = 0.634$$
$$g_{n2}^2 = 2.366$$
$$\beta_{21} = 2 - 0.634 = 1.366$$
$$\beta_{22} = 2 - 2.366 = -0.366$$

Using Equation 5.14b,

(a) $P_1\beta_{11} = 42(1) = 42$ lb

$P_2\beta_{21} = 65(1.366) = 88.8$ lb

$P_{n1} = 42 \twoheadrightarrow 88.8 = 98.2$ lb

(b) $P_1\beta_{12} = 42(1) = 42$ lb
$P_2\beta_{22} = 65(-0.366) = -23.8$ lb
$P_{n2} = 42 \nrightarrow 23.8 = 48.3$ lb

EXAMPLE 5.5. The damped free–free system is at resonance, forced by the 60 lb excitation. Find:

(a) The amplitude of each mass.
(b) The amplitude of the vibratory spring force.

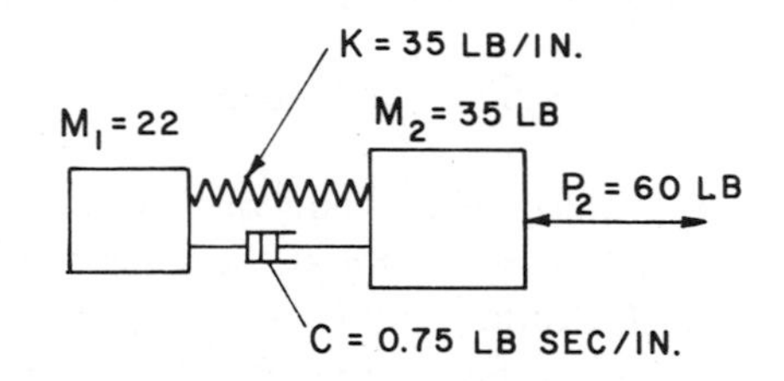

Solution

(a) For *system* natural frequency (Table 5.4*a*), with $C = 0$,

$$\mathsf{g}_n = 1.30 \qquad \omega_1 = 24.78 \text{ rad/sec}$$

$$\mathsf{g} = \mathsf{g}_n \qquad \zeta = 0.27$$

$$B = 0.69 \qquad C = 0$$

$$D = -2.87 \qquad A_1 = -0.69$$

$$\mathsf{a}_1 = 0.73 \text{ in.} \qquad \mathsf{a}_2 = 0.58 \text{ in.}$$

(b)

$$\mathsf{g}^2 = 1.69$$

$$\mathsf{a}_{12} = 1.01 \text{ in.}$$

$$F_k = 35.4 \text{ lb}$$

EXAMPLE 5.6. The system in the Holzer calculation (Figure 5.20) has the following dimensional data:

$K_1 = 50{,}000$ lb/in. $\quad M_1 = 2$ lb $\quad C_1 = 4.6$ lb sec/in.
$K_2 = 180{,}000 \quad M_2 = 1 \quad C_2 = 7.4$
$\mathsf{a}_3 = 0.040$ in. at 400 Hz

Find the forced amplitudes of each mass:

(a) Using the Holzer values.

(b) Using Table 5.4*c*.

Solution

(a) Using Section 5.13 factors,

$$(a_1/a_3) = 4.47 \qquad a_1 = 0.179 \text{ in.}$$

$$(a_2/a_3) = 1.81 \qquad a_2 = 0.072 \text{ in.}$$

(b) In Table 5.4*c*, $A = 3.51$, $A_1 = 1.16$, $B = 0.744$, $B_1 = 0.593$, $C = 0.388$, $D = 0.733$. Then $a_1 = 4.47(0.04) = 0.179$ in. and $a_2 = 1.81(0.04) = 0.072$ in.

6 TORSIONAL VIBRATION OF CRANKSHAFTS

A multicylinder engine is an excellent example of a mechanism that develops various types of vibratory excitations by virtue of its normal operation. One is the cyclic torque generated on the crank in the plane of each cylinder, which can cause serious crankshaft torsional vibration and which is analyzed in this chapter. Another is the inertia unbalance developed by the motions of the various rotating and reciprocating parts. The latter excite the engine body as a total rigid mass.

Historically, failures of crankshafts were something of a mystery to engineers in the very early days of the internal-combustion engine, and these catastrophic failures became more frequent as the lengths of the crankshafts increased. The problem was identified by World War I, as German U-boats were found to have red lines painted on tachometers warning of resonant speeds. By the advent of World War II analytical methods of predicting resonance were well understood and widely used. Also in this era the centrifugal pendulum absorber (Chapter 7), received much attention and was developed and installed in various configurations. Damped devices were also incorporated to mitigate the engine torsional problem.

This chapter draws on much material previously presented; however, several new concepts are introduced, including the kinematics of the slider-crank mechanism and harmonic analysis by the Fourier series for periodic excitation. Modal response, developed in Sections 5.8 and 5.9, is especially pertinent, as it is impractical and unnecessary to attempt to obtain total response with multiple excitation frequencies, phase relations, and degrees of freedom.

6.1. KINEMATICS OF THE SLIDER CRANK

The slider crank (Figure 6.1) is a fundamental and widely used linkage. It is the common denominator of nearly all reciprocating engines, converting trans-

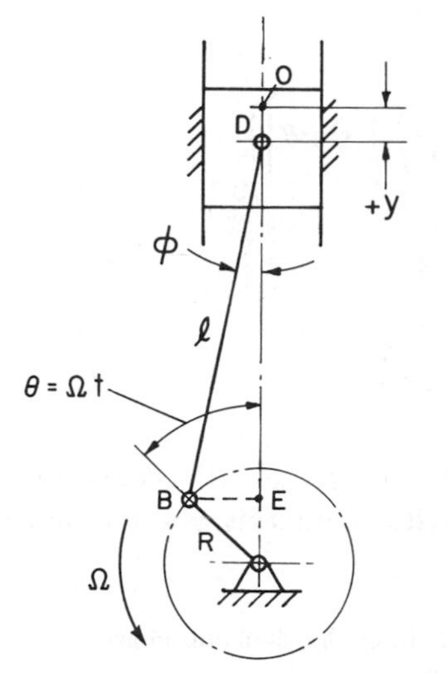

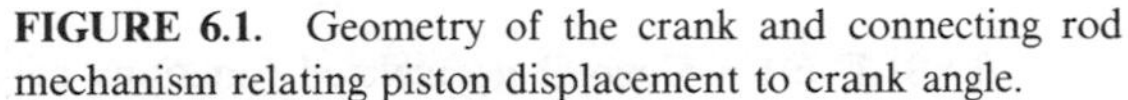
FIGURE 6.1. Geometry of the crank and connecting rod mechanism relating piston displacement to crank angle.

lational quantities to rotational. For constant angular velocity of the crank the piston displacement approximates simple harmonic motion, but there is distortion because of the finite length of the connecting rod. This distortion is often referred to as the *angularity effect*.

An expression for the slider or piston displacement in terms of crank angle, measured from top dead center, can be derived from geometric relations (Figure 6.1):

$$y = (R + l) - (l\cos\phi + R\cos\theta)$$

$$= R(1 - \cos\theta) + l(1 - \cos\phi)$$

$$= R(1 - \cos\theta) + l\left[1 - \sqrt{1 - \left(\frac{R}{l}\right)^2 \sin^2\theta}\right]$$

since $R\sin\theta = l\sin\phi$. Letting $(R/l)^2\sin^2\theta = \delta$, a small dimensionless term,

$$\sqrt{1 - \delta} \approx \left(1 - \frac{\delta}{2}\right)$$

because

$$\left(1 - \frac{\delta}{2}\right)^2 = \left(1 - \delta + \frac{\delta^2}{4}\right)$$

The $(\delta^2/4)$ term is negligible, as with a ratio of R/l in the order of $1/4$ it is a maximum of only $1/1024$ with respect to unity. With this slight approximation,

$$y = R(1 - \cos\theta) + l\left\{1 - \left[1 - \frac{1}{2}\left(\frac{R}{l}\right)^2 \sin^2\theta\right]\right\}$$

$$= R(1 - \cos\theta) + \frac{R^2}{2l}\left(\frac{1 - \cos 2}{2}\right)$$

$$= R\left[\left(1 + \frac{R}{4l}\right) - \left(\cos\theta + \frac{R}{4l}\cos 2\theta\right)\right] \qquad (6.1)$$

Both terms in Equation 6.1 involving the R/l ratio decrease with increasing rod length, approaching zero. They represent the distortion terms, and the sum

TABLE 6.1. General normalized functions describing the dynamic characteristics of the crank and connecting rod mechanism.

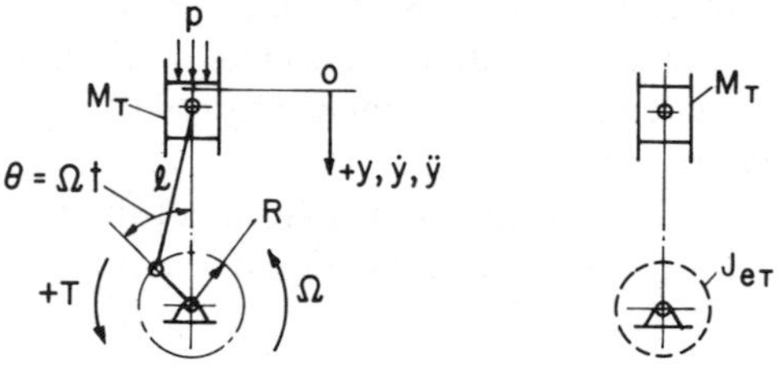

DISPLACEMENT	$\frac{y}{R}$	$\left(1+\frac{R}{4\ell}\right) - \left(\cos\theta + \frac{R}{4\ell}\cos 2\theta\right)$
VELOCITY	$\frac{\dot{y}}{R\Omega}$	$\sin\theta + \frac{R}{2\ell}\sin 2\theta$
ACCELERATION	$\frac{\ddot{y}}{R\Omega^2}$	$\cos\theta + \frac{R}{\ell}\cos 2\theta$
PRESSURE TORQUE	$\frac{T_p}{pAR}$	$\sin\theta + \frac{R}{2\ell}\sin 2\theta$
INERTIA TORQUE	$\frac{T_i}{\left(\frac{M_T R^2 \Omega^2}{4}\right)}$	$\left(\frac{R}{\ell}\right)\sin\theta - 2\sin 2\theta - 3\left(\frac{R}{\ell}\right)\sin 3\theta - \frac{1}{4}\left(\frac{R}{\ell}\right)^2 \sin 4\theta$
EQUIV. INERTIA OF RECIP. MASS	$\frac{J_{eT}}{\left(\frac{M_T R^2}{2}\right)}$	$[1 + \frac{1}{4}\left(\frac{R}{\ell}\right)^2] + [\left(\frac{R}{\ell}\right)\cos\theta - \cos 2\theta - \left(\frac{R}{\ell}\right)\cos 3\theta - \frac{1}{4}\left(\frac{R}{\ell}\right)^2\cos 4\theta]$
AVERAGE EQUIVALENT DISK	$\frac{J_{eT}}{\left(\frac{M_T R^2}{2}\right)}$	$[1 + \frac{1}{4}\left(\frac{R}{\ell}\right)^2] \approx 1$

is always positive, corresponding to a piston displacement greater than the basic harmonic motion, except at top and bottom dead center. The frequency of the distortion term is double the rotative frequency or 2Ω. Higher frequency terms exist, but are negligible, and have been eliminated by the radical approximation in the derivation.

Piston velocity results from the first derivative with respect to time (Table 6.1). Angularity causes a maximum velocity before the 90° crank angle when the rod and crank are approximately perpendicular. At 90° the piston velocity is exactly equal to the tangential velocity of the crankpin B, or $R\Omega$, both in magnitude and direction. At this instant the rod has simple translation.

Acceleration is particularly important, since it determines the inertia effects of the reciprocating parts. By a second differentiation we have the expression in Table 6.1. The terms represent *primary* and *secondary* accelerations, both proportional to the square of the rotative velocity.

6.2. MASS MODELING THE CONNECTING ROD

Moving masses in the mechanism (Figure 6.2) include the piston assembly, which moves the pure translation; the crankshaft, which moves with simple rotation; and the connecting rod, which has composite motion. We now must resolve the effects of the dynamics of the rod.

As a rigid link for our purposes, the rod has mass, a centroid, and a mass moment of inertia about the centroid. The link is similar to the compound pendulum in these respects (see Section 8.2). For dynamic equivalence we substitute two concentrated masses connected by a stiff, weightless bar. If one of the two masses is located at the wristpin M_0, the other will lie a calculated

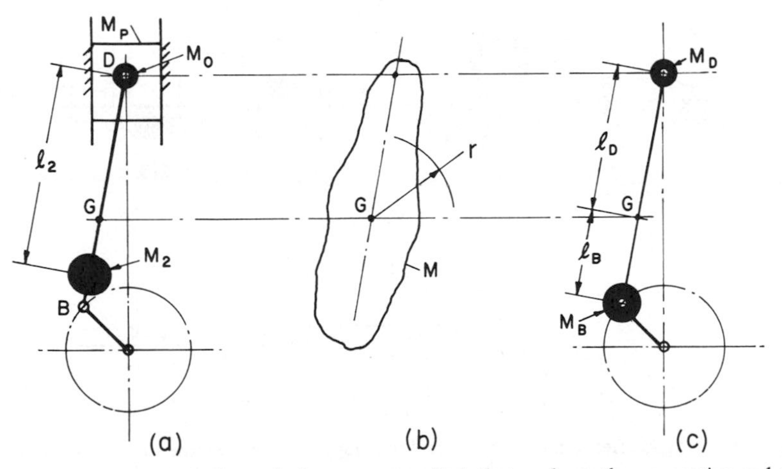

FIGURE 6.2. The dynamically equivalent two-mass link that replaces the connecting rod (a) is slightly compromised to the more convenient two-mass link (c).

distance from this mass of l_2. The probability is negligible that the second mass M_2 will coincide with the crankpin (Figure 6.2a). Generally it falls short.

With only a slight error, the precise two-mass system of Section 8.2 is approximated with part of the total rod mass at each end, or pin (Figure 6.2c). Then M_D has pure translation and M_B pure rotation. We maintain the same total mass and center of gravity by

$$\begin{cases} M_B + M_D = M \\ M_B l_B = M_D l_D \end{cases} \tag{6.2}$$

The inertia condition (Equation 8.3c) is not used. We then add M_D to the other reciprocating mass M_P, including the piston, rings, and wristpin, to obtain the *total translational mass* M_T.

$$M_T = M_D + M_P \tag{6.3}$$

6.3. EQUIVALENT TORSIONAL INERTIA

Determination of natural frequencies requires that we reduce the entire cylinder mechanism to a single equivalent rotating inertia for the discrete model, shown in Figure 6.3, including rotational and reciprocating effects. The former are obvious conceptually, as the crankshaft rotates about a fixed bearing axis and can be evaluated by numerical integration for the contributions of the

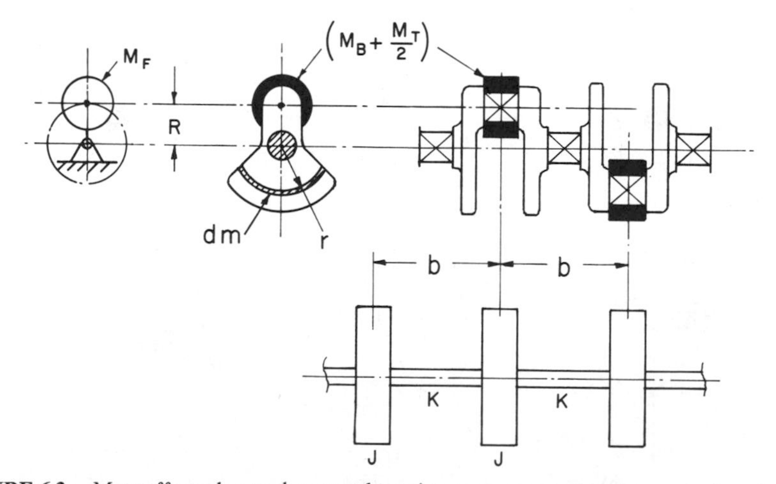

FIGURE 6.3. Mass effects due to the several moving parts are combined to a single polar inertia at each cylinder. An equivalent spring is also obtained for the crankshaft.

crankcheeks, crankpin, and counterweights. For one span this is

$$J_C = \int r^2 dm \tag{6.4}$$

with integration of the solid extending from the center to the extreme radius (Figure 6.3).

Reciprocating mass is completely uncoupled for dead-center crank positions, but partially to fully coupled for other crank angles. Equating rotational and translational kinetic energies,

$$\frac{1}{2} J_{eT} \Omega^2 = \frac{1}{2} M_T (\dot{y})^2 = \frac{1}{2} M_T R^2 \Omega^2 \left(\sin\theta + \frac{R}{2l} \sin 2\theta \right)^2 \tag{6.5}$$

By expansion and trigonometric substitution this becomes a cosine series (Table 6.1). Since all cosine terms have zero average value over a complete cycle or revolution, the constant terms represent the *average* disk effect. Furthermore, the $\frac{1}{4}(R/l)^2$ factor is negligible with respect to unity and we can take the effective disk to be $\frac{1}{2} M_T R^2$. This can be interpreted as an average of a maximum of $M_T R^2$ and a minimum of zero. We can also view this result to indicate the variable disk as equivalent to an actual disk of radius R and mass M_T at the crankshaft axis. The polar inertia of such a disk is Mr^2, but the radius of gyration squared is $\frac{1}{2} R^2$.

For the total effective inertia at one cylinder, we finally combine all contributions,

$$J_e = J_C + M_B R^2 + \tfrac{1}{2} M_T R^2 = M_F R^2 \tag{6.6}$$

where M_F = the equivalent total concentrated mass at B.

6.4. EQUIVALENT TORSIONAL SPRING

In Figure 6.3 we see the inertia per cylinder applying to the central plane of a cylinder, with the interval b. Similarly, the torsional spring has the span of b, but the crankshaft is usually highly irregular geometrically. Theoretical elastic behavior about the torque axis is a complex problem involving three-dimensional stresses, but as a first approximation we can use the basic torsional equation from Table 1.4a, substituting the span b for the length and the main journal diameter for the shaft diameter. Real flexibilities will be greater or less than this nominal value, but we have a first approximation.

Another approach is experimental. If an actual crankshaft is available it can be torqued in the laboratory with angular deflections accurately measured. Variations in materials, particularly in cast crankshafts, which affect the shear modulus G, can also cause deviations in K and in calculated frequencies;

however, this is only a problem when a running speed is very close to resonance.

6.5. CONVERSION OF PISTON FORCE TO TORQUE

A downward force on the piston due to gas pressure (Figure 6.4) is $F_p = pA$, with A the area of the piston. This varying force can be converted to instantaneous torque at the crank by various means, the simplest by using the principle of conservation of energy.

Considering an incremental displacement of the mechanism, the work done at the piston must equal the work done on the crank.

$$F_p\, dy = T_p\, d\theta$$

$$T_p = F_p\left(\frac{dy}{d\theta}\right)$$

Taking the derivative of Equation 6.1 with respect to crank angle rather than time,

$$T_p = F_p R\left(\sin\theta + \frac{R}{2l}\sin 2\theta\right) \tag{6.7}$$

We note that the actual torque on the crank is a nominal $(F_p R)$ multiplied by a dimensionless factor depending on crank angle. This factor is zero at dead-center positions, and provides proper sign directions, becoming negative between $\theta = 180°$ and $360°$.

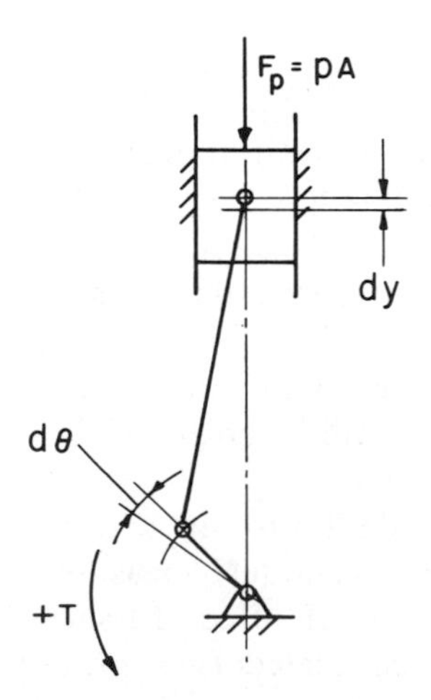

FIGURE 6.4. The pressure force on the piston is converted to turning effort on the crank.

6.6. INERTIA TORQUE DUE TO RECIPROCATING MASS

The translational mass M_T develops torque on the crank because of the related cyclic inertia effects. Replacing the pressure force in Equation 6.7 by the corresponding inertia force at the piston,

$$T_i = \left[-M_T R\Omega^2\left(\cos\theta + \frac{R}{l}\cos 2\theta\right)\right]\left[R\left(\sin\theta + \frac{R}{2l}\sin 2\theta\right)\right] \tag{6.8}$$

In the resulting expression (Table 6.1), there are only sine terms, with the second or double-frequency term the largest component. This factor, the only one independent of R/l, represents the conversion of a simple harmonic inertia force to a torque through a simple harmonic factor. With increasing rod length, R/l approaches zero and the inertia torque is simply

$$\begin{aligned} T_i &= (-M_T R\Omega^2 \cos\theta)(R\sin\theta) \\ &= -M_T R^2\Omega^2\left(\frac{\sin 2\theta}{2}\right) \end{aligned} \tag{6.9}$$

Inertia torque can be predicted accurately by the equation given for any combination of mass, geometry, and speed. It increases as the square of both the crank radius and the speed, with a mean value of zero.

6.7. THE FOURIER TRIGONOMETRIC SERIES

If a function continuously repeats itself in the same interval of time, it is cyclic, and equivalent to the sum of component sinusoidal functions at integer multiples of the fundamental frequency, plus a possible constant.

$$\begin{aligned} f(\theta) &= A_1\sin\theta + A_2\sin 2\theta + \cdots + A_n\sin n\theta \\ &\quad + B_0 + B_1\cos\theta + B_2\cos 2\theta + \cdots + B_n\cos n\theta \\ &= \sum_{n=1}^{n=\infty} A_n\sin n\theta + \sum_{n=0}^{n=\infty} B_n\cos n\theta \end{aligned} \tag{6.10}$$

where A and B = amplitudes of the Fourier components in proper units, that is, of the function

n = all successive integers

$\theta = \omega t$ = function angle

ω = fundamental circular frequency

As seen in Figure 6.5, ordinates of the components add to the ordinate of the function at any instant, the approximation becoming closer as more terms are taken. The coefficients are evaluated by harmonic analysis using the following relations:

$$A_n = \frac{1}{\pi}\int_0^{2\pi} f(\theta)\sin n\theta\, d\theta \tag{6.11a}$$

$$B_n = \frac{1}{\pi}\int_0^{2\pi} f(\theta)\cos n\theta\, d\theta \tag{6.11b}$$

$$B_0 = \frac{1}{2\pi}\int_0^{2\pi} f(\theta)\, d\theta \tag{6.11c}$$

The series actually contains only one sinusoidal function at each integer frequency, obtained by combination of the sine and cosine terms.

$$f(\theta) = C_0 + C_1\cos(\theta - \phi_1) + C_2\cos(2\theta - \phi_2) + \cdots \tag{6.12}$$

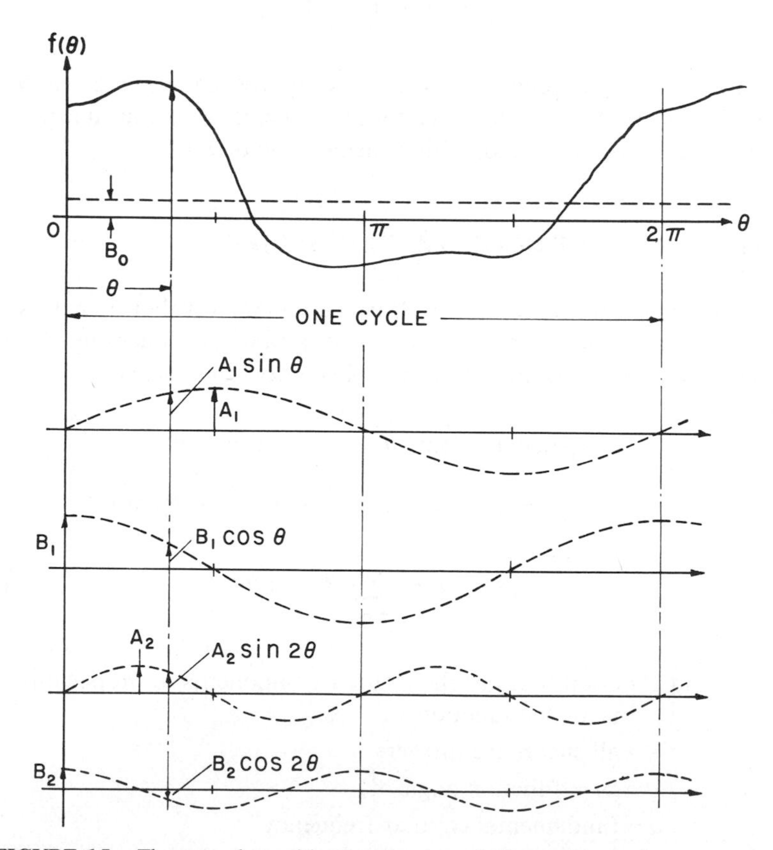

FIGURE 6.5. The general repetitive function is equivalent to harmonic components.

where $C_0 = B_0$

$C_n = \sqrt{A_n^2 + B_n^2}$

$\phi_n = \tan^{-1}(A_n/B_n)$

The Fourier series, although a mathematical concept, has many important physical implications. Our study of vibratory systems has largely revolved about sinusoidal behavior of linear systems with excitation at constant frequency. We now observe in Equation 6.12 with $\theta = \omega t$ and in Figure 6.5 that there are, in addition to the fundamental frequency ω, exciting frequencies of 2ω, 3ω, $4\omega, \ldots, n\omega$. Thus a vibratory system with cyclic excitation will respond to and can become resonant with any integer multiple of the fundamental frequency. In fact, the system responds simultaneously to all component excitations resulting in superposition of all responses.

This profound result complicates the analytical solution but fortunately does not preempt application of the basic relations of Table 2.1. Superposition of the multiple responses is not by simple algebraic addition, as we have rotating vectors in the complex plane at different velocities. Two such vectors will only add algebraically if they cross the real axis simultaneously. We can, however, consider the sum of the absolute vectors as a maximum potential total amplitude.

6.8. TORSIONAL EXCITING FREQUENCIES

A single-cylinder engine operating in steady state subjects the crank to cyclic pressure torque, which can be reduced to harmonic components as indicated. The most common engine is the four-cycle, requiring two revolutions for a complete cycle. The two-cycle is repetitive every revolution (Figure 6.6). In the Fourier analysis, completion of an engine cycle corresponds to $\theta = 2\pi$, regardless of the revolutions required.

Exciting frequencies developed relate to engine speed and are defined in terms of the *order*, N, as a ratio of exciting frequency to rotative frequency; that is, the number of excitation cycles per revolution. We then have

Two-cycle engine: $N = \dfrac{\omega}{\Omega} = n \qquad \omega = n\Omega$

Four-cycle engine: $N = \dfrac{\omega}{\Omega} = \dfrac{n}{2} \qquad \omega = \left(\dfrac{n}{2}\right)\Omega$

where n = any integer in the Fourier series

ω = exciting frequency

Ω = rotative frequency

N = exciting order

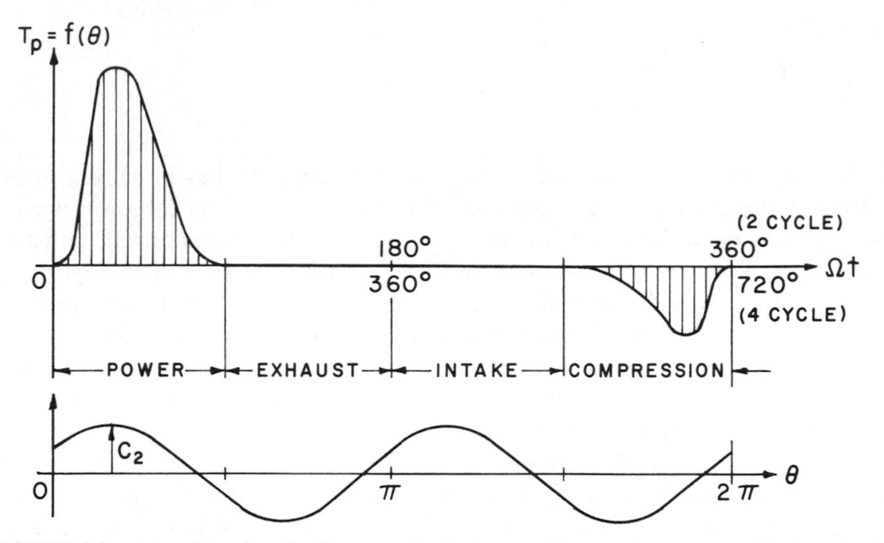

FIGURE 6.6. An exciting function for an engine is the repetitive pressure torque, with the related second harmonic illustrated.

For example, a two-cycle engine operating at 2500 RPM has Fourier frequencies at 2500, 5000, 7500,..., CPM. A four-cycle engine at the same speed has frequencies of 1250, 2500, 3750,..., CPM.

6.9. RESONANT TORSIONAL SPEEDS

Increase of torsional exciting frequencies with rotational speed is shown in Figure 6.7. The slope of a straight line is numerically an order of excitation, with the ordinate to a straight line the excitation frequency at a particular speed. If now we superimpose natural frequencies on this diagram we mate a crankshaft torsional system with means for its excitation, noting that the engine frequencies are independent of rotation and therefore horizontal lines.

From this combination in Figure 6.7, a number of intersections are observed. Each represents a resonant condition in which a particular harmonic torque component frequency coincides with a modal frequency, and is identified by both order and mode. Projection of an intersection down to the axis determines the speed at which the resonance occurs. A maximum speed limitation is indicated.

Figure 6.7 displays the resonant problem graphically and we are appalled by the magnitude of the difficulties that have arisen. There are many resonances to be expected when operating the engine through the speed range, and the spectrum is now defined. Fortunately the vibratory problem is not as serious as it appears, but is limited as a rule to several potentially dangerous resonances that can be avoided or controlled.

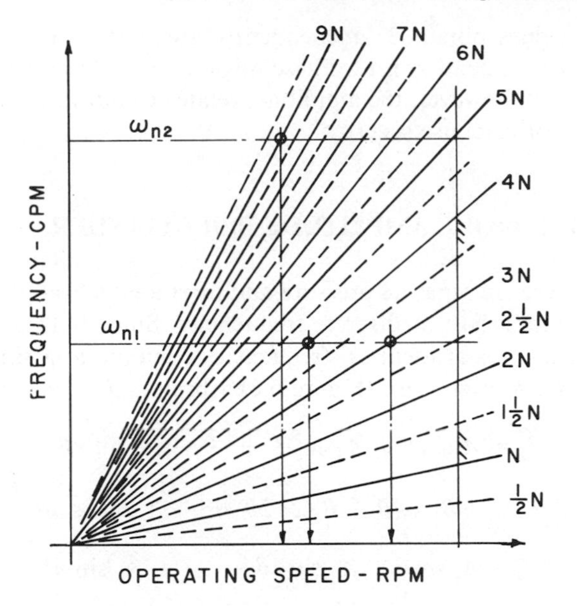

FIGURE 6.7. The spectrum of resonant operating speeds is determined by the intersections of the Fourier frequencies and the modal frequencies.

6.10. AMPLITUDE OF PRESSURE TORQUE EXCITATION

Quantification of expected operational amplitudes at a resonance point requires an evaluation of the exciting torques at the various discrete frequencies. For pressure components it is necessary to know the cyclic variation in cylinder pressure, and pressure–volume data can be reduced to torque angle variation using Equation 6.7. A typical pressure torque cycle is shown in Figure 6.6.

Since the torque function is not sinusoidal, it contains all the Fourier components (Equation 6.10). The component amplitudes can be calculated from Equation 6.11, which requires multiplication of the torque ordinates by the respective sine and cosine functions. For instance,

$$A_5 = \frac{1}{\pi}\int_0^{2\pi} f(\theta)\sin 5\theta \, d\theta$$

$$B_5 = \frac{1}{\pi}\int_0^{2\pi} f(\theta)\cos 5\theta \, d\theta$$

when determining the fifth harmonic.

Ordinates under the integral represent a product function, and the area under this curve, with due regard for algebraic sign, evaluates the definite integral in torque units. This can be done by numerical or analog methods. Division by π then yields the torque coefficients.

A_n and B_n values obtained independently are then combined to a single amplitude C_n from Equation 6.12. Phase angle is of little physical importance and is disregarded; however, the amplitude relates to the strength of excitation and is an index of response severity.

6.11. TOTAL TORQUE AMPLITUDE PER CYLINDER

As indicated, there are separate pressure and inertia effects applying sinusoidal torques to the crankshaft at discrete frequencies. Since both are multiples of rotative speed, torques at identical frequencies combine to a single excitation. For the simpler two-cycle case (Figure 6.6) vibratory torques are (Figure 6.8)

$$T_p = A_1 \sin\theta + A_2 \sin 2\theta + \cdots + A_n \sin n\theta$$

$$+ B_1 \cos\theta + B_2 \cos 2\theta + \cdots + B_n \cos n\theta$$

$$T_i = A_{1i} \sin\theta + A_{2i} \sin 2\theta + \cdots + A_{ni} \sin n\theta$$

$$\Sigma T = \left(\sqrt{(A_1 + A_{1i})^2 + B_1^2}\right) \sin(\theta + \phi_1)$$

$$+ \left(\sqrt{(A_2 + A_{2i})^2 + B_2^2}\right) \sin(2\theta + \phi_2) + \cdots$$

$$= T_N \sin\theta + T_{2N} \sin 2\theta + \cdots \tag{6.13}$$

where T_N is the total torque amplitude at order N.

In Equation 6.13 phase angles are neglected, although the T_N terms are essentially complex variables at the respective frequencies; however, as will be shown in Section 6.12, it is only necessary to define the absolute torque amplitudes. Above $4N$ only pressure torques occur.

The four-cycle engine similarly requires combination at 1, 2, 3, and $4N$, and has pressure orders of $\frac{1}{2}$, $1\frac{1}{2}$, $2\frac{1}{2}$, $3\frac{1}{2}$, $4\frac{1}{2}$, 5, $5\frac{1}{2}$, $6N, \ldots$.

6.12. STAR DIAGRAMS

It was shown in Section 4.6 that a lumped spring–mass system can be reduced to modal springs, usually at the end of the system. In combination with damping, excitation, and forced frequency ratio, these springs will predict modal response amplitudes (Equation 5.28).

In a multicylinder engine each cylinder develops multiple-order torques as indicated, presumably identical at each individual cylinder, with the resultant effect on a mode determined by considering the phase differences between the

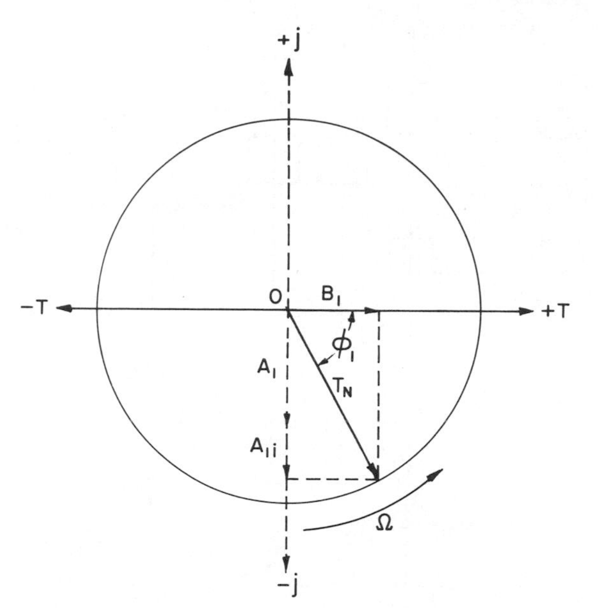

FIGURE 6.8. Torque excitations due to reciprocating masses and to cyclic pressure are combined in the complex plane at common orders $1N$, $2N$, $3N$, and $4N$, with $1N$ illustrated.

torques (Section 5.10). We take for example a three-cylinder, two-cycle engine with firing order 1, 3, 2, 1,... (Figure 6.9). Normalized modal data are shown in Figure 6.10. Considering the lowest exciting frequency, $1N$, there are three sinusoidal torques distributed axially, each T_N. These torques are complex variables at a common frequency, and they can be combined or added vectorially (Figure 1.3). There are constant phase angles between the rotating vectors of 120°, corresponding to the end view of the crankshaft in Figure 6.9*e*, with No. 3 lagging the reference No. 1 by 120°, and No. 2 lagging No. 3 by 120°. With equal amplitudes the sum of the vectors is zero, and there is apparent cancelation of excitation. But this is not true with respect to modal excitation, as No. 3 has less effect on the mode (Figure 6.9*c*), than the other cylinders because of lesser modal amplitude. This results from energy considerations; however, intuitively, if a cylinder is at a node, it can provide no excitation to that mode. As indicated in Equation 5.14, the excitation referred to M_{ni} is *directly proportional* to the modal amplitude at which the excitation is applied.

Introducing this modification, Figure 6.9*e* becomes Figure 6.9*f*. Factoring the T_N term, the resultant equivalent single torque at No. 1 is obtained by transferring the phase relations to the modal amplitudes β_i, or

$$\Sigma T_N \beta_{qi} = T_N \beta_{1i} + T_N \beta_{2i} + T_N \beta_{3i} + \cdots$$

$$= (T_N \beta)_i = T_N (\beta_1 \nrightarrow \beta_2 \nrightarrow \beta_3 \nrightarrow \cdots)_i \qquad (6.14)$$

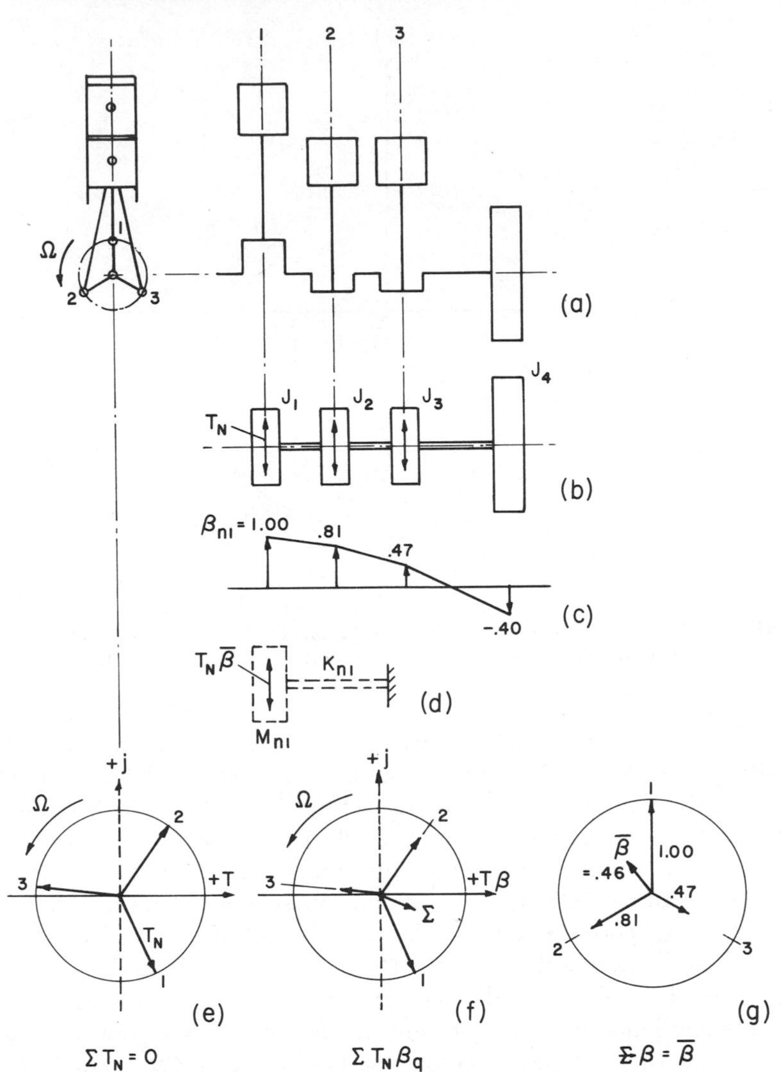

FIGURE 6.9. A three-cylinder, two-cycle engine excited by first-order out-of-phase torques leads to an equivalent simplified result (d).

where numerical subscripts refer to the individual cylinders and

β_q = normal mode amplitudes at the cylinders
T_N = absolute torque amplitude at order N per cylinder
$\overline{\beta}$ = vector resultant in star diagram, dimensionless
i = mode

The final star diagram, Figure 6.9g, is similar to (f), but with phase disregarded we start the construction with unit radius and vertically upward for β_1.

$g_{n1}^2 = 0.190$						
q	m	β	Σ	k	$\Sigma m\beta$	$m\beta^2$
1	1	1.000	.190	1	1.000	1.000
2	1	.810	.343	1	1.810	.656
3	1	.466	.431	.5	2.276	.217
4	5.67	−.400	0	—	0	.907
					$m_{n1} = \Sigma$	2.780

FIGURE 6.10. Normalized converged Holzer values for the first mode of the Figure 6.9 engine.

With distributed torques reduced to $T_N\bar{\beta}$ at J_1 and with the distributed mass-elastic system reduced to the modal spring–mass, we have in Figure 6.9*d* simplified a complex situation to the basic single-degree case of Table 2.1*a*.

6.13. MAJOR ORDERS OF EXCITATION

In the sample star diagram constructions (Figure 6.11), ψ_1 for $N = 1$ is a direct substitution, but at $2N$ the complex torque vectors rotate at twice the velocity, or 2Ω. They then travel through twice the angle of the crank during successive cylinder dead-center positions, or $\psi_{2N} = 240°$. To generalize the phase angle in

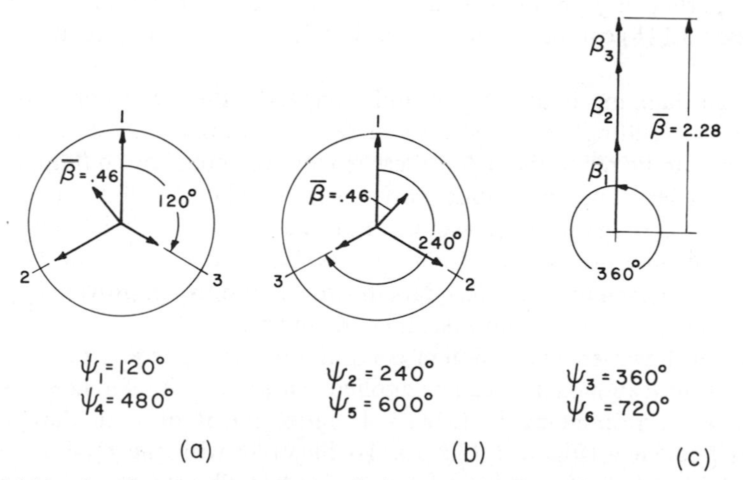

FIGURE 6.11. Star diagrams for the three-cylinder example indicate phase angles, and only two possible resultant $\bar{\beta}$ factors.

the star diagram,

$$\psi_N = N\alpha \tag{6.15}$$

where α is the crank angle between successive cylinder firing. Star diagrams can thus be developed using this phase angle and the modal amplitudes β_{qi}.

The most profound result in terms of β occurs when $\psi_N = 360°$. For usual positive values of β at all cylinders in the first mode, $\bar{\beta}$ is the summation of all modal amplitudes at the cylinders, maximizing the excitation combination. Orders of this type ar particularly serious and are termed *major orders*. These are easily identified, depending only on the number of cylinders and whether an engine is two or four cycle. The major orders are

Two-cycle engine: $N = n_c, 2n_c, 3n_c, \ldots$

Four-cycle engine: $N = \frac{1}{2}n_c, n_c, \frac{3}{2}n_c, \ldots$

where n_c is the number of cylinders.

These relations apply to any type of engine—in-line, radial, or vee. For instance an eight-cylinder, four-cycle vee engine has major orders of 4, 8, 12, 16N,... . For a five-cylinder, two-cycle engine they are 5, 10, 15N,... . The major-order summation problem cannot be avoided by changes in crankshaft configuration or firing order.

6.14. MODAL RESPONSE

Having reduced our problem to an equivalent P_1 in Equation 5.28, we must now have a value for damping ratio in order to determine the magnification factor, particularly in the vicinity of resonance. Rolling mode amplitude (Equation 5.12b) can be calculated, but involves neither damping nor torsional stress.

System damping is distributed and complex, with the majority of energy dissipation resulting from hysteresis in reversed stressing of the crankshaft.[11] Damping can be relatively high if there is vibratory coupling to fluid, as with a ship propeller. The magnitude varies with modal distribution, materials, clearances, and many other factors, but the resonant magnification factor usually falls in a broad range from 5 to 100. We can assume a mean value of 20–30, recognizing that only test data during operation can provide a quantitative answer to this important system characteristic.

We can, however, make system computations independent of damping to which magnification factors can be applied. These furnish reference *equilibrium amplitudes* corresponding to (MF) = 1, independent of both damping and forcing frequency (Figure 6.12b, c). To illustrate with the torsional data in Figure 6.10 and absolute values of $K_1 = 1.7(10)^5$ in.lb/rad and a 6N excitation

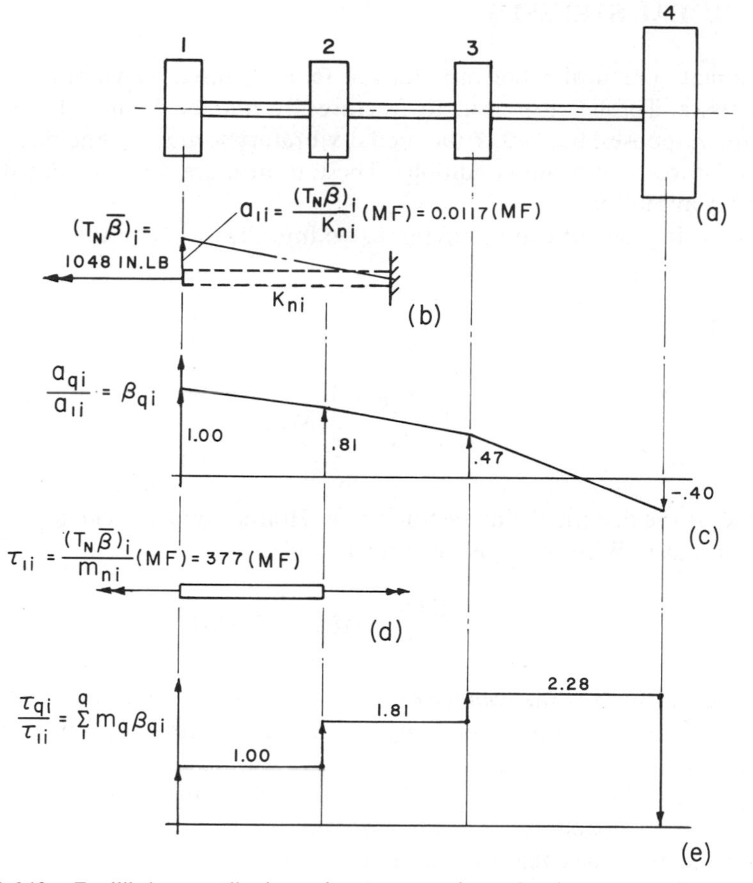

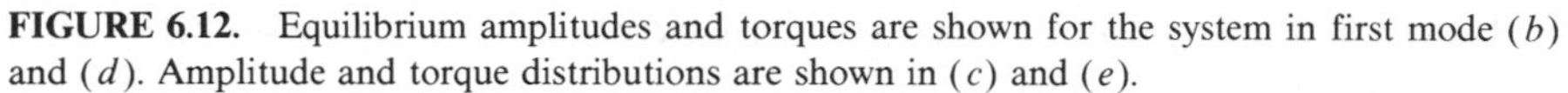
FIGURE 6.12. Equilibrium amplitudes and torques are shown for the system in first mode (*b*) and (*d*). Amplitude and torque distributions are shown in (*c*) and (*e*).

of 460 in.lb/cylinder, for first mode,

$$K_{n1} = K_1 \mathsf{g}_{n1}^2 \Sigma m_q \beta_{qi}^2 = 1.7(10)^5(0.190)(2.78) = 0.90(10)^5$$

$$(\mathsf{a}_{N1})_i = \left[\frac{(T_N\beta)_i}{K_{ni}}\right](\mathrm{MF}) \tag{6.16}$$

$$(\mathsf{a}_{61})_1 = \left[\frac{(460)(2.28)}{0.90(10)^5}\right](\mathrm{MF}) = \left[\frac{1048}{0.9(10)^5}\right](\mathrm{MF}) = 0.0117\ (\mathrm{MF})$$

where the 0.0117 is the static twist of K_{n1} (radians) in Figure 6.12*b*.

In addition to the equilibrium amplitude at 1 we are studying a modal response with known distribution (Figure 6.12*c*). As shown, $\mathsf{a}_{q1} = \beta_{q1}(\mathsf{a}_{11})$, providing equilibrium system amplitudes.

6.15. MODAL STRESSES

The ultimate practical problem is fatigue failure caused by vibratory torques and stresses. Torques that lead to fracture are not the Fourier torques that drive the responses, but rather the modal vibratory torques as the disks of the system approach resonant conditions. These in turn are directly related to the Σ column in Holzer.

Again using the left end as reference, we find the torque in the first interval to be

$$(\tau_{N1})_i = (a_{N1})_i\left[K_1\left(m_1 g_{ni}^2 \beta_{n1}\right)\right]$$
$$= \left[\frac{(T_N \bar{\beta})_i}{m_{ni}}\right](\mathrm{MF}) \tag{6.17}$$

where K_1 is the denormalizing factor for the Holzer table and $(m_1 g_{ni}^2 \beta_{n1}) = g_{ni}^2$ at the first disk. With previous numerical values,

$$(\tau_{61})_1 = \left(\frac{1048}{2.78}\right)(\mathrm{MF}) = 377(\mathrm{MF})$$

where the 377 in.lb is the *equilibrium torque* between J_1 and J_2 caused by $6N$ excitation in the first mode, occurring in the actual spring element K_1 (Figure 6.12*d*).

Torque distribution in Figure 6.12*e* follows from the Σ column in Figure 6.10, and is analogous to a shear diagram for a beam. Maximum torque typically exists in the span containing the node.

Torsional vibratory shear stress can be calculated on a nominal basis as for a simple shaft.

$$S_{sq} = \frac{T}{J/r} = \frac{(\tau_{Nq})_i}{\left(\pi D_q^3/16\right)} \tag{6.18}$$

where the stress is to the right of disk q for a particular order and mode, and D is the minimum shaft diameter subjected to the torque. Effects of stress concentration and related allowable fatigue stresses are the final practical considerations as to the feasibility of operation at a resonant or near resonant speed. With resonant operation, only damping limits amplitudes and stresses, and we have $(\mathrm{MF}) = 1/2\zeta_1$.

Stresses are directly proportional to amplitudes in a given mode, Figure 6.12. Engine tests using a torsiograph (Figure 16.11) at the end J_1 to measure operational amplitudes can be interpreted as shown in terms of modal stress distribution throughout the crankshaft. If an elastically coupled load is connected, torsionally, computations will apply to the complete torsional system

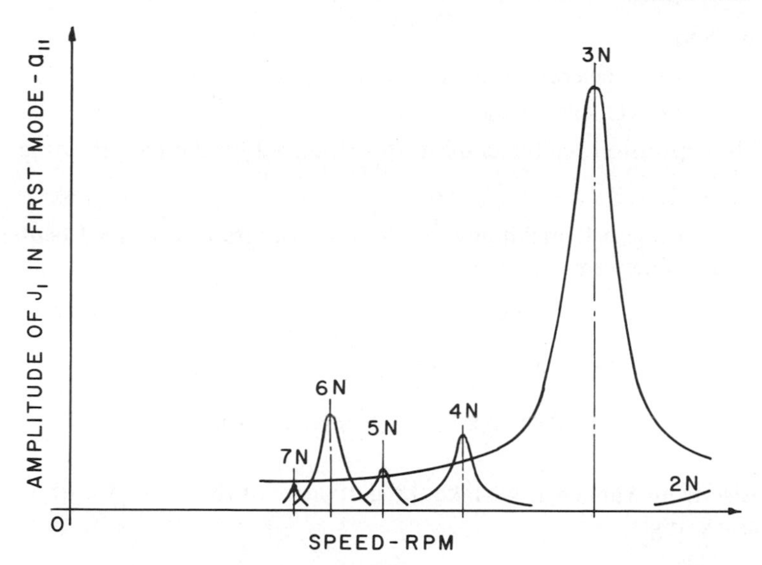

FIGURE 6.13. Total crankshaft response is composed of multiple discrete resonances with additive effects.

and stresses can be determined similarly in these extensions. Excessive stress combinations can be suppressed by means of auxiliary dampers or absorbers as explained in Chapter 7.

6.16. MODAL SUPERPOSITION

Since a number of resonances will be encountered in an operational speed range (Figure 6.7) vibratory amplitude of the end disk predicted by Equation 6.16 will consist of a series of independent single-degree modal peaks at discrete speeds. With the simple case of Figure 6.9 related qualitatively to Figure 6.7, we have the resonances shown in Figure 6.13. Major orders $3N$ and $6N$ amplify these peaks. Order excitations lower than $2\frac{1}{2}N$ fall above the maximum speed limitation. Second and higher modes have higher modal stiffness and resonate at the higher orders, which tend to become progressively smaller in amplitude. Thus modes above the first can often be neglected analytically; however, in the more complex systems a number of higher modes can require consideration.

EXAMPLES

EXAMPLE 6.1. A small $3\frac{1}{8} \times 2\frac{3}{4}$ in. (bore and stroke) engine operates at 3600 RPM. The connecting rod length is 4.50 in. and the reciprocating mass is

0.80 lb. Find:

(a) Maximum acceleration of the piston.

(b) Maximum velocity of the piston.

(c) The expression for the crank torque induced by the reciprocating mass.

Solution

(a) From Table 6.1, maximum acceleration occurs at top dead center when both cosine terms are 1.00.

$$R = 1.38 \text{ in.}$$

$$\Omega = 377 \text{ rad/sec}$$

$$\ddot{y} = 195{,}000(1 + (R/l)) = 255{,}000 \text{ in./sec}^2$$

(b) Also from Table 6.1, we take the derivative of the velocity with respect to crank angle,

$$\frac{d\dot{y}}{d\theta} = \cos\theta + (R/l)\cos 2\theta = 0$$

$$2\cos^2\theta + (R_l)\cos\theta - 1 = 0$$

$$\theta = 75°$$

$$\text{Maximum } \dot{y} = (1.38)(377)(1.12) = 580 \text{ in./sec}$$

(c) From Table 6.1 for inertia torque,

$$\frac{M_T R^2 \Omega^2}{4} = 139 \text{ in.lb}$$

$$T_i = 139(0.306\sin\theta - 2\sin 2\theta - 0.917\sin 3\theta - 0.023\sin 4\theta)$$

$$= 42.5\sin\theta - 278\sin 2\theta - 127\sin 3\theta - 3\sin 4\theta$$

EXAMPLE 6.2. A function consists of a reversing constant value as shown. Find the Fourier series that describes the function.

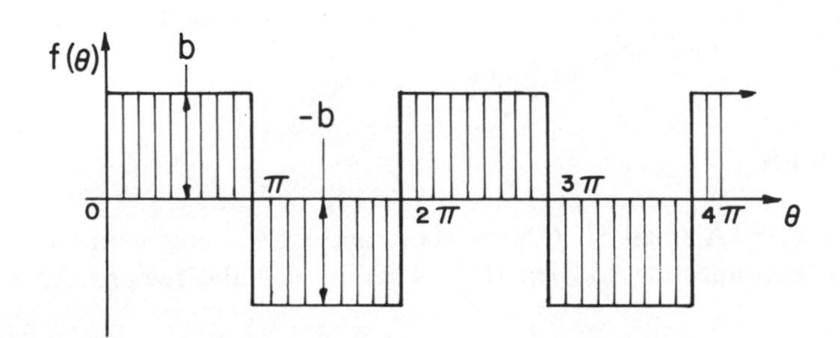

Solution. From Equation 6.11, the average or constant term $B_0 = 0$, and,

$$A_1 = \frac{1}{\pi}\int_0^{\pi} b \sin\theta \, d\theta - \frac{1}{\pi}\int_0^{\pi} b \sin\theta \, d\theta$$

$$= \frac{b}{\pi}\left[-\cos\theta|_0^{\pi} + \cos\theta|_{\pi}^{2\pi} = \left(\frac{4}{\pi}\right)b\right.$$

$$A_2 = \frac{b}{\pi}\int_0^{\pi} \sin 2\theta \, d\theta - \frac{b}{\pi}\int_{\pi}^{2\pi} \sin 2\theta \, d\theta = 0$$

$$A_3 = \frac{b}{\pi}\int_0^{\pi} \sin 3\theta \, d\theta - \frac{b}{\pi}\int_{\pi}^{2\pi} \sin 3\theta \, d\theta = \left(\frac{4}{3\pi}\right)b$$

$$B_1 = \frac{b}{\pi}\int_0^{\pi} \cos\theta \, D\theta - \frac{b}{\pi}\int_{\pi}^{2\pi} \cos\theta \, d\theta = 0$$

$$= B_2 = B_3 \cdots = 0$$

$$f(\theta) = \left(\frac{4}{\pi}\right)b\left(\sin\theta + \frac{1}{3}\sin 3\theta + \frac{1}{5}\sin 5\theta + \cdots\right)$$

Only the odd sine terms are present.

EXAMPLE 6.3. A repetitive function exists as a constant pulse for the first one-quarter of the cycle. Determine the first three Fourier components.

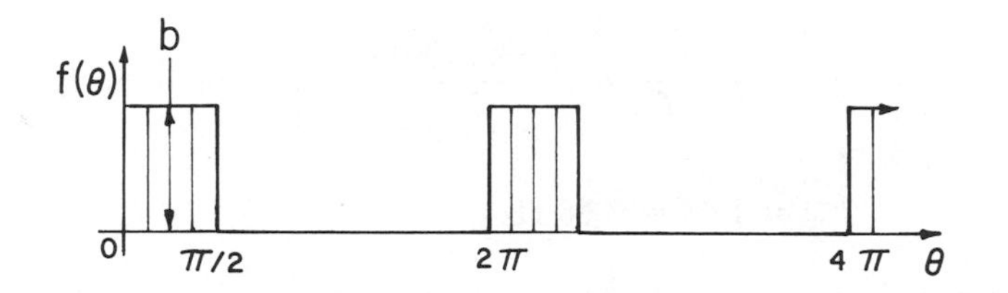

Solution

$$A_1 = \frac{b}{\pi}\int_0^{\pi/2} \sin\theta \, d\theta = \frac{b}{\pi}(-\cos\theta \, d\theta)\Big|_0^{\pi/2} = \frac{b}{\pi}$$

$$A_2 = \frac{b}{\pi}\int_0^{\pi/2} \sin 2\theta \, d\theta = \frac{b}{\pi}$$

$$A_3 = \frac{b}{\pi}\int_0^{\pi/2} \sin 3\theta \, d\theta = \frac{b}{3\pi}$$

$$B_1 = \frac{b}{\pi}\int_0^{\pi/2} \cos d\theta = \frac{b}{\pi}(\sin\theta)\Big|_0^{\pi/2} = \frac{b}{\pi}$$

$$B_2 = \frac{b}{\pi}\int_0^{\pi/2} \cos 2\theta \, d\theta = 0$$

$$B_3 = \frac{b}{\pi}\int_0^{\pi/2} \cos 3\theta \, d\theta = -\frac{b}{3\pi}$$

$$B_0 = \frac{1}{2\pi}\int_0^{\pi/2} b \, d\theta = \frac{b}{4}$$

EXAMPLE 6.4. The pulse function of Example 6.3 represents a displacement excitation of an undamped single-degree system with $b = 2.9$ cm. Fundamental cycle time for the function is 0.35 sec. Calculate:

(a) The component amplitudes of the mass in response to the first three harmonics.

(b) Required amplitude of the driving force with the largest component amplitude.

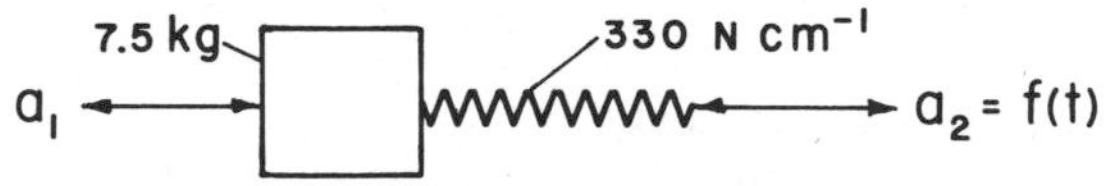

Solution

(a) From Table 2.1*c*,

$$\omega_1 = \sqrt{\frac{K_1}{M_1}} = \sqrt{\frac{330(100)}{7.5}} = 66.3, \text{ or } 10.56 \text{ Hz}$$

$$\omega = 1/T = 2.86 \text{ Hz}$$

For $n = 1$,

$$g = (2.86/10.56) = 0.27$$

$$\left(\frac{a_{11}}{a_{21}}\right) = \left(\frac{1}{1 - g^2}\right) = 1.08$$

$$a_{21} = \sqrt{\left(\frac{2.9}{\pi}\right)^2 + \left(\frac{2.9}{\pi}\right)^2} = 1.31 \text{ cm}$$

$$a_{11} = 1.08(1.31) = 1.41 \text{ cm at } 2.86 \text{ Hz}$$

For $n = 2$,

$$\mathsf{g} = 2(0.27) = 0.54$$

$$\left(\frac{\mathsf{a}_{12}}{\mathsf{a}_{22}}\right) = 1.42$$

$$\mathsf{a}_{12} = 1.42(2.9/\pi) = 1.31 \text{ cm at } 5.72 \text{ Hz}$$

For $n = 3$,

$$\mathsf{g} = 3(0.27) = 0.81$$

$$\left(\frac{\mathsf{a}_{13}}{\mathsf{a}_{23}}\right) = 2.94$$

$$\mathsf{a}_{13} = 2.94(0.435) = 1.28 \text{ cm at } 8.58 \text{ Hz}$$

(b) The largest product of $\mathsf{a}_2[\mathsf{g}^2/(1 - \mathsf{g}^2)]$ from Table 2.1*c* occurs at $n = 3$, and $\mathsf{g}^2 = 0.66$.

$$\left(\frac{P_2}{K_1\mathsf{a}_2}\right) = \left(\frac{0.66}{1 - 0.66}\right) = 1.93$$

$$P_2 = (33{,}000)\left(\frac{0.435}{100}\right)(1.93) = 277 \text{ N at } 8.58 \text{ Hz}$$

EXAMPLE 6.5. The four-cycle engine of Example 6.1 is found by harmonic analysis of test data to have the following Fourier pressure torque components:

n	*A*	*B*	*n*	*A*	*B*
1	475 in.lb	150 in.lb	4	185 in.lb	−40 in.lb
2	450	−20	5	115	−45
3	255	−45	6	50	−30

Find the resultant exciting torque amplitudes due to the combination of pressure and inertia effects at 3600 RPM.

Solution. Using inertia torque from Example 6.1 and Equation 6.13,

$$T_{1/2} = \sqrt{A_{1/2}^2 + B_{1/2}^2} = \sqrt{(475)^2 + (150)^2} = 498 \text{ in. lb}$$

$$T_1 = \sqrt{(450 + 42.5)^2 + (-20)^2} = 493$$

$$T_{1\,1/2} = \sqrt{(255)^2 + (-45)^2} = 259$$

$$T_2 = \sqrt{(185 - 278)^2 + (-40)^2} = 101$$

$$T_{2\,1/2} = \sqrt{(115)^2 + (-45)^2} = 124$$

$$T_3 = \sqrt{(50 - 127)^2 + (-30)^2} = 83$$

7 VIBRATION ABSORBERS AND DAMPERS

Vibration suppression is often both necessary and desirable to avoid fatigue failure, excessive roughness, or noise. A family of vibratory devices can be applied to reduce the severity of a vibratory condition, usually at or near resonance. These include the *absorber*, which is an auxiliary tuned spring–mass system that in rotation becomes a centrifugal pendulum. The absorber alters the vibratory response with negligible damping, and is usually quite sensitive to tuning. Also, spring–mass–dashpot or just mass–dashpot combinations can be employed. These are termed *dynamic dampers* or simply *dampers*, respectively, and they introduce beneficial inertia effects and increase substantially the equivalent system damping. Dampers and absorbers as a group characteristically are employed to reduce resonant peaks or to shift resonant peaks from an operating condition. There is in all cases an element of tuning involved that tends to magnify the effectiveness of a given auxiliary mass.

7.1. FRAHM VIBRATION ABSORBER

In the simplest form a tuned spring–mass system is a *Frahm vibration absorber* when superimposed on a forced main vibratory system (Table 7.1*a*). The main system is reference, with the original main system natural frequency $\omega_1 = \sqrt{K_1/M_1}$ the reference frequency, and the exciting force on M_1. The auxiliary system is K_0 and M_0, and there is no significant damping in either of the component systems.

This arrangement is similar to that of Table 5.1*a*, with P_2 excitation, except that the notation has been modified to be more meaningful for the interpretation of absorber behavior. A tuning factor f is introduced to relate the two independent natural frequencies, and is the ratio of the auxiliary to the main system, $\mathsf{f} = \omega_0/\omega_1$, where $\omega_0 = \sqrt{K_0/M_0}$. This factor represents the proximity of the absorber to main natural frequency. Also, the reference is now the

TABLE 7.1. Comparative response factors and optimum parameters for three basic forms of auxiliary systems.

	FRAHM ABSORBER	DYNAMIC DAMPER	VISCOUS DAMPER
	a: 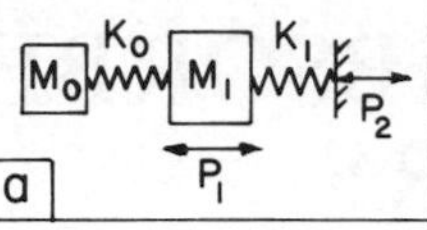 M_0, K_0, M_1, K_1, P_2, P_1	b: M_0, K_0, C_0, M_1, K_1, P_2, P_1 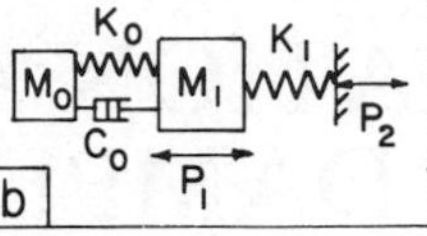	c: M_0, C_0, M_1, K_1, P_2, P_1 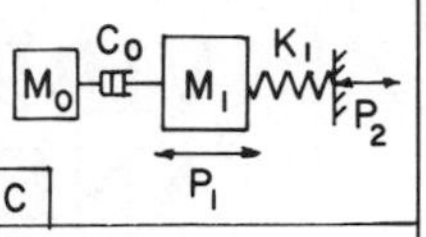
$\frac{a_1}{P_1/K_1}$	$\frac{A}{C}$	$\frac{A+jB}{C+jD}$	$\frac{-A+jB}{C+jD}$
$\frac{a_0}{P_1/K_1}$	$\frac{f^2}{C}$	$\frac{f^2+jB}{C+jD}$	$\frac{jB}{C+jD}$
$\frac{a_{01}}{P_1/K_1}$	$\frac{g^2}{C}$	$\frac{g^2}{C+jD}$	$\frac{A}{C+jD}$
$\frac{P_2}{P_1}$	$\frac{-A}{C}$	$\frac{-(A+jB)}{C+jD}$	$\frac{A-jB}{C+jD}$
C	$g^4-(1+f^2[1+m_0])g^2+f^2$		$m_0 g^2(g^2-1)$
A	f^2-g^2		$m_0 g^2$
B	0	$2\zeta_0 f g$	$2\zeta_0 g$
D	0	$B(1-g^2[1+m_0])$	
ζ_0	0	$\frac{C_0}{2M_0\omega_0}$	$\frac{C_0}{2M_1\omega_1}$
OPTIMUM $\frac{a_1}{P_1/K_1}$	0	$\sqrt{1+\frac{2}{m_0}}$	$1+\frac{2}{m_0}$
OPTIMUM f	g	$\frac{1}{1+m_0}$	—
OPTIMUM ζ_0	0	$\sqrt{\frac{3}{8}\left(\frac{m_0}{1+m_0}\right)}$	$\frac{m_0}{\sqrt{2(1+m_0)(2+m_0)}}$

$f^2=\left(\frac{\omega_0}{\omega_1}\right)^2=\frac{k_0}{m_0}$, $\quad k_0=\frac{K_0}{K_1}$, $\quad m_0=\frac{M_0}{M_1}$, $\quad \omega_1=\sqrt{\frac{K_1}{M_1}}$, $\quad g=\frac{\omega}{\omega_1}$

second spring–mass system, left to right, rather than the first, as has been the previous practice.

7.2. FRAHM EQUATIONS

Considering the Table 7.1*a* system, and writing equations for the auxiliary and total systems in steady-state sinusoidal equilibrium,

$$\begin{cases} M_0\omega^2 a_0 + K_0(a_1 - a_0) = 0 & (7.1a) \\ M_0\omega^2 a_0 + M_1\omega^2 a_1 - K_1 a_1 + P_1 = 0 & (7.1b) \end{cases}$$

Dividing Equation 7.1a by K_0 and 7.1b by K_1,

$$\begin{cases} a_0\left[\left(\dfrac{\omega}{\omega_0}\right)^2 - 1\right] + a_1 = 0 & (7.2a) \\ -a_0(m_0 g^2) + a_1(1 - g^2) = \left(\dfrac{P_1}{K_1}\right) & (7.2b) \end{cases}$$

Equation 7.2a can be further reduced to

$$a_0(g^2 - f^2) + f^2 a_1 = 0 \tag{7.3}$$

where $f = \omega_0/\omega_1 = \sqrt{K_0 M_1 / M_0 K_1} = \sqrt{k_0/m_0}$ = ratio of absorber to main natural frequency

$g = \omega/\omega_1$ = ratio of forcing to main-system natural frequency

$m_0 = M_0/M_1$ = mass ratio

Simultaneous solution of Equations 7.2b and 7.3 yields the response results in Table 7.1a.

7.3. FRAHM RESPONSE

The two-degree system has two natural frequencies resulting from the coupling of the two single-degree systems, obtained by setting $C = 0$. For exact tuning, $f = 1$, and

$$g_n^4 - (2 + m_0)g_n^2 + 1 = 0 \tag{7.4}$$

Main-system amplitude becomes zero if $f^2 = g^2$, or

$$f = g = \left(\frac{\omega_0}{\omega_1}\right) = \left(\frac{\omega}{\omega_1}\right) \text{ or } \omega_0 = \omega$$

That is, when the exciting frequency at M_1 becomes equal to the auxiliary system natural frequency, M_1 becomes a node, representing the optimum reduction. This is the ideal operational condition, resulting in a stationary M_1, no stress in K_1, and no vibratory force transmitted to ground, P_2. In Figure 7.1 at $\mathsf{g} = 1$, the null condition, the amplitude a_1 changes from out of phase to in phase with the exciting force.

In this example there is exact tuning, $m_0 = 0.20$, and the auxiliary amplitude at $\mathsf{g} = 1$ has a magnification factor of -5. Auxiliary amplitude reaches a minimum between the two resonances, and is out of phase with P_1. Although this example is for exact tuning, nodal operation will occur whenever $\omega = \omega_0 = \sqrt{K_0/M_0}$, or whenever $\mathsf{f} = \mathsf{g}$. Thus the absorber can be tuned *to any objectionable exciting frequency, independent of any consideration of main-system natural frequency*; however, typically the problem is with respect to a main-system resonance.

At nodal conditions P_1 is exactly balanced by the inertia effect of the absorber, with $\mathsf{a}_1 = 0$ in Equation 7.2b.

$$\frac{\mathsf{a}_0}{P_1/K_1} = -\frac{1}{m_0\mathsf{g}^2} = -\frac{1}{m_0\mathsf{f}^2} \tag{7.5}$$

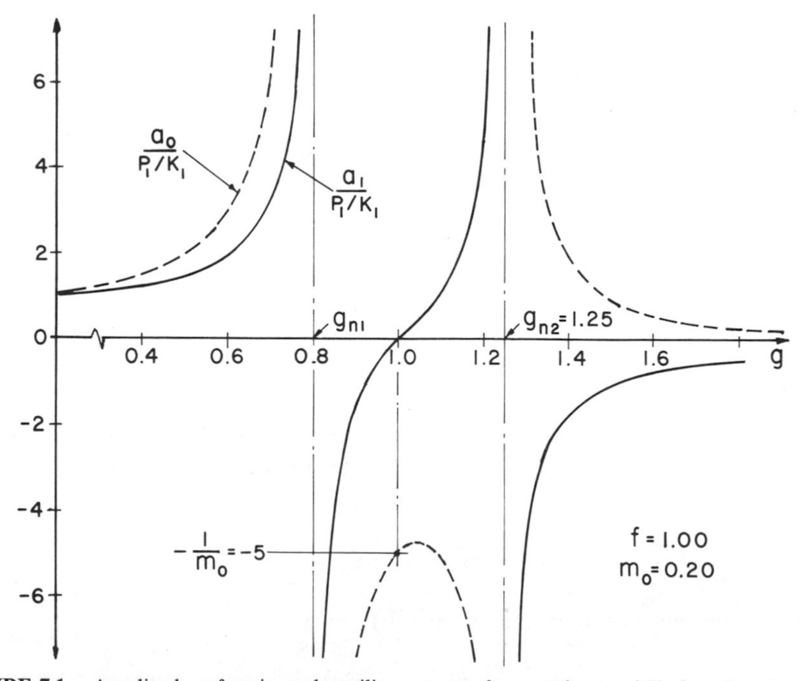

FIGURE 7.1. Amplitudes of main and auxiliary masses for exactly tuned Frahm absorber with main mass subjected to sinusoidally forced excitation.

For equilibrium at $a_1 = 0$,

$$P_1 = -M_0\omega^2 a_0 \tag{7.6}$$

The absorber action therefore provides a cancelation effect. By balancing the exciting force the main mass is motionless because it lacks an excitation. The Frahm absorber literally provides a counterexcitation.

7.4. MASS RATIO EFFECTS

Although a small Frahm absorber can theoretically suppress a large system and a large excitation, absorber amplitude must increase as the absorber mass decreases (Equation 7.6). Additionally, tuning becomes more critical at small values of m_0, so it is usually desirable to separate the two system resonances as much as possible. This peak spread is shown in Figure 7.2, with exact tuning assumed. Resonances are roughly equally spaced above and below the original frequency, $g = 1$.

From a design standpoint the absorber spring must be capable of carrying the fatigue load P_1, and must also be capable of the required relative deflection

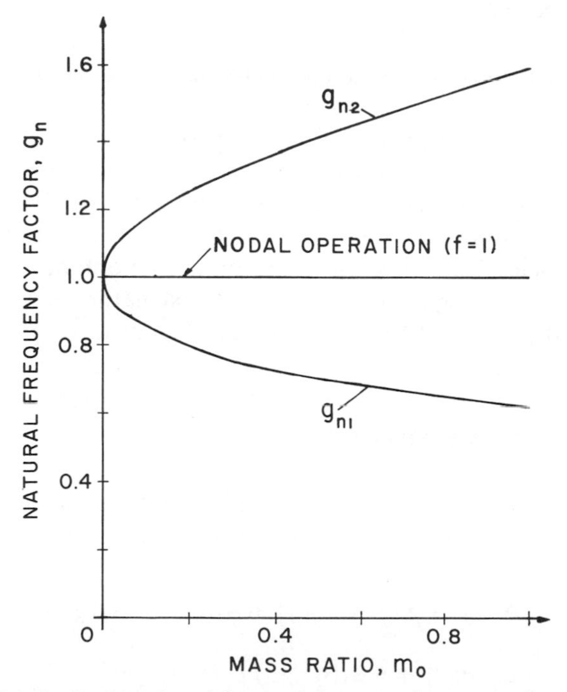

FIGURE 7.2. The distribution of modal natural frequency factors as a function of mass ratio shows nearly symmetrical disposition.

a_{01}. The spring force is equal to P_1 at nodal operation, and the deflection a_{01} of K_0 is inversely proportional to M_0. Also, considering ω^2 in Equation 7.6, the Frahm mass tends to be small and feasible at high operational frequencies.

7.5. PRACTICAL CONSIDERATIONS

The simple Frahm vibration absorber is uniquely successful in achieving a main system null at a particular forcing frequency; however, if this frequency varies the situation can be hazardous. The absorber–main combination has two resonant frequencies, where originally there was only one (Figure 7.1). Also in starting and stopping we must always pass through the lower resonance. Amplitudes of both main and absorber masses can increase dramatically if the steady-state forcing frequency deviates slightly from the null condition.

Thus the Frahm absorber has applications limited to those cases in which the operational frequency is substantially constant, for instance in alternating current devices governed by line frequency. The topic is presented to illustrate the fundamental nature of most types of auxiliary suppression devices.

7.6. THE DYNAMIC DAMPER

The Frahm restrictions can be largely overcome by the deliberate introduction of relative damping between the main and auxiliary masses, specifically by the dashpot C_0 (Table 7.1*b*). Natural frequency solutions are identical with the Frahm, but substantial damping effects alter the suppression mechanism. Whereas the Frahm is restricted to one frequency, the dynamic damper is adaptable to a significant response reduction over the complete range of exciting frequencies. The reduction behavior can be optimized with respect to both tuning and damping parameters.[3] In fact, the nature of the response curve can be drastically modified by damper design (Figure 7.3).

7.7. EQUATIONS OF MOTION

The analysis relates to Equations 7.1, 7.2, and 7.3, with a factor f again used as the tuning parameter. Normalized equations with damping included are

$$\begin{cases} \mathsf{a}_0(\mathsf{g}^2 - \mathsf{f}^2 - j2\zeta_0\mathsf{fg}) + \mathsf{a}_1(\mathsf{f}^2 + j2\zeta_0\mathsf{fg}) = 0 \\ \mathsf{a}_0(m_0\mathsf{g}^2) + \mathsf{a}_1(\mathsf{g}^2 - 1) = -\left(\dfrac{P_1}{K_1}\right) \end{cases} \tag{7.7}$$

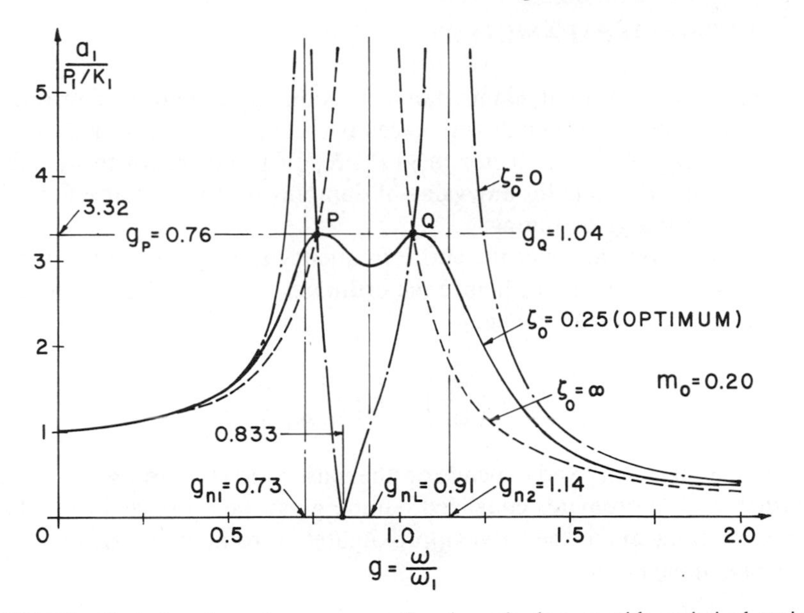

FIGURE 7.3. Typical main system responses for dynamic damper with optimized tuning at various damping ratios, and ordinates at *P* and *Q* equalized.

Finalized expressions for displacements of the masses, relative motion, and net reaction to ground are indicated in Table 7.1*b*. They permit calculation of responses for any combination of mass ratio, tuning, and damping. It should be noted that ζ_0 represents the true damping ratio of the isolated auxiliary system.

In addition to the system natural frequencies, the concept of the single natural frequency resulting from physically locking the auxiliary to the main mass is of interest for test purposes. The locked frequency can be compared with the original frequency for the determination of effective mass ratio.

$$\omega_{nL} = \sqrt{\frac{K_1}{M_0 + M_1}} \tag{7.8}$$

$$\left(\frac{\omega_{nL}}{\omega_1}\right)^2 = g_{nL}^2 = \left(\frac{1}{1 + m_0}\right) \tag{7.9}$$

The larger the mass ratio, the greater the reduction in locked frequency with respect to the original frequency.

7.8. OPTIMIZED PARAMETERS

In Figure 7.3 we see the dynamic damper producing a main-system response with two finite peaks, although with excessive damping this can degenerate to one central peak. Also, the factor $[a_1/(P_1/K_1)]$ can be shown to intersect at two fixed points P and Q for any value of damping, in this instance for $\zeta_0 = 0$, 0.25, and ∞, for a given tuning.

If we wish to operate over the entire frequency range and to limit the peak response, the first step is to equalize the ordinates at the fixed points P and Q. This requires an *optimum tuning*,

$$f_0 = \left(\frac{\omega_0}{\omega_1}\right) = \left(\frac{1}{1 + m_0}\right) \tag{7.10}$$

which is less than unity and depends on the mass ratio (Figure 7.4).

With P and Q ordinates equalized damping can be optimized to make the slope of the tangents to the curve approximately zero at P and Q, resulting in an *optimum damping ratio*,

$$\zeta_0 = \sqrt{\frac{3}{8}\left(\frac{m_0}{1 + m_0}\right)} \tag{7.11}$$

also a function of m_0 only (Figure 7.4).

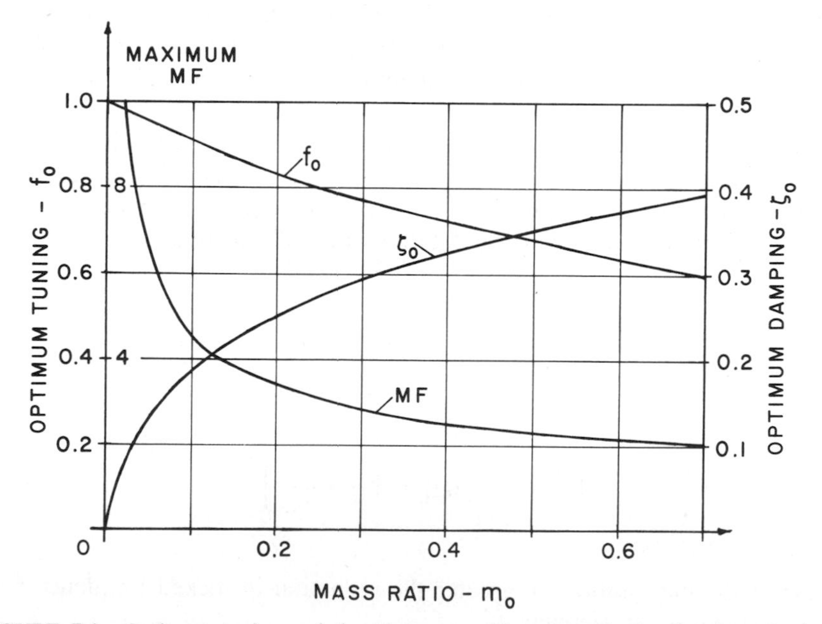

FIGURE 7.4. Optimum tuning and damping vary with mass ratio for the dynamic damper. Minimized magnification factors are also indicated.

Given optimum tuning and damping, the optimized response over the entire range is limited to two maxima at P and Q.

$$\left(\frac{a_1}{P_1/K_1}\right)_{P,Q} = \left(1 + \frac{2}{m_0}\right) \tag{7.12}$$

As seen in Figure 7.4, the limiting factor drops rather rapidly with increasing m_0, but then levels off.

7.9. TUNING EFFECTS

With optimum tuning the P and Q points straddle the locked peak (Figure 7.5). As with the Frahm (Figure 7.2), the larger mass ratios produce a larger spread between peaks; however, with the dynamic damper there is a decreasing trend in all frequencies. The equation for these P and Q frequencies is given by the quadratic

$$(2 + m_0)g_{P,Q}^4 - 2[1 + f^2(1 + m_0)]g_{P,Q}^2 + 2f^2 = 0 \tag{7.13}$$

Optimum tuning may not be obtainable,[5] or we might deliberately skew the peaks. For instance, with undertuning a lower magnification factor will occur at P and a higher at Q than with optimum. This effect is indicated in Figure 7.6. For $m_0 = 0.20$ we have equal factors of 3.32 at optimum tuning, or

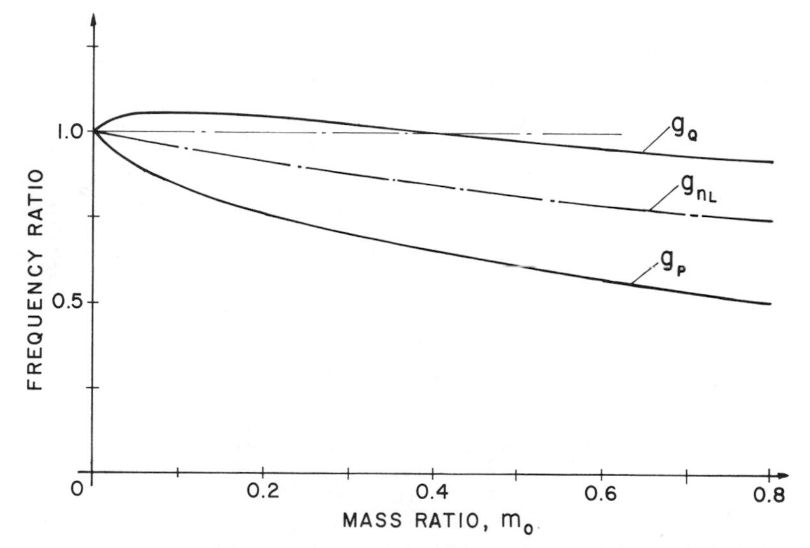

FIGURE 7.5. Decreasing frequency factors at P and Q are shown relative to the locked resonant frequency.

$(f/f_0) = 1$, but as the damper natural frequency is decreased, $(MF)_P$ drops rapidly to approximately one-half at 70% of the optimum; however, the peak at Q doubles. In this situation we would obviously plan to operate well below the Q peak.

7.10. DAMPING EFFECTS

Optimum damping with optimum tuning results in a slight dip in the response curve between P and Q (Figure 7.3). Greater damping will cause a peak between P and Q greater than at these points. Insufficient damping results in two higher peaks lying symmetrically outside the P–Q interval. These characteristics provide a guide for experimental damping correction.

As indicated in Section 7.9, $(MF)_P$ and $(MF)_Q$ will be unequal for tuning other than optimum, and the extent of this skewing is shown in Figure 7.6 for $m_0 = 0.20$. This effect is advantageous if we expect operational frequency ranges usually below or above $g = 1$. If below, we should undertune to involve the lesser magnification factors associated with P, and vice versa. With this type of intentional detuning we can further reduce the maximum main system response at P or Q by *increasing* the damping beyond optimum, shown by the dashed curve in Figure 7.6. If so, we will insure that the slope of the response

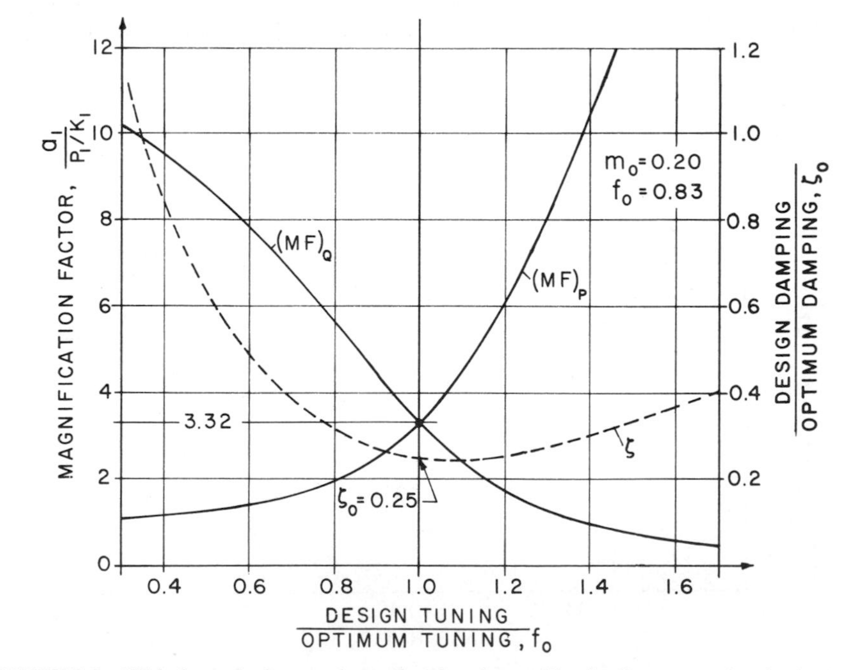

FIGURE 7.6. With dynamic damper detuned, reduced magnification factors can be obtained at P or Q. The damping ratio must be increased as shown by the dashed curve.

curve at *P or Q* will be zero at *P or Q*, and the ordinate at *P or Q* will also be the maximum response of M_1. In effect we have then optimized damping with respect to the lower of the two peaks, rather than with respect to both *P* and *Q* peaks.

As the dynamic damper has relatively heavy damping, the dissipation rate may be of importance with respect to heat continuously generated, and it is given by

$$E_d = \pi\left(C_0\omega \mathsf{a}_{01}^2\right)\frac{\omega}{2\pi} \tag{7.14}$$

Using the relative displacement equation from Table 7.1*b* we can obtain the normalized form

$$\frac{E_d}{\left(P_1^2\omega_1/K_1\right)} = \left(\frac{m_0\zeta_0 \mathsf{f}\mathsf{g}^6}{C^2 + D^2}\right) \tag{7.15}$$

The energy rate, or power level, becomes maximum in the vicinity of *Q* with optimum conditions.

7.11. DYNAMIC DAMPER CONSTRUCTION

Although the analysis has been illustrated by classical translational systems, torsional applications are probably more common. The only difference is in the use of angular units in the equations; for instance, radius of gyration squared entering the mass terms.

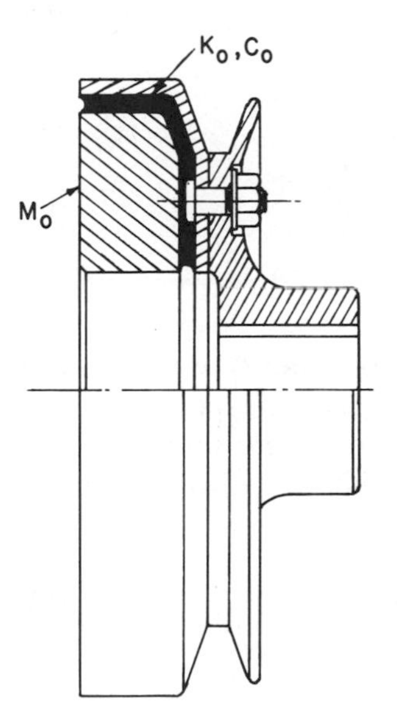

FIGURE 7.7. Typical automotive engine torsional dynamic damper uses bonded rubber to provide both spring and dashpot coupling.

Spring and dashpot elements are often combined into a single element, namely a highly damped elastomer material that mounts the damper mass to the main mass. Both the rubber geometry and compounding collectively determine the effective stiffness and damping of the coupling. This arrangement is shown in Figure 7.7 as an automotive engine crankshaft damper for torsional vibratory control. The outer shell connects rigidly to the main crankshaft with the annular auxiliary mass suspended internally by means of bonded rubber, subjected to vibratory shear.

7.12. THE VISCOUS DAMPER

If the spring coupling element of the Frahm absorber is replaced by a dashpot element, a quite different behavior results.[13] The viscous damper of this type operates to dissipate vibratory energy, and although less effective than related devices for a given size, has the advantage of utilizing a simple fluid surface in shear. Thus the moving element is not a potential source of wear or fatigue failure. The auxiliary mass–dashpot combination is shown in Table 2.5*c*; however, as we now have a main system with a spring, a critical damping can now be defined using K_1 and M_1 of the main system in conjunction with the auxiliary dashpot for a reference damping ratio.

Simultaneous equations for the auxiliary and complete systems (Table 7.1*c*) are

$$\begin{cases} M_0\omega^2\mathsf{a}_0 + jC_0\omega(\mathsf{a}_1 - \mathsf{a}_0) = 0 & (7.16\text{a}) \\ M_0\omega^2\mathsf{a}_0 + M_1\omega^2\mathsf{a}_1 - K\mathsf{a}_1 + P_1 = 0 & (7.16\text{b}) \end{cases}$$

Normalized, the equations become,

$$\begin{cases} \mathsf{a}_0(m_0\mathsf{g}^2 - j2\zeta_0\mathsf{g}) + j2\zeta_0\mathsf{g}\mathsf{a}_1 = 0 & (7.17\text{a}) \\ -\mathsf{a}_0(m_0\mathsf{g}^2) + (1 - \mathsf{g}^2)\mathsf{a}_1 = \dfrac{P_1}{K_1} & (7.17\text{b}) \end{cases}$$

Solution results are given in Table 7.1*c*.

Whereas the dynamic damper has two fixed points P and Q at which there is independence of damping, the viscous damper has one peak and one intersection point P (Figure 7.8). Optimization can only occur with respect to ζ_0, and this desired ratio is again obtained by maintaining a horizontal slope to the response curve tangent at P, for which

$$\zeta_0 = \frac{m_0}{\sqrt{2(1 + m_0)(2 + m_0)}} \qquad (7.18)$$

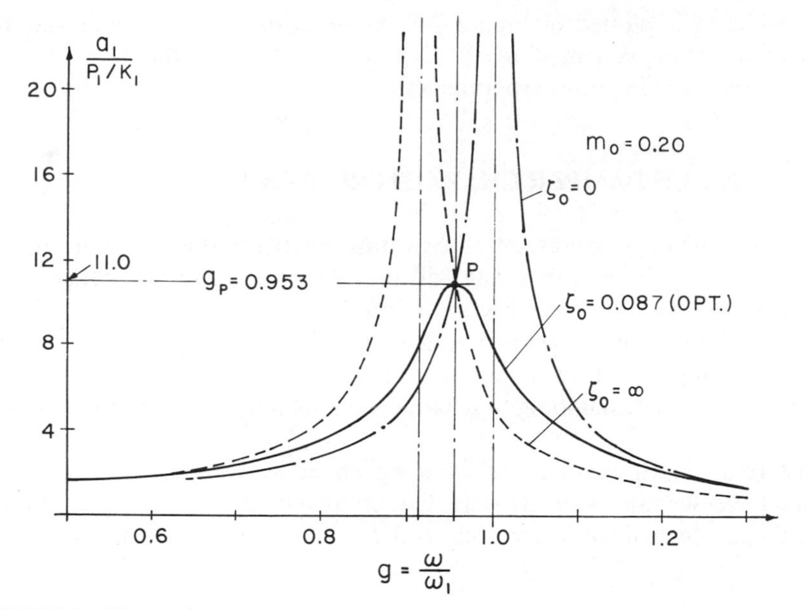

FIGURE 7.8. Viscous damper on a single-degree main system produces only one peak, optimized at *P*.

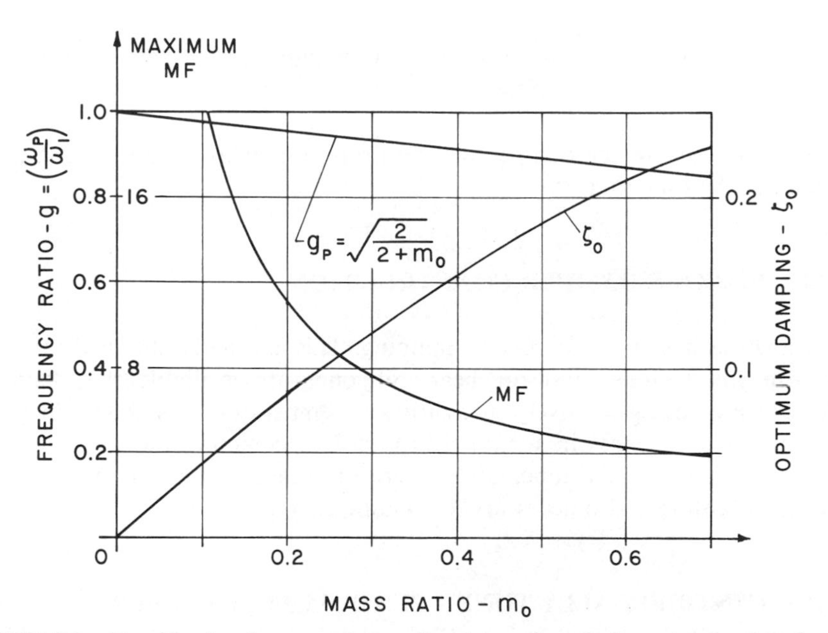

FIGURE 7.9. Magnification factor, optimum damping, and peak frequency for the viscous damper as a function of mass ratio.

This variation is plotted in Figure 7.9. As defined, ζ_0 is the actual damping ratio of the main system if m_0 is fixed to ground, and this is a means of studying the damping ratio experimentally.

7.13. VISCOUS DAMPER CHARACTERISTICS

The viscous damper operates by permitting relative motion of the auxiliary mass. Obviously, infinite damping will lock M_0 to M_1 and no energy will be dissipated, so the response becomes infinite at g_{nL}. Similarly, with zero damping relative motion occurs, but no energy is involved with a zero dashpot. Optimum damping is then a compromise in which a usable motion combines with an intermediate damping to maximize dissipation and minimize main-system response.

The result is an exaggerated damping effect with energy dissipated at a constant average rate in steady state. Quantitatively the power level is obtained from the product of work per cycle and frequency, as in Equations 7.14 and 7.15.

$$E_d = \pi(C_0\omega \mathsf{a}_{01})\mathsf{a}_{01}\left(\frac{\omega}{2\pi}\right) \tag{7.19}$$

With equations from Table 7.1*c* the dimensionless rate is

$$\frac{E_d}{\left(P_1^2\omega_1/K_1\right)} = \left(\frac{\zeta_0 m_0^2 \mathsf{g}^6}{C^2 + D^2}\right) \tag{7.20}$$

Heat generation in the damper will be proportional to E_d, increasing as the square of the excitation P_1.

7.14. VISCOUS DAMPER CONSTRUCTION

The analysis has assumed viscous damping. This is usually obtained by means of a highly viscous fluid in shear. Silicones are available that have the advantage of being relatively insensitive to temperature. As shown in Figure 7.10 the damper mass in torsion has annular geometry, and control of the clearances between the inner and outer surfaces is critical to the damping. Hermetic sealing is also necessary to contain the fluid.

7.15. CENTRIFUGALLY TUNED PENDULUM ABSORBER

The Frahm absorber can be adapted to rotating systems with torsional modes so that it does not suffer from the constant exciting frequency limitation

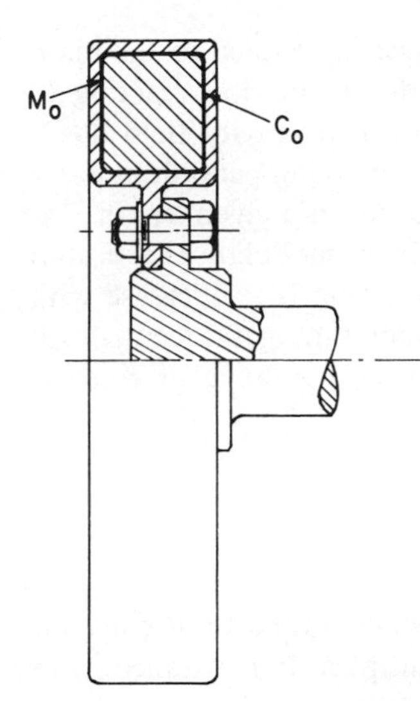

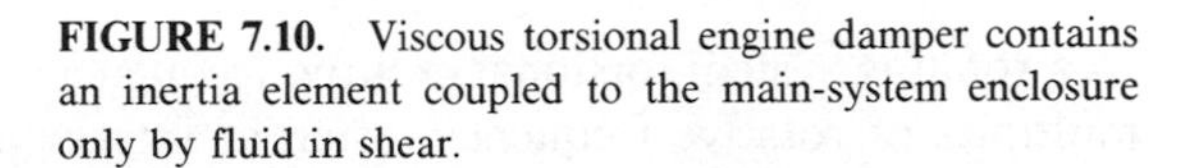

FIGURE 7.10. Viscous torsional engine damper contains an inertia element coupled to the main-system enclosure only by fluid in shear.

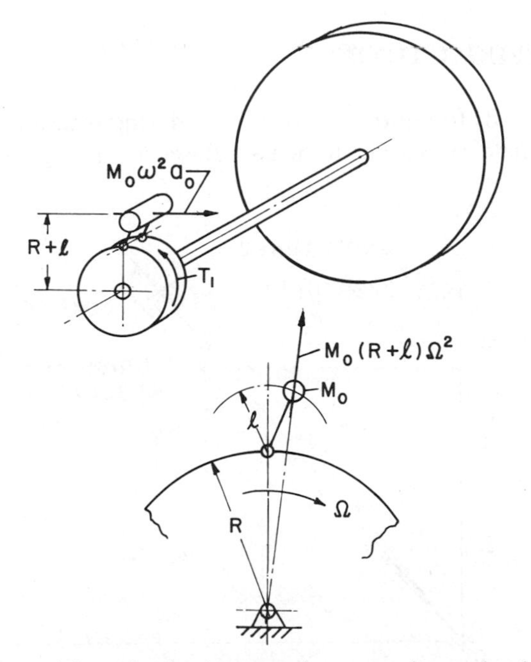

FIGURE 7.11. The pendulum absorber is mounted on a rotating torsional system to suppress torque excitation by means of a tangential inertia force component.

previously indicated. In Figure 7.11 the auxiliary mass is coupled to the main mass by a pendular suspension, with angular pendulum motion reacting to oppose the torsional excitation. A simple pendulum in a gravity field has a constant frequency, depending on the length, but a revolving pendulum has a natural frequency directly proportional to rotative speed for a given length.

With this unique tuning characteristic we effectively modulate the Frahm frequency with speed. Thus if the nature of the excitation is to increase with speed, we have a matching situation. Given an exact tuning the centrifugal pendulum will provide a Frahm-type null over the entire range of rotative speed.

7.16. ORDER EXCITATION

In a rotating system torsional exciting frequencies usually do occur at constant multiples of rotative frequency, often at integer multiples. For instance, with third-*order* excitation $N = 3$, $\omega = N\Omega = 3\Omega$, where ω is the exciting frequency and Ω is the rotative frequency in the same units, as in Chapter 6.

As shown in Figure 7.12, N represents the slope of the excitation frequency and the exciting frequency is the product of the order and the speed. If the torsional system has a natural frequency ω_1, an order will produce resonance at the rotational speed determined by the intersection, $\omega_1 = N\Omega$.

7.17. CENTRIFUGAL TUNING

The pendulum tends to return to midposition during rotation at Ω by virtue of the centrifugal field acting on the mass.[6] Based on this geometry, and using

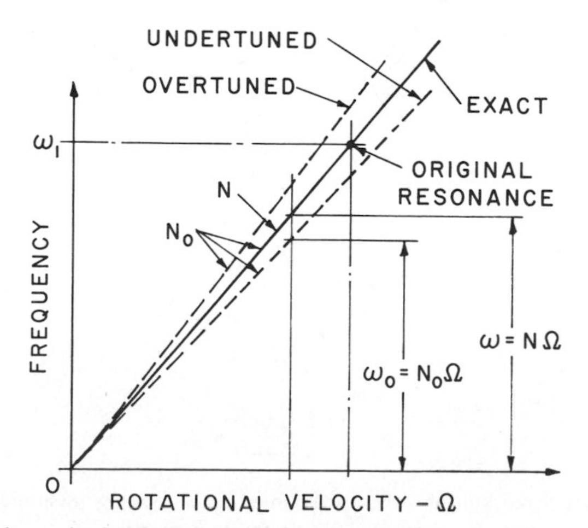

FIGURE 7.12. Order excitation frequency produces resonance at the intersection with the natural frequency. Pendulum tuning order N_0 can be greater or less than N.

force equilibrium, the equivalent tangential restoring spring at M_0 is

$$K_c = \frac{M_0 R \Omega^2}{l} \tag{7.21}$$

The natural frequency of the pendulum is then

$$\omega_0 = \sqrt{\frac{K_c}{M_0}} = \Omega\sqrt{\frac{R}{l}}$$

$$N_0 = \frac{\omega_0}{\Omega} = \sqrt{\frac{R}{l}} \tag{7.22}$$

where N_0 is the tuning order of the absorber (Figure 7.12). N_0 is not necessarily coincident with N, although in most cases this is the intention. Deviations can be due to manufacturing tolerances, or there can be intentional detuning. If $N_0 > N$ the pendulum is shorter than nominal and it is *overtuned*. Conversely if the length is greater than nominal the pendulum is *undertuned*. Whatever the geometric condition, the ratio (ω_0/Ω) remains constant at any speed.

In practice amplitudes are not usually small, and Equation 7.22 will not apply exactly. The greater the amplitude, the greater the deviation, and the greater the tendency to nonlinear behavior. Actual pendulum frequency is always slightly less than calculated, and compensation requires a pendulum that is 5% or more shorter than the nominal, as an order of magnitude.

7.18. EQUIVALENT TANGENTIAL SYSTEM

Although the pendulum is usually treated in angular coordinates with an inertia torque caused by pendulum displacement, we will convert the torsional to an equivalent translational system referred to the rotational radius of the center of gravity of the pendulum. This permits us to relate the pendulum to the Frahm equations, and we have another two-degree system similar to those in Table 7.1.

The conversion applies to the main system, which is referred to as follows (see Table 7.2):

1. M_1 has an equivalent mass moment of inertia about the main axis of rotation, or

$$M_1 = \frac{J_1}{(R + l)^2}$$

2. To maintain the same main-system natural frequency, $K_1 = M_1\omega_1^2$.

TABLE 7.2. Response equations for centrifugally tuned absorber apply to equivalent two-degree translational system.

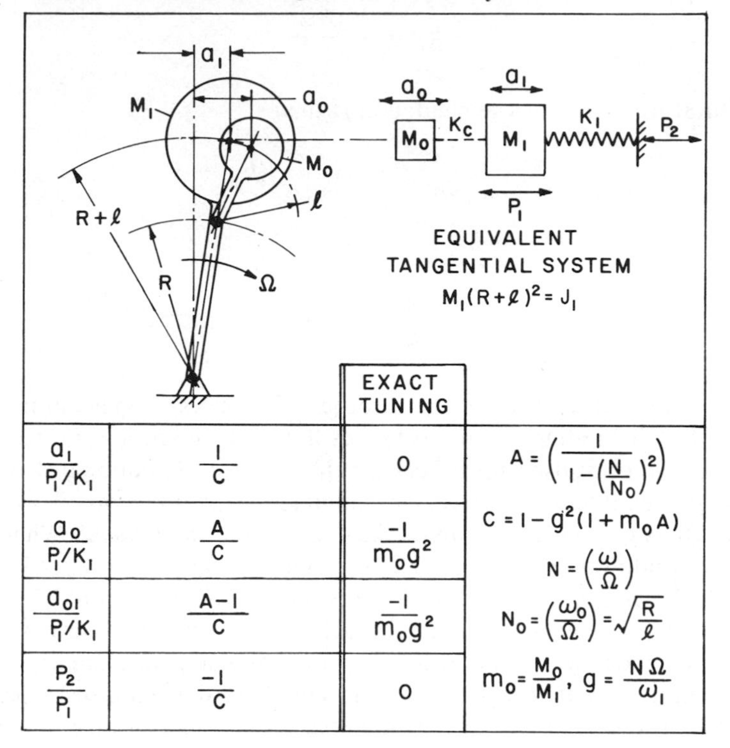

		EXACT TUNING	
$\frac{a_1}{P_1/K_1}$	$\frac{1}{C}$	0	$A = \left(\frac{1}{1-\left(\frac{N}{N_0}\right)^2}\right)$
$\frac{a_0}{P_1/K_1}$	$\frac{A}{C}$	$\frac{-1}{m_0 g^2}$	$C = 1 - g^2(1 + m_0 A)$ $N = \left(\frac{\omega}{\Omega}\right)$
$\frac{a_{01}}{P_1/K_1}$	$\frac{A-1}{C}$	$\frac{-1}{m_0 g^2}$	$N_0 = \left(\frac{\omega_0}{\Omega}\right) = \sqrt{\frac{R}{\ell}}$
$\frac{P_2}{P_1}$	$\frac{-1}{C}$	0	$m_0 = \frac{M_0}{M_1},\ g = \frac{N\Omega}{\omega_1}$

3. P_1 must be obtained from the torque excitation, or

$$P_1 = \frac{T_1}{(R + l)}$$

Vibratory inertia forces at M_0 and M_1 as well as P_1 and the spring force $K_1 a_1$ are now all tangential and colinear.

7.19. SYSTEM RESPONSE

Frahm equations are now applicable modified at K_0 to K_c.

$$f = \sqrt{\frac{R}{l}}\left(\frac{\Omega}{\omega_1}\right) = \left(\frac{N_0\Omega}{\omega_1}\right) \quad \text{and} \quad g = \left(\frac{N\Omega}{\omega_1}\right)$$

With these substitutions the pendulum-governed response reduces to

$$\frac{\mathsf{a}_1}{(P_1/K_1)} = \left\{ \frac{1}{1 - \mathsf{g}^2 \left[1 + \dfrac{m_0}{1 - (N/N_0)^2} \right]} \right\} \tag{7.23}$$

which we interpret as a modified single-degree response, with a factor applied to g^2, and only one resonant peak possible regardless of tuning. The ratio (N/N_0) is the reciprocal of pendulum tuning for a given exciting order.

It is difficult to visualize the variation of $\mathsf{a}_1/(P_1/K_1)$ from Equation 7.23, but this is demonstrated by numerical examples in Figure 7.13 assuming $m_0 = 0.10$. For a low tuning of $(N_0/N) = 0.90$, the relation becomes

$$\frac{\mathsf{a}_1}{(P_1/K_1)} = \left(\frac{1}{1 - 0.574\mathsf{g}^2} \right)$$

which corresponds to a single-degree resonance with the peak at $\mathsf{g}_n = 1.32$, or C as shown.

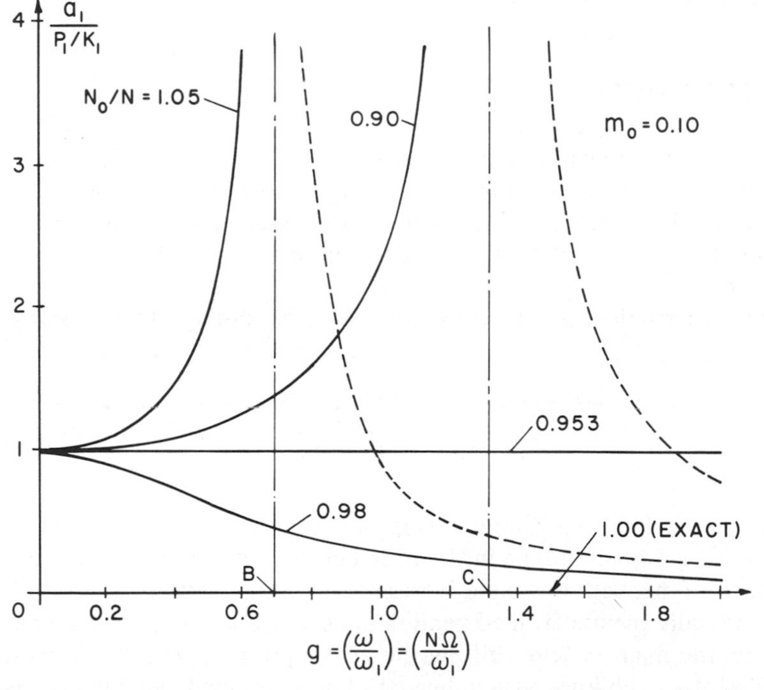

FIGURE 7.13. Main-system response is altered dramatically by the centrifugal pendulum, depending on the tuning ratio N_0/N.

At a higher tuning of $(N_0/N) = 1.05$,

$$\frac{a_1}{(P_1/K_1)} = \left(\frac{1}{1 - 2.075g^2}\right)$$

or a resonant system at $g_n = 0.694$, at B.

If $(N_0/N) = 0.953$, $[a_1/(P_1/K_1)] = 1$, and we have a constant magnification factor of unity over the entire speed range.

With tuning between 0.953 and 1.000, say 0.98,

$$\frac{a_1}{(P_1/K_1)} = \left(\frac{1}{1 + 2.43g^2}\right)$$

leading to the gradually decreasing factor shown in Figure 7.13.

Finally, at exact tuning, $(N_0/N) = 1$, and the main-system amplitude is zero, providing a Frahm null result not at one condition, but *over the complete speed range* with respect to a particular excitation order. By any standards this is a spectacular achievement.

We have seen that the response curve can be drastically manipulated by relatively small tuning changes, and this is a powerful option; however, similarly minor errors in tuning can produce vibratory results that are entirely different from those intended.

7.20. PEAK SHIFT

Equating the denominator of Equation 7.23 to zero we can find the location of the g_n factor as a function of m_0 and (N_0/N), providing a generalized concept of the peak shift phenomenon, (Figure 7.14). Resonance exists for any mass ratio for all overtuned cases, with the peak increased from the original resonant speed and frequency.

With undertuning, a resonance may or may not exist. The transitional tuning is

$$\frac{N_0}{N} = \sqrt{\frac{1}{1 + m_0}} \tag{7.24}$$

As indicated by the horizontal asymptotes in Figure 7.14, there will be no response peak for any tuning in the zone between the tuning of Equation 7.24 and exact tuning, with this band increasing with mass ratio.

Functionally the undertuned pendulum acts as a highly magnified auxiliary spring on the main system, shifting the peak upward in speed, except for the nonresonant possibilities just indicated. The overtuned pendulum acts as a highly magnified (resonant) auxiliary mass to shift the peak downward.

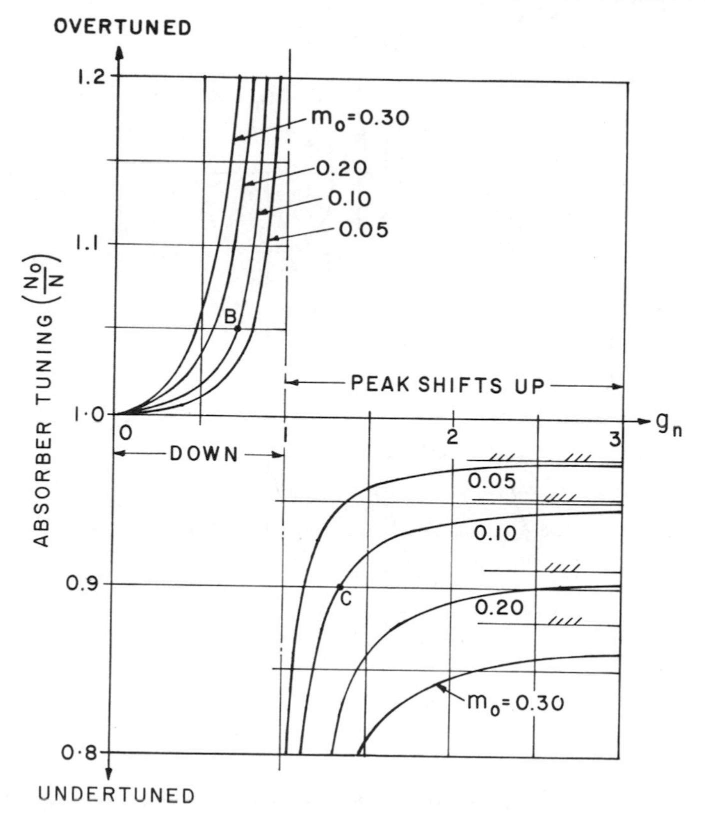

FIGURE 7.14. Resonant locations in Figure 7.13 are defined for mass ratio m_0 and tuning ratio N_0/N. None can occur for tuning between the hatched asymptotes and exact tuning.

7.21. BIFILAR SUSPENSION

One of the characteristic difficulties with the pendulum absorber is that the combination of geometry and exciting order usually leads to a pendulum length too short to be achieved with a simple pendulum configuration. One solution to this is the *bifilar* suspension (Figure 7.15), using two cylindrical pins interposed between two pairs of cylindrical holes. With this geometry all points in the suspended mass have circular motion, but the mass moves with translation. Thus a simple or mathematical pendulum is obtained with no angular inertia effects.

The pendulum radius is $l = (D_1 - D_2)$. By increasing the pin diameter relative to the holes, the length can be made to approach zero for high tuning orders.

As shown in Figure 7.15, the centroid of M_0 rotates about an apparent center D; therefore the radius R to be used in the tuning equation pertains to

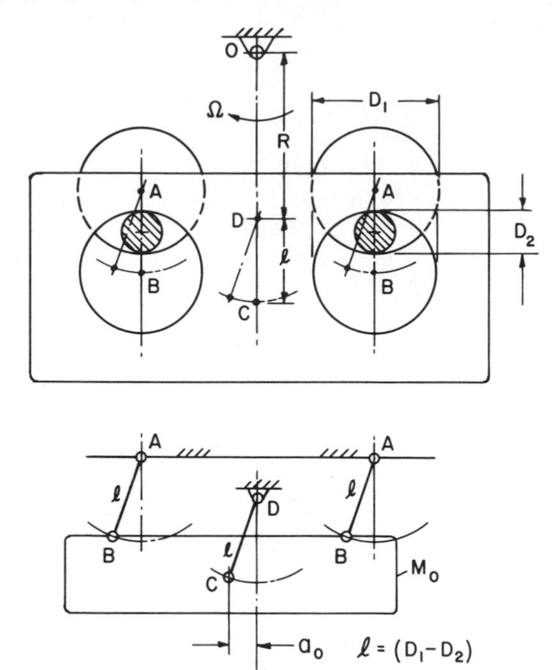

FIGURE 7.15. The bifilar suspension consists of a pair of rotating support holes A supporting a mass by cylindrical pins. An equivalent gravity bifilar pendulum is shown.

the distance from the center of rotation to this equivalent fulcrum. Note that this dimension is unrelated to the radial position of the holes and pins.

7.22. PUCK ABSORBER

A simpler arrangement for effecting small pendulum lengths is shown in Figure 7.16. The cylindrical mass or puck is confined in a cylindrical cavity, which is part of the main rotating vibratory system. The puck is held in the outward radial position by centrifugal effects. As a pendulum, B swings about A, and the length is the difference in the radii, or $l = (D_1 - D_2)/2$. Again, as the puck diameter approaches the hole diameter, the difference and the pendulum length approach zero.

Puck-style absorbers are rugged and simple, but involve a possible complication. If contact friction is sufficient, the puck will rotate as the centroid displaces. Such rotation introduces polar inertia effects and kinetic energy, reducing the pendulum natural frequency and subtracting from the useful inertia effect. Fortunately, in most cases with relatively high frequency and puck inertia, and with lubricated surfaces, puck motion will tend to sliding or skidding rather than rolling. Then simple mathematical pendulum action results. This problem is not present in the bifilar absorber, as the pins are constrained to roll.

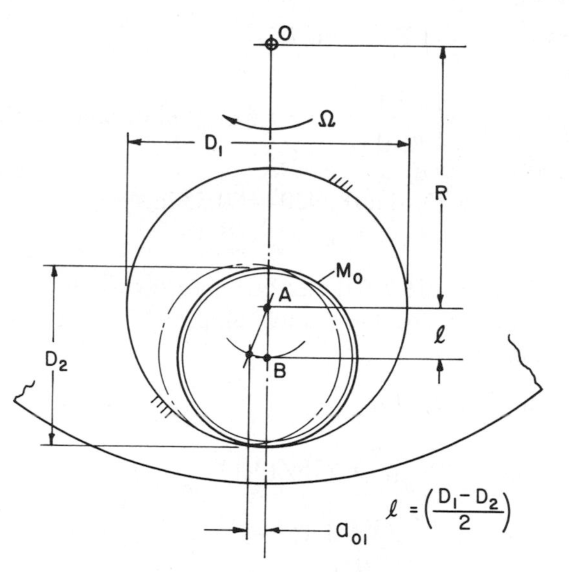

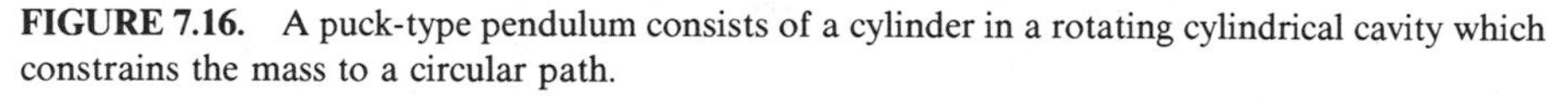
FIGURE 7.16. A puck-type pendulum consists of a cylinder in a rotating cylindrical cavity which constrains the mass to a circular path.

EXAMPLES

EXAMPLE 7.1. A Frahm absorber is applied at a mass ratio of 0.10 to suppress a main system completely, and

$$K_1 = 420 \text{ lb/in.}$$
$$M_1 = 24 \text{ lb} \qquad P_1 = 18 \text{ lb at } 15 \text{ Hz}$$

Determine the auxiliary spring and the auxiliary amplitude.

Solution. Using Table 7.1*a*,

$$\omega_1 = \sqrt{\frac{420(386)}{24}} = 82.2 \text{ r/s} = 13.1 \text{ Hz}$$

$$\mathsf{g} = 15/13.1 = 1.15 = \sqrt{k_0/0.10}$$

$$k_0 = 0.132 = K_0/K_1$$

$$K_0 = 0.132(420) = 55.2 \text{ lb/in.}$$

Check: $\omega_0 = \sqrt{\dfrac{55.2(386)}{2.4}} = 94.3 \text{ r/s} = 15 \text{ Hz}$

At nodal operation (From Equation 7.5),

$$\left(\frac{\mathsf{a}_0}{P_1/K_1}\right) = \left(\frac{-1}{m_0\mathsf{g}^2}\right) = -7.61 \quad \text{(out of phase with } P_1\text{)}$$

$$\mathsf{a}_0 = (7.61)(18/420) = 0.33 \text{ in.}$$

EXAMPLE 7.2. In Example 7.1 the frequency rises to 16 Hz. Find the amplitude of the main system and the relative amplitude of the absorber mass with respect to the main mass.

Solution. With Table 7.1*a*,

$$\mathsf{g}^2 = (16/13.1)^2 = 1.50$$

$$\mathsf{f}^2 = 1.15^2 = 1.32,$$

$$C = -0.108,$$

$$\left(\frac{\mathsf{a}_1}{P_1/K_1}\right) = \left(\frac{-0.18}{-0.108}\right) = 1.67$$

$$\mathsf{a}_1 = 1.67(18/420) = 0.07 \text{ in.}$$

$$\frac{\mathsf{a}_{01}}{(P_1/K_1)} = \left(\frac{1.50}{-0.108}\right) = -13.9,$$

$$\mathsf{a}_{01} = 0.60 \text{ in.}$$

EXAMPLE 7.3. The system of Example 7.1 is to be equipped with a dynamic damper instead of a Frahm. Mass ratio is also 0.10, and tuning and damping are optimum. Calculate the auxiliary spring and damping ratio.

Solution. From Table 7.1*b*,

$$\mathsf{f} = \left(\frac{1}{1.10}\right) = 0.91$$

$$\omega_0 = 0.91(82.2) = 74.7 \text{ r/s} = \sqrt{\frac{K_0 386}{2.4}}$$

$$K_0 = 34.7 \text{ lb/in.}$$

$$\zeta_0 = \sqrt{\left(\frac{3}{8}\right)\left(\frac{0.10}{1.10}\right)} = 0.19$$

EXAMPLE 7.4. The system of Example 7.1 is fitted with a viscous damper. Determine the maximum main-system amplitude and the rate of energy dissipation for optimum damping.

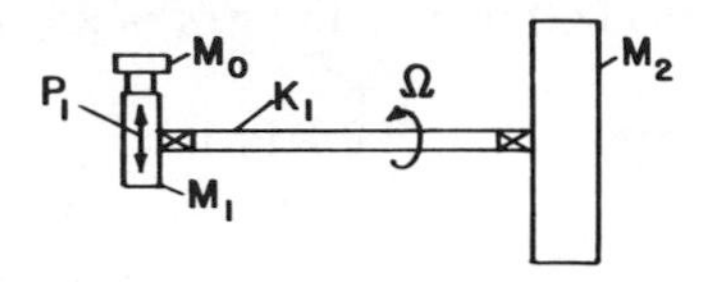

Solution. Using Table 7.1*c*,

$$\left(\frac{\mathsf{a}_1}{P_1/K_1}\right) = \left(1 + \frac{2}{0.10}\right) = 21$$

$$\mathsf{a}_1 = 21(18/420) = 0.90 \text{ in.}$$

From Equation 7.20,

$$\frac{E_d}{(P_1^2\omega_1/K_1)} = \left[\frac{0.046(0.01)(0.976)^6}{(0.0045)^2 + (0.0042)^2}\right] = 10.23$$

Note that this is the factor at $\mathsf{g}_P = \sqrt{2/(2 + m_0)}$, and is slightly less than the maximum rate, which occurs at a somewhat higher frequency than g_P.

$$E_d = \frac{(18)^2(82.2)}{420}(10.23) = 648 \text{ in. lb/sec}$$

$$= 54 \text{ ft lb/sec}$$

$$= \frac{54}{550} = 0.10 \text{ HP}$$

EXAMPLE 7.5. A bifilar pendulum acts on a rotational system with M_2 so much larger than M_1 that it is assumed to enforce a node.

$K_1 = 160{,}000$ N · m	$P_1 = 250$ N · m
$M_1 = 6.5$ kg · m^2	$N = 3$
$m_0 = 0.08$	$R + l = 26$ cm

Find:

(a) The resonant speed of the original rotating system.

(b) The length of the pendulum for exact tuning.

(c) The pendulum mass.
(d) Angular amplitude of the pendulum at the original resonant speed.

Solution. From Table 7.2,

(a)
$$\omega_1 = 160{,}000/6.5 = 157\ \text{r/s} = 25\ \text{Hz}$$
$$\Omega_3 = 25(60)/3 = 500\ \text{RPM}$$

(b)
$$N_0^2 = N^2 = 9 = R/l \quad \text{and} \quad (R + l) = 26$$

from which $l = 2.60$ cm.

(c) Referring the main system to the $(R + l)$ radius,

$$M_1 = \left[6.50/(0.26)^2\right] = 96.2\ \text{kg}$$
$$M_0 = 0.08(96.2) = 7.70\ \text{kg}$$

(d)
$$\frac{\mathsf{a}_{01}}{P_1/K_1} = \frac{-1}{(0.08)(1)^2} = -12.5$$
$$(P_1/K_1) = (250/160{,}000) = 0.00156\ \text{r}$$

The corresponding linear displacement at $(R + l)$,

$$\mathsf{a}_0 = 26(0.00156)(12.5) = 0.51\ \text{cm}$$
$$\theta_0 = (\mathsf{a}_0/l) = (0.51/2.6) = 0.19\ \text{r} = 11.2°$$

Note that the absolute and relative *tangential* amplitudes are identical, since the main system is nodal.

8 COMBINED LINEAR AND ANGULAR COORDINATES

Previous chapters have discussed only translational and torsional systems. In the former the inertia effect relates to scalar mass; in the latter the inertia effect is proportional to scalar mass moment of inertia about a fixed axis of rotation. Corresponding coordinates have been exclusively translational or torsional for the particular system analyzed. For the more general planar case, however, an elastically supported rigid mass can have simultaneous linear and angular vibratory motions, with system coordinates specified by two different transverse linear motions or a simultaneous linear and angular displacement. Small amplitudes are assumed with respect to a reference axis. Axial displacement is not considered, although it is a possible third degree of freedom.

8.1. THE COMPOUND PENDULUM

A rigid mass, pivoted to oscillate in a vertical plane, responds when displaced at a natural frequency (Figure 8.1). Although a single-degree case, this is a situation in which the body has both mass and moment of inertia effects, the latter distinguishing it from the simple pendulum.

In the position of maximum displacement shown there is a restoring moment about the pivot due to gravity analogous to a spring effect, an inertia force at the centroid related to translation, and an inertia couple related to mass rotation. As an equilibrium equation for the free vibration,

$$\Sigma T_0 = -lM\ddot{x} - Mr^2\ddot{\theta} - Wx = 0 \tag{8.1}$$

where x = horizontal displacement of centroid
M = total mass of pendulum
W = weight of pendulum = Mg_c

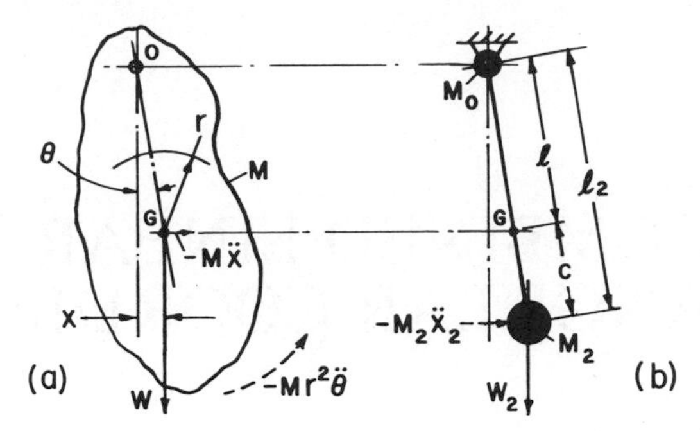

FIGURE 8.1. The compound pendulum involves an inertia couple as well as a force. It is equivalent to a simple pendulum of increased length (*b*).

r = radius of gyration about centroid
l = distance from fulcrum to centroid
$\theta = x/l$ for small angles
g_c = acceleration of gravity

Converting to complex form for steady-state oscillation, and with a representing horizontal motion of the centroid,

$$lM\omega_n^2\mathsf{a} + Mr^2\omega_n^2\frac{\mathsf{a}}{l} - W\mathsf{a} = 0$$

$$M\omega_n^2\left[1 + \left(\frac{r}{l}\right)^2\right] = \frac{W}{l}$$

$$\omega_n = \sqrt{\left[\frac{1}{1 + (r/l)^2}\right]\left(\frac{g_c}{l}\right)} = \mathsf{g}_n\omega_1$$

$$= \sqrt{\frac{1}{\left[1 + (r/l)^2\right]}}\ \omega_1 \tag{8.2}$$

Equation 8.2 gives the natural frequency of the compound pendulum, with $\omega_1 = \sqrt{g_c/l}$ the reference natural frequency of the simple pendulum for which $r = 0$ and the mass is concentrated at the centroid G. The larger the radius of gyration the lower the natural frequency factor g_n for the compound pendulum (Figure 8.2). Natural frequency is independent of the mass and the amplitude, and depends only on the r and l dimensions.

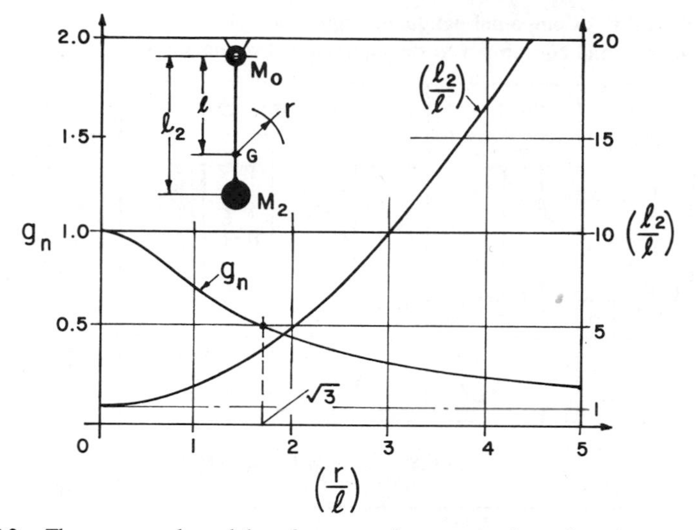

FIGURE 8.2. The compound pendulum frequency decreases as the radius of gyration increases for a given centroidal distance, corresponding to an increase in length of the equivalent simple pendulum, l_2.

8.2. EQUIVALENT TWO-MASS PENDULUM

As shown in Figure 8.1*b*, the distributed rigid mass can be replaced by a dynamically equivalent system consisting of two concentrated masses connected by a stiff, weightless bar for analytical purposes. By locating M_0 at the pivot, an equivalent simple pendulum results. Equations for the two-mass system that must be satisfied are

$$\begin{cases} M_0 + M_2 = M & (8.3a) \\ M_0 l = M_2 c & (8.3b) \\ M_0 l^2 + M_2 c^2 = M r^2 & (8.3c) \end{cases}$$

In combination these equations produce an important result.

$$lc = r^2 \qquad \left(\frac{c}{r}\right) = \left(\frac{r}{l}\right) \tag{8.4}$$

If l_2 is the length of the equivalent simple pendulum and the distance between concentrated masses, $(l + c)$, and if M_2 is determined from Equation 8.4, the complete equivalent simple pendulum is known. See Table 8.1 for several simple but useful cases.

TABLE 8.1. Compound pendulum rigid body characteristics of common geometries and conversion to the equivalent two-mass system.

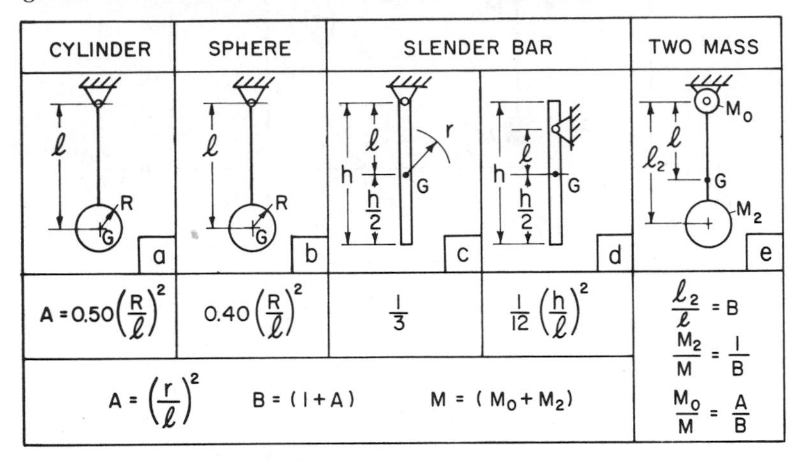

8.3. THE FORCED PENDULUM

Addition of sinusoidal forcing at a radius b (Table 8.2a) results in modification of Equation 8.2, adding the P_3 term,

$$\Sigma T_0 = lM\omega^2\mathrm{a} + Mr^2\omega^2\left(\frac{\mathrm{a}}{l}\right) - W\mathrm{a} + bP_3 = 0 \tag{8.5}$$

which is normalized by division by W to yield the magnification factor shown in terms of the complex variable θ. The reference frequency ω_1 is that of the simple pendulum on which the frequency ratio g is based. Resonance and infinite response ratio occur when the exciting frequency corresponds to the natural frequency, or when $B\mathrm{g}^2 = 1$.

Ground reaction P_0 at the fulcrum involves Equation 8.6 terms as they relate to the horizontal effects.

$$\Sigma F_x = M\omega^2\mathrm{a} + P_3 + P_0 = 0 \tag{8.6}$$

We note that both the inertia couple and the weight term are absent. Dividing by P_3 and substituting for a, we get the transmissibility factor given, which will be zero if the numerator is zero. Torque excitation about the pivot (Table 8.2b), is similar, but P_0 is not affected directly by the excitation, as in Table 8.2a.

Table 8.2 is based on actual centroidal distance l and total mass M, which define the pendulum directly; however, damping is more conveniently introduced using the equivalent system l_2 and M_2. In Table 8.3 we then locate the viscous dashpot at the l_2 radius and define the critical damping coefficient as

TABLE 8.2. Compound pendulum with transverse force excitation (*a*), couple excitation (*b*), and transverse displacement excitation of the fulcrum (*c*).

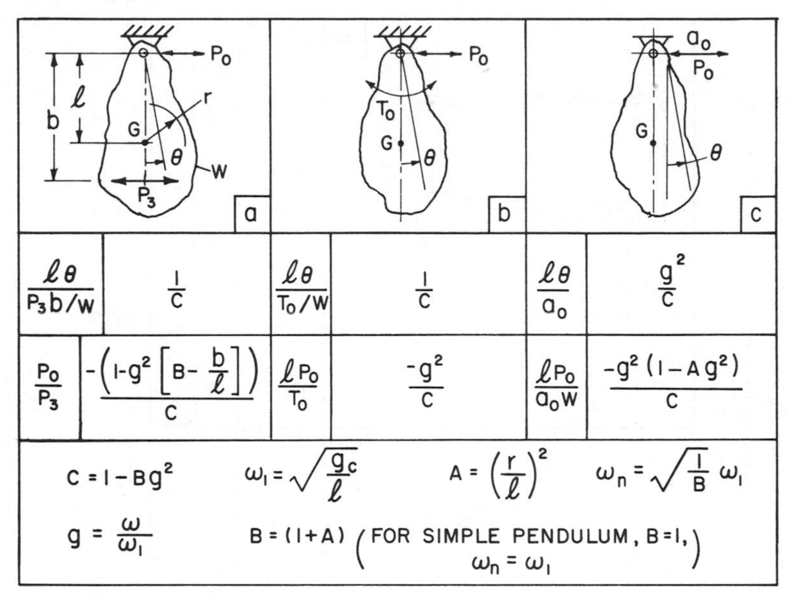

a		b		c	
$\frac{\ell\theta}{P_3 b/W}$	$\frac{1}{C}$	$\frac{\ell\theta}{T_0/W}$	$\frac{1}{C}$	$\frac{\ell\theta}{a_0}$	$\frac{g^2}{C}$
$\frac{P_0}{P_3}$	$\frac{-\left(1-g^2\left[B-\frac{b}{\ell}\right]\right)}{C}$	$\frac{\ell P_0}{T_0}$	$\frac{-g^2}{C}$	$\frac{\ell P_0}{a_0 W}$	$\frac{-g^2(1-Ag^2)}{C}$

$C = 1 - Bg^2$ $\omega_1 = \sqrt{\frac{g_c}{\ell}}$ $A = \left(\frac{r}{\ell}\right)^2$ $\omega_n = \sqrt{\frac{1}{B}}\,\omega_1$

$g = \frac{\omega}{\omega_1}$ $B = (1 + A)$ (FOR SIMPLE PENDULUM, $B = 1$, $\omega_n = \omega_1$)

$2M_2\omega_n$. Relative damping (Table 8.3*c*) with excitation by pivot displacement corresponds to viscous damping at the fulcrum.

8.4. FORCED PENDULUM MODES

The equivalent pendulum of Table 8.3*c* exhibits the classical single-degree response, as in Table 2.1*c*. Although this is a two-degree-of-freedom system, the a_0 coordinate corresponds to the impressed sinusoidal displacement. Forced undamped modes are shown in Figure 8.3, as forcing frequency increases.

Approaching zero frequency the pendulum tends to remain vertical with the node at infinity. Below resonance, a_2 is in phase with a_0 and greater than a_0, with the node above the fulcrum. At resonance, finite a_2 results from zero excitation, with the node at the fulcrum. Above resonance, the a_2/a_0 ratio is negative, with the node between the fulcrum and M_2. In the limit at high exciting frequency, M_2 becomes seismic, tending to remain stationary with the node at M_2. Note that the forced node can range above M_2 but never below.

With viscous damping (Table 8.3*c*), a phase relationship exists between a_2 and a_0, and a fixed node does not exist. As seen in Figure 8.4, velocities at these two points are represented by synchronized rotating vectors in a complex plane with $a_0\omega$ leading $a_2\omega$ by a phase angle ϕ. At some instant of time, projections on the real axis will be equal, and the pendular rigid body has

TABLE 8.3. **Compound pendulum with viscous damping and force excitation P_3 (*a*), couple excitation T_0 (*b*), and displacement excitation with relative damping (*c*). Body has been reduced to the equivalent two-mass system as in Table 8.1*e*.**

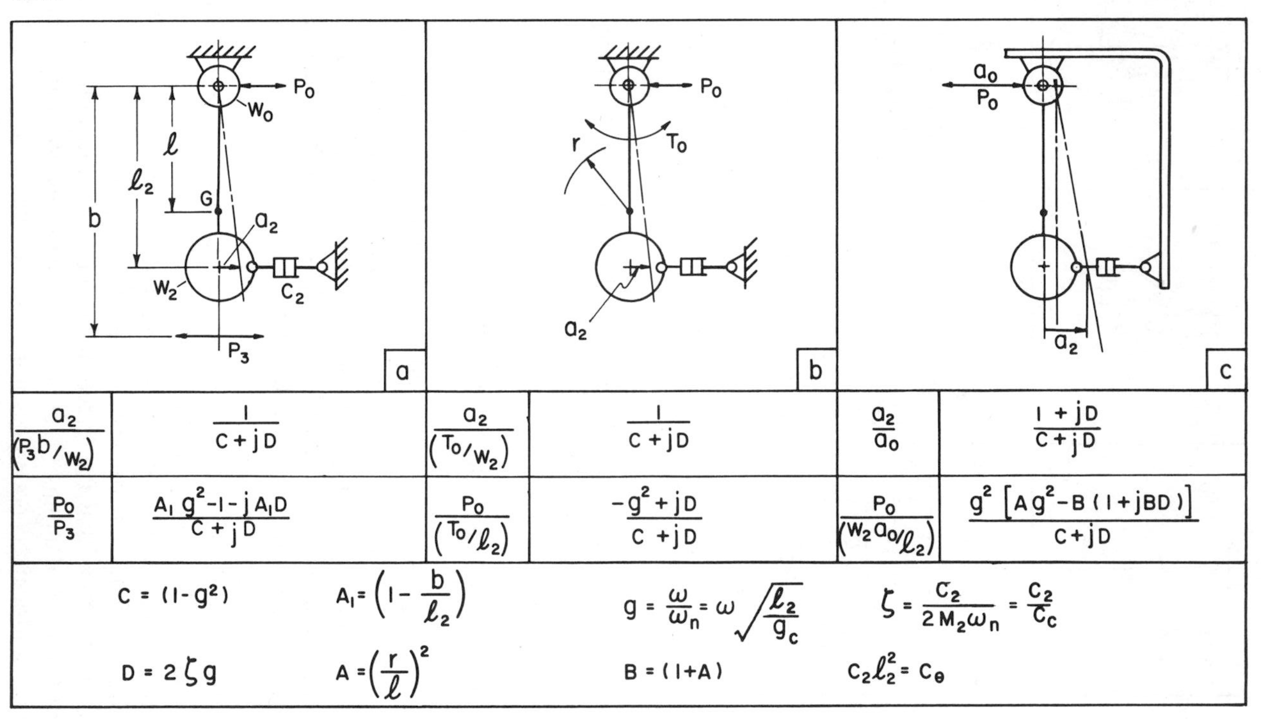

	a		b		c
$\frac{a_2}{(P_3 b / W_2)}$	$\frac{1}{C + jD}$	$\frac{a_2}{(T_0 / W_2)}$	$\frac{1}{C + jD}$	$\frac{a_2}{a_0}$	$\frac{1 + jD}{C + jD}$
$\frac{P_0}{P_3}$	$\frac{A_1 g^2 - 1 - jA_1 D}{C + jD}$	$\frac{P_0}{(T_0 / \ell_2)}$	$\frac{-g^2 + jD}{C + jD}$	$\frac{P_0}{(W_2 a_0 / \ell_2)}$	$\frac{g^2 [A g^2 - B(1 + jBD)]}{C + jD}$

$C = (1 - g^2)$ $\quad A_1 = \left(1 - \frac{b}{\ell_2}\right)$ $\quad g = \frac{\omega}{\omega_n} = \omega \sqrt{\frac{\ell_2}{g_c}}$ $\quad \zeta = \frac{C_2}{2 M_2 \omega_n} = \frac{C_2}{C_c}$

$D = 2\zeta g$ $\quad A = \left(\frac{r}{\ell}\right)^2$ $\quad B = (1 + A)$ $\quad C_2 \ell_2^2 = C_\theta$

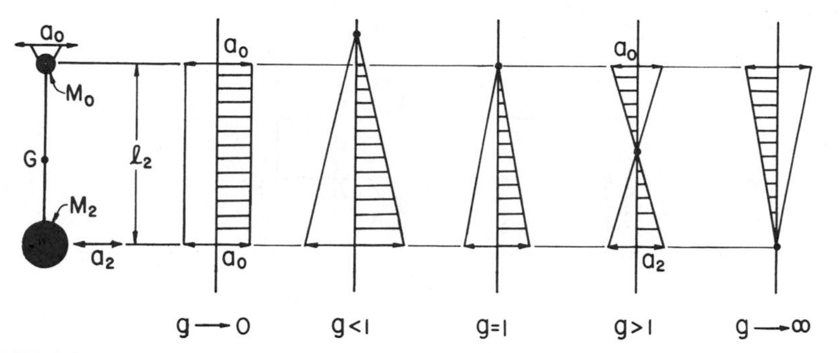

FIGURE 8.3. Pendulum forced at fulcrum has a modal variation (Table 8.3c), for zero damping. The node progresses downward with increasing frequency, but cannot exist below M_2.

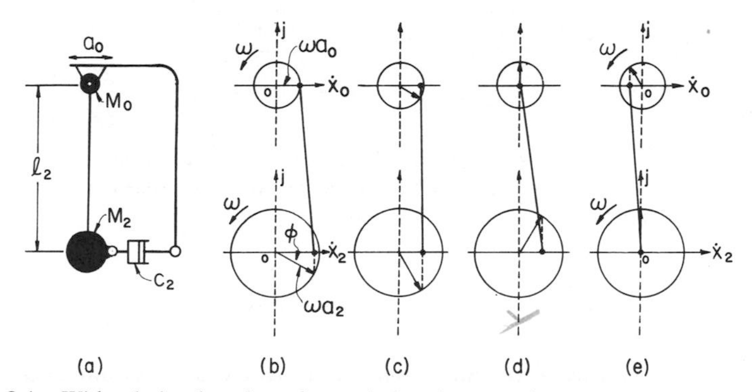

FIGURE 8.4. With relative damping, phase relations between the motions of the two masses can be visualized in terms of the respective complex vectors.

instantaneous horizontal *translation* (Figure 8.4c). At a different time, the $\mathbf{a}_0\omega$ vector crosses the j axis and the pendulum has simple *rotation* about the fulcrum (d). Another possibility occurs when $\mathbf{a}_2\omega$ crosses its j axis, and we observe rotation about M_2 as an instantaneous center (e).

8.5. DEFINITION OF THE PLANAR SPRING-MOUNTED BODY

A rigid body (Figure 8.5a) is mounted on two equal springs and lacks the fulcrum constraint of the pendulum. It has two degrees of freedom in a plane; coordinates can involve two transverse displacements, y_1 and y_2, or a transverse and an angular coordinate, y and θ. Total spring rate is K with respect to a force at E, the elastic center. Mass, as with the pendulum, is defined by the total mass M, by radius of gyration r about the centroid, and by the axial location of the centroid G.

Although the Figure 8.5a system appears restricted to the special case of equal springs, the more general system with unequal springs (Figure 8.6) can

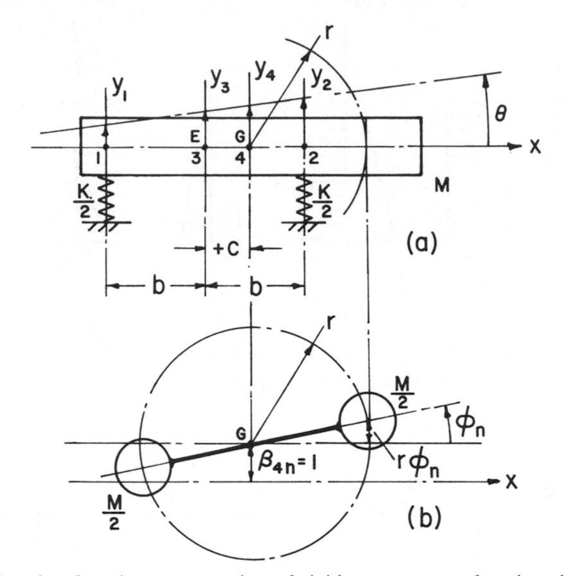

FIGURE 8.5. Notation for planar suspension of rigid mass on equal springs includes the elastic center E and the center of gravity G. The equivalent two-mass system and normal mode coordinates are shown in (b).

be reduced to the case of equivalent equal springs on the basis of static loads and deflections.

If we apply a couple T (Figure 8.6a), it is resisted by equal and opposite forces at A and B, with respective spring deflections of

$$y_{AT} = \left(\frac{T}{lK_A}\right) \qquad y_{BT} = \left(\frac{T}{lK_B}\right)$$

defining the static node E with respect to angular loading. Axial distances l_A and l_B are directly proportional to the spring deflections. Then if a transverse force F is applied at E (Figure 8.6b), spring forces for equilibrium are

$$F_A = \left(\frac{l_B}{l}\right)F \qquad F_B\left(\frac{l_A}{l}\right)F$$

with a deflection

$$y_F = \left(\frac{F_A}{K_A}\right) = \left(\frac{l_B}{l}\right)\left(\frac{F}{K_A}\right) = \left(\frac{F_B}{K_B}\right) = \left(\frac{l_A}{l}\right)\left(\frac{F}{K_B}\right)$$

Thus the point E determined by the applied couple is indeed the elastic center of the system, since a force at E produces only simple translation (Figure

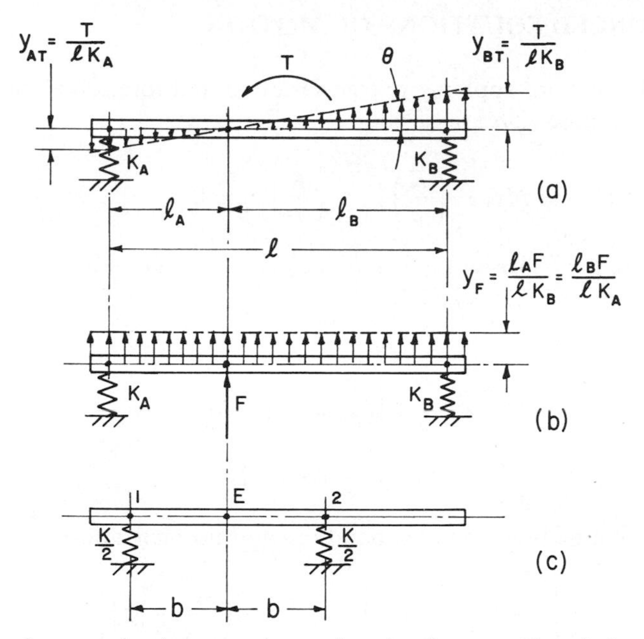

FIGURE 8.6. An unequal spring suspension can be reduced to one with equivalent equal springs, and by determining the proper equivalent span, b, (c).

8.6*b*). With E identified, it is possible to convert the original system to the simplified case (c), or Figure 8.5*a* with equal springs. For equivalent translational and rotational flexibility,

$$K = K_A + K_B \tag{8.7a}$$

$$Kb^2 = K_A l_A^2 + K_B l_B^2 = 2\left(\frac{K}{2} b^2\right) \tag{8.7b}$$

These two relations can be combined into an expression for the required horizontal spacing of the equal equivalent springs.

$$\frac{b}{l} = \frac{\sqrt{K_A K_B}}{(K_A + K_B)} = \frac{\sqrt{K_A K_B}}{K} \tag{8.8a}$$

And from Figure 8.6a we locate the midpoint E from

$$\frac{l_A}{l} = \frac{1}{1 + K_A/K_B} \tag{8.8b}$$

8.6. UNFORCED EQUATIONS OF MOTION

Differential equations applying to translation of and rotation about the center of gravity in Figure 8.5*a* are

$$\begin{cases} \Sigma F_y = -M\ddot{y}_4 - \left(\frac{K}{2}\right)y_1 - \left(\frac{K}{2}\right)y_2 = 0 & \text{(8.9a)} \\ \Sigma T_4 = -Mr^2\ddot{\theta} + \left(\frac{K}{2}\right)(b+c)y_1 - \left(\frac{K}{2}\right)(b-c)y_2 = 0 & \text{(8.9b)} \end{cases}$$

where

$$\theta = [(y_2 - y_1)/2b]$$

$$y_4 = y_1 + (b+c)\theta$$

With sinusoidal motion, complex notation, and normalization,

$$\begin{cases} \left[1 - \left(1 - \frac{c}{b}\right)\mathsf{g}_n^2\right]\mathsf{a}_1 + \left[1 - \left(1 + \frac{c}{b}\right)\mathsf{g}_n^2\right]\mathsf{a}_2 = 0 & \text{(8.10a)} \\ \left[1 + \left(\frac{c}{b}\right) - \left(\frac{r}{b}\right)^2\mathsf{g}_n^2\right]\mathsf{a}_1 - \left[1 - \left(\frac{c}{b}\right) - \left(\frac{r}{b}\right)^2\mathsf{g}_n^2\right]\mathsf{a}_2 = 0 & \text{(8.10b)} \end{cases}$$

Equating the determinant to zero, the generalized natural frequency quadratic becomes

$$\left(\frac{r}{b}\right)^2\mathsf{g}_n^4 - \left[1 + \left(\frac{c}{b}\right)^2 + \left(\frac{r}{b}\right)^2\right]\mathsf{g}_n^2 + 1 = 0 \qquad \text{(8.11)}$$

where

$$\mathsf{g}_n = (\omega_n/\omega_1)$$

$$\omega_1 = \sqrt{K/M}$$

In the symmetrical case (Figure 8.7), $c = 0$ and Equation 8.11 reduces to

$$\left(\frac{r}{b}\right)^2\mathsf{g}_n^4 - \left[1 + \left(\frac{r}{b}\right)^2\right]\mathsf{g}_n^2 + 1 = 0. \qquad \text{(8.12)}$$

with roots of $\mathsf{g}_n = 1$ and (b/r). The former corresponds to simple translation; the latter, to rotation about the centroid G. A further symmetrical case exists if $(b/r) = 1$, or $b = r$. Physically we now have $M/2$ at each spring, resulting in

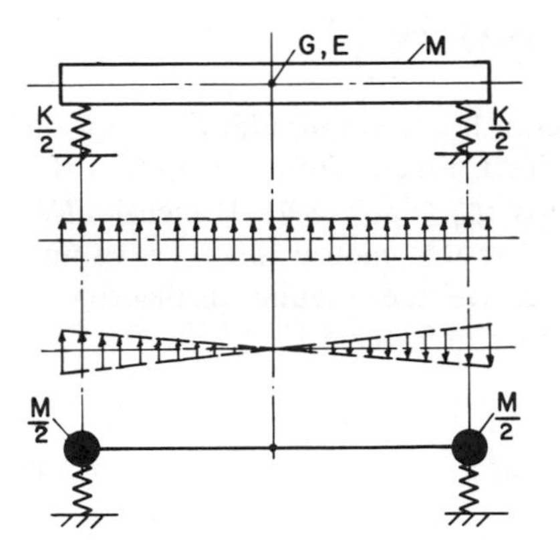

FIGURE 8.7. The modes for a symmetrical system are simple translation and rotation. If $r = b$, there are two equal and independent single-degree systems.

two independent, uncoupled single-degree systems, each with a natural frequency of $\sqrt{K/M}$.

Another simplified case results for the concentrated mass, $r = 0$, supported on a stiff weightless bar (Figure 8.8). Then Equation 8.11 becomes

$$\mathsf{g}_n^2 = \frac{1}{1 + (c/b)^2} \tag{8.13}$$

Now if $c = \pm b$, the concentrated mass is supported directly by one spring or the other and

$$\mathsf{g}_n^2 = \frac{1}{1+1} = \frac{1}{2}$$

$$\omega_n = \sqrt{\frac{K}{2M}}$$

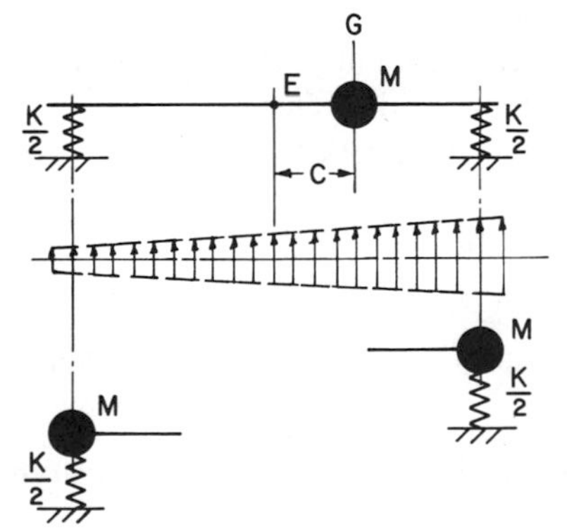

FIGURE 8.8. A concentrated mass on a stiff, weightless bar has a single mode. If G is located at either spring, simple spring–mass systems result.

8.7. NORMAL NODES AND ORTHOGONALITY

Free vibratory modes are described by the coordinate amplitudes β_1 and β_2 at the two natural frequencies. These are obtained directly from Equation 8.10*a* evaluated at the two modal frequency ratios g_{ni} (Table 8.4). Dimensionless frequency and modal ratios are functions of the geometric ratios (c/b) and (r/b) only. Modal amplitudes can also involve the angular displacement, resulting in a simple expression similar to those in Table 4.1.

$$\left(\frac{c\phi}{\beta_4}\right)_n = \left(1 - \mathsf{g}_n^2\right) \tag{8.14}$$

Using the radius of gyration to define the modal amplitude,

$$\left(\frac{r\phi}{\beta_4}\right)_n = \left(\frac{r}{c}\right)\left(1 - \mathsf{g}_n^2\right) \tag{8.15}$$

As seen in Figure 8.5*b*, the $r\phi_n$ displacement represents the physical motion of the two equivalent masses, $M/2$.

Orthogonal relationships connect the normal modes, and as shown in Figure 8.5*b*, for the double mass with two modes,

$$\left(M\beta_i\beta_j\right) + M(r\phi_n)_i(r\phi_n)_j = 0$$

$$1 + \left(\phi_i\phi_j\right)r^2 = 0$$

$$\phi_i\phi_j = -\left(\frac{1}{r^2}\right) \tag{8.16}$$

Although Equation 8.16 appears inconsistent dimensionally, the unit numerator represents the square of the unit displacements at the centroid G.

Modal amplitudes at the springs also satisfy orthogonality.

$$M\beta_i\beta_{1j} + M\beta_{2i}\beta_{2j} = 0$$

$$1 + \beta_{2i}\beta_{2j} = 0 \tag{8.17}$$

Although normal mode expressions are shown in different forms at the bottom of Table 8.4, both result in identical modes.

TABLE 8.4. Rigid body on equal springs with central forcing (*a*), and couple forcing (*c*). There is translational displacement excitation in (*b*) and angular displacement excitation about a central pivot in (*d*).

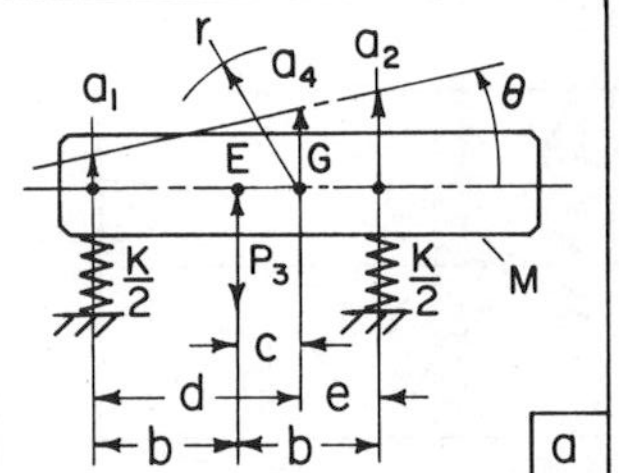 a		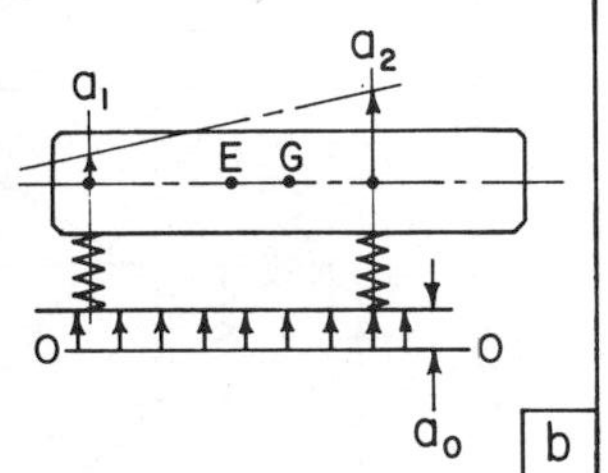 b	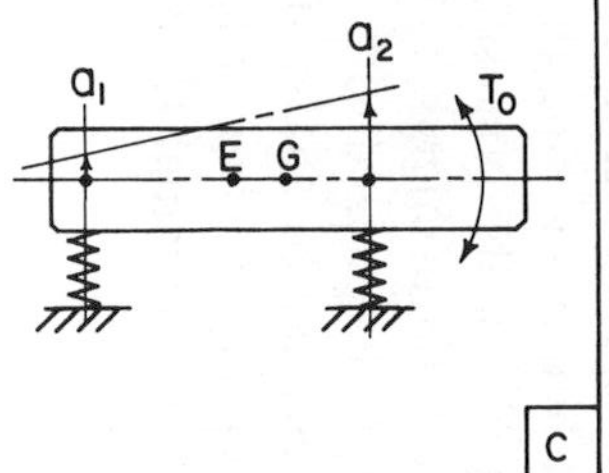 c		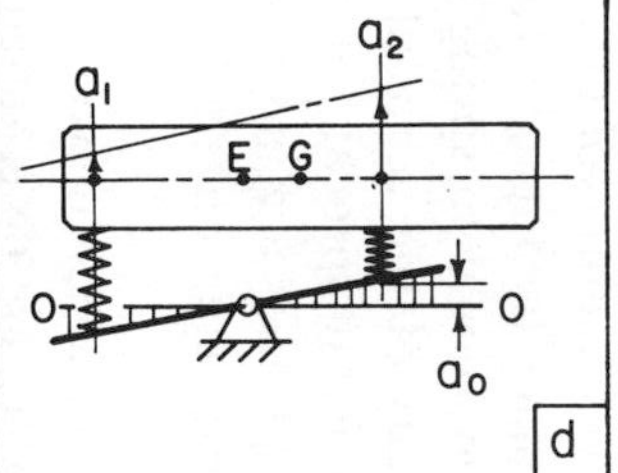 d
$\dfrac{a_1}{P_3/K}$	$\dfrac{1-g^2\left[\left(\frac{c}{b}\right)\left(\frac{d}{b}\right)+\left(\frac{r}{b}\right)^2\right]}{c_1}$	$\dfrac{a_1}{a_0}$	$\dfrac{a_1}{T_0/Kb}$	$\dfrac{\left[\left(\frac{d}{b}\right)g^2-1\right]}{c_1}$	$\dfrac{a_1}{a_0}$
$\dfrac{a_2}{P_3/Kb}$	$\dfrac{1+g^2\left[\left(\frac{c}{b}\right)\left(\frac{e}{b}\right)-\left(\frac{r}{b}\right)^2\right]}{c_1}$	$\dfrac{a_2}{a_0}$	$\dfrac{a_2}{T_0/Kb}$	$\dfrac{\left[1-\left(\frac{e}{b}\right)g^2\right]}{c_1}$	$\dfrac{a_2}{a_0}$
$\dfrac{b\theta}{P_3/K}$	$\dfrac{\left(\frac{c}{b}\right)g^2}{c_1}$	$\dfrac{b\theta}{a_0}$	$\dfrac{b\theta}{T_0/Kb}$	$\dfrac{1-g^2}{c_1}$	$\dfrac{b\theta}{a_0}$

$$c_1=\left(\frac{r}{b}\right)^2 g^4-\left[1+\left(\frac{c}{b}\right)^2+\left(\frac{r}{b}\right)^2\right]g^2+1, \quad g=\frac{\omega}{\sqrt{K/M}}$$

$$\left(\frac{\beta_2}{\beta_1}\right)_n=\frac{1-\left(\frac{e}{b}\right)g_n^2}{\left(\frac{d}{b}\right)g_n^2-1}$$

$$\left(\frac{r\phi}{\beta_4}\right)_n=\left(\frac{r}{c}\right)\left(1-g_n^2\right)$$

8.8. NATURAL FREQUENCY BEHAVIOR

Figure 8.5*a* represents a basic, or classical, planar spring-mounted body, with angular and translational freedom, and typifies a number of practical systems. A plot similar to Figures 4.2 and 4.3 can be made for this case from Equation 8.11 (Figure 8.9). The horizontal axis is proportional to spring spacing. The vertical axis is proportional to the eccentricity of the centroid *C*. Both are normalized by the radius of gyration and are reciprocals of the ratios in Equation 8.11. Higher and lower natural frequencies are indicated relative to the reference translational frequency ω_1. Intercepts on the horizontal axis correspond to the symmetrical case, $c = 0$.

Figure 8.10 is an extension of Figure 8.9 in which constant ratios of the higher to the lower natural frequency factors g_{ni} are determined from Figure 8.9, and are exact semicircles.[12] The dashed line connecting the tops of the circles represents, at a given spring spacing (b/r), the corresponding offset ratio (c/r) which will result in the minimum ratio of higher to lower modal frequency. Or, conversely, the lower frequency is maximized with respect to the higher. Thus relative coupled natural frequencies can be analyzed on the basis of the linear dimensions representing lateral spring spacing, centroidal offset, and radius of gyration for any planar system having equal transverse springs using these plots.

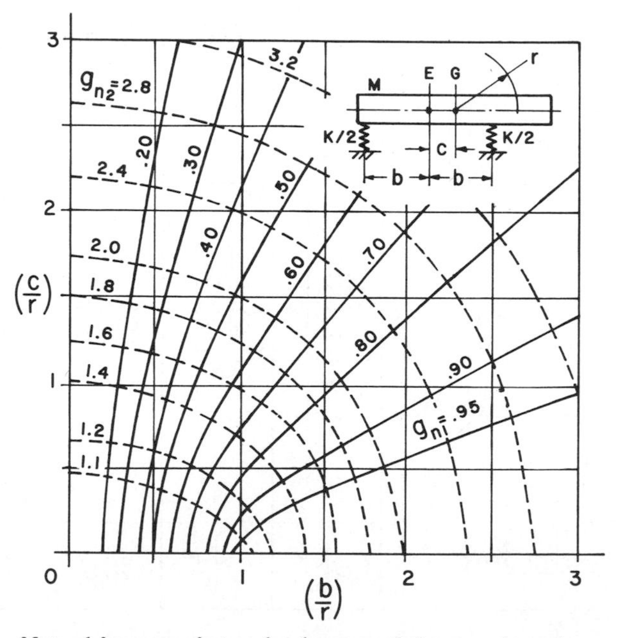

FIGURE 8.9. Natural frequency factors for the suspended system depend on centroidal position and spring spacing relative to the radius of gyration. Solid curves indicate the lower mode and dashed curves indicate the higher.

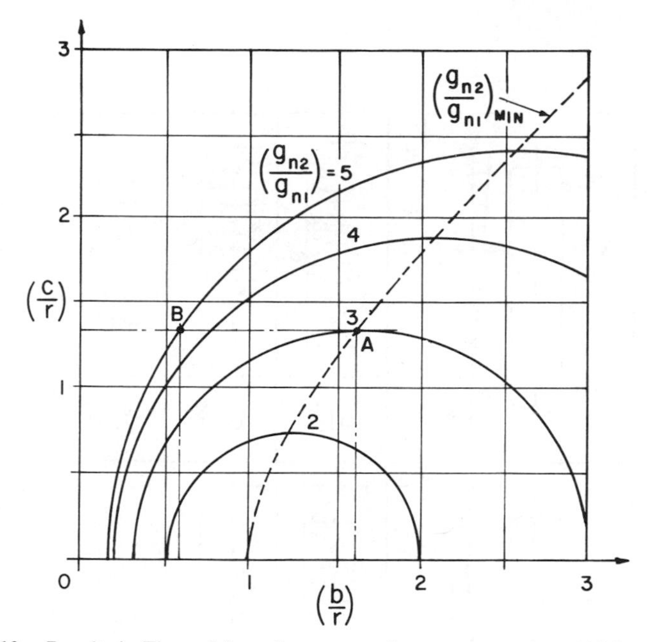

FIGURE 8.10. Results in Figure 8.9 can be converted to constant ratios of higher to lower modal frequencies.

8.9. FORCED VIBRATION

The general example of sinusoidal excitation is shown in Table 8.5*a*, with P_5 applied at a distance f from the elastic center. Equations 8.9*a* and *b* applied to this case become

$$\Sigma F_y = -M\ddot{y}_4 - \left(\frac{K}{2}\right)(y_1 + y_2) + P_5 = 0 \tag{8.18a}$$

$$\Sigma T_4 = -Mr^2\ddot{\theta} + \left(\frac{K}{2}\right)(d)y_1 - \left(\frac{K}{2}\right)(e)y_2 + P_5(f - c) = 0 \tag{8.18b}$$

After normalizing, these results are tabulated with the special case of $f = 0$ shown in Table 8.4*a*. Couple forcing is indicated in Table 8.4*c*. Table 8.5*a* is a combination of 8.4*a* and 8.4*c* results, as an eccentric forcing is equivalent to the force referred to E plus the moment effect, $T_0 = P_4 f$. As with other systems, the excitation can be a specified sinusoidal displacement. In Table 8.4*b* it is translatory, and equal at both springs. In Table 8.4*d* it is angular relative to E, in 8.5*b* at 1 only, and in 8.5*c* at 2 only.

TABLE 8.5. **In (*a*) force excitation is applied transversely at any axial position. There is displacement excitation at the left and right springs respectively in (*b*) and (*c*).**

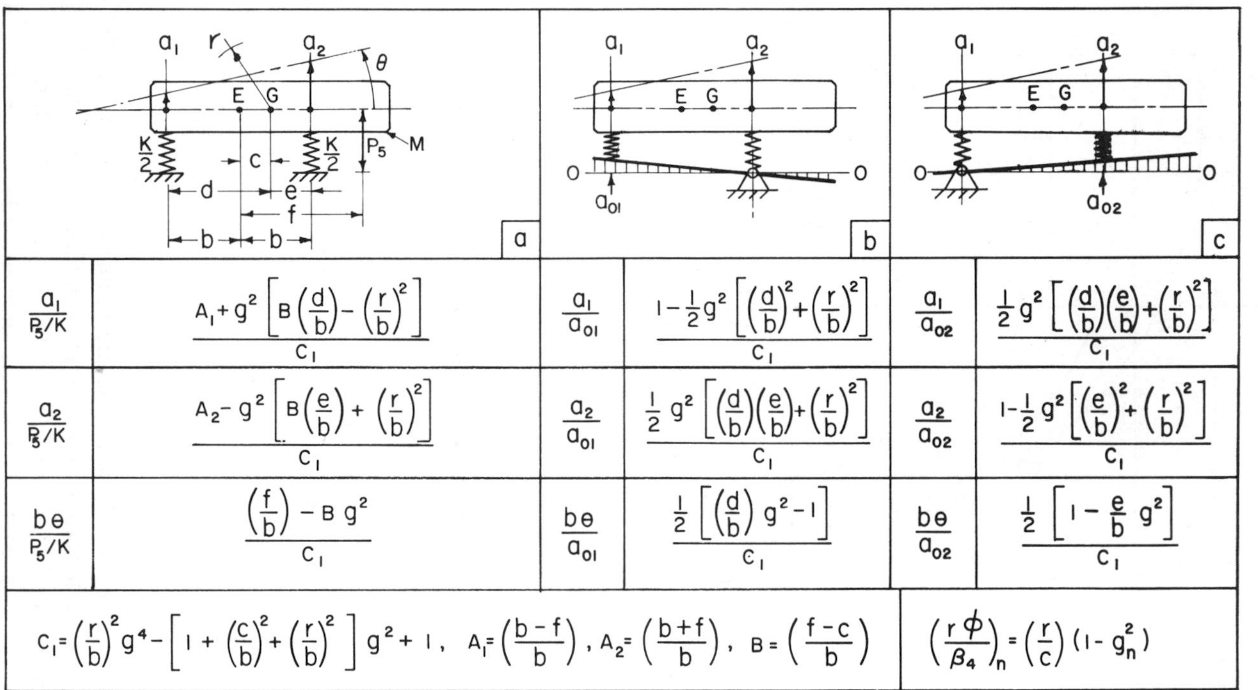

(a)		(b)		(c)	
$\frac{a_1}{P_5/K}$	$\frac{A_1 + g^2\left[B\left(\frac{d}{b}\right) - \left(\frac{r}{b}\right)^2\right]}{C_1}$	$\frac{a_1}{a_{01}}$	$\frac{1 - \frac{1}{2}g^2\left[\left(\frac{d}{b}\right)^2 + \left(\frac{r}{b}\right)^2\right]}{C_1}$	$\frac{a_1}{a_{02}}$	$\frac{\frac{1}{2}g^2\left[\left(\frac{d}{b}\right)\left(\frac{e}{b}\right) + \left(\frac{r}{b}\right)^2\right]}{C_1}$
$\frac{a_2}{P_5/K}$	$\frac{A_2 - g^2\left[B\left(\frac{e}{b}\right) + \left(\frac{r}{b}\right)^2\right]}{C_1}$	$\frac{a_2}{a_{01}}$	$\frac{\frac{1}{2}g^2\left[\left(\frac{d}{b}\right)\left(\frac{e}{b}\right) + \left(\frac{r}{b}\right)^2\right]}{C_1}$	$\frac{a_2}{a_{02}}$	$\frac{1 - \frac{1}{2}g^2\left[\left(\frac{e}{b}\right)^2 + \left(\frac{r}{b}\right)^2\right]}{C_1}$
$\frac{b\theta}{P_5/K}$	$\frac{\left(\frac{f}{b}\right) - Bg^2}{C_1}$	$\frac{b\theta}{a_{01}}$	$\frac{\frac{1}{2}\left[\left(\frac{d}{b}\right)g^2 - 1\right]}{C_1}$	$\frac{b\theta}{a_{02}}$	$\frac{\frac{1}{2}\left[1 - \frac{e}{b}g^2\right]}{C_1}$

$$C_1 = \left(\frac{r}{b}\right)^2 g^4 - \left[1 + \left(\frac{c}{b}\right)^2 + \left(\frac{r}{b}\right)^2\right] g^2 + 1, \quad A_1 = \left(\frac{b-f}{b}\right), \quad A_2 = \left(\frac{b+f}{b}\right), \quad B = \left(\frac{f-c}{b}\right)$$

$$\left(\frac{r\phi}{\beta_4}\right)_n = \left(\frac{r}{c}\right)(1 - g_n^2)$$

8.10. OPTIMIZED VIBRATION ISOLATION

The planar system (Figure 8.5*a*) is representative of a motor or engine mounting in which the actual three-dimensional arrangement is symmetrical about the center plane shown; that is, there are four springs, each $K/4$, with two pairs disposed equidistant from the center elevation shown. With the centroid also in the central plane, all tabulations given apply.

Transmissibility, discussed for the single-degree case in Section 2.8, requires a low natural frequency relative to a forcing frequency, and low damping, for minimizing the transmission of an exciting force to the base that supports the spring. In the coupled case, $c \neq 0$ and $f \neq 0$, the same objective applies, but two natural frequencies and modes are involved. Furthermore, as seen in the tabulated solutions, forced amplitudes are generally unequal at the two springs, causing the isolation problem to be more complex than for colinear systems. Nevertheless the broad fundamental concept of desired minimized natural frequencies still applies. We must reduce both modal frequencies to a certain fraction of the exciting frequency, particularly the higher. Relative frequencies are readily determined using Figures 8.9 and 8.10.

If the higher mode is suppressed, the lower mode can present another problem. If this natural frequency is considerably less than the higher, the mounting may become extremely soft or unstable. On this basis it is desirable to maintain the lower frequency as close as possible to the higher.

For instance, a design condition at B (Figure 8.10) has a lower frequency that is one-fifth the higher. But at A, by increasing the spring span ratio from 0.60 to 1.65, the lower frequency is one-third the higher. The frequency gain is thus 0.33/0.20, an improvement factor of 1.67.

8.11. RESPONSE TO UNBALANCE

Table 2.1*b* indicates single-degree response to an eccentric rotating mass. The body (Table 8.5*a*), can similarly be excited by a single unbalance mounted on the body producing excitation in a plane perpendicular to the axis shown. The entire body can also be rotating about its axis, in which case an unbalance develops a centrifugal force rotating in space. Introduction of $me\omega^2$ for P_5 in Equations 8.18*a* and *b* leads to modification of the amplitude expressions in Table 8.5*a*.

$$\mathsf{a}_{15} = \left[\frac{(me\omega^2)_5}{K}\right]\left\{\frac{A_1 + \mathsf{g}^2\left[B(d/b) - (r/b)^2\right]}{C_1}\right\}$$

$$= \left[\frac{(me)_5\omega^2}{M(K/M)}\right]\left(\frac{N_{15}}{C_1}\right) = \left[\frac{(me)_5}{M}\right]\mathsf{g}^2\left(\frac{N_{15}}{C_1}\right)$$

$$= \left[\frac{(me)_5}{M}\right]A_{15} \tag{8.19a}$$

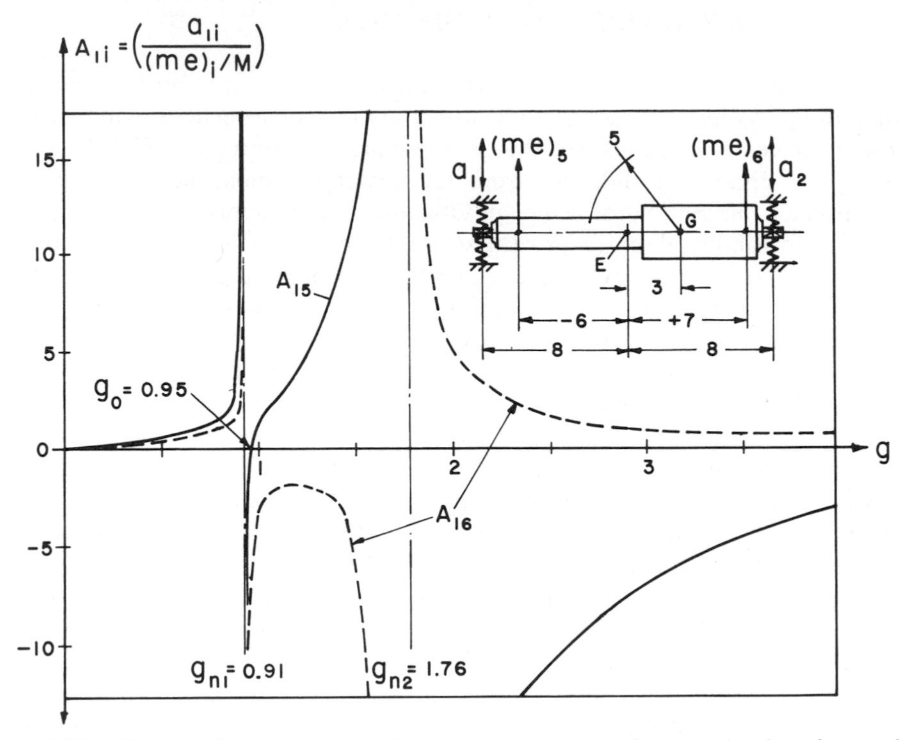

FIGURE 8.11. Amplitude response a_1 due to unbalance $(me)_5$ is proportional to the speed dependent factor A_{15} in Equation 8.19*a*, shown by the solid curve. Numerical values are for the assumed dimensions.

and similarly

$$a_{25} = \left[\frac{(me)_5}{M}\right] g^2 \left(\frac{N_{25}}{C_1}\right) = \left[\frac{(me)_5}{M}\right] A_{25} \tag{8.19b}$$

In this form we see that the response amplitude at either spring is directly proportional to the unbalance in a given plane and to the magnification factor at a particular rotative speed. See Figure 8.11 for a typical variation of A_{15}. Amplitude is also inversely proportional to the total mass. For zero damping the two amplitudes will be in or out of phase with the unbalance vector, $(me)_5$. Additionally a_{15} and a_{25} will be in or out of phase with each other.

8.12. RESPONSE TO DYNAMIC UNBALANCE

The single-plane unbalance just described is termed *force* or *static unbalance*. In the more general case of *dynamic unbalance*, two independent centrifugal forces occur in distinct, axially spaced planes (Figure 8.12). The two unbal-

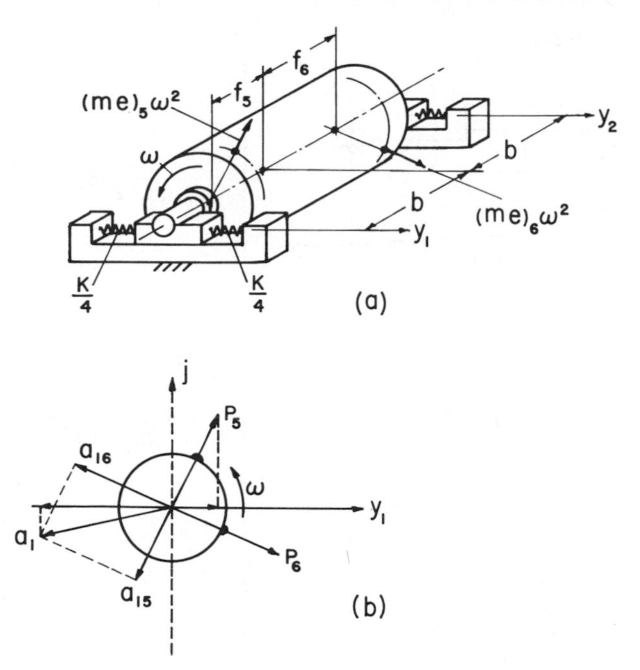

FIGURE 8.12. With dynamic unbalance total amplitude $\mathbf{a}_1$ is the resultant of two vector components $\mathbf{a}_{15}$ and $\mathbf{a}_{16}$ due to the unbalances $(me)_5$ and $(me)_6$ respectively. The real axis y_1 corresponds to the direction of planar oscillation.

ances, and the forces developed in these two planes, are the resultants of all the mass eccentricities existing in the rigid rotor with respect to the axis of rotation. They thus represent total rotor unbalance in terms of two vectors having magnitude and direction with respect to the rotor.

As each unbalance corresponds to a separate excitation, and since all frequencies are at rotative velocity, we have vector addition or superposition of responses (Figure 8.12*b*).

$$\mathbf{a}_1 = \mathbf{a}_{15} + \mathbf{a}_{16} = \left[\frac{(me)_5}{M}\right] A_{15} + \left[\frac{(me)_6}{M}\right] A_{16} \tag{8.20a}$$

$$\mathbf{a}_2 = \mathbf{a}_{25} + \mathbf{a}_{26} = \left[\frac{(me)_5}{M}\right] A_{25} + \left[\frac{(me)_6}{M}\right] A_{26} \tag{8.20b}$$

As shown, the rotor oscillates in the horizontal plane, the direction of the real axis. There is no motion in the vertical or imaginary direction, and exciting forces are vectors in phase with their respective unbalances. Responses are both assumed out of phase with the respective force vectors. The resultant $\mathbf{a}_1$ vector revolves with the rotor, its projection on the horizontal axis or plane corresponding to the instantaneous harmonic displacement.

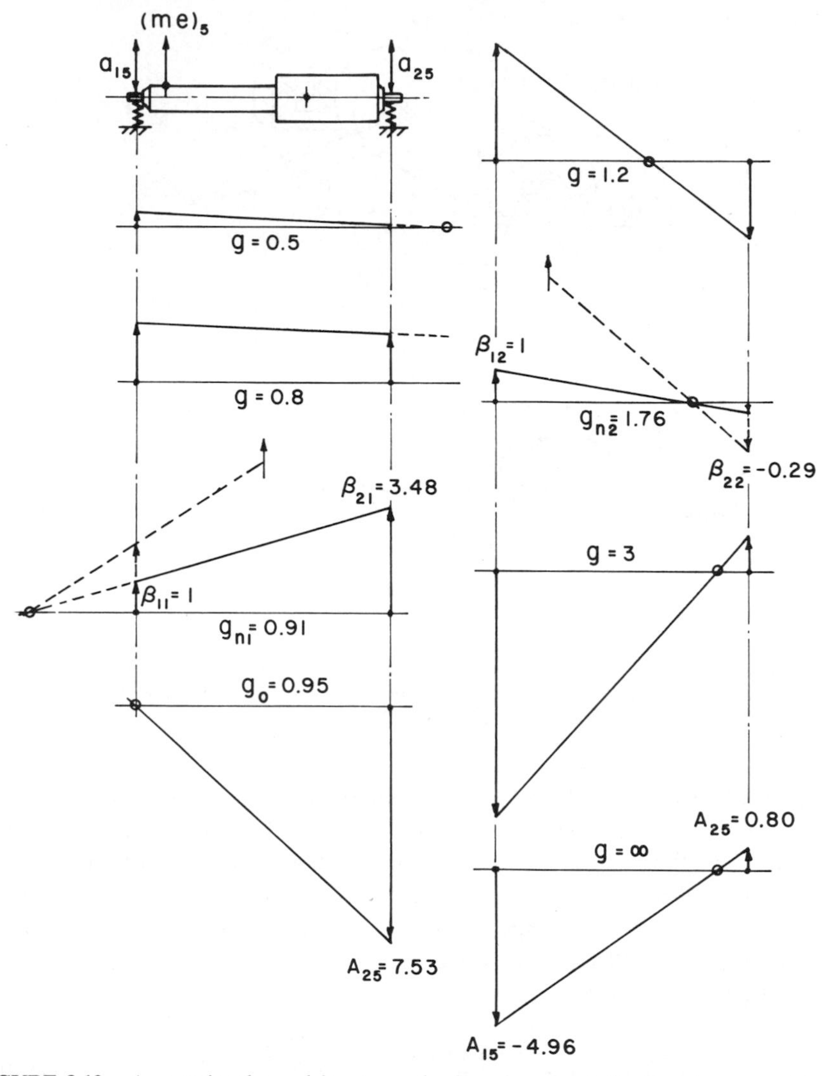

FIGURE 8.13. As rotational speed increases, the forced mode of the flexibly supported rotor (Figure 8.12) progresses through the two resonances, becoming asymptotic to the mode shown for $\mathsf{g} = \infty$ at high speeds. Unbalance excitation is in plane 5.

Figure 8.11 is a plot of the variation of influence factors A_{15} and A_{16} with speed, or frequency, for one end of the rotor. Both factors become infinite at the modal resonances. As A_{15} crosses the axis at $\mathsf{g} = \mathsf{g}_0 = 0.95$, there is no component amplitude at 1 due to the proximate unbalance, $(me)_5$. Both factors approach asymptotic values at high speeds. These can be determined by taking $\mathsf{g} = \infty$ in the response equation, say for a_{15}. Using Equation 8.19*a* and

Table 8.5 expressions,

$$\frac{\mathsf{a}_{15}}{[(me)_5/M]} = \frac{\mathsf{g}^2\{A_1 + \mathsf{g}^2[A - (r/b)^2]\}}{(r/b)^2\mathsf{g}^4 - [1 + (c/b)^2 + (r/b)^2]\mathsf{g}^2 + 1}$$

In the high-speed limit, only the g^4 terms in the numerator and in the denominator govern the asymptotic amplitude.

$$\frac{\mathsf{a}_{15}}{[(me)^5/M]} = \frac{\mathsf{g}^4[A - (r/b)^2]}{\mathsf{g}^4(r/b)^2} = \left[\frac{A}{(r/b)^2} - 1\right] = \left[\left(\frac{b}{r} + \frac{c}{r}\right)\left(\frac{d}{r} - \frac{c}{r}\right) - 1\right] \tag{8.21}$$

The effect of increasing rotor speed on the forced mode shape is indicated in Figure 8.13, again due only to an unbalance in plane 5. The left amplitude factor corresponds to A_{15} in Figure 8.11. Nodal position shifts from outboard right to outboard left, to plane 1. It then progresses to the right, with the limiting mode shown for $\mathsf{g} = \infty$.

8.13. MEASUREMENT OF DYNAMIC UNBALANCE

Since rotors, no matter how carefully manufactured, contain geometric and material variations that produce unbalance forces during rotation, detection and correction in selected balancing planes becomes a required procedure for many rotating parts. A number of types of machines have been designed and built for this purpose. Their primary function is to measure vibratory response of a revolving part in bearings and to interpret this response in terms of the existing dynamic unbalance. Practically, this requires the generation of the data indicating the magnitudes and phase positions of the unbalances in two preselected balancing planes if the rotor is rigid.

Early machines did this rather directly, but crudely. The rotor was mounted in a stiff cradle, pivoted mechanically to oscillate in a vertical plane with a fulcrum located below the cradle in one of the balancing planes (Figure 8.14). This reduces the system to single degree, excited by only one unbalance, with response at a particular speed calibrated in terms of the unbalance in the plane being measured. There are disadvantages to this arrangement:

1. It is inconvenient to stop rotation and to shift the fulcrum mechanically.
2. The cradle constitutes parasitic or additional mass moment of inertia, reducing response and sensitivity.
3. The fulcrum may introduce undesirable damping and phase shift.

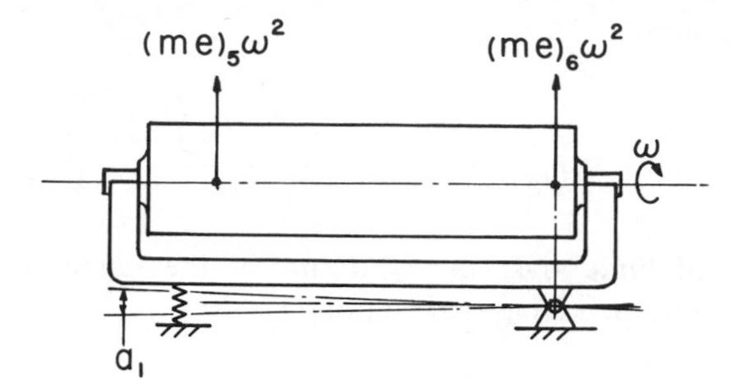

FIGURE 8.14. Schematic side view of dynamic balancing machine using fulcrum located in plane 6 to cancel vibratory effects of $(me)_6$. Responses at 1 and 2 are proportional to $(me)_5$ only.

If the rotor is suspended elastically so both bearings oscillate freely in a plane (Figure 8.15), simultaneous measurements can be made of the end amplitudes, magnitude and phase, providing complete $\mathbf{a}_1$ and $\mathbf{a}_2$ information. Vector solution of Equations 8.20*a* and 8.20*b* then can be achieved by an electrical dividing circuit that nulls the effect of first one unbalance and then the other.

Another scheme for separating the effects of the combined unbalances is suggested by Figure 8.11. At $\mathbf{g}_0$ there is no amplitude at 1 due to an unbalance at 5; therefore, this is a unique rotative speed at which the amplitude at 1 is proportional to and out of phase with the unbalance $(me)_6$.[7] A similar speed exists for the reverse condition, in which $A_{26} = 0$.

In addition, the modes shown in Figure 8.13 suggest another alternative for obtaining response due to one unbalance at a time. All modes caused by a single exciting unbalance contain a node, often within the supports. Thus a vibration pickup attached parallel to the rotor and to the end supports (Figure 8.15), can be located axially at a forced node to null the effect of a component unbalance. The forced mode, superimposed by the second unbalance, however, will provide a measure at this nodal position of the magnitude and phase of the second unbalance.

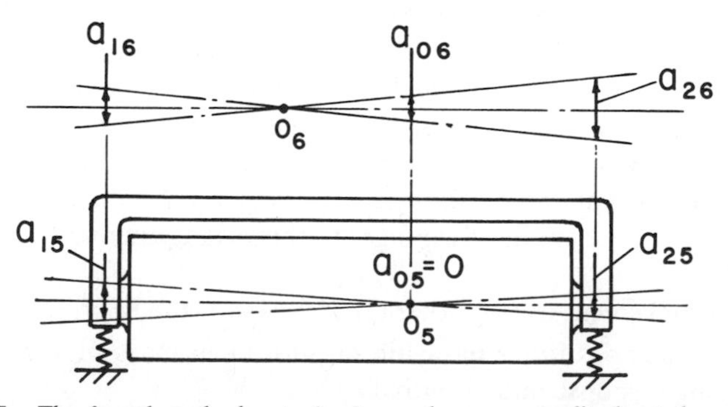

FIGURE 8.15. The forced mode due to $(me)_5$ produces no amplitude at the modal node, as $\mathbf{a}_{05} = 0$, but the forced mode due to $(me)_6$ is $\mathbf{a}_{06}$, which thus measures $(me)_6$ only. Both bearings oscillate freely in the horizontal plane shown.

8.14. ADDITIONAL CASES

Relations are provided for treating the vertically suspended rigid body with symmetrical vertical flexibility K_1 and horizontal flexibility K_2. In Table 8.6*a* the excitation P_2 is aligned with K_2, and in Table 8.6*b* a couple T_0 provides the excitation. A horizontal excitation at a different height can be reduced to an

TABLE 8.6. Symmetrically mounted body with unequal stiffness horizontally and vertically excited by a force P_2 at 2 or a couple T_0. Horizontal forcing at a different height can be converted to an equivalent P_2 and T_0.

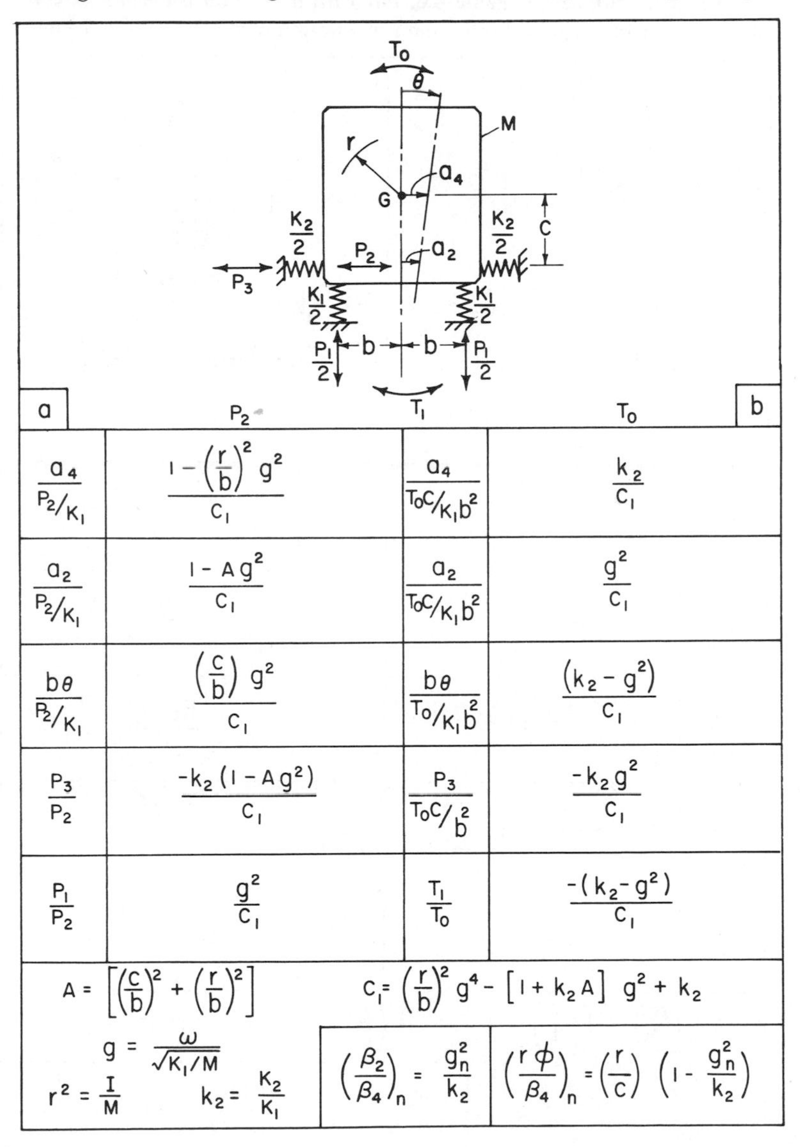

a	P_2	T_1	T_0 (b)
$\dfrac{a_4}{P_2/K_1}$	$\dfrac{1-\left(\frac{r}{b}\right)^2 g^2}{C_1}$	$\dfrac{a_4}{T_0 c/K_1 b^2}$	$\dfrac{k_2}{C_1}$
$\dfrac{a_2}{P_2/K_1}$	$\dfrac{1 - A g^2}{C_1}$	$\dfrac{a_2}{T_0 c/K_1 b^2}$	$\dfrac{g^2}{C_1}$
$\dfrac{b\theta}{P_2/K_1}$	$\dfrac{\left(\frac{c}{b}\right) g^2}{C_1}$	$\dfrac{b\theta}{T_0/K_1 b^2}$	$\dfrac{(k_2 - g^2)}{C_1}$
$\dfrac{P_3}{P_2}$	$\dfrac{-k_2(1 - A g^2)}{C_1}$	$\dfrac{P_3}{T_0 c/b^2}$	$\dfrac{-k_2 g^2}{C_1}$
$\dfrac{P_1}{P_2}$	$\dfrac{g^2}{C_1}$	$\dfrac{T_1}{T_0}$	$\dfrac{-(k_2 - g^2)}{C_1}$

$$A = \left[\left(\frac{c}{b}\right)^2 + \left(\frac{r}{b}\right)^2\right] \qquad C_1 = \left(\frac{r}{b}\right)^2 g^4 - [1 + k_2 A]\, g^2 + k_2$$

$$g = \frac{\omega}{\sqrt{K_1/M}} \qquad r^2 = \frac{I}{M} \qquad k_2 = \frac{K_2}{K_1}$$

$$\left(\frac{\beta_2}{\beta_4}\right)_n = \frac{g_n^2}{k_2} \qquad \left(\frac{r\phi}{\beta_4}\right)_n = \left(\frac{r}{c}\right)\left(1 - \frac{g_n^2}{k_2}\right)$$

equivalent P_2 and T_0. Translational and angular displacement excitations are given in Tables 8.7*a* and 8.7*b* for the same system. Although shown as planar, this could be a three-dimensional problem with eight springs, provided the springs, centroid, and excitation are symmetrically disposed in the axial direction. Otherwise the case is much more complexly coupled.

The free rigid mass, in Table 8.8, is ungrounded and could represent a body floating on a liquid, in space, or suspended on extremely flexible springs. Planar responses to planar sinusoidal excitations are given.

TABLE 8.7. Similar to Table 8.6, but with horizontal translatory excitation (*a*), or with angular displacement excitation about the axis at 2 (*b*).

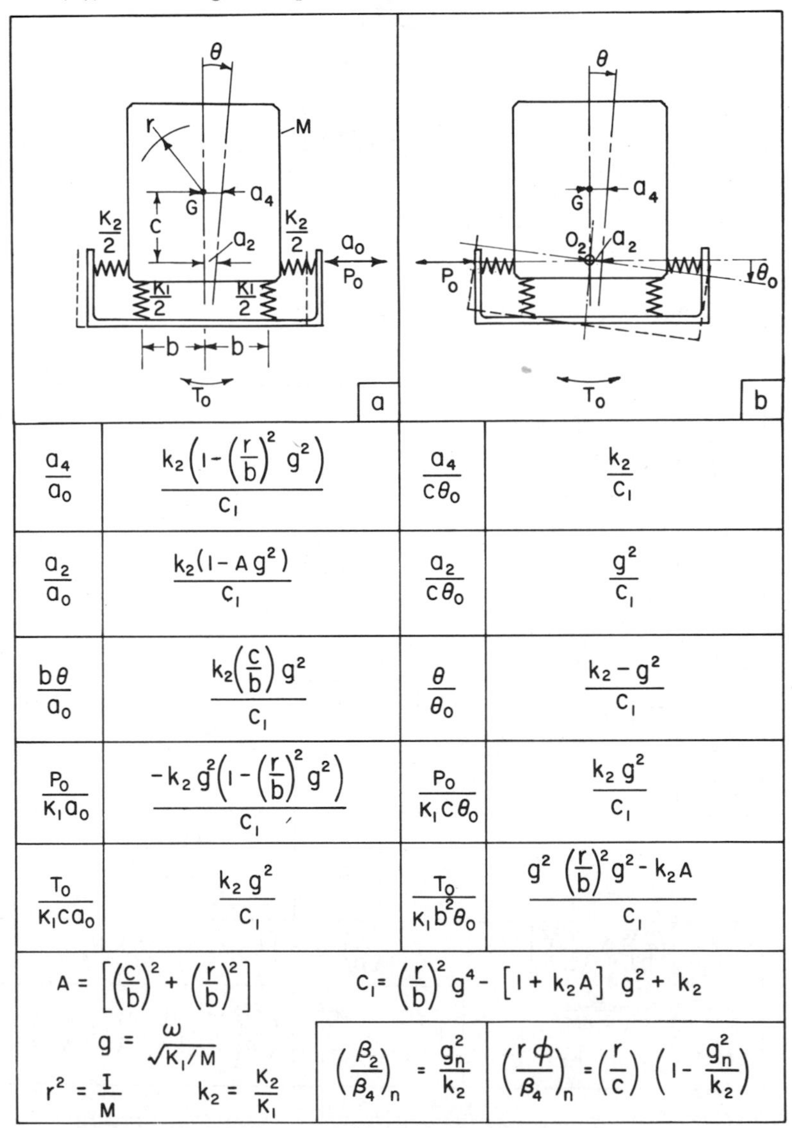

(a)		(b)	
$\frac{a_4}{a_0}$	$\frac{k_2\left(1-\left(\frac{r}{b}\right)^2 g^2\right)}{C_1}$	$\frac{a_4}{c\theta_0}$	$\frac{k_2}{C_1}$
$\frac{a_2}{a_0}$	$\frac{k_2(1-Ag^2)}{C_1}$	$\frac{a_2}{c\theta_0}$	$\frac{g^2}{C_1}$
$\frac{b\theta}{a_0}$	$\frac{k_2\left(\frac{c}{b}\right)g^2}{C_1}$	$\frac{\theta}{\theta_0}$	$\frac{k_2-g^2}{C_1}$
$\frac{P_0}{K_1 a_0}$	$\frac{-k_2 g^2\left(1-\left(\frac{r}{b}\right)^2 g^2\right)}{C_1}$	$\frac{P_0}{K_1 c\theta_0}$	$\frac{k_2 g^2}{C_1}$
$\frac{T_0}{K_1 c a_0}$	$\frac{k_2 g^2}{C_1}$	$\frac{T_0}{K_1 b^2\theta_0}$	$\frac{g^2\left(\frac{r}{b}\right)^2 g^2 - k_2 A}{C_1}$

$$A = \left[\left(\frac{c}{b}\right)^2 + \left(\frac{r}{b}\right)^2\right] \qquad C_1 = \left(\frac{r}{b}\right)^2 g^4 - [1 + k_2 A]\, g^2 + k_2$$

$$g = \frac{\omega}{\sqrt{K_1/M}} \qquad r^2 = \frac{I}{M} \qquad k_2 = \frac{K_2}{K_1}$$

$$\left(\frac{\beta_2}{\beta_4}\right)_n = \frac{g_n^2}{k_2} \qquad \left(\frac{r\phi}{\beta_4}\right)_n = \left(\frac{r}{c}\right)\left(1 - \frac{g_n^2}{k_2}\right)$$

TABLE 8.8. Free planar body with sinusoidal force excitation at any point (*a*), and with couple excitation (*b*).

TABLE 8.9. Slender pivoted bar with cantilever spring to ground, force-excited at outboard end of bar by P_2 (*a*), and with transverse displacement excitation at either or both support points at the same frequency (*b*).

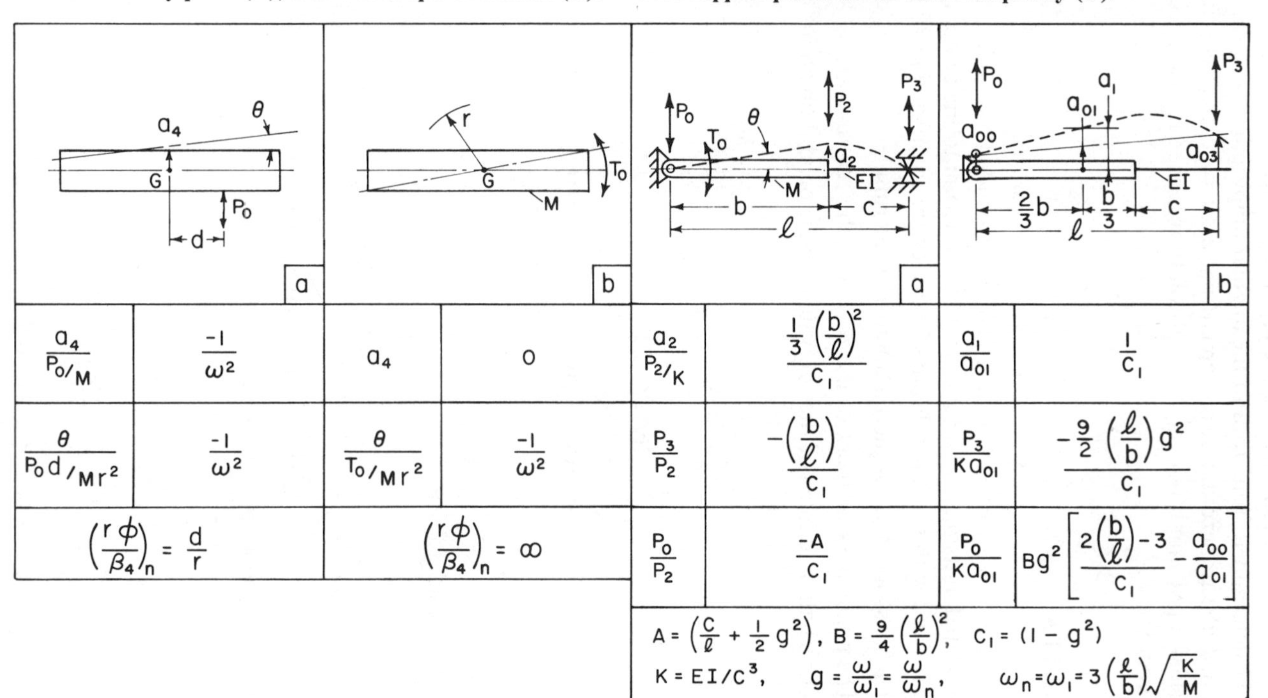

Table 8.8 (a)		Table 8.8 (b)	
$\frac{a_4}{P_0/M}$	$\frac{-1}{\omega^2}$	a_4	0
$\frac{\theta}{P_0 d/Mr^2}$	$\frac{-1}{\omega^2}$	$\frac{\theta}{T_0/Mr^2}$	$\frac{-1}{\omega^2}$
$\left(\frac{r\phi}{\beta_4}\right)_n = \frac{d}{r}$		$\left(\frac{r\phi}{\beta_4}\right)_n = \infty$	

Table 8.9 (a)		Table 8.9 (b)	
$\frac{a_2}{P_2/K}$	$\frac{\frac{1}{3}\left(\frac{b}{\ell}\right)^2}{C_1}$	$\frac{a_1}{a_{01}}$	$\frac{1}{C_1}$
$\frac{P_3}{P_2}$	$\frac{-\left(\frac{b}{\ell}\right)}{C_1}$	$\frac{P_3}{Ka_{01}}$	$\frac{-\frac{9}{2}\left(\frac{\ell}{b}\right)g^2}{C_1}$
$\frac{P_0}{P_2}$	$\frac{-A}{C_1}$	$\frac{P_0}{Ka_{01}}$	$Bg^2\left[\frac{2\left(\frac{b}{\ell}\right)-3}{C_1} - \frac{a_{00}}{a_{01}}\right]$

$A = \left(\frac{c}{\ell} + \frac{1}{2}g^2\right)$, $B = \frac{9}{4}\left(\frac{\ell}{b}\right)^2$, $C_1 = (1 - g^2)$

$K = EI/c^3$, $g = \frac{\omega}{\omega_1} = \frac{\omega}{\omega_n}$, $\omega_n = \omega_1 = 3\left(\frac{\ell}{b}\right)\sqrt{\frac{K}{M}}$

In Table 8.9 the pivoted rigid slender bar is supported to ground by a constant cantilevered beam spring of negligible mass. Small amplitudes are again assumed. In Table 8.9*a* the excitation P_2 is referred to the radius b, which could be determined from T_0, and the angular amplitude θ is a_2/b. Similarly in Table 8.9*b* the trapezoidal displacement pattern $(a_{00} - a_{03})$ is referred to the sinusoidal input amplitude a_{01}. The absolute response is also specified at the $2b/3$ coordinate.

EXAMPLES

EXAMPLE 8.1. A flywheel weighing 44 lb is suspended on a fulcrum at the rim and oscillated in its plane. Distance from the fulcrum to the center of the wheel is 6.70 in., and with small amplitudes 25 cycles are completed in 30.8 sec. Find:

(a) The mass moment of inertia about the fulcrum.

(b) The polar mass moment of inertia about the central axis.

(c) The length and mass of the equivalent single pendulum about the fulcrum.

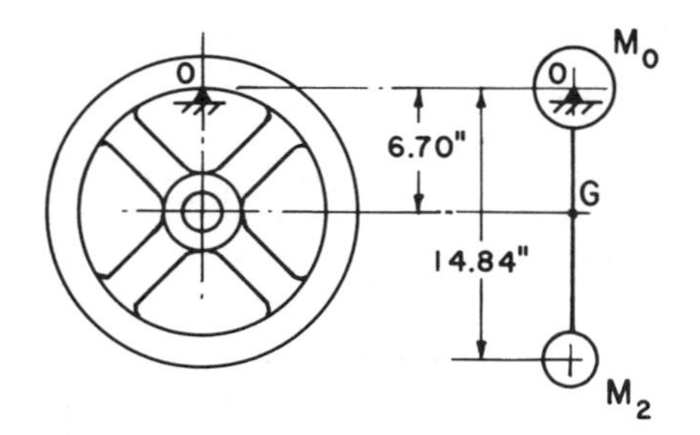

Solution

(a) Using Tables 8.1 and 8.2,

$$\omega_1 = \sqrt{\frac{386}{6.70}} = 7.59 \text{ rad/sec} = 1.208 \text{ Hz}$$

$$\omega_n = 25/30.8 = 0.812 \text{ Hz} = 5.10 \text{ rad/sec}$$

$$(5.10)^2 = \frac{1}{B}(7.59)^2$$

$$B = 2.215$$

$$A = 1.215 = (r/l)^2$$

$$r^2 = 1.215(6.70)^2 = 54.5 \text{ in.}^2$$

$$I_O = 44(54.5 + 6.7^2) = 4370 \text{ lb in.}^2$$

(b)

$$I_G = 44(54.5) = 2400 \text{ lb in.}^2$$

(c) Using Table 8.1,

$$l_2/l = B = 2.215$$

$$l_2 = 14.84 \text{ in.}$$

$$M_2/M = 1/B = 0.451$$

$$M_2 = 19.86 \text{ lb}$$

EXAMPLE 8.2. A 1.50 in. diamater steel shaft 42 in. long is to be pivoted at a point on its longitudinal axis as a vertical pendulum. Find:

(a) The position of the fulcrum if the natural frequency is 0.60 Hz.

(b) The maximum possible natural frequency.

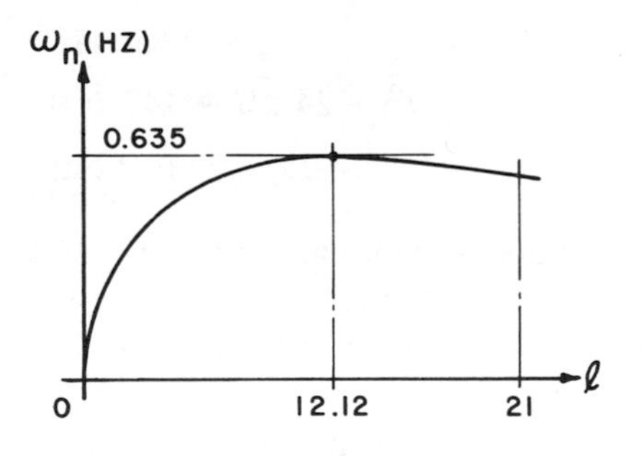

Solution

(a) Using Table 8.2,

$$\omega_n^2 = \frac{1}{B}\omega_1^2 = [0.60(2\pi)]^2 = 14.21$$

$$\frac{1}{B}\left(\frac{386}{l}\right) = 14.21$$

$$Bl = 27.16$$

Using Table 8.1*d*,

$$B = 1 + \left(\frac{1}{12}\right)\left(\frac{42}{l}\right)^2 = 1 + \left(\frac{147}{l^2}\right)$$

$$l^2 - 27.16l + 147 = 0$$

$$l = \frac{27.16 \pm \sqrt{737.6 - 588}}{2}$$

$$l = 7.46 \text{ in. and } 19.70 \text{ in.}$$

Note that there are two positions for this frequency.

(b) From part a, the 147 factor is constant, but the 27.16 factor in general terms is

$$Bl = \left(\frac{386}{\omega_n^2}\right)$$

$$l^2 - \left(\frac{386}{\omega_n^2}\right)l + 147 = 0$$

$$l = \frac{(386/\omega_n^2) \pm \sqrt{(386/\omega_n^2)^2 - 588}}{2}$$

The largest value of ω_n that will retain a positive radical is

$$(386/\omega_n^2) = \sqrt{588}$$

$$\omega_n^2 = 15.92$$

$$\omega_n = 0.635 \text{ Hz}$$

$$l^2 - 24.25l + 147 = 0$$

$$l = 24.25/2 = 12.12 \text{ in.}$$

EXAMPLE 8.3. A compound pendulum weighing 15 kg measures 32 cm from the fulcrum to the centroid. Radius of gyration about the centroid is 11.5 cm. Horizontal sinusoidal displacement of the pivot is 1.6 cm at 2 Hz. Damping is negligible Find:

(a) Vibratory amplitude of the centroid.
(b) Driving force at the fulcrum.
(c) Angular amplitude.

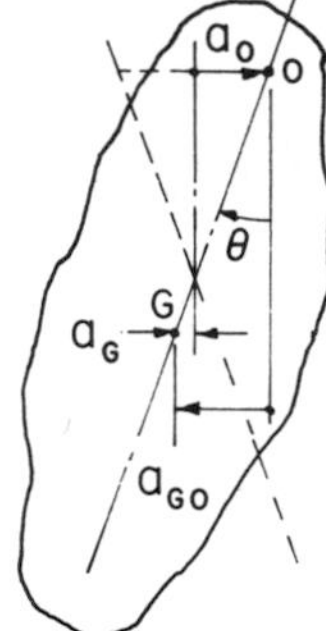

Solution

(a) Using Table 8.2c, it is unnecessary to convert to the equivalent two-mass system, and

$$A = (11.5/32)^2 = 0.129$$

$$B = 1.129$$

$$\omega_1 = \sqrt{\frac{980}{32}} = 5.53 \text{ rad/sec} = 0.88 \text{ Hz}$$

$$\mathsf{g}^2 = 5.145$$

$$\frac{32\theta}{1.6} = \frac{\mathsf{a}_{GO}}{1.6} = \left(\frac{5.145}{-4.82}\right) = -1.067$$

$$\mathsf{a}_{GO} = -1.71 \text{ cm} \quad \text{(relative)}$$

$$\mathsf{a}_G = \mathsf{a}_O + \mathsf{a}_{GO} = 1.60 - 1.71 = -0.11 \text{ cm} \quad \text{(absolute)}$$

(b) $$\left[\frac{P_O}{(1.6/32)(15)(9.80)}\right] = \left[\frac{-5.145(1 - 0.664)}{-482}\right] = +0.0036$$

$$P_O = 0.0026 \text{ N}$$

(c) $$\theta = \left(\frac{-1.71}{32}\right) = 0.053 \text{ rad} = 3.06°$$

EXAMPLE 8.4. A rigid body is supported on three springs, as shown. Only vertical motion is to be considered, symmetrical about the *CG* axis. Find:

(a) The natural frequencies.

(b) The normal modes.

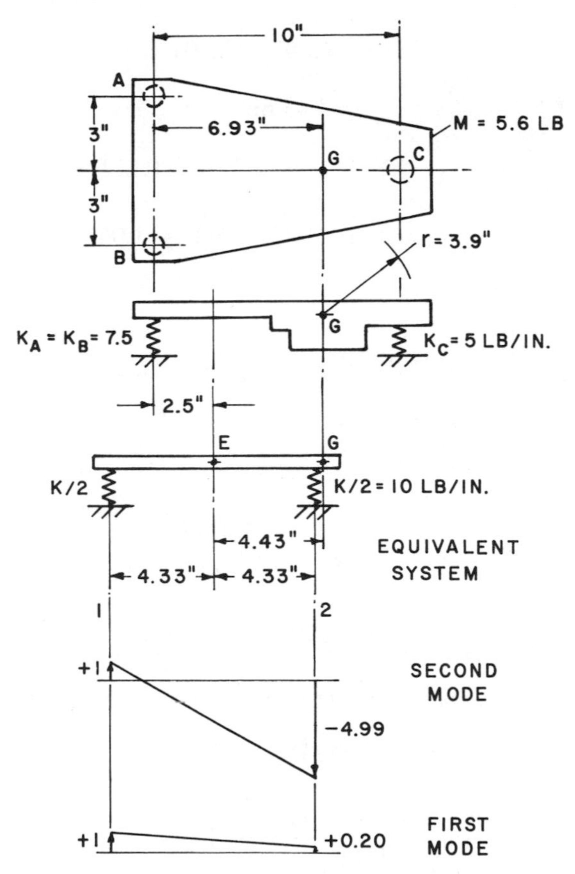

Solution

(a) Using Equation 8.7*a*,

$$\frac{K}{2} = \left(\frac{7.5 + 7.5 + 5}{2}\right)$$

$$= 10 \text{ lb/in.}$$

Using Equation 8.8a,

$$\frac{b}{l} = \left(\frac{\sqrt{15(5)}}{20}\right) = 0.433$$

$$b = 4.33 \text{ in.}$$

Using Equation 8.8b for E,

$$l_A/l = \frac{1}{1 + 15/5} = 0.25$$

$$l_A = 2.50 \text{ in.}$$

$$c = (6.93 - 2.50)$$

$$= 4.43 \text{ in.}$$

$$(r/b)^2 = (3.9/4.33)^2 = 0.811$$

$$(c/b)^2 = (4.43/4.33)^2 = 1.0467$$

From Table 8.4,

$$C_1 = 0.81\mathsf{g}_n^4 - 2.857\mathsf{g}_n^2 + 1 = 0$$

$$\mathsf{g}_{n1}^2 = 0.394 \qquad \mathsf{g}_{n2}^2 = 3.133$$

$$\mathsf{g}_{n1} = 0.628 \qquad \mathsf{g}_{n2} = 1.77$$

$$\omega_1 = \sqrt{\frac{K}{M}} = \sqrt{\frac{20(386)}{5.6}}$$

$$= 37.13/2\pi = 5.91 \text{ Hz}$$

$$\omega_{n1} = 3.71 \text{ Hz}$$

$$\omega_{n2} = 10.46 \text{ Hz}$$

(b)

$$\left(\frac{\beta_2}{\beta_1}\right)_n = \left(\frac{1 + 0.023\mathsf{g}_n^2}{2.023\mathsf{g}_n^2 - 1}\right)$$

$$\left(\frac{\beta_2}{\beta_1}\right)_{n1} = -4.99$$

$$(\beta_2/\beta_1)_{n2} = +0.201$$

EXAMPLE 8.5. The Example 8.4 system is excited by a vertical sinusoidal translational external motion. Amplitude is 0.042 in. at 15 Hz. Calculate the response.

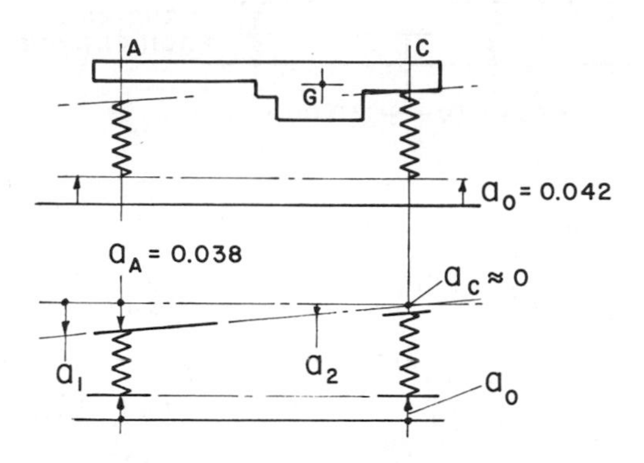

Solution. Using Table 8.4*b*,

$$g^2 = (15/5.91)^2 = 6.44$$

$$c/b = 1.023, \; d/b = 2.023$$

$$\frac{a_1}{a_0} = \frac{1 - 6.44(2.88)}{16.24} = -1.08$$

$$a_1 = (-1.08)(0.042) = -0.045 \text{ in.}$$

$$\frac{a_2}{a_0} = \frac{1 - 6.44(0.913)}{16.24} = -0.30$$

$$a_2 = (-0.30)(0.042) = -0.013 \text{ in.}$$

Having obtained forced amplitudes for the equivalent system, we determine displacements at *A*, *B* and *C* using the geometry of the straight-line mode. All displacements are absolute.

EXAMPLE 8.6. A single-cylinder engine-driven generator is mounted on four springs. Reciprocating inertia effects subject the unit to a sinusoidal vertical excitation of 260 N at engine frequency, with operation at 900 RPM. Total mounted mass, including base, is 33 kg, and radius of gyration about the centroid is 90 cm. Assume symmetry about the central plane through the axis of rotation. Find:

(a) The vertical amplitude at *A*.

(b) The angular amplitude in the vertical plane.

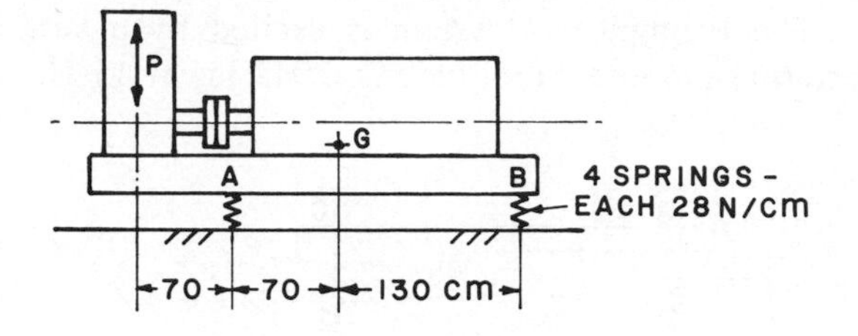

Solution

(a) Using Table 8.5a,

$$\left(\frac{f}{b}\right) = \left(\frac{-170}{100}\right) = -1.7 \qquad \left(\frac{d}{b}\right) = \left(\frac{70}{100}\right) = 0.7$$

$$\omega_1 = \sqrt{\frac{11{,}200}{33}} = 18.4 \qquad \left(\frac{c}{b}\right)^2 = \left(\frac{-30}{100}\right)^2 = 0.09$$

$$\mathsf{g}^2 = \left(\frac{15}{2.93}\right)^2 = 26.2 \qquad \left(\frac{r}{b}\right)^2 = \left(\frac{90}{100}\right)^2 = 0.81$$

$$\left[\frac{\mathsf{a}_1}{(260/11{,}200)}\right] = \left[\frac{2.7 + (26.21)(-0.98 - 0.81)}{507.6}\right] = -0.087$$

$$\mathsf{a}_1 = \mathsf{a}_A = -0.202 \text{ cm}$$

(b)
$$\theta = \left(\frac{260}{11{,}200(1)}\right)\left[\frac{-1.7 + (1.4)(26.2)}{507.6}\right] = +0.0016r$$

$$= +0.092°$$

9 SINGLE-MASS SYSTEMS WITH BEAM ELASTICITY

In Chapter 8 we analyzed single rigid masses with various types of spring suspensions and planar vibratory motions. They involved translational and rotational coordinates, mass moment of inertia, and two natural frequencies and normal modes. In this chapter an analogous situation will be studied in which the elastic behavior relates not to placement of simple springs but to the bending characteristics of beams. This type of system typifies many mechanical structures in which the primary elastic element is a beam or shaft. Although the results are illustrated by basic examples, such as statically determinate beams of constant cross section, methods are general and can be extended as necessary. In this chapter the single-mass element is not rotating and is assumed sufficiently large that the mass of the elastic element can be neglected.

9.1. ELASTIC DEFLECTION OF BEAMS

A beam subjected to a transverse load (Figure 9.1*a*) develops an elastic deflection curve consisting of distributed transverse deflection and slope functions. Similarly, a beam loaded statically by a couple (Figure 9.1*b*) has distributed linear and angular deflection. Within the elastic limit of the material and with typical supports beam deflections are linear, or proportional to the concentrated loads imposed.

In this analysis we are interested only in the deflection characteristics or factors at the single loading point on the beam, the center of gravity of the mass. At this point the deflection rates can be interpreted as springs, either linear or angular; however, the reciprocal form is more convenient and permits us to relate the cross-effects, between force-slope and moment-deflection.

These *compliance* or *influence* factors are defined at a point as

$\alpha = (y/F) =$ linear deflection due to unit load

$\beta = (\theta/F) =$ angular deflection due to unit load

$= (y/T) =$ linear deflection due to unit moment (equal by Maxwell's theorem of reciprocal displacements)

$\gamma = (\theta/T) =$ angular deflection due to unit moment

Thus for a given beam there are four characteristic elastic influence factors at a point, but by reciprocity only three factors must be determined. These are functions of beam geometry, area moment of inertia, and modulus of elasticity.

The factors can be reduced to a dimensional term involving the absolute length and EI, and a dimensionless factor involving load placement (Table 9.1). These terms occur similarly for more complex beams but in less simple form.

Several other important deflection properties, given in Table 9.2, relate to the stiff longitudinal bar with spring supports, as in Chapter 8. For instance, (b) and (c) can be used to account for flexibility in bearing supports. We superimpose beam elasticity upon support elasticity if both are present, by addition of the α, β, and γ factors algebraically. Unequal spring cases are included in Table 9.3.

This brief discussion concerns *elastic coupling*. We have seen that a given load produces a direct and an indirect deflection; α and γ are direct effects, and β represents a cross-coupling effect.

9.2. EQUATIONS INCLUDING MASS MOMENT OF INERTIA

The oscillating rigid mass (Figure 9.2a) develops a transverse inertia force resulting in a transverse deflection, but the associated slope also produces

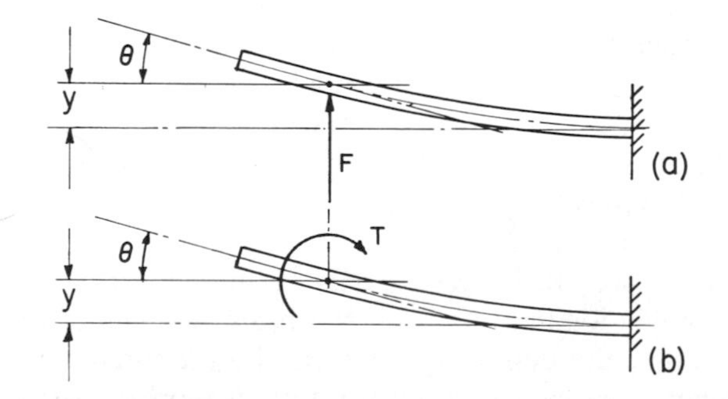

FIGURE 9.1. A static transverse force on a beam produces bending resulting in transverse deflection and slope at a point (a). A couple at this point similarly causes linear and angular deflections (b).

TABLE 9.1. Static elastic compliance factors for basic beams of constant section with clamped or simple supports. Also shown are ratio combinations for use in vibratory equations.

		a (T, θ; F, y; ℓ)	b (T, θ; F, y; a; b; ℓ)	c (T, θ; F, y; a; ℓ)
α	$\frac{y}{F}$	$\frac{\ell^3}{3EI}$	$\frac{\ell^3}{3EI}\left(\frac{a}{\ell}\right)^2\left(\frac{b}{\ell}\right)^2$	$\frac{\ell^3}{3EI}\left(\frac{a}{\ell}\right)^2\left[1+\frac{a}{\ell}\right]$
β	$\frac{\theta}{F}=\frac{y}{T}$	$\frac{\ell^2}{2EI}$	$\frac{\ell^2}{3EI}\left(\frac{a}{\ell}\right)\left(\frac{b}{\ell}\right)\left[2\frac{a}{\ell}-1\right]$	$\frac{\ell^2}{3EI}\left(\frac{a}{\ell}\right)\left[1+\frac{3}{2}\frac{a}{\ell}\right]$
γ	$\frac{\theta}{T}$	$\frac{\ell}{EI}$	$\frac{\ell}{3EI}\left[\left(\frac{a}{\ell}\right)^3+\left(\frac{b}{\ell}\right)^3\right]$	$\frac{\ell}{3EI}\left[1+3\frac{a}{\ell}\right]$
$\ell\psi_1$	$\left(\frac{\beta}{\alpha}\right)\ell$	$\left(\frac{3}{2}\right)$	$\frac{\left(2\frac{a}{\ell}-1\right)}{\left(\frac{a}{\ell}\right)\left(\frac{b}{\ell}\right)}$	$\frac{\left(1+\frac{3}{2}\frac{a}{\ell}\right)}{\left(\frac{a}{\ell}\right)\left(1+\frac{a}{\ell}\right)}$
$\ell^2\psi_2$	$\left(\frac{\gamma}{\alpha}\right)\ell^2$	3	$\frac{\left(1-3\left(\frac{a}{\ell}\right)+3\left(\frac{a}{\ell}\right)^2\right)}{\left(\frac{a}{\ell}\right)^2\left(\frac{b}{\ell}\right)^2}$	$\frac{\left(1+3\frac{a}{\ell}\right)}{\left(\frac{a}{\ell}\right)^2\left(1+\frac{a}{\ell}\right)}$
$\ell^2(\psi_2-\psi_1^2)$		$\left(\frac{3}{4}\right)$	$\frac{1}{\left(\frac{a}{\ell}\right)\left(\frac{b}{\ell}\right)}$	$\frac{\left(1+\frac{3}{4}\frac{a}{\ell}\right)}{\left(\frac{a}{\ell}\right)\left(1+\frac{a}{\ell}\right)^2}$

TABLE 9.2. Elastic factors for a rigid bar supported by spring elements. Both (*b*) and (*c*) represent typical bearing or support flexibilities.

		a (T, θ; F, y; K_θ; ℓ)	b (T, θ; F, y; K/2; K/2; a; b; ℓ)	c (T, θ; F, y; K/4; K/4; K/2; K/2; a; ℓ)
α	$\frac{y}{F}$	$\frac{\ell^2}{K_\theta}$	$\frac{2}{K}\left[\left(\frac{a}{\ell}\right)^2+\left(\frac{b}{\ell}\right)^2\right]$	$\frac{2}{K}\left[1+2\left(\frac{a}{\ell}\right)+2\left(\frac{a}{\ell}\right)^2\right]$
β	$\frac{\theta}{F}=\frac{y}{T}$	$\frac{\ell}{K_\theta}$	$\frac{2}{K\ell}\left[\frac{b}{\ell}-\frac{a}{\ell}\right]$	$\frac{2}{K\ell}\left[1+2\left(\frac{a}{\ell}\right)\right]$
γ	$\frac{\theta}{T}$	$\frac{1}{K_\theta}$	$\frac{4}{K\ell^2}$	$\frac{4}{K\ell^2}$
$\ell\psi_1$	$\left(\frac{\beta}{\alpha}\right)\ell$	1	$\frac{\left(\frac{b}{\ell}\right)-\left(\frac{a}{\ell}\right)}{\left(\frac{b}{\ell}\right)^2+\left(\frac{a}{\ell}\right)^2}$	$\frac{1+2\left(\frac{a}{\ell}\right)}{1+2\left(\frac{a}{\ell}\right)+2\left(\frac{a}{\ell}\right)^2}$
$\ell^2\psi_2$	$\left(\frac{\gamma}{\alpha}\right)\ell^2$	1	$\frac{2}{\left(\frac{b}{\ell}\right)^2+\left(\frac{a}{\ell}\right)^2}$	$\frac{2}{1+2\left(\frac{a}{\ell}\right)+2\left(\frac{a}{\ell}\right)^2}$
$\ell^2(\psi_2-\psi_1^2)$		0	$\frac{1}{\left[\left(\frac{b}{\ell}\right)^2+\left(\frac{a}{\ell}\right)^2\right]^2}$	$\frac{1}{\left[1+2\left(\frac{a}{\ell}\right)+2\left(\frac{a}{\ell}\right)^2\right]^2}$

TABLE 9.3. Compliance coefficients for a beam with a rigid section: (*a*) a rigid overhung element with simple bearing supports and bending between the bearings; (*b*) a rigid body with unequal spring stiffness at each end; and (*c*) superimposed conditions from (*a*) and (*b*).

	T,θ; F,y; EI; a; ℓ — a	T,θ; F,y; $K_1/2$; $K_2/2$; $K_1/2$; $K_2/2$; a; ℓ — b	T,θ; F,y; $K_1/2$; $K_2/2$; EI; $K_1/2$; $K_2/2$; a; ℓ — c
α	$\frac{\ell^3}{3EI}\left(\frac{a}{\ell}\right)^2$	$\frac{1}{K_1}\left[1+\left(\frac{a}{\ell}\right)\right]^2+\frac{1}{K_2}\left(\frac{a}{\ell}\right)^2$	$\frac{B}{K_2}$
β	$\frac{\ell^2}{3EI}\left(\frac{a}{\ell}\right)$	$\frac{1}{K_1\ell}\left[1+\left(\frac{a}{\ell}\right)\right]+\frac{1}{K_2\ell}\left(\frac{a}{\ell}\right)$	$\frac{(a/\ell)A+k_2}{K_2\ell}$
γ	$\frac{\ell}{3EI}$	$\frac{1}{K_1\ell^2}+\frac{1}{K_2\ell^2}$	$\frac{A}{K_2\ell^2}$
$\ell\psi_1$	$\left(\frac{\ell}{a}\right)$	$\frac{k_2(1+a/\ell)+a/\ell}{C}$	$\frac{(a/\ell)A+k_2}{B}$
$\ell^2\psi_2$	$\left(\frac{\ell}{a}\right)^2$	$\frac{1+k_2}{C}$	$\frac{A}{B}$

$A=\left(\frac{K_2\ell^3}{3EI}+k_2+1\right)$ $\quad B=\left[\left(\frac{a}{\ell}\right)^2A+k_2\left(1+2\frac{a}{\ell}\right)\right]$ $\quad C=\left[k_2\left(1+\frac{a}{\ell}\right)^2+\left(\frac{a}{\ell}\right)^2\right]$

$k_2=\frac{K_2}{K_1}$

sinusoidal rotational accelerations. When multiplied by the mass moment of inertia about the transverse axis at the mass centroid, this becomes an inertia couple in the direction of angular displacement, equal to $(Mr^2)(\omega^2\theta_1)$.

Two equilibrium equations can be written for the extreme position using the influence coefficients:

$$\begin{cases} \mathsf{a}_1 = \alpha\left(M\omega^2\mathsf{a}_1\right) + \beta\left(Mr^2\omega^2\theta_1\right) & (9.1a) \\ \theta_1 = \beta\left(M\omega^2\mathsf{a}_1\right) + \gamma\left(Mr^2\omega^2\theta_1\right) & (9.1b) \end{cases}$$

where a_1 and θ_1 are the complex variables corresponding to the transverse and the angular displacements, respectively. Mr^2 is mass moment of inertia, with r the radius of gyration about the centroidal transverse axis.

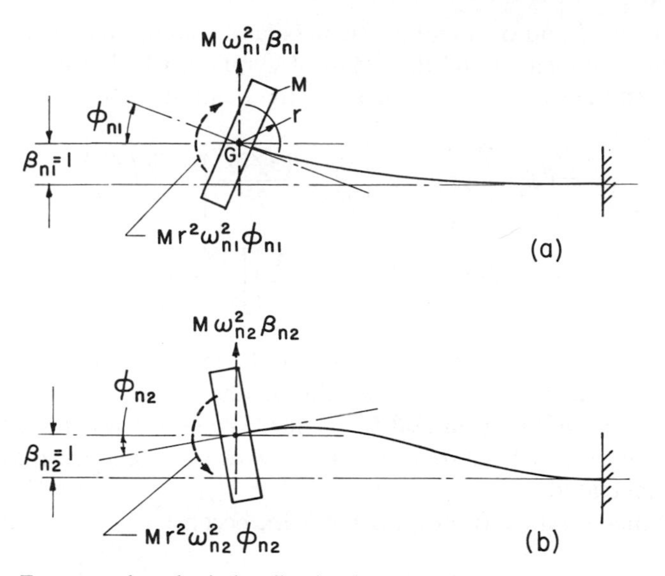

FIGURE 9.2. Two normal modes in bending involve reversal of the angular displacement with respect to the transverse. Corresponding inertia loads are indicated.

The simultaneous equations normalize to

$$\begin{cases} \mathsf{a}_1(1 - \mathsf{g}^2) - \theta_1\left[\left(\dfrac{\beta}{\alpha}\right)r^2\mathsf{g}^2\right] = 0 & (9.2\text{a}) \\ -\mathsf{a}_1\left(\dfrac{\beta}{\alpha}\mathsf{g}^2\right) + \theta_1\left[1 - \left(\dfrac{\gamma}{\alpha}\right)r^2\mathsf{g}^2\right] = 0 & (9.2\text{b}) \end{cases}$$

where $\mathsf{g} = (\omega/\omega_1)$ and $\omega_1 = \sqrt{1/\alpha M}$.

We can consolidate the elastic factors by letting

$$\psi_1 = \left(\frac{\beta}{\alpha}\right) \qquad \psi_2 = \left(\frac{\gamma}{\alpha}\right)$$

Neither factor is dimensionless, as ψ_1 involves a length, and ψ_2 a length squared (Table 9.1). Finally,

$$\begin{cases} \mathsf{a}_1(1 - \mathsf{g}^2) - \theta_1(\psi_1 r^2\mathsf{g}^2) = 0 & (9.3\text{a}) \\ -\mathsf{a}_1(\psi_1\mathsf{g}^2) + \theta_1(1 - \psi_2 r^2\mathsf{g}^2) = 0 & (9.3\text{b}) \end{cases}$$

The first equation has dimensions of displacement and the second has angular dimensions, but this creates no problem in the simultaneous solution. The natural frequency quadratic becomes

$$r^2(\psi_2 - \psi_1^2)\mathsf{g}_n^4 - (1 + r^2\psi_2)\mathsf{g}_n^2 + 1 = 0 \tag{9.4}$$

Coefficients of g_n^4 and g_n^2 are dimensionless, and contain mass-elastic relations between the beam length and the radius of gyration, which are geometric. For instance, with the cantilever (Table 9.1a), the quadratic coefficients are

$$r^2(\psi_2 - \psi_1^2) = r^2\left[\frac{3}{l^2} - \frac{(3/2)^2}{l^2}\right] = \frac{3}{4}\left(\frac{r}{l}\right)^2 \tag{9.5a}$$

$$(1 + r^2\psi_2) = \left[1 + r^2\left(\frac{3}{l^2}\right)\right] = 1 + 3\left(\frac{r}{l}\right)^2 \tag{9.5b}$$

Thus, in this generalized form, if we double the length of the cantilever as we also double the radius of gyration, we alter neither the g_n factors nor the ratio of the higher to the lower natural frequency. The larger system, however, will have lower absolute natural frequencies as α varies as l^3 and the mass may also have been increased.

Normal modal ratios, from Equation 9.3a, become

$$\left(\frac{r\phi}{\beta}\right)_n = \left(\frac{1 - \mathsf{g}_n^2}{r\psi_1 \mathsf{g}_n^2}\right) \tag{9.6}$$

Note that β_n is a normal modal amplitude, not to be confused with the influence factor β. Also in Equations 9.1, 9.2, and 9.3, since they are equated to zero, natural frequency conditions are implied, and technically $\mathsf{a}_1 = \beta_n$, $\theta_1 = \phi_n$, and $\mathsf{g} = \mathsf{g}_n$.

Normal modes are orthogonal (Equation 8.16).

9.3. BEHAVIOR OF NATURAL FREQUENCIES

For given elastic suspension factors α, β, and γ, the remaining characteristic is the radius of gyration when considering generalized frequencies and modes. As shown in Equation 9.5a and 9.5b for the cantilever, frequency coefficients reduce to a single geometric system parameter, (r/l). In Figure 9.3 the natural frequency roots g_n are plotted against this parameter.

At zero the mass is concentrated or we have infinite beam length; neither is possible physically. However, as the axis is approached, $\mathsf{g}_{n1} = 1$, and we have only the basic reference frequency with g_{n2} at infinity. As (r/l) increases, first modal frequency decreases, gradually approaching zero. The higher frequency falls from infinity, approaching a minimum of $\mathsf{g}_{n2} = 2$ for large r or small l. No natural frequency factor g_{ni} can exist between 1.00 and 2.00.

Thus both modal frequencies decrease as (r/l) increases. At small (r/l) we have essentially only first mode. At large (r/l) we have only the second, and the latter frequency will be approximately twice the former.

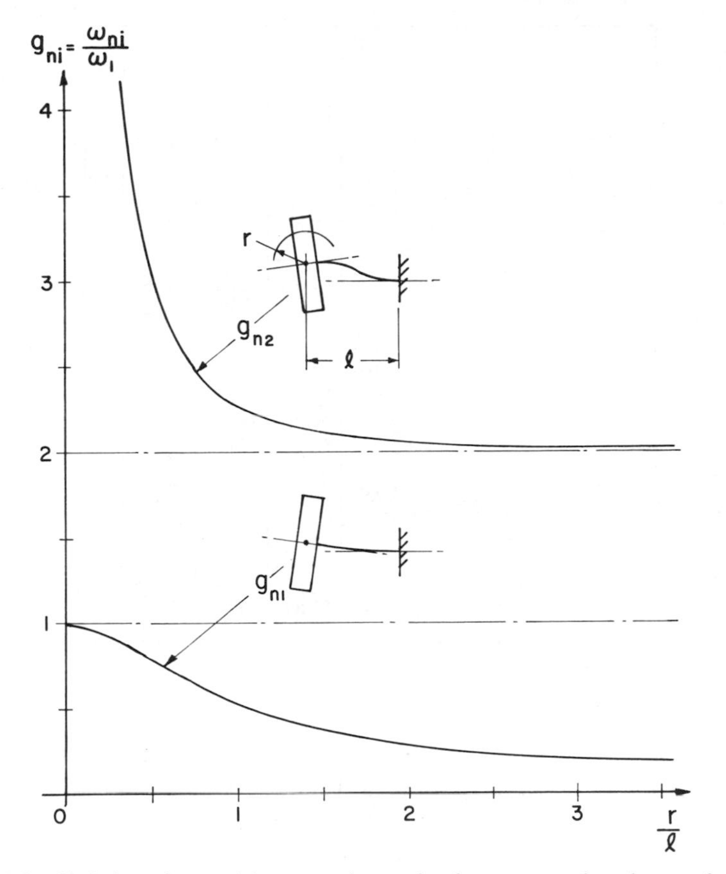

FIGURE 9.3. Variation of natural frequency factors for the two normal modes as a function of the geometric ratio r/l, with $C_n = 0$. Values are for a simple cantilever support.

9.4. FORCED TOTAL RESPONSE

Harmonic excitation on the mass can occur as a sinusoidal force or couple. With a force on the mass at the centroid, we add $P_1\alpha$ and $P_1\beta$ to Equations 9.3a and b respectively. For a couple we similarly add $T_1\beta$ and $T_1\gamma$. Results in Table 9.4*a* and *b* apply to the general case of a beam system elastically supported to ground with compliance in only the transverse direction. Reactions to the ground support are P_2 and T_2.

These reactions are given with P_1 acting colinearly with P_2, as with a single-degree system. Actual sinusoidal forces at the ground points are determined from static equilibrium. For instance, with a shaft supported at the ends (Figure 9.4*a*), these are the sinusoidal forces, P_A and P_B. The calculated P_2 applied at G results in two bearing reactions (*b*). The couple T_2 creates two equal and opposite forces (*c*). By addition we obtain the superimposed loads.

TABLE 9.4. General steady-state response for rigid mass and elastically coupled mounting, all forces and linear amplitudes transverse to the axis at the centroid: (*a*) and (*b*) excited by P_1 and T_1; (*c*) and (*d*) displacement excited by $\mathbf{a}_0$ and θ_0 from the structure.

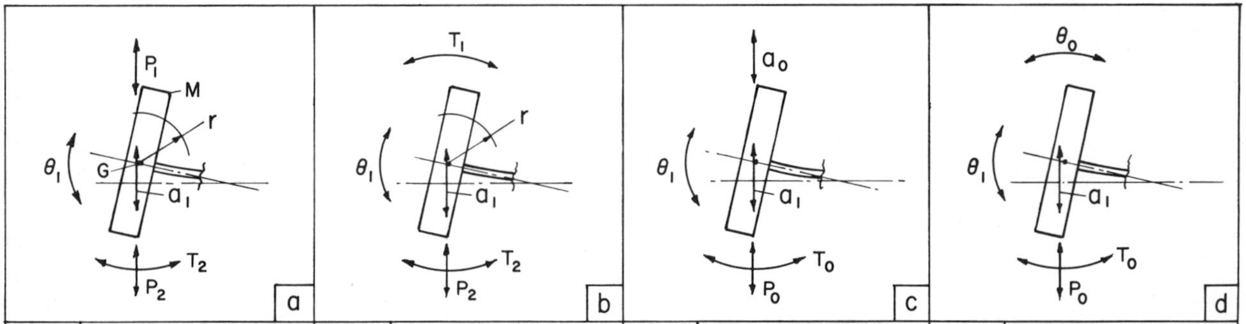

a		b		c		d	
$\frac{a_1}{P_1\alpha}$	$\frac{1-r^2Ag^2}{C}$	$\frac{a_1}{T_1\alpha/r}$	$\frac{r\psi_1}{C}$	$\frac{a_1}{a_0}$	$\frac{1-r^2\psi_2 g^2}{C}$	$\frac{a_1}{\theta_0 r}$	$\frac{r\psi_1 g^2}{C}$
$\frac{\theta_1}{P_1\alpha/r}$	$\frac{r\psi_1}{C}$	$\frac{\theta_1}{T_1\alpha/r^2}$	$\frac{r^2(\psi_2-Ag^2)}{C}$	$\frac{\theta_1}{a_0/r}$	$\frac{r\psi_1 g^2}{C}$	$\frac{\theta_1}{\theta_0}$	$\frac{(1-g^2)}{C}$
$\frac{P_2}{P_1}$	$\frac{(r^2\psi_2 g^2-1)}{C}$	$\frac{P_2}{T_1/r}$	$\frac{-r\psi_1 g^2}{C}$	$\frac{P_0}{a_0/\alpha}$	$\frac{g^2(r^2\psi_2 g^2-1)}{C}$	$\frac{P_0}{\theta_0 r/\alpha}$	$\frac{-r\psi_1 g^4}{C}$
$\frac{T_2}{P_1 r}$	$\frac{-r\psi_1 g^2}{C}$	$\frac{T_2}{T_1}$	$\frac{(g^2-1)}{C}$	$\frac{T_0}{a_0 r/\alpha}$	$\frac{-r\psi_1 g^4}{C}$	$\frac{T_0}{\theta_0 r^2/\alpha}$	$\frac{g^2(g^2-1)}{C}$

$A = (\psi_2 - \psi_1^2)$, $g = \omega\sqrt{\alpha M}$, $r^2 = \frac{I}{M}$, $C = r^2Ag^4 - (1 + r^2\psi_2)g^2 + 1$

$$\left(\frac{r\phi}{\beta}\right)_n = \frac{1-g_n^2}{r\psi_1 g_n^2}$$

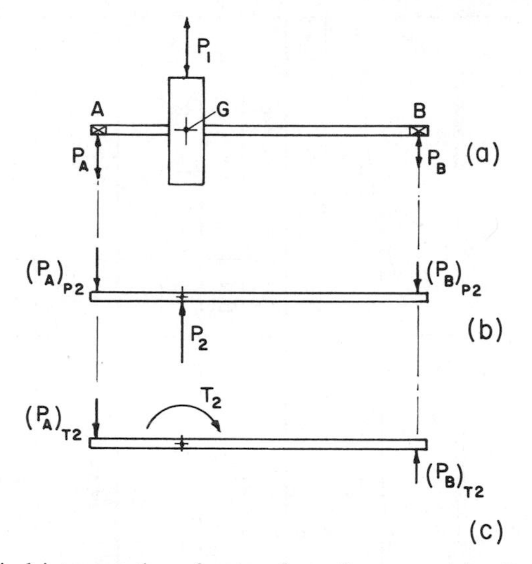

FIGURE 9.4. Physical interpretation of external reactions caused by P_2 and T_2 (Table 9.4*a*) illustrated by a simple beam, static equilibrium, and superposition.

It is necessary to take P_2 and T_2 as having a static direction relative to P_1 because the addition is algebraic. Vibrationwise, we thus obtain the actual phase relationships between P_A, P_B, and P_1.

With displacement forcing, a_0, or θ_0, Table 9.4*c* and *d* apply. Equations result from adding these terms directly on the right in Equations 9.3a and b, respectively. P_0 and T_0 are the driving functions associated with the displacement excitations.

The cantilevered single mass is a special but important case for which the elastic factors from Table 9.1*a* can be used to reduce the total response equations to the results in Table 9.5 for the four basic excitation possibilities.

9.5. EQUIVALENT ELASTIC SUSPENSIONS

This chapter has dealt with the effects of inertia in systems in which the characteristic compliances are defined at a point by α, β, and γ. A resemblance should be noted between the systems of Chapter 8 and those now discussed. The former were transverse rigid bodies with elasticity analyzed in terms of the axial dimensions and simple springs; however, rigid masses with planar motion are elastically supported, and influence factors can be developed for the Chapter 8 cases. Thus although these analyses could be merged, they are treated separately to expedite solutions and to illustrate different conceptual approaches.

TABLE 9.5. Response of a cantilevered mass with transverse inertia to sinusoidal excitation: (*a*) and (*b*) force P_1 or couple T_1; (*c*) and (*d*) translation of the base, $\mathbf{a}_3$, or rotation of the base, θ_3.

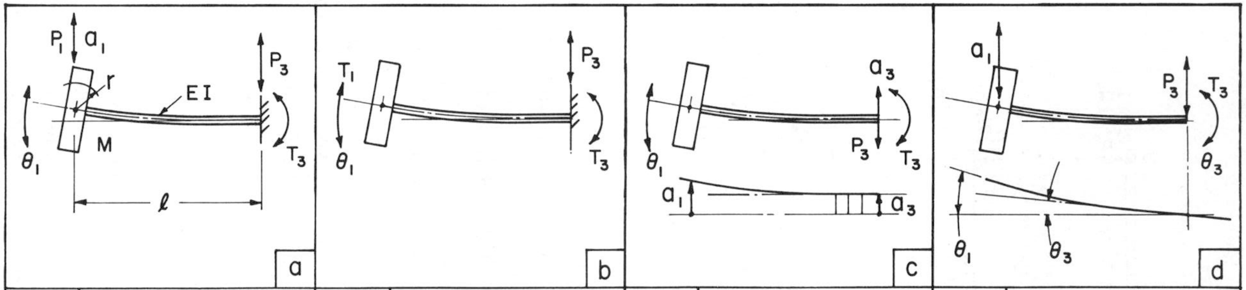

a		b		c		d	
$\frac{a_1}{P_1\alpha}$	$\frac{1-\frac{3}{4}A}{C}$	$\frac{a_1}{T_1\beta}$	$\frac{1}{C}$	$\frac{a_1}{a_3}$	$\frac{1-3A}{C}$	$\frac{a_1}{\theta_3\ell}$	$\frac{1-\frac{3}{2}A}{C}$
$\frac{\theta_1}{P_1\beta}$	$\frac{1}{C}$	$\frac{\theta_1}{T_1\gamma}$	$\frac{1-\frac{1}{4}g^2}{C}$	$\frac{\theta_1}{a_3/\ell}$	$\frac{\frac{3}{2}g^2}{C}$	$\frac{\theta_1}{\theta_3}$	$\frac{1+\frac{1}{2}g^2}{C}$
$\frac{P_3}{P_1}$	$\frac{-1+3A}{C}$	$\frac{P_3}{T_1/\ell}$	$\frac{\frac{3}{2}g^2}{C}$	$\frac{P_3}{a_3/\alpha}$	$\frac{3A-g^2}{C}$	$\frac{P_3}{\theta_3\ell/\alpha}$	$\frac{g^2(\frac{3}{2}A-1)}{C}$
$\frac{T_3}{P_1\ell}$	$\frac{-1+\frac{3}{2}A}{C}$	$\frac{T_3}{T_1}$	$\frac{-(1+\frac{1}{2}g^2)}{C}$	$\frac{T_3}{a_3\ell/\alpha}$	$\frac{g^2(\frac{3}{2}A-1)}{C}$	$\frac{T_3}{\theta_3\ell^2/\alpha}$	$\frac{A(g^2-1)-g^2}{C}$

$A=\left(\frac{r}{\ell}\right)^2 g^2,\quad C=\frac{3}{4}\left(\frac{r}{\ell}\right)^2 g^4-\left[1+3\left(\frac{r}{\ell}\right)^2\right]g^2+1,\quad g^2=\omega^2(\alpha M),\quad \alpha=\frac{\ell^3}{3EI},\ \beta=\frac{\ell^2}{2EI},\ \gamma=\frac{\ell}{EI}$

$\left(\frac{r\phi}{\beta}\right)_n=\frac{1-g_n^2}{\frac{3}{2}\left(\frac{r}{\ell}\right)g_n^2}$

In transferring the system of Table 8.4a, for instance, to Table 9.4a, or transforming to elastic influence factors, the relations are

$$\alpha = \left[\frac{1 + (c/b)^2}{K}\right] \qquad \beta = \left[\frac{(c/b^2)}{K}\right] \qquad \gamma = \left[\frac{(1/b^2)}{K}\right]$$

$$\psi_1 = \left(\frac{c}{b^2 + c^2}\right) \qquad \psi_2 = \left(\frac{1}{b^2 + c^2}\right) \tag{9.7}$$

9.6. EQUIVALENT MODAL SYSTEMS

As shown in Section 4.6, multimass translational and torsional systems have vibrationally equivalent spring and mass systems based on natural frequencies and normal modes. This principle is now extended to the present coupled coordinates, which include rotation of the mass. To maintain generality and permit couple excitation, both a translational and rotational spring–mass equivalent are required for each mode.

In translation (Figure 9.5b), summing kinetic energies at maximum velocities as in Equation 4.9, the equivalent mass becomes

$$\tfrac{1}{2}M_n(\beta_n\omega_n)^2 = \tfrac{1}{2}M(\beta_n\omega_n)^2 + \tfrac{1}{2}Mr^2(\phi_n\omega_n)^2$$

$$m_{ni} = \left(\frac{M_n}{M}\right) = 1 + \left(\frac{r\phi}{\beta}\right)_n^2 \tag{9.8}$$

The second term on the right represents the increase in modal mass due to angular effects and involves the normal mode from Equation 9.6. The equivalent modal spring is then

$$K_n = M_n\omega_n^2 = \left[1 + \left(\frac{r\phi}{\beta}\right)_n^2\right] M\omega_n^2$$

$$k_n = \left(\frac{K_n}{K}\right) = \left[1 + \left(\frac{r\phi}{\beta}\right)_n^2\right] g_n^2 \tag{9.9}$$

where

$$K = (1/\alpha)$$

$$g_n = (\omega_n/\omega_1)$$

In rotation (Figure 9.5c), taking rotational kinetic energies as in Equation

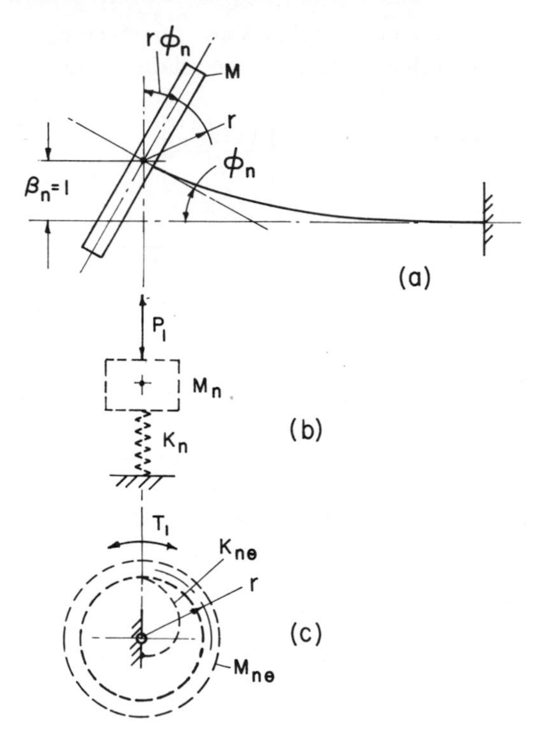

FIGURE 9.5. Replacement of the actual system by equivalent modal single-degree systems requires two for each mode. A translational system simulates lateral modal stiffness, as seen by P_1. A torsional modal spring resists the excitation couple T_1.

9.8 the angular modal mass becomes

$$\tfrac{1}{2}M_{n\theta}r^2(\phi_n\omega_n)^2 = \tfrac{1}{2}M(\beta_n\omega_n)^2 + \tfrac{1}{2}Mr^2(\phi_n\omega_n)^2$$

$$m_{n\theta} = \left(\frac{M_{n\theta}}{M}\right) = 1 + \left(\frac{\beta}{r\phi}\right)_n^2 \tag{9.10}$$

And for the modal torsional spring,

$$k_{n\theta} = \left(\frac{K_{n\theta}}{Kr^2}\right) = \left[1 + \left(\frac{\beta}{r\phi}\right)_n^2\right]g_n^2 \tag{9.11}$$

9.7. FORCED MODAL RESPONSE

In the simplest possible case (Table 2.1*a*), without damping, the magnification factor involves the forced frequency ratio, and the equilibrium amplitude is

TABLE 9.6. Modal characteristics and steady-state responses: (*a*) and (*b*) forced by P_1 and T_1; (*c*) and (*d*) displacement excited in translation and rotation.

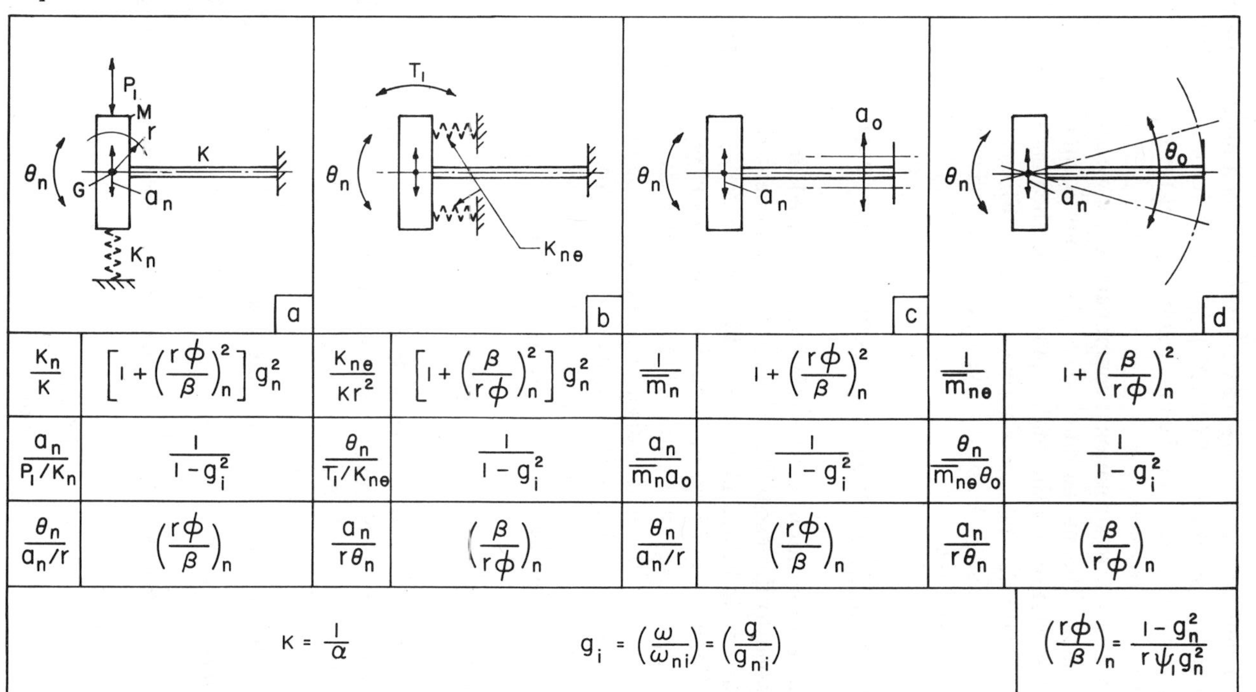

(a)		(b)		(c)		(d)	
$\frac{K_n}{K}$	$\left[1+\left(\frac{r\phi}{\beta}\right)_n^2\right] g_n^2$	$\frac{K_{n\theta}}{Kr^2}$	$\left[1+\left(\frac{\beta}{r\phi}\right)_n^2\right] g_n^2$	$\frac{1}{\bar{m}_n}$	$1+\left(\frac{r\phi}{\beta}\right)_n^2$	$\frac{1}{\bar{m}_{n\theta}}$	$1+\left(\frac{\beta}{r\phi}\right)_n^2$
$\frac{a_n}{P_1/K_n}$	$\frac{1}{1-g_i^2}$	$\frac{\theta_n}{T_1/K_{n\theta}}$	$\frac{1}{1-g_i^2}$	$\frac{a_n}{\bar{m}_n a_o}$	$\frac{1}{1-g_i^2}$	$\frac{\theta_n}{\bar{m}_{n\theta}\theta_o}$	$\frac{1}{1-g_i^2}$
$\frac{\theta_n}{a_n/r}$	$\left(\frac{r\phi}{\beta}\right)_n$	$\frac{a_n}{r\theta_n}$	$\left(\frac{\beta}{r\phi}\right)_n$	$\frac{\theta_n}{a_n/r}$	$\left(\frac{r\phi}{\beta}\right)_n$	$\frac{a_n}{r\theta_n}$	$\left(\frac{\beta}{r\phi}\right)_n$

$$K = \frac{1}{\alpha} \qquad g_i = \left(\frac{\omega}{\omega_{ni}}\right) = \left(\frac{g}{g_{ni}}\right) \qquad \left(\frac{r\phi}{\beta}\right)_n = \frac{1-g_n^2}{r\psi_1 g_n^2}$$

(P_1/K_1). Similarly, in Table 9.6*a* we have the modal magnification factor and an equilibrium amplitude of (P_1/K_1). With this information it is a relatively simple matter to determine the linear amplitude of the mass in a given mode at a given forcing frequency.

But we also have a rotational component, since we are isolating a particular normal mode. This modal ratio is $(r\phi/\beta)_n$ from Equation 9.6.

If the excitation is a sinusoidal couple T_1 on the mass (Table 9.6*b*), we must find the rotational static displacement $T_1/K_{n\theta}$. From this we have the modal angular response θ_n. As before, we find the associated linear component a_n from the modal ratio.

9.8. MODAL RESPONSE TO DISPLACEMENT EXCITATION

With base translation, inertia effects are induced at the mass centroid that correspond to the excitation. For modal component response an $\overline{m}$ factor is necessary, as indicated in Section 5.10, for relative or absolute modal amplitudes. Results are shown in Table 9.6*c*, and *d* for $\overline{m}_n$ and $\overline{m}_{n\theta}$ from energy considerations used in Section 9.6. Modal responses are then proportional to the $\overline{m}$ factors, to the driving displacement, and to the magnification factor. The complementary modal displacements also follow from modal ratios as shown.

We note that the angular base excitation in Table 9.6*d* is unusual, but it is the only type that generates simple rotation of M.

EXAMPLES

EXAMPLE 9.1. A uniform beam, 75 cm between simple supports, supports a static couple of 155 N · m at a distance of 15 cm from one support. The beam is steel and 4 cm in diameter. Find:

(a) The slope at the couple.

(b) The transverse deflection at the couple.

Solution

$$E = 200(10)^9 \text{ N/m}^2$$

$$I = \frac{\pi d^4}{64} = \frac{\pi(4)^4}{64} = 12.57 \text{ cm}^4 = 0.1257(10)^{-6} \text{ m}^4$$

$$EI = 25{,}120 \text{ N} \cdot \text{m}^2$$

From Table 9.1*b*,

(a) $$\gamma = \frac{\theta}{T} = \frac{0.75}{3EI}\left[(0.20)^3 + (0.80)^3\right] = 5.18(10)^{-6} \text{ r/N} \cdot \text{m}$$

$$\theta = 155\left[5.18(10)^{-6}\right]$$

$$= 0.0008 \text{ r}$$

(b) $$\beta = \frac{y}{T} = \frac{(0.75)^2}{3EI}(0.20)(0.80)(0.40 - 1) = -0.72(10)^{-6}N^{-1}$$

$$y = 155[-0.72(10)^{-6}]$$
$$= -0.011 \text{ cm}$$

EXAMPLE 9.2. A cantilevered steel shaft supports a thin cylindrical disk as shown. For the sinusoidal couple excitation at the disk, determine:

(a) The shear reaction at the fixed end.

(b) Amplitude of the moment reaction.

(c) The natural frequencies.

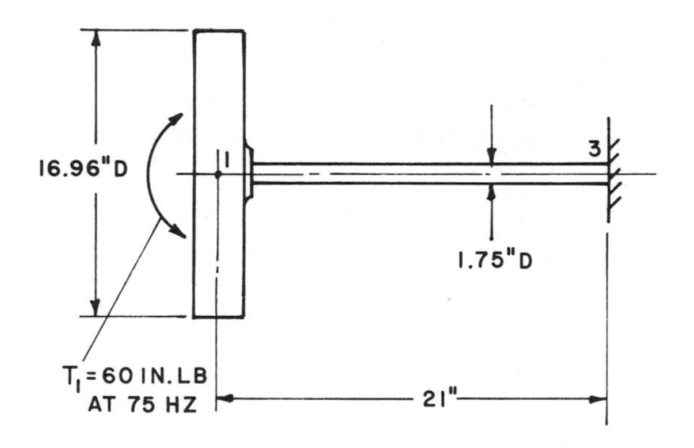

Solution. From Tables 9.1*a* and 9.5*b*,

$$\alpha = \frac{(21)^3}{3(30)(10)^6(0.460)} = 224(10)^{-6} \text{ in./lb}$$

$$r^2A = \left(\frac{3}{4}\right)\left(\frac{r}{l}\right)^2 \qquad r^2 = \frac{R^2}{4} = 17.98 \text{ in.}^2$$

$$r\psi_1 = \frac{3}{2}\left(\frac{r}{l}\right) \qquad r = 4.24 \text{ in.}$$

$$r^2\psi_2 = 3\left(\frac{r}{l}\right)^2 \qquad \left(\frac{r}{l}\right) = 0.202$$

$$\omega_1 = \sqrt{\frac{386}{224(10)^{-6}42}} = 202.6 \text{ rad/sec} = 32.2 \text{ Hz}$$

$$g^2 = \left(\frac{75}{32.2}\right)^2 = 5.41$$

$$C = 0.0306g^4 - 1.122g^2 + 1 = -4.18$$

(a) $$P_2 = \left(\frac{60}{4.24}\right)\left[\frac{1.5(0.202)5.41}{-4.18}\right] = +5.55 \text{ lb}$$

Note that P_2 is the required reaction in the plane of the disk. Also, we could have more easily used Table 9.5*b* to solve this problem, from which the shear amplitude at the base is $P_3 = -5.55$ lb.

(b)
$$T_2 = 60\left(\frac{5.41 - 1}{-4.18}\right) = -63.3 \text{ in. lb}$$

Taking summation of moments about an axis at the base,

$$\Sigma T_3 = T_3 + T_2 + P_2 l = 0$$

$$T_3 = 63.3 - 5.55(21) = -53.2 \text{ in. lb} \quad (\text{out of phase with } T_1)$$

(c)
$$C_n = 0.0306\mathsf{g}_n^4 - (1.122)\mathsf{g}_n^2 + 1$$

$$\mathsf{g}_{n1}^2 = 0.907 \qquad \mathsf{g}_{n2}^2 = 35.76$$
$$\mathsf{g}_{n1} = 0.952 \qquad \mathsf{g}_{n2} = 5.98$$

$$\omega_{n1} = (0.952)(32.2) = 30.7 \text{ Hz}$$

$$\omega_{n2} = (5.98)(32.2) = 193 \text{ Hz}$$

EXAMPLE 9.3. The disk of Example 9.2 is supported in a single span as shown, with the shaft again 1.75 in. in diameter. Find:

(a) The natural frequencies.
(b) The normal modes.

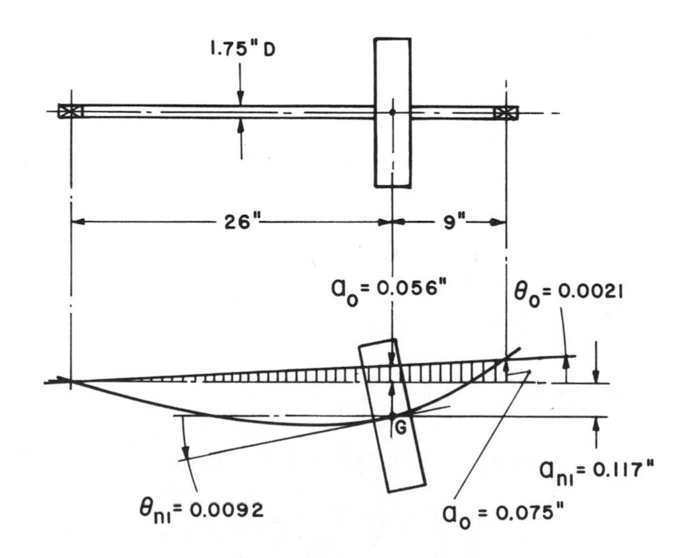

Solution. Using Table 9.1*b* for elastic characteristics,

$$\frac{a}{l} = \left(\frac{26}{35}\right) = 0.743$$

$$\frac{b}{l} = \left(\frac{9}{35}\right) = 0.257$$

$$\alpha = \left(\frac{l^3}{3EI}\right)\left(\frac{a}{l}\right)^2\left(\frac{b}{l}\right)^2$$

$$= 37.8(10)^{-6} \text{ in./lb}$$

$$l\psi_1 = \left[\frac{2(0.743) - 1}{(0.743)(0.257)}\right] = 2.545$$

$$l^2\psi_2 = \left[\frac{1 - 3(0.743) + 3(0.743)^2}{(0.743)^2(0.247)^2}\right] = 11.715$$

$$l^2(\psi_2 - \psi_1^2) = \left[\frac{1}{(0.743)(0.257)}\right] = 5.237$$

Note that all elasticity factors and g_n values are identical if $a = 26$ and $b = 9$, or, for the reversed system, if $a = 9$ and $b = 26$.

(a) From Table 9.4,

$$(r/l)^2 = (4.24/35)^2 = (0.121)^2 = 0.0147$$

$$C_n = (0.077)\mathsf{g}_n^4 - (1.172)\mathsf{g}_n^2 + 1 = 0$$

$$\mathsf{g}_{n1}^2 = 0.907 \qquad \mathsf{g}_{n1} = 0.952$$

$$\mathsf{g}_{n2}^2 = 14.31 \qquad \mathsf{g}_{n2} = 3.78$$

$$\omega_1 = \sqrt{\frac{1}{\alpha M}} = \sqrt{\frac{386}{37.8(10)^{-6}42}} = 493 \quad \text{or} \quad 78.5 \text{ Hz}$$

$$\omega_{n1} = 0.952(78.5) = 74.7 \text{ Hz}$$

$$\omega_{n2} = 3.78(78.5) = 297 \text{ Hz}$$

(b)

$$\left(\frac{r\phi}{\beta}\right)_{n1} = \left[\frac{1 - 0.907}{(4.24)(0.0727)(0.907)}\right] = +0.33$$

$$\left(\frac{r\phi}{\beta}\right)_{n2} = -3.02$$

EXAMPLE 9.4. The shaft of Example 9.3 is subjected to a sinusoidal vertical displacement of 0.075 in. at 90 Hz at the right bearing. The left bearing has zero motion. Determine the first modal response.

Solution. Using Table 9.6c and d, the respective excitational displacements in a translational and a rotational sense are

$$\mathsf{a}_0 = \left(\frac{26}{35}\right)(0.075) = 0.0557$$

$$\theta_0 = \frac{0.075}{35} = 0.00214 \text{ rad.}$$

In translation (Table 9.6c),

$$\overline{m}_{n1} = \frac{1}{1 + (0.333)^2} = 0.900$$

$$\mathsf{g} = \left(\frac{90}{74.7}\right)$$

$$\mathsf{a}_{n1} = (0.900)(0.0557)(-2.214) = -0.111 \text{ in.}$$

$$\theta_{n1} = \left(\frac{-0.111}{4.24}\right)(0.333) = -0.0087 \text{ rad}$$

With angular excitation (Table 9.6d),

$$\overline{m}_{n\theta} = \left[\frac{1}{1 + (1/0.333)^2}\right] = 0.100$$

$$\theta_{n1} = (0.100)(0.00214)(-2.214) = -0.00048 \text{ rad}$$

$$\mathsf{a}_{n1} = (-4.24)(0.00048)\left(\frac{1}{0.333}\right) = -0.0061 \text{ in.}$$

$$\Sigma\mathsf{a}_{n1} = -0.111 - 0.006 = -0.117 \text{ in.}$$

$$\Sigma\theta_{n1} = -0.0087 - 0.0005 = -0.0092 \text{ rad}$$

Note that the amplitudes are absolute and out of phase with the displacement excitation.

10 WHIRL OF THE SINGLE-MASS SYSTEM

Whirl refers to an orbital motion of a mass-elastic system during rotation. This contrasts with the normal system with sinusoidal motions in a plane, but the two cases are closely related. In whirl a circular displacement exists in space and the radial vector projected on a plane results in simple harmonic motion. Rotative frequency corresponds to circular cyclic frequency. Similarly, critical speeds in whirl are linked to the planar natural frequencies.

Physically, if we take a cantilever-mounted mass, displace it radially, and release it, we observe the planar vibration. But if we give the same mass an orbital start, an orbital action ensues at exactly the same frequency. A simple pendulum can also be excited in a planar or circular orbit, the latter a spherical pendulum.

Gyroscopic effects involving the spin velocity about the axis and the polar mass moment of inertia distinguish the analysis in whirl from the equations in Chapter 9; however, those tabulated results are generally applicable to whirl when modified by the gyroscopic factors.

There are specialized types of whirl, some depending on bearing clearances, some induced by gravity, and some induced by lubrication effects in sleeve bearings. The current treatment is confined to classical cases with linear elasticity and a simple rotating whirl excitation at constant speed.

10.1. THE ROTATING BALANCED MASS

In the simplest arrangement, a rotating mass (Figure 10.1) is in perfect balance, with the center of gravity coinciding exactly with the axis of rotation. The elastic shaft has deflection characteristics (Table 9.1*b*) for constant diameter; however, only the factor α is of interest because no slope or angular inertia effects are present. Considering static equilibrium in the rotating plane, or the plane of the paper instantaneously, the only two forces acting are the inertia

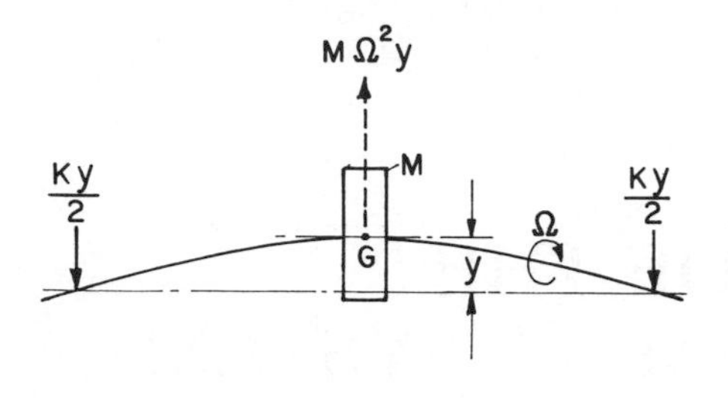

FIGURE 10.1. A central, balanced mass in rotation is loaded by the centrifugal force resulting from elastic deflection; however, this only occurs at the critical speed.

and the spring on the mass.

$$\Sigma F_y = -M\ddot{y} - Ky = 0 \tag{10.1a}$$

$$M\Omega^2 y - \left(\frac{1}{\alpha}\right)y = 0$$

$$\Omega_c = \sqrt{\frac{1}{\alpha M}} = \omega_1 \tag{10.1b}$$

where y is the transverse elastic deflection of the shaft in a rotating plane containing the shaft axis.

This solution indicates that deflection only occurs at a particular speed satisfying Equation 10.1b. This is the *critical speed*, equal to the natural transverse frequency with y undetermined at this condition. Thus the system is unstable with the centrifugal force at any radius equal to the elastic restraining force. The rotating mass is then potentially in equilibrium over the entire range of radius from zero to infinity. Stability and zero radius exist at all other speeds.

10.2. STATIC UNBALANCE EXCITATION

If the mass includes an unbalance, providing an eccentricity between the mass centroid and the axis of rotation, there is a centrifugal force of $M\Omega^2 e_0$ if the shaft is stiff. With shaft flexure, the radius and the force increase (Figure 10.2), and Equation 10.1a becomes

$$\Sigma F_y = M\Omega^2(y + e_0) - \left(\frac{1}{\alpha}\right)y = 0$$

$$\left(\frac{y}{e_0}\right) = \left[\frac{(\Omega/\omega_1)^2}{1 - (\Omega/\omega_1)^2}\right] = \left(\frac{g^2}{1 - g^2}\right) \tag{10.2}$$

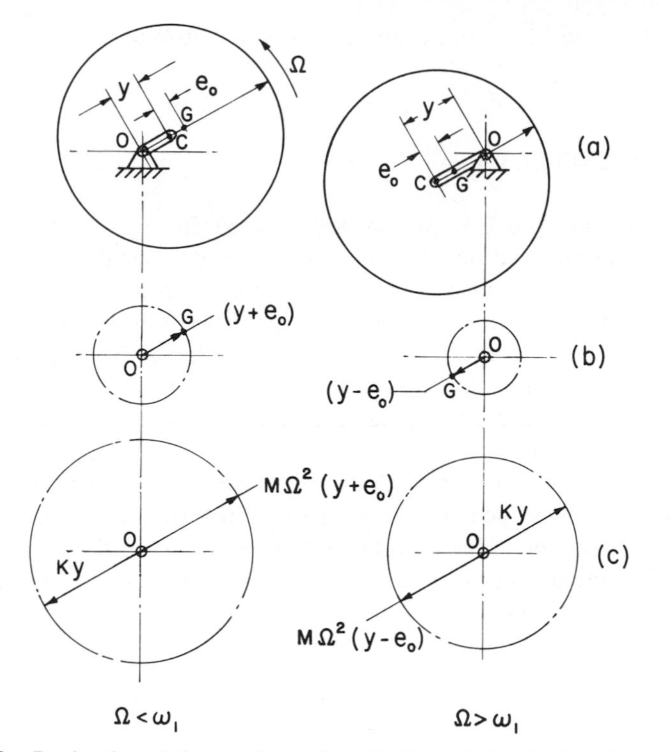

FIGURE 10.2. Passing from below to above the critical speed, the elastic deflection y and the radius of unbalance shift from in phase to out of phase.

where y = radial elastic deflection of shaft

e_0 = static eccentricity of mass center of gravity

$\mathsf{g} = \Omega/\omega_1$ = ratio of rotative (forcing) frequency to natural frequency, or to critical speed

$\omega_1 = \sqrt{1/\alpha M} = \Omega_c$ = natural frequency and critical speed

Equation 10.2 is a magnification effect, corresponding to Table 2.1*b* for zero damping. In whirl the displacement excitation is a displacement e_0 rotating in space, rather than the vector a_2, which rotates in the complex plane.

The mass centroid whirls at a radius of $(y + e_0)$, in phase below resonance, and out of phase above resonance, with total radius given by

$$\left(\frac{\mathsf{a}_1}{e_0}\right) = \left(\frac{y + e_0}{e_0}\right) = \left[\frac{1}{1 - (\Omega/\omega_1)^2}\right] = \left(\frac{1}{1 - \mathsf{g}^2}\right) \qquad (10.3)$$

This corresponds to the $(\mathsf{a}_1/\mathsf{a}_2)$ ratio of Table 2.1*c*. We note that e_0 is analogous to a_2 and $(y + e_0)$ is analogous to the absolute displacement of the mass, a_1.

10.3. SPEED EFFECTS

Equation 10.2 and 10.3 responses are plotted in Figure 10.3. With zero damping the phase relation of y to e_0 is algebraic. Instability occurs at the critical speed and radial displacement of the mass centroid diverges with time, more drastically with large unbalance, e_0. At speeds well above the critical, (y/e_0) approaches -1 and y approaches $-e_0$. The shaft deflection therefore compensates for the unbalance such that the mass tends ultimately to rotate about its centroid, thus reducing the developed centrifugal force asymptotically to zero.

This interesting feature resembles the seismic mounting of a vibratory system but its achievement requires passing through the critical speed, both starting and stopping. This is often possible if the instability is traversed rapidly and if the mass is accurately balanced. Stops can also be used to limit radial amplitudes temporarily.

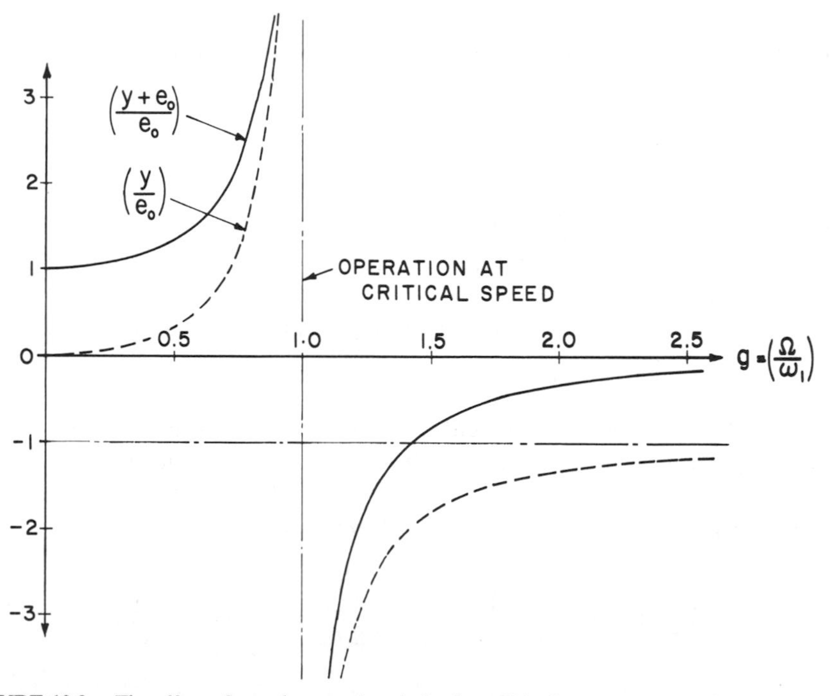

FIGURE 10.3. The effect of speed on total and elastic radii indicates usual resonant characteristics, tending to infinity at the critical speed.

10.4. THE GYROSCOPIC COUPLE

In special cases, such as in Figure 10.1, there is no concern with mass moment of inertia because of symmetry, but in general this factor creates a vibratory couple in the plane of bending (Figure 9.2). A further complication arises when the mass is rotating, or *spinning*, which results in a second inertia couple, adding to or subtracting to the former.[2]

As shown in Figure 10.4, the disk rotates at a *spin velocity* Ω about BG, a principal axis of the mass. There is a simultaneous orbital conical displacement of BG, with cone angle θ, a small angle representing the slope of a deflected shaft. Constant *orbital velocity* ω occurs about the central axis OB.

In the vertical plane (Figure 10.4*b*), we note the presence of the inertia couple T_I, as in Figure 9.2. This results from the tilt of the disk and acts about the transverse diametral axis perpendicular to the plane OGB and passing through the centroid G. A second reversed effective couple T_J applied about this same axis will now be determined. T_J is a gyroscopic effect superimposed if the disk rotates or spins about its polar axis. The source is a constant change in the angular direction of the momentum vector $J\Omega$, which coincides with spin axis GB.

To obtain T_J we use Figure 10.5, with double-headed vectors representing rotational quantities. With this convention the variable turns about the vector axis and lies in a plane perpendicular to the vector; it has a clockwise sense

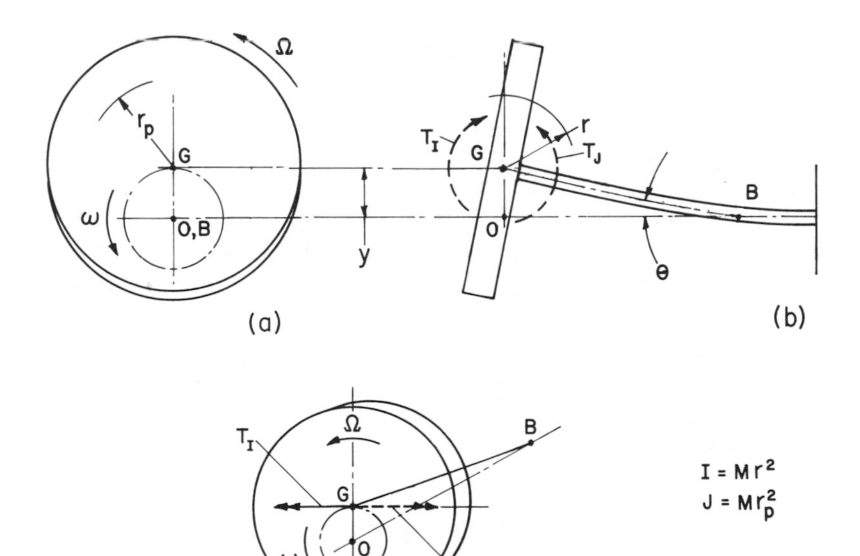

FIGURE 10.4. Conical whirl displacement of a mass creates both the inertia couple of Chapter 9 and a gyroscopic couple, T_J. The latter is proportional to rotative speed Ω and the polar mass moment of inertia J.

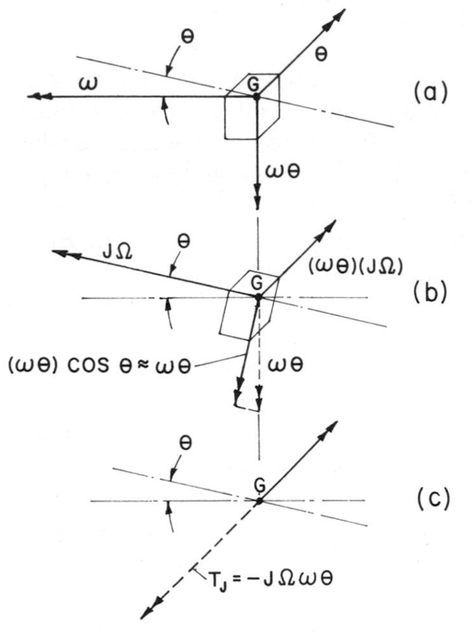

FIGURE 10.5. Vector components that lead to the gyroscopic couple result from two vector cross-products. Views relate to Figure 10.4*b*, but include double-headed vectors θ, $(\omega\theta)$ $(J\Omega)$, and T_J, which are shown isometrically, perpendicular to the basic plane *OGB*.

when viewed from the tail to the vector. The sequence is as follows:

1. The whirl velocity ω multiplies the tilt vector θ, resulting in an angular velocity $\omega\theta$ coincident with *OG*.
2. We now multiply the vectors $(\omega\theta)$ and $(J\Omega)$ to obtain the vector product, which is the vector rate of change of momentum of the disk. It exists in the plane *OGB*, directed clockwise in Figure 10.5*b*; however, a slight complication arises with respect to the direction of $\omega\theta$, which is perpendicular to *OB* rather than *OG*. But with small θ, $\cos\theta \approx 1$, and we neglect the sine component. This approximation is satisfactory in most whirl situations.
3. The gyroscopic couple required on the mass is provided by the shaft. Thus the inertia load on the shaft is equal and opposite, as with D'Alembert, and we have by reversal the magnitude and direction of T_J for the assumed directions of θ, ω, and Ω.

In the vector multiplications shown in Figure 10.5, the product is perpendicular to the plane of the factors, and all three vectors are mutually perpendicular. Direction of the product is consistent with the multiplication of factors in the order indicated by the products. This procedure resembles that of Section 1.3, in which we used multiplication by two successive operators to convert displacement to acceleration in the complex plane.

The gyroscopic couple ($J\Omega\omega\theta$) is directly proportional to each factor. If any of these parameters is zero, the couple vanishes.

10.5. EQUATIONS OF THE UNFORCED SYSTEM

Given a system (Figure 10.4), in a steady-state whirling condition, the equations are similar to Equations 9.1, but include T_J.

$$
\begin{cases}
y_1 = \alpha\left(M\omega^2 y_1\right) + \beta\left(I\omega^2 - J\Omega\omega\right)\theta_1 & (10.4a) \\
\theta_1 = \beta\left(M\omega^2 y_1\right) + \gamma\left(I\omega^2 - J\Omega\omega\right)\theta_1 & (10.4b)
\end{cases}
$$

These can be normalized to

$$
\begin{cases}
\left(1 - \mathsf{g}^2\right)y_1 - \left(\rho r^2 \psi_1 \mathsf{g}^2\right)\theta_1 = 0 & (10.5a) \\
-\left(\psi_1 \mathsf{g}^2\right)y_1 + \left(1 - \rho r^2 \psi_2 \mathsf{g}^2\right)\theta_1 = 0 & (10.5b)
\end{cases}
$$

TABLE 10.1. Natural frequencies and normal modes of the single mass spinning about a principal axis with general elastic support (*a*), and cantilever support (*b*).

TABLE 10.2. Whirl response to static (force) unbalance (*a*) and to dynamic (couple) unbalance (*b*), with $N = +1$.

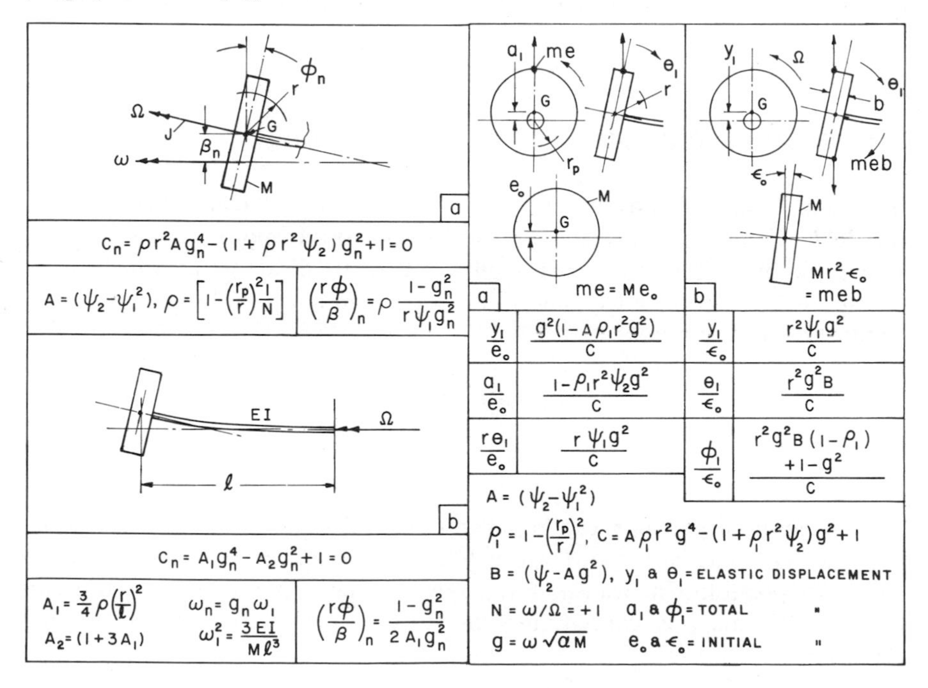

where $N = (\omega/\Omega)$ = order of orbital velocity, or whirl cycles per revolution
$\omega_1 = 1/\sqrt{\alpha M}$ = reference natural frequency
$\mathrm{g} = N\Omega/\omega_1$ = ratio of orbital to reference frequency
$\rho = (1 - J\Omega/I\omega) = [1 - (r_p/r)^2(1/N)]$ = a dimensionless parameter
$r^2 = (I/M)$ = radius of gyration about a transverse axis
$r_p^2 = (J/M)$ = radius of gyration about polar (spin) axis
$\psi_1 = (\beta/\alpha)$
$\psi_2 = (\gamma/\alpha)$

(See Chapter 9 for a discussion of elastic coefficients.)

Equating the determinant to zero, we have the natural frequency quadratic, Table 10.1a,

$$\rho r^2(\psi_2 - \psi_1^2)\mathrm{g}_n^4 - (1 + \rho r^2\psi_2)\mathrm{g}_n^2 + 1 = 0 \tag{10.6}$$

and the normal modes are given by

$$\left(\frac{r\phi}{\beta}\right)_n = \left(\frac{1 - \mathrm{g}_n^2}{\rho r\psi_1\mathrm{g}_n^2}\right) \tag{10.7}$$

Orthogonality is satisfied if

$$\left(\frac{r\phi}{\beta}\right)_{n1}\left(\frac{r\phi}{\beta}\right)_{n2} = -\left(\frac{1}{\rho}\right) \tag{10.8}$$

10.6. BEHAVIOR OF FREQUENCIES AND MODES

In Figure 10.4, with positive N, T_J subtracts from the conventional inertia couple T_I. Conversely, with spin and whirl velocities of opposite sense, T_J and T_I are additive. This leads to an equivalent T_I, or disk inertia, which incorporates the gyroscopic effect, and to an equivalent single radius of gyration r_e about the transverse axis.

$$I_e = Mr_e^2 = M\rho r^2$$

$$r_e = r\sqrt{\rho} \tag{10.9}$$

Since ρ can be negative, r_e can be imaginary, and Equation 10.6 then has one real and one imaginary root (Figure 10.6). There is then only one frequency and mode possible. The transition from two to one at $\rho = 0$ and $r_e = 0$ occurs if either the mass is concentrated or if $N = (r_p/r)^2$. The latter is when $J\Omega = I\omega$.

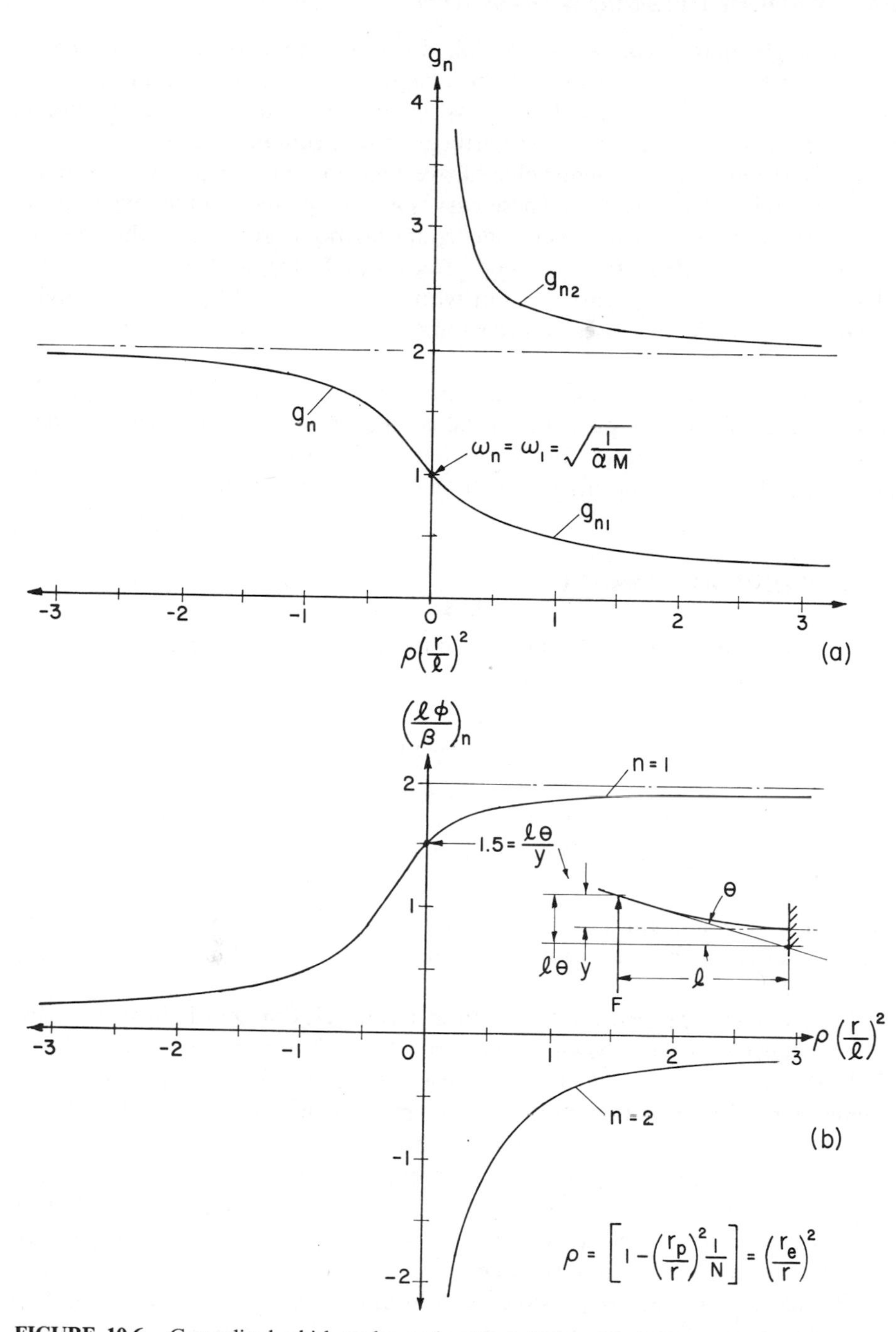

FIGURE 10.6. Generalized whirl modes and natural frequencies for any spinning disk with cantilever elastic support. A concentrated mass corresponds to the vertical axis, for which the mode is static-elastic (*b*).

The single-mode result seems to violate orthogonality, since we only have one mode; however, calculation of the imaginary root and mode produces a complementary mode. Although not possible physically, this mode, in conjunction with the real mode, will satisfy orthogonality (Equation 10.8).

To illustrate how a system behaves, we take the basic cantilever support elasticity and plot Figure 10.6. These curves are completely general, with r/l a geometric ratio, and ρ a parameter incorporating both order and polar inertia effects. In Figure 10.6*a* the first mode frequency factor, and the second, are identical with those of Figure 9.3, but with ρ introduced. As ρ is increasingly negative, the single root rises, approaching 2 for very small positive values of N.

In Figure 10.6*b* the modal characteristics are shown. The ordinate is converted from r to l for purposes of the cantilever plot. Normal modal angles are always positive, except for the higher mode, when it exists. The static elastic cantilever mode occurs at $\rho = 0$ or $r = 0$, as shown inset.

10.7. WHIRL RESONANCE

Figure 10.7 illustrates graphically the interactions between spin, whirl, and natural frequencies. Three rotational velocity ratios are involved:

$g = (\omega/\omega_1) =$ angular velocity of whirl, or orbital velocity relative to reference critical speed

$G = (\Omega/\Omega_1) =$ spin velocity relative to reference critical speed

$N = (g/G) = (\omega/\Omega) =$ order of whirl

with the reference natural frequency the reference critical speed, or $\sqrt{1/\alpha M}$. In this plot elastic constants for the cantilever are used, a thin disk is assumed, and $r = l$.

Solid curves represent the two natural frequencies or speeds in whirl, with nonrotative frequencies shown on the vertical axis. The negative horizontal axis indicates spin and whirl of opposite sense. The dashed straight lines show the increase in whirl velocity with increasing spin velocity for a constant multiple, or order. All intersections constitute a resonant condition, say A, for which the dimensionless speed is read on the horizontal axis, B.

For the conditions of Figure 10.7, whirl resonance cannot exist for orders of $+2$ or less in the second mode, but the first mode is possible. All negative orders intersect both natural frequencies.

We have been viewing possible equilibrium combinations of whirl as free orbital motions. There are infinite possibilities since N need not be an integer in free steady state; however, Figure 10.7 also applies to the forced condition. Then the straight lines represent orbital excitation frequency ratios. For instance, with disk unbalance, a force vector rotates with the mass, and $\omega = \Omega$,

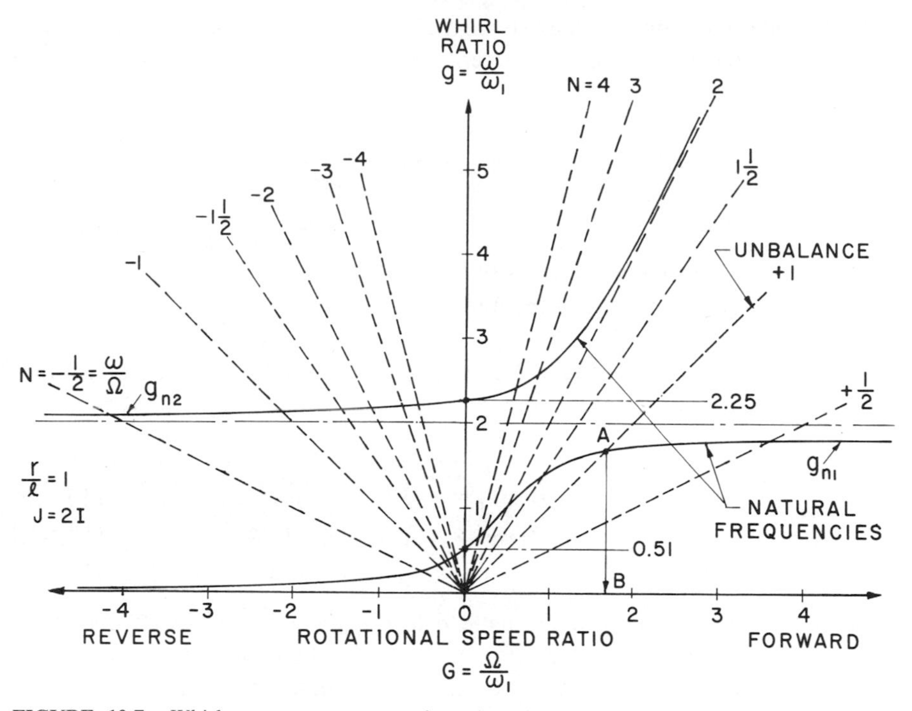

FIGURE 10.7. Whirl resonance occurs when the whirl order N intersects a whirl natural frequency curve. Cantilever elasticity is assumed.

or $N = +1$. Since operation is confined to this line, there is one resonant condition at A. Forcing excitation orders tend to be integers; for instance $1N$ for unbalance, $2N$ for secondary engine unbalance and so on.

10.8. WHIRL EXCITED BY FORCE UNBALANCE

Excitation by unbalance at $N = +1$ is so important that we develop special expressions for this response. There are two possibilities, force and couple unbalance, which in combination describe the total unbalance of any rigid rotor. This case is, incidentally, the only order excitation that induces no fatigue stress in the rotating beam as the system rotates in a constant flexed condition (Figure 10.2). Also for this reason, no hysteretic damping is present in the shaft at $N = +1$, because of a lack of stress reversal. The dynamic damper is similarly ineffective on $1N$ whirl unbalance, as there is no relative motion or damping with respect to the auxiliary system.

With no relative motion at $N = +1$, we conclude that there is a planetary effect between the whirl and spin motions. A second-order whirl, $N = +2$, involves a relative velocity of $(2 - 1) = +1$, with the disk orbiting one more turn than it rotates per revolution. If $N = -4$, this becomes $(-4 - 1) = -5$.

Static, or force, unbalance exists in the plane of the mass centroid and can be defined in terms of total mass eccentricity of an equivalent smaller mass at an arbitrary radius:

$$Me_0 = me$$

where e_0 = centroidal eccentricity of the entire mass
me = equivalent mass unbalance

Introducing $M\Omega^2 e_0 = M\omega^2 e_0$ in Equations 10.4a and 10.4b as a static unbalance additive to inertia force,

$$\begin{cases} y_1 = \alpha M\Omega^2 (y_1 + e_0) + \beta M r^2 \rho \Omega^2 \theta_1 & (10.10a) \\ \theta_1 = \beta M\Omega^2 (y_1 + e_0) + \gamma M r^2 \rho \Omega^2 \theta_1 & (10.10b) \end{cases}$$

After normalization these equations lead to the general results in Table 10.2*a*. Note that y_1 is just the elastic portion of the radial deflection of G. Total radius a_1 is the algebraic sum of y_1 and e_0, as in Equation 10.3. The factor ρ_1 designates ρ at first-order excitation.

10.9. WHIRL EXCITED BY DYNAMIC UNBALANCE

Although dynamic unbalance often refers to both force and moment effects, we are using the term here to indicate two equal and opposite unbalances, spaced apart axially to produce a rotating couple. The term *dynamic* implies that this unbalance cannot be detected statically, but is manifested during rotation.

In Table 10.2*b* the equal and opposite rotating unbalances create a couple, $T_D = (me\Omega^2)b$, which rotates with the mass. Adding this effect to the other inertia couples, Equations 10.4a and 10.4b can be written

$$\begin{cases} y_1 = (\alpha M\Omega^2) y_1 + (\beta \rho r^2 M\Omega^2)\theta_1 + \beta(meb)\Omega^2 & (10.11a) \\ \theta_1 = (\beta M\Omega^2) y_1 + (\gamma \rho r^2 M\Omega^2)\theta_1 + \gamma(meb)\Omega^2 & (10.11b) \end{cases}$$

Normalized results in Table 10.2*b* include a simplifying term, $\varepsilon_0 = meb/Mr^2$. It is an angular deflection of the disk defining the couple unbalance.

10.10. GENERAL COMMENTS

To solve for natural frequencies and normal modes quantitatively it is necessary to have values for all terms in Table 10.1*a*. The mass must be defined, including the centroid, principal axes, and mass moments of inertia about these axes. For cylindrical rotors these determinations usually are relatively simple.

Elasticity factors can be more difficult. Tables 9.1–9.3 are useful for basic cases. As indicated in Chapter 9, deflections at a point are additive with compliances in series, such as bearing deflections combined with shaft deflections. The component α, β, and γ terms are then summed from the various sources. *It is incorrect to sum* ψ_1 *and* ψ_2 *before having obtained the total* α, β, *and* γ *compliance factors*.

With forced whirl, we have four possible types of excitation, as in Tables 9.4 and 9.5, and we can use these tables for whirl by simply substituting ρr^2 for r^2. Where r has no exponent, the ρ factor is not inserted. Obviously the gyroscopic factor ρ will always influence the denominator, and thereby the natural frequencies and modes.

In practice multiple disks and coupling effects can arise. The principles outlined here should serve as guidelines, and are intended to provide insights into the more complex cases; however, the single rigid rotating disk is a frequent problem.

EXAMPLES

EXAMPLE 10.1. A rotating weight of 36 lb is carried at the center of a steel shaft 1.75 in. in diameter. The shaft is simply supported in bearings with a span of 21 in. Operating speed is 2500 RPM and mass eccentricity is 0.042 in. Mass of the shaft is neglected. Find:

(a) The critical speed of rotation.

(b) The steady-state radial elastic deflection of the shaft at the mass.

(c) The orbital radius of the mass centroid.

Solution. From Table 1.3*k*,

$$\alpha = \frac{1}{K} = \frac{l^3}{48EI} = \left[\frac{(21)^3}{48(30)(10)^6(0.46)}\right] = 13.97(10)^{-6}\text{ in./lb}$$

$$\text{(a)}\quad \omega_1 = \Omega_c = \sqrt{\frac{1}{\alpha M}} = 876 \text{ or } 8366 \text{ RPM}$$

(b) From Equation 10.2,

$$\left(\frac{y}{e_0}\right) = \left(\frac{\mathrm{g}^2}{1 - \mathrm{g}^2}\right)$$

$$\mathrm{g}^2 = \left(\frac{\omega}{\omega_1}\right)^2 = \left(\frac{2500}{8366}\right)^2 = 0.0893$$

$$\left(\frac{y}{e_0}\right) = 0.0981$$

$$y = (0.098)(0.042) = 0.004 \text{ in.}$$

(c) $\mathrm{a}_1 = y + e_0 = 0.004 + 0.042 = 0.046$ in.

EXAMPLE 10.2. An overhung flywheel is supported on a 3.75 cm diameter steel shaft and is excited by the rotating unbalance of the mass. Find:

(a) The nonrotative natural frequencies.
(b) The critical speed at which the unbalance results in a whirl resonance.
(c) The normal mode corresponding to the resonance.

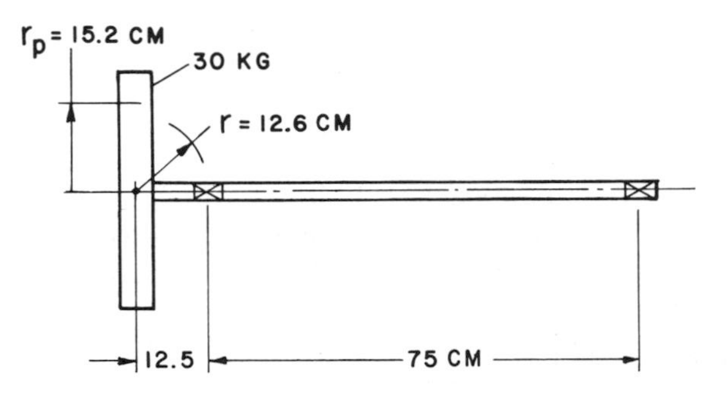

Solution. Applicable elastic coefficients from Table 9.1*c* are as follows:

$$\alpha = 23.5(10)^{-6} \text{ cm/N} \qquad l^2\psi_2 = 46.3$$

$$l\psi_1 = 6.43 \qquad l^2(\psi_2 - \psi_1^2) = 4.96$$

The geometric ratio is

$$\frac{r}{l} = \left(\frac{12.6}{75}\right) = 0.168$$

(a) Taking $C = 0$ from Table 9.4, the frequency quadratic is

$$0.14\mathsf{g}_n^4 - 2.306\mathsf{g}_n^2 + 1 = 0$$

$$\mathsf{g}_{n1}^2 = 0.445 \qquad \mathsf{g}_{n2}^2 = 16.04$$
$$\mathsf{g}_{n1} = 0.67 \qquad \mathsf{g}_{n2} = 4.01$$

$$\omega_1 = 10\sqrt{\frac{(10)^6}{23.5(30)}} = 377 \text{ r/s} = 3600 \text{ RPM}$$

$$\omega_{n1} = 0.67(3600) = 2400 \text{ RPM}$$

$$\omega_{n2} = 4.01(3600) = 14{,}400 \text{ RPM}$$

(b) Introducing ρ for gyroscopic effects (Table 10.1*a*),

$$\rho = \left[1 - \left(\frac{15.2}{12.6}\right)^2\left(\frac{1}{1}\right)\right] = -0.455$$

$$(-0.455)(0.140)\mathsf{g}_n^4 - (1 - (0.455)(1.306))\mathsf{g}_n^2 + 1 = 0$$

$$-0.0637\mathsf{g}_n^4 - 0.406\mathsf{g}_n^2 + 1 = 0$$

$$\mathsf{g}_{n1}^2 = 1.90$$

$$\mathsf{g}_{n1} = 1.38$$

$$\Omega_{c1} = 1.38(3600) = 4960 \text{ RPM}$$

Note that the speed has approximately doubled from 2400 RPM by the gyroscopic effect.

(c) Also from Table 10.1*a*,

$$\left(\frac{r\phi}{\beta}\right)_{n1} = \left[\frac{1 - 1.90}{(-0.455)(12.6)(6.43/75)1.90}\right] = +0.96$$

If $\beta_{n1} = 1$ cm,

$$\phi_{n1} = \left(\frac{0.96}{12.6}\right) = 0.076 \text{ r or } 4.4°$$

EXAMPLE 10.3. The rotor of Example 10.2 operates at 5200 RPM, with a dynamic (couple) unbalance of 3000 g · cm^2. Find the orbital radius of the mass centroid.

Solution. Using Table 10.2*b* for the 1*N* excitation,

$$\mathsf{g}^2 = \left(\frac{5200}{3600}\right)^2 = 2.086$$

$$y_1 = \left(\frac{3000}{M}\right)\left(\frac{\psi_1 \mathsf{g}^2}{C}\right) = \left(\frac{3}{30}\right)\left(\frac{6.43}{75}\right)\left(\frac{2.09}{-0.124}\right) = 0.14 \text{ cm}$$

This elastic orbital radius of *G* is the total radius. The couple unbalance corresponds to an angular misalignment of the axis of the disk, with no initial radius.

11 MULTIMASS SYSTEMS WITH BEAM ELASTICITY

This chapter relates to bending or rotational whirl of systems having two or more concentrated masses supported on a straight elastic element in flexure. Beam mass effects are assumed small, and negligible with respect to the masses. Specific tabulations are limited to the three-degree case, but the organization of the simultaneous linear equations based on elastic influence coefficients for higher-order systems follows from the methods indicated. Rayleigh and Stodola methods introduced now have been associated historically with beam and rotor problems and are applicable to systems with many masses.

There is considerable interchangeability among most of the basic conceptual methods of solutions for natural frequencies and modes, and this is one of the fascinations of working in the vibration field. We have many options in attacking a problem and can choose a favorite. The adventurous reader is invited to verify the equivalence of the results now presented and those obtained from methods given in previous chapters, particularly Chapter 4.

As determination of the influence factors is often a major portion of the beam-coupled vibratory problem, dimensionless tabulations are provided to facilitate this aspect for beams of constant cross section.

11.1. TRANSVERSE EFFECTS IN BEAMS

With several masses located along a beam element (Figure 11.1), there are two modes and lateral coupling effects as the system vibrates in a principal or natural mode. Masses are concentrated, (shown spherically), as slope and angular inertia factors are considered negligible. This approximation has not been necessary in previous chapters. Vibratory behavior is tied closely to the static elastic characteristics that must be determined. In the idealized case a constant section or diameter is shown, otherwise deflection behavior must be

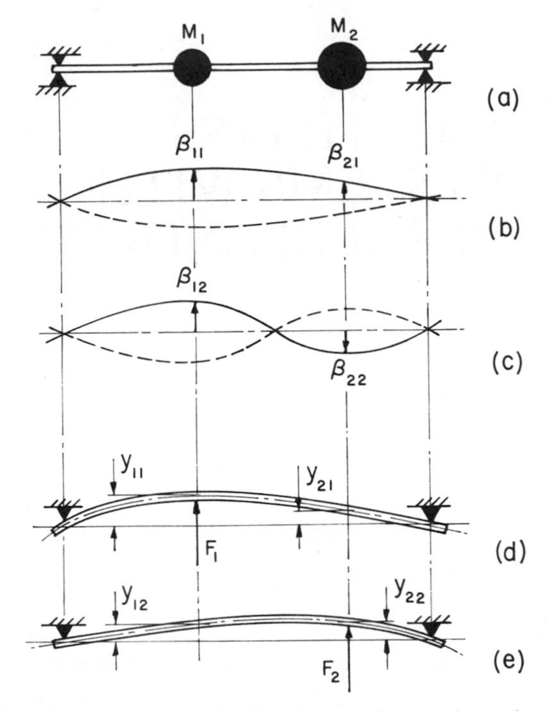

FIGURE 11.1. The two-mass system in bending has two frequencies and modes, related to the elastic behavior at 1 and 2 when forces are applied at these points.

calculated by more complex means. The beam can contain various sections, and the supports can introduce indeterminate restraints or can contribute flexibilities. We must assume, however, in the following development that the system elasticity is linear. We also limit flexibilities to those perpendicular to the straight central beam axis.

In Figure 11.1*d* and *e*, a force at 1 produces deflection not only at its point of application y_{11} due to F_1, but at all other points on the elastic curve, including y_{21}, the deflection at 2 due to a load at 1. In a two-degree system there are thus four transverse deflection rates of importance, but these reduce to three by Maxwell's theorem of reciprocal displacements.

$$\alpha_{11} = \left(\frac{y_{11}}{F_1}\right) \qquad \alpha_{12} = \left(\frac{y_{12}}{F_2}\right) = \alpha_{21} = \left(\frac{y_{21}}{F_1}\right) \qquad \alpha_{22} = \left(\frac{y_{22}}{F_2}\right)$$

All factors are in terms of displacement per unit load. They are similar to the influence coefficients of Section 9.1, which applied to linear and angular deflections of a single point. Present factors include coupling effects between points, and can be calculated for basic beams from Tables 11.1–11.3. Angular factors are also provided, but are not applied in this chapter.

TABLE 11.1. Nomenclature for the nine compliance factors involved in three transverse coordinates, and the five corresponding normalized ratios.

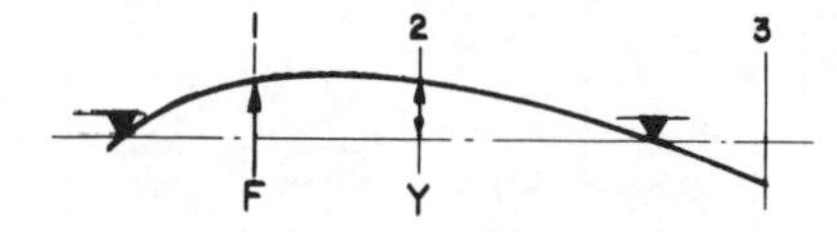

α_{11}	$\frac{Y_1}{F_1}$	α_{22}	$\frac{Y_2}{F_2}$	α_{33}	$\frac{Y_3}{F_3}$
α_{21}	$\frac{Y_2}{F_1}$	α_{12}	$\frac{Y_1}{F_2}$	α_{13}	$\frac{Y_1}{F_3}$
α_{31}	$\frac{Y_3}{F_1}$	α_{32}	$\frac{Y_3}{F_2}$	α_{23}	$\frac{Y_2}{F_3}$

ψ_1	$\frac{\alpha_{21}}{\alpha_{11}}$	$\frac{\alpha_{12}}{\alpha_{11}}$
ψ_2	$\frac{\alpha_{22}}{\alpha_{11}}$	
ψ_3	$\frac{\alpha_{23}}{\alpha_{11}}$	$\frac{\alpha_{32}}{\alpha_{11}}$
ψ_4	$\frac{\alpha_{33}}{\alpha_{11}}$	
ψ_5	$\frac{\alpha_{31}}{\alpha_{11}}$	$\frac{\alpha_{13}}{\alpha_{11}}$

11.2. EQUATIONS OF MOTION

With steady-state sinusoidal vibration of the unforced system (Figure 11.1), the equations are

$$\begin{cases} \mathsf{a}_1 = (\alpha_{11} M_1 \omega^2)\mathsf{a}_1 + (\alpha_{12} M_2 \omega^2)\mathsf{a}_2 & (11.1a) \\ \mathsf{a}_2 = (\alpha_{21} M_1 \omega^2)\mathsf{a}_1 + (\alpha_{22} M_2 \omega^2)\mathsf{a}_2 & (11.1b) \end{cases}$$

In normalized form,

$$\begin{cases} (1 - \mathsf{g}^2)\mathsf{a}_1 - \left[m_2 \left(\frac{\alpha_{12}}{\alpha_{11}} \right) \mathsf{g}^2 \right] \mathsf{a}_2 = 0 & (11.2a) \\ - \left[\left(\frac{\alpha_{21}}{\alpha_{11}} \right) \mathsf{g}^2 \right] \mathsf{a}_1 + \left[1 - m_2 \left(\frac{\alpha_{22}}{\alpha_{11}} \right) \mathsf{g}^2 \right] \mathsf{a}_2 = 0 & (11.2b) \end{cases}$$

Simplification of the notation is obtained by letting

$$\left(\frac{\alpha_{12}}{\alpha_{11}} \right) = \left(\frac{\alpha_{21}}{\alpha_{11}} \right) = \psi_1 \qquad \left(\frac{\alpha_{22}}{\alpha_{11}} \right) = \psi_2$$

Thus ψ_1 and ψ_2 are dimensionless ratios of influence factors, and ψ_1 carries algebraic sign.

Equations 11.2*a* and 11.2*b* then become

$$\begin{cases} (1 - \mathsf{g}^2)\mathsf{a}_1 - (m_2 \psi_1 \mathsf{g}^2)\mathsf{a}_2 = 0 & (11.3a) \\ - (\psi_1 \mathsf{g}^2)\mathsf{a}_1 + (1 - m_2 \psi_2 \mathsf{g}^2)\mathsf{a}_2 = 0 & (11.3b) \end{cases}$$

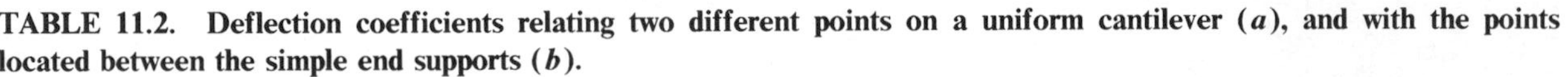

TABLE 11.2. Deflection coefficients relating two different points on a uniform cantilever (*a*), and with the points located between the simple end supports (*b*).

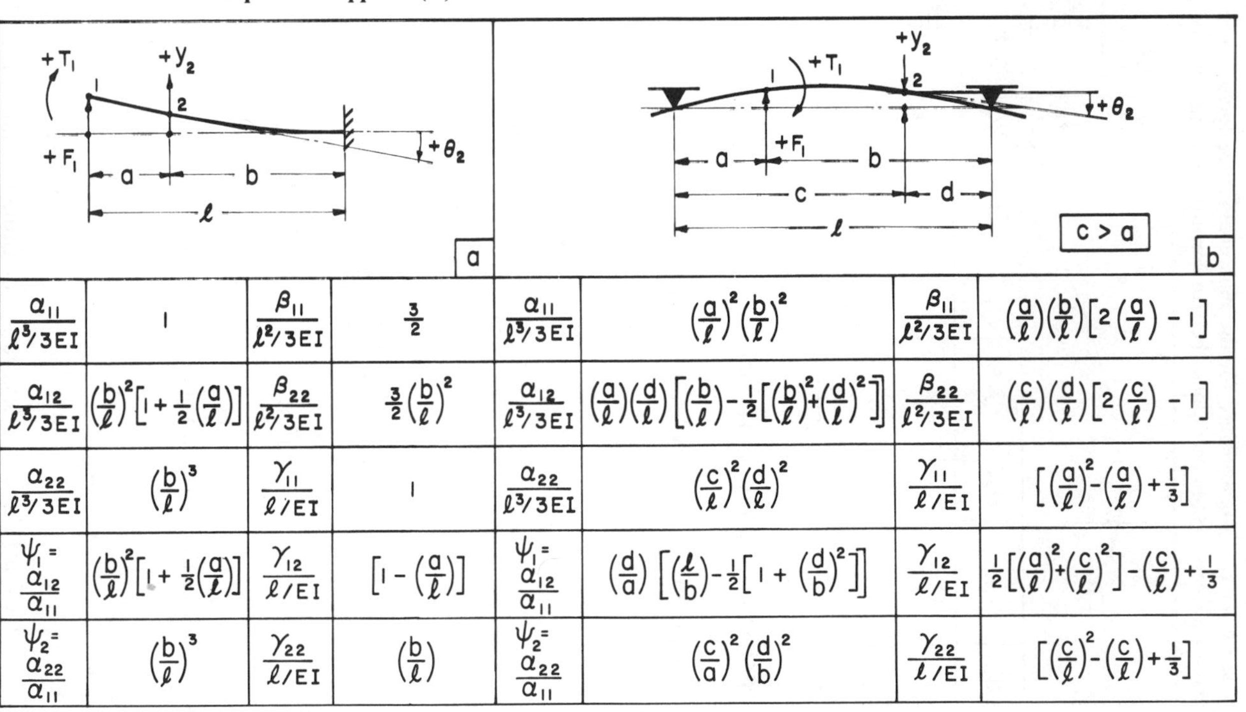

(a)				(b)			
$\frac{\alpha_{11}}{\ell^3/3EI}$	1	$\frac{\beta_{11}}{\ell^2/3EI}$	$\frac{3}{2}$	$\frac{\alpha_{11}}{\ell^3/3EI}$	$\left(\frac{a}{\ell}\right)^2\left(\frac{b}{\ell}\right)^2$	$\frac{\beta_{11}}{\ell^2/3EI}$	$\left(\frac{a}{\ell}\right)\left(\frac{b}{\ell}\right)\left[2\left(\frac{a}{\ell}\right)-1\right]$
$\frac{\alpha_{12}}{\ell^3/3EI}$	$\left(\frac{b}{\ell}\right)^2\left[1+\frac{1}{2}\left(\frac{a}{\ell}\right)\right]$	$\frac{\beta_{22}}{\ell^2/3EI}$	$\frac{3}{2}\left(\frac{b}{\ell}\right)^2$	$\frac{\alpha_{12}}{\ell^3/3EI}$	$\left(\frac{a}{\ell}\right)\left(\frac{d}{\ell}\right)\left[\left(\frac{b}{\ell}\right)-\frac{1}{2}\left[\left(\frac{b}{\ell}\right)^2+\left(\frac{d}{\ell}\right)^2\right]\right]$	$\frac{\beta_{22}}{\ell^2/3EI}$	$\left(\frac{c}{\ell}\right)\left(\frac{d}{\ell}\right)\left[2\left(\frac{c}{\ell}\right)-1\right]$
$\frac{\alpha_{22}}{\ell^3/3EI}$	$\left(\frac{b}{\ell}\right)^3$	$\frac{\gamma_{11}}{\ell/EI}$	1	$\frac{\alpha_{22}}{\ell^3/3EI}$	$\left(\frac{c}{\ell}\right)^2\left(\frac{d}{\ell}\right)^2$	$\frac{\gamma_{11}}{\ell/EI}$	$\left[\left(\frac{a}{\ell}\right)^2-\left(\frac{a}{\ell}\right)+\frac{1}{3}\right]$
$\psi_1=\frac{\alpha_{12}}{\alpha_{11}}$	$\left(\frac{b}{\ell}\right)^2\left[1+\frac{1}{2}\left(\frac{a}{\ell}\right)\right]$	$\frac{\gamma_{12}}{\ell/EI}$	$\left[1-\left(\frac{a}{\ell}\right)\right]$	$\psi_1=\frac{\alpha_{12}}{\alpha_{11}}$	$\left(\frac{d}{a}\right)\left[\left(\frac{\ell}{b}\right)-\frac{1}{2}\left[1+\left(\frac{d}{b}\right)^2\right]\right]$	$\frac{\gamma_{12}}{\ell/EI}$	$\frac{1}{2}\left[\left(\frac{a}{\ell}\right)^2+\left(\frac{c}{\ell}\right)^2\right]-\left(\frac{c}{\ell}\right)+\frac{1}{3}$
$\psi_2=\frac{\alpha_{22}}{\alpha_{11}}$	$\left(\frac{b}{\ell}\right)^3$	$\frac{\gamma_{22}}{\ell/EI}$	$\left(\frac{b}{\ell}\right)$	$\psi_2=\frac{\alpha_{22}}{\alpha_{11}}$	$\left(\frac{c}{a}\right)^2\left(\frac{d}{b}\right)^2$	$\frac{\gamma_{22}}{\ell/EI}$	$\left[\left(\frac{c}{\ell}\right)^2-\left(\frac{c}{\ell}\right)+\frac{1}{3}\right]$

TABLE 11.3. Similar to Table 11.2, but including points outboard of the end supports.

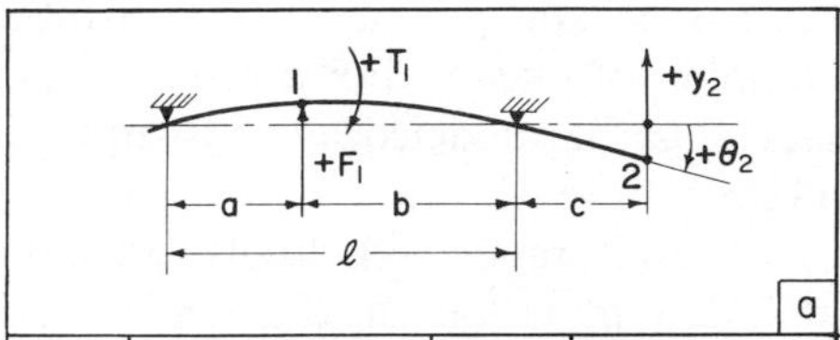

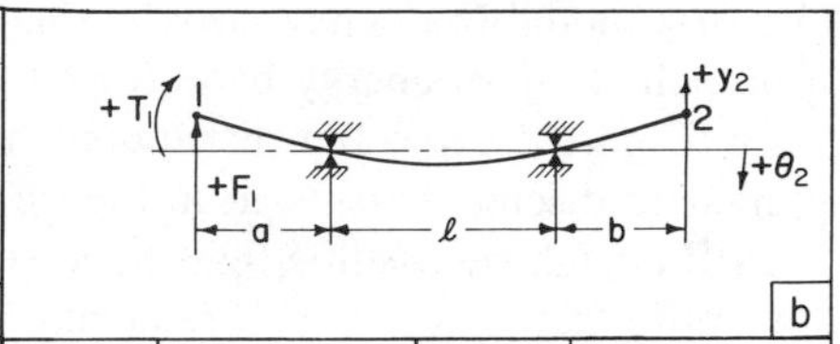

a				b			
$\dfrac{\alpha_{11}}{\ell^3/3EI}$	$\left(\frac{a}{\ell}\right)^2\left(\frac{b}{\ell}\right)^2$	$\dfrac{\beta_{11}}{\ell^2/3EI}$	$\left(\frac{a}{\ell}\right)\left(\frac{b}{\ell}\right)\left[2\left(\frac{a}{\ell}\right)-1\right]$	$\dfrac{\alpha_{11}}{\ell^3/3EI}$	$\left(\frac{a}{\ell}\right)^2\left[1+\frac{a}{\ell}\right]$	$\dfrac{\beta_{11}}{\ell^2/3EI}$	$\left(\frac{a}{\ell}\right)\left[1+\frac{3}{2}\left(\frac{a}{\ell}\right)\right]$
$\dfrac{\alpha_{12}}{\ell^3/3EI}$	$-\frac{1}{2}\left(\frac{a}{\ell}\right)\left(\frac{b}{\ell}\right)\left(\frac{c}{\ell}\right)\left[1+\frac{a}{\ell}\right]$	$\dfrac{\beta_{22}}{\ell^2/3EI}$	$-\left(\frac{c}{\ell}\right)\left[1+\frac{3}{2}\left(\frac{c}{\ell}\right)\right]$	$\dfrac{\alpha_{12}}{\ell^3/3EI}$	$\frac{1}{2}\left(\frac{b}{\ell}\right)\left(\frac{a}{\ell}\right)$	$\dfrac{\beta_{22}}{\ell^2/3EI}$	$-\left(\frac{b}{\ell}\right)\left[1+\frac{3}{2}\left(\frac{b}{\ell}\right)\right]$
$\dfrac{\alpha_{22}}{\ell^3/3EI}$	$\left(\frac{c}{\ell}\right)^2\left[1+\left(\frac{c}{\ell}\right)\right]$	$\dfrac{\gamma_{11}}{\ell/EI}$	$\left[\left(\frac{a}{\ell}\right)^2-\left(\frac{a}{\ell}\right)+\frac{1}{3}\right]$	$\dfrac{\alpha_{22}}{\ell^3/3EI}$	$\left(\frac{b}{\ell}\right)^2\left[1+\frac{b}{\ell}\right]$	$\dfrac{\gamma_{11}}{\ell/EI}$	$\left[\frac{1}{3}+\frac{a}{\ell}\right]$
$\psi_1 = \left(\frac{\alpha_{12}}{\alpha_{11}}\right)$	$-\frac{1}{2}\left(\frac{c}{b}\right)\left[1+\left(\frac{\ell}{a}\right)\right]$	$\dfrac{\gamma_{12}}{\ell/EI}$	$\left[\frac{1}{2}\left(\frac{a}{\ell}\right)^2-\frac{1}{6}\right]$	$\psi_1 = \left(\frac{\alpha_{12}}{\alpha_{11}}\right)$	$\frac{1}{2}\left(\frac{b}{a}\right)\left[\frac{(\ell/a)}{1+\ell/a}\right]$	$\dfrac{\gamma_{12}}{\ell/EI}$	$-\frac{1}{6}$
$\psi_2 = \left(\frac{\alpha_{22}}{\alpha_{11}}\right)$	$\left(\frac{c}{a}\right)\left(\frac{c}{b}\right)\left(\frac{\ell}{a}\right)\left(\frac{\ell}{b}\right)\left[1+\frac{c}{\ell}\right]$	$\dfrac{\gamma_{22}}{\ell/EI}$	$\left[\frac{c}{\ell}+\frac{1}{3}\right]$	$\psi_2 = \left(\frac{\alpha_{22}}{\alpha_{11}}\right)$	$\left(\frac{b}{a}\right)^2\left[\frac{a+\ell}{b+\ell}\right]$	$\dfrac{\gamma_{22}}{\ell/EI}$	$\left[\frac{1}{3}+\frac{b}{\ell}\right]$

Equating the denominator of the determinant to zero results in the natural frequency quadratic,

$$A m_2 \mathsf{g}_n^4 - (1 + \psi_2 m_2)\mathsf{g}_n^2 + 1 = 0 \tag{11.4}$$

where

$$A = (\psi_2 - \psi_1^2)$$

$$m_2 = (M_2/M_1)$$

$$\mathsf{g}_n^2 = (\omega_n/\omega_1)^2$$

$$\omega_1 = \sqrt{\frac{1}{\alpha_{11} M_1}}$$

Normal mode amplitudes are, from Equation 11.3a,

$$\left(\frac{\beta_2}{\beta_1}\right)_n = \left(\frac{1 - \mathsf{g}_n^2}{\psi_1 m_2 \mathsf{g}_n^2}\right) \tag{11.5}$$

11.3. RAYLEIGH APPROXIMATION

The first modal frequency, usually the most important practically, can also be approached on an energy basis for any number of masses. For a general type rotor (Figure 13.1*a*), the distributed mass must be modeled as consisting of equivalent discrete masses, as in Figure 11.2.

In Rayleigh we assume that the lowest vibratory mode will closely resemble the static deflection curve requiring the determination of the *total* static deflection at each mass due to simultaneous weight loading using strength of materials methods.

One method of obtaining this total deflection is the superposition of the deflections at a given mass caused by that mass, plus all other loads multiplied by their cross-coefficients, as indicated in Figure 11.2*c*. We can also use this method in the Stodola solution to convert loads to total deflections.

In Figure 11.2 static potential energy (PE) of the deflected system stored as strain energy in beam bending is

$$\mathrm{PE} = \tfrac{1}{2}W_1\delta_1 + \tfrac{1}{2}W_2\delta_2 + \tfrac{1}{2}W_3\delta_3 \tag{11.6}$$

This corresponds in the vibratory case to the system potential energy at maximum displacement.

With simultaneous harmonic motion of the masses, total kinetic energy (KE) at the midposition is

$$\mathrm{KE} = \tfrac{1}{2}M_1\omega^2\delta_1^2 + \tfrac{1}{2}M_2\omega^2\delta_2^2 + \tfrac{1}{2}M_3\omega^2\delta_3^2 \tag{11.7}$$

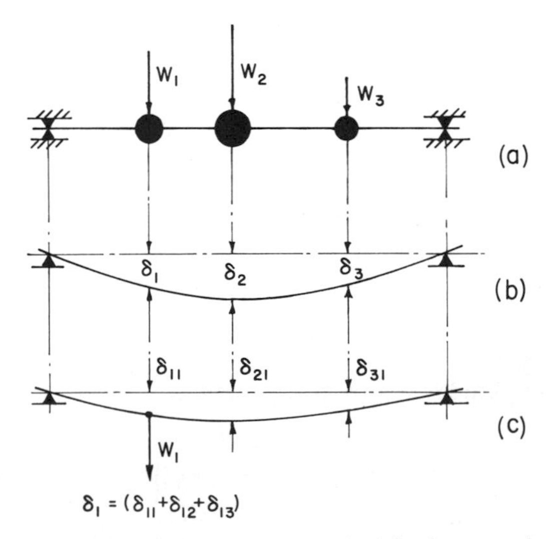

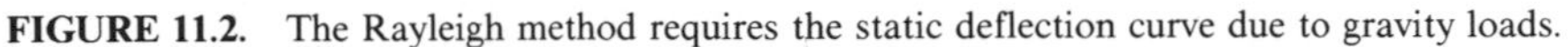

FIGURE 11.2. The Rayleigh method requires the static deflection curve due to gravity loads.

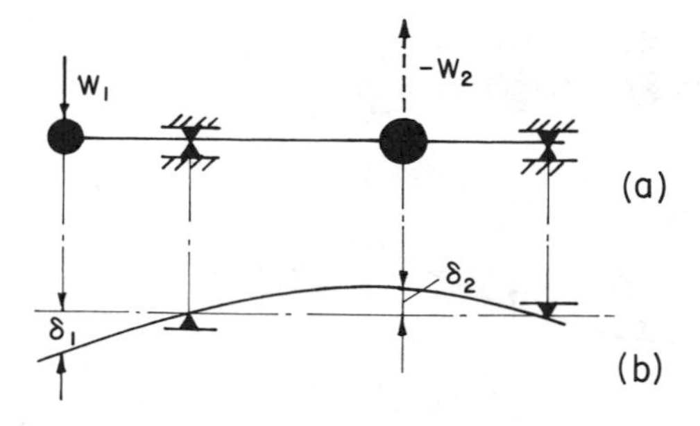

FIGURE 11.3. With the Rayleigh method an overhung mass requires reversal of the gravity force to approximate the first vibratory mode. This concept is also applicable when using the Stodola method or matrix iteration.

Equating these energies provides the frequency relation, with $M = W/g_c$.

$$\omega_{n1} = \sqrt{\frac{g_c \Sigma W\delta}{\Sigma W\delta^2}} \tag{11.8}$$

And in normalized form,

$$\omega_{n1} = \sqrt{\frac{g_c}{\delta_1}} \left[\sqrt{\frac{1 + m_2(\delta_2/\delta_1) + \cdots}{1 + m_2(\delta_2/\delta_1)^2 + \cdots}} \right] \tag{11.9}$$

The Rayleigh calculation gives frequencies that are slightly high in all cases.

If the masses overhang the supports (Figure 11.3), static loading tends to violate the first-mode flexural pattern. As shown, this is rectified by reversal of one gravity load. The resulting elastic curve then agrees with the nature of the mode as visualized intuitively.

11.4. THE STODOLA METHOD

In the Stodola method both a modal deflection curve and a vibratory frequency are assumed. Resulting inertia loads act on the beam simultaneously at all masses, causing a beam deflection that can be evaluated numerically at each mass coordinate. If the *shape* of the deflection curve agrees with the assumed curve, a normal mode has been identified.

For example, in Figure 11.4 $F_1 = M_1\omega^2\beta_1$ is the inertia load at 1, but by assuming $\omega = 1$, the force is numerically $M_1\beta_1$. M is in mass units, and we take β_1 as 1.00. The downward loads are taken as static, and correspond to the maximum vibratory loading. The deflection analysis is not detailed here, but the classical method requires four successive integrations to convert loading to shear, shear to moment, moment to slope, and slope to deflection.

When the y values are determined, Figure 11.4d, we note the ratios of the ordinates, β_1/y_1, β_2/y_2, and β_3/y_3, which should be roughly equal in the first

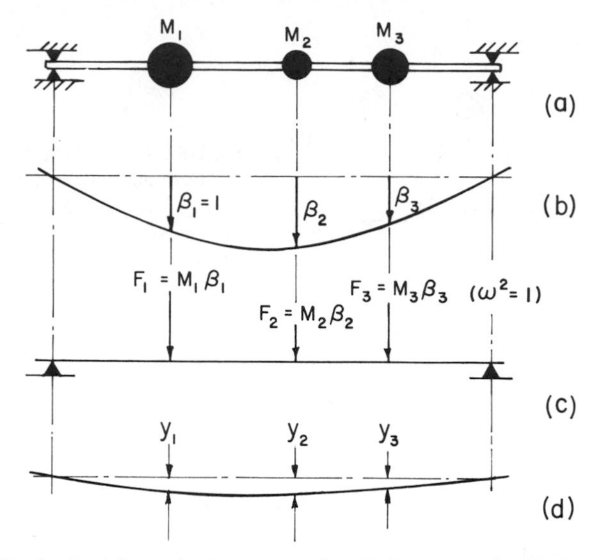

FIGURE 11.4. In the Stodola method an assumed mode is compared with the actual deflection curve resulting from vibratory loading in the assumed mode.

convergence. The ratios will be large, however, indicating that the assumed β mode is not close to the y mode obtained.

This should not be too discouraging, as we have made a highly unrealistic natural frequency estimate of $\omega^2 = 1$, and this is only $1/2\pi$, or 0.16 Hz. Corresponding inertia loads and resulting beam deflections have suffered from this poor judgment; however, it is not too late for reconciliation.

We do this by finding a frequency, $\omega^2 = \beta_1/y_1$. If this factor is applied to our previous loads, $M_1\beta_1$, $M_2\beta_2$, and $M_3\beta_3$, we increase all beam loadings and deflections by ω^2, $\beta_1 = y_1 = 1$, and our assumed and determined modes intersect at M_1. We then conclude that the numerical value $1/y_1$ as determined is the first approximation to the actual first-mode natural frequency.

For the next convergence we take for the assumed mode, $\beta_1 = 1$, $\beta_2 = y_2/y_1$, and $\beta_3 = y_3/y_1$, from the previously calculated deflection curve, revert to $\omega^2 = 1$, and find a new set of deflections using inertia loads $F_q = M_q\omega^2\beta_q$. By comparing the new ratio $(1/y_1)$ with the previous ratio $(1/y_1)$, we can judge the decrease in ω_n^2. When this change is negligible we have sufficient convergence, and the normal mode has been obtained. As with the Holzer method (see Chapter 4), we have satisfied physical conditions, and assumptions are now factual.

With Holzer it is necessary to assume a frequency, but not a mode, and this frequency must be modified by judgment or computer program. Lack of automatic convergence is an advantage in Holzer if we are seeking a higher mode. Loadings, which determine spring stress distributions, are represented by the Σ column in Holzer. With a Stodola calculation, modal stress distribution in the beam relates to the moment distribution, $M(x)$, obtained in calculating $y(x)$ from $\beta(x)$.

The Holzer method has been adapted to beam systems, moving incrementally from mass to mass, left to right. Although logical, and based on the mechanics of beam flexure, the number of parameters involved tends to make this approach unwieldy.

11.5. MATRIX ITERATION

The Stodola concept of assuming a modal distribution and then converging successively upon the actual mode and frequency can be systematized in matrix notation and solved by matrix iteration. This method also builds on the simultaneous equations (11.1) and requires evaluation of all elastic influence coefficients. Normalization simplifies both the notation and the computations, as all terms are ratios.

For the simple two-mass case (Figure 11.1), the matrix format becomes

$$\begin{bmatrix} \beta_1 \\ \beta_2 \end{bmatrix} = \mathsf{g}^2 \begin{bmatrix} 1 & \psi_1 m_2 \\ \psi_1 & \psi_2 m_2 \end{bmatrix} \begin{bmatrix} \beta_1 \\ \beta_2 \end{bmatrix}$$

Assuming unit normal mode,

$$\begin{bmatrix} 1 \\ 1 \end{bmatrix} = \mathsf{g}^2 \begin{bmatrix} 1 & \psi_1 m_2 \\ \psi_1 & \psi_2 m_2 \end{bmatrix} \begin{bmatrix} 1 \\ 1 \end{bmatrix}$$

$$= \mathsf{g}^2 \begin{bmatrix} (1 + \psi_1 m_2) \\ (\psi_1 + \psi_2 m_2) \end{bmatrix}$$

$$= (1 + \psi_1 m_2)\mathsf{g}^2 \begin{bmatrix} 1 \\ \left(\dfrac{\psi_1 + \psi_2 m_2}{1 + \psi_1 m_2} \right) \end{bmatrix} \tag{11.10}$$

The first approximation for the natural frequency ratio is

$$\mathsf{g}_n^2 = \frac{1}{1 + \psi_1 m_2} \tag{11.11}$$

And for the normal mode, we have the terms in brackets.

A numerical example is less cumbersome. If we take $m_2 = 4$, $\psi_1 = 0.75$, and $\psi_2 = 1.2$, and start with equal modal amplitudes,

$$\begin{bmatrix} 1 \\ 1 \end{bmatrix} = g^2 \begin{bmatrix} 1 & 3 \\ 0.75 & 4.8 \end{bmatrix} \begin{bmatrix} 1 \\ 1 \end{bmatrix} = g^2 \begin{bmatrix} 4 \\ 5.55 \end{bmatrix} = 4g^2 \begin{bmatrix} 1 \\ 1.388 \end{bmatrix}$$

Using the new mode,

$$\begin{bmatrix} 1 \\ 1.388 \end{bmatrix} = g^2 \begin{bmatrix} 1 & 3 \\ 0.75 & 4.8 \end{bmatrix} \begin{bmatrix} 1 \\ 1.388 \end{bmatrix} = g^2 \begin{bmatrix} 5.16 \\ 7.41 \end{bmatrix} = 5.16g^2 \begin{bmatrix} 1 \\ 1.436 \end{bmatrix}$$

The final convergence is

$$\begin{bmatrix} 1 \\ 1.436 \end{bmatrix} = g^2 \begin{bmatrix} 1 & 3 \\ 0.75 & 4.8 \end{bmatrix} \begin{bmatrix} 1 \\ 1.436 \end{bmatrix} = g^2 \begin{bmatrix} 5.31 \\ 7.64 \end{bmatrix} = 5.31g^2 \begin{bmatrix} 1 \\ 1.440 \end{bmatrix}$$

or

$$g_{n1}^2 = \frac{1}{5.31} = 0.188$$

Exact values from the quadratic solution are $g_{n1}^2 = 0.1879$ and $\beta_{21} = 1.440$. We see that the convergence is rapid, and the numerical or algebraic effort is minimal.

Considering the three-degree system, the notation of Table 11.1 is applicable, and we can use the dimensionless factors to write

$$\begin{bmatrix} \beta_1 \\ \beta_2 \\ \beta_3 \end{bmatrix} = g^2 \begin{bmatrix} 1 & \psi_1 m_2 & \psi_5 m_3 \\ \psi_1 & \psi_2 m_2 & \psi_3 m_3 \\ \psi_5 & \psi_3 m_2 & \psi_4 m_3 \end{bmatrix} \begin{bmatrix} \beta_1 \\ \beta_2 \\ \beta_3 \end{bmatrix} \tag{11.12a}$$

The general formulation for the larger system becomes

$$\begin{bmatrix} \beta_1 \\ \beta_2 \\ \beta_3 \\ \vdots \\ \beta_q \end{bmatrix} = g^2 \begin{bmatrix} 1 & & & \\ \left(\frac{\alpha_{21}}{\alpha_{11}}\right) & \left(\frac{\alpha_{22}}{\alpha_{11}}\right) m_2 & \left(\frac{\alpha_{23}}{\alpha_{11}}\right) m_3 & \cdots \\ \left(\frac{\alpha_{31}}{\alpha_{11}}\right) & \vdots & \vdots & \cdots \\ \vdots & & & \cdots \\ \left(\frac{\alpha_{q1}}{\alpha_{11}}\right) & \left(\frac{\alpha_{q2}}{\alpha_{11}}\right) m_2 & \left(\frac{\alpha_{q3}}{\alpha_{11}}\right) m_3 & \cdots \end{bmatrix} \begin{bmatrix} \beta_1 \\ \beta_2 \\ \beta_3 \\ \vdots \\ \beta_q \end{bmatrix} \tag{11.12b}$$

11.6. HIGHER MODES

Since Stodola converges on the first mode, if we desire the second mode or higher, we must circumvent this feature. As a procedure we assume a mode β_q (Figure 11.5*a*), shaped as a second mode, and in addition we remove from our assumption all first-mode characteristics. Having first determined β_{n1} values accurately, we next calculate a first modal component factor of the assumed second mode from

$$\overline{m}_1 = \left(\frac{m_1 \beta_{q1} \beta_{11} + m_2 \beta_{q2} \beta_{21} + \cdots}{m_1 \beta_{11}^2 + m_2 \beta_{21}^2 + \cdots} \right) \tag{11.13}$$

where the denominator is the normalized modal mass. The basis for this relation is indicated in Section 13.4.

As shown in Figure 11.5*b*, the first normal mode is reduced by the factor $\overline{m}_1$. The resulting dashed curve represents the portion of the β_q curve that would have produced responses in the y ordinates (Figure 11.4*d*). By subtraction (Figure 11.5*a*), the revised dashed assumed mode β_p is a *purified* curve, having been purged of first-mode contamination. Convergence will now focus on second mode.

When using matrix iteration there is a similar problem with first-mode convergence. Then the relationship that permits elimination of lower modes is basic orthogonality of normal modes. From these equations the matrix format can be voided of lower modes. In fact, with a two-degree system, orthogonality permits us to solve directly for the exact second mode.

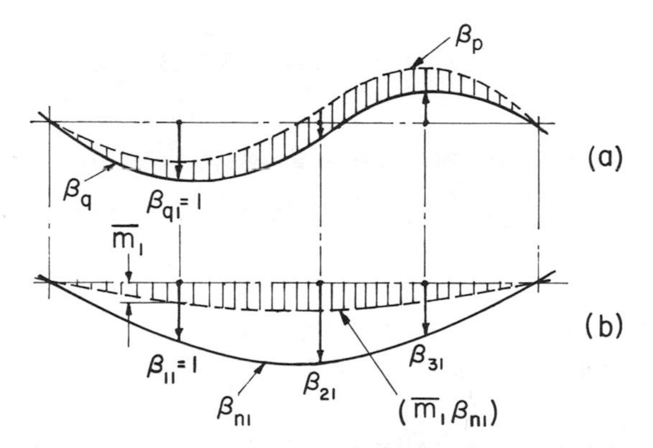

FIGURE 11.5. To find a second mode by the Stodola method, the assumed mode must be purified by subtracting the latent first modal component, as shown by the hatching.

11.7. TYPICAL CASES

Table 11.4 indicates natural frequencies and normal modes with equally spaced equal masses on a constant cross-section beam and simple supports. Table 11.5*b* completely describes the free vibration of the general three-mass system, for which six influence coefficients are required. The frequency equation is a cubic in g_n^2, from which the three roots can be calculated. Equations apply to beams with masses inboard or outboard of the bearings, to cantilever suspension, and to spring-coupled discrete systems.

If a system is not constrained to ground, it can have vibratory characteristics and rigid modes, as in the simpler systems (see Section 4.4). Such cases are possible in a structure floating on a liquid, or suspended to ground by extremely soft springs or shock cords. Analysis is different from grounded systems with defined influence factors, and with lumped masses three is the minimum number for which a vibratory bending mode can exist. Conditions of force and moment equilibrium must be satisfied under the action of simultaneous vibratory inertia loads, leading first to the modal amplitudes (Table 11.5*a*). Second, with known amplitudes, the kinetic energy at midposition is equated to the potential energy at full amplitude. The latter is based on the strain energy in the two constant beams with cantilever bending stress. From the energies the natural frequency is determined.

This three-degree system, having only one vibratory mode, seems to lack the required three modes; however, these are present in the translation and rotation of the rigid system. In translation, all amplitudes are +1. In rotation, the node is the center of gravity. Orthogonality of modes is satisfied by any combination of rigid and vibratory modes. Rigid-mode natural frequencies are zero.

11.8. TOTAL FORCED RESPONSE

Addition of a forcing term on M_1 in Equations 11.3*a* and 11.3*b* results in

$$\begin{cases} (1 - g^2)a_1 - (m_2\psi_1 g^2)a_2 = P_1\alpha_{11} & (11.14a) \\ -(\psi_1 g^2)a_1 + (1 - m_2\psi_2 g^2)a_2 = P_1\alpha_{12} & (11.14b) \end{cases}$$

where $g = \omega/\omega_1$ = forced frequency ratio

$P_1\alpha_{11}$ = reference amplitude at 1

Similar modifications to the free equations can be used to introduce P_2. Resulting equations are shown in Table 11.6*a*.

TABLE 11.4. Natural frequency factors, normal modes, and normalized compliance factors for systems with equal masses equally spaced on a simply supported constant beam.

TABLE 11.5. Frequency and modal data for the three-mass flexurally supported case; free–free in (a) and simply supported in (b).

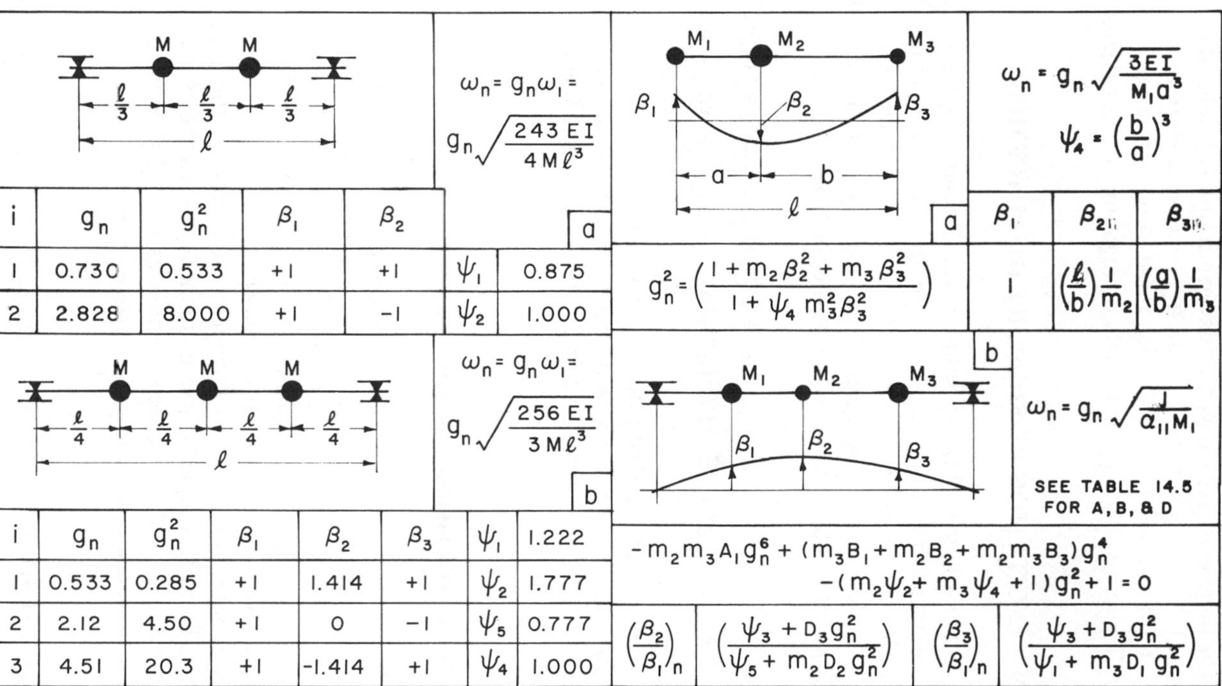

Table 11.4 (a) — two masses M at spacing $\ell/3$, span ℓ

$$\omega_n = g_n \omega_1 = g_n \sqrt{\frac{243\,EI}{4M\ell^3}}$$

i	g_n	g_n^2	β_1	β_2		
1	0.730	0.533	+1	+1	ψ_1	0.875
2	2.828	8.000	+1	−1	ψ_2	1.000

Table 11.4 (b) — three masses M at spacing $\ell/4$, span ℓ

$$\omega_n = g_n \omega_1 = g_n \sqrt{\frac{256\,EI}{3M\ell^3}}$$

i	g_n	g_n^2	β_1	β_2	β_3	ψ_1	1.222
1	0.533	0.285	+1	1.414	+1	ψ_2	1.777
2	2.12	4.50	+1	0	−1	ψ_5	0.777
3	4.51	20.3	+1	−1.414	+1	ψ_4	1.000

Table 11.5 (a) — masses M_1, M_2, M_3; distances a, b, ℓ

$$\omega_n = g_n \sqrt{\frac{3EI}{M_1 a^3}} \qquad \psi_4 = \left(\frac{b}{a}\right)^3$$

$$g_n^2 = \left(\frac{1 + m_2\beta_2^2 + m_3\beta_3^2}{1 + \psi_4 m_3^2 \beta_3^2}\right)$$

β_1	β_2	β_3
1	$\left(\frac{\ell}{b}\right)\frac{1}{m_2}$	$\left(\frac{a}{b}\right)\frac{1}{m_3}$

Table 11.5 (b) — masses M_1, M_2, M_3 simply supported

$$\omega_n = g_n \sqrt{\frac{1}{\alpha_{11} M_1}}$$

SEE TABLE 14.5 FOR A, B, & D

$$-m_2 m_3 A_1 g_n^6 + (m_3 B_1 + m_2 B_2 + m_2 m_3 B_3) g_n^4 - (m_2\psi_2 + m_3\psi_4 + 1) g_n^2 + 1 = 0$$

$$\left(\frac{\beta_2}{\beta_1}\right)_n = \left(\frac{\psi_3 + D_3 g_n^2}{\psi_5 + m_2 D_2 g_n^2}\right) \qquad \left(\frac{\beta_3}{\beta_1}\right)_n = \left(\frac{\psi_3 + D_3 g_n^2}{\psi_1 + m_3 D_1 g_n^2}\right)$$

The three-mass system with transverse coupling is related to the C_a frequency cubic (Table 14.5), with a_1, a_2, and a_3 corresponding to the coordinates a_x, a_y, and a_z; however, there are now three different masses applicable to the coupling terms. In Chapter 14 only a single mass is considered.

11.9. EXCITATION BY SUPPORT DISPLACEMENT

Table 11.6*b* is the vibratory solution for a beam-coupled system with inertia excitation arising from the motion of the beam axis induced by sinusoidal support amplitudes at a common frequency, producing planar motion of the masses. Response is determined by Equations 11.1a and 11.1b with the addition of inertia forcing effects at the masses,

$$\begin{cases} \mathsf{a}_1^r = \alpha_{11} M_1 \omega^2 (\mathsf{a}_1^r + \mathsf{a}_{01}) + \alpha_{12} M_2 \omega^2 (\mathsf{a}_2^r + \mathsf{a}_{02}) & (11.15\text{a}) \\ \mathsf{a}_2^r = \alpha_{21} M_1 \omega^2 (\mathsf{a}_1^r + \mathsf{a}_{01}) + \alpha_{22} M_2 \omega^2 (\mathsf{a}_2^r + \mathsf{a}_{02}) & (11.15\text{b}) \end{cases}$$

where a_1^r and a_2^r are elastic deflection amplitudes relative to the beam axis and a_{01} and a_{02} are the in-phase or out-of-phase displacement excitations at the masses. Responses from the two excitations are given separately, and can be combined algebraically.

11.10. ROTATIONAL WHIRL

If the previous two-mass case is rotating with unbalances at either mass, we will have synchronous whirl, $N = +1$. Equations are similar to 11.15, with e replacing a_0, or with 11.15*a* the planar projection of 11.15*b*. Additionally, the supports in (a) can be orbited at fixed radii, a_{00} and a_{03}, producing orbital, nonrotative displacements at the masses.

It is interesting to note the identical nature of the expressions in Table 11.6*b*. There is complete correspondence between the excitation displacements at a mass, a_0 and e_0, between responses a^r and y, and between the exciting frequencies ω and Ω.

11.11. MODAL RESPONSE TO DISPLACEMENT EXCITATION

In simply coupled systems (see Figure 5.14), a sinusoidal displacement at the right spring subjects all masses to the same alternating acceleration field, $\omega^2 \mathsf{a}_3$, as the root of the induced inertia forces in the several masses. Modal excitation factor $\overline{m}_i$ is defined in Equation 5.16, and its components are $m_q \beta_{qi}$; that is, the excitation effect of a mass is directly proportional to the mass and the modal amplitude.

TABLE 11.6. Response of the two-mass beam coupled system to sinusoidal forcing at either mass (*a*) and to displacement forcing in (*b*). The latter includes the rotating or whirl condition with forcing by eccentricity.

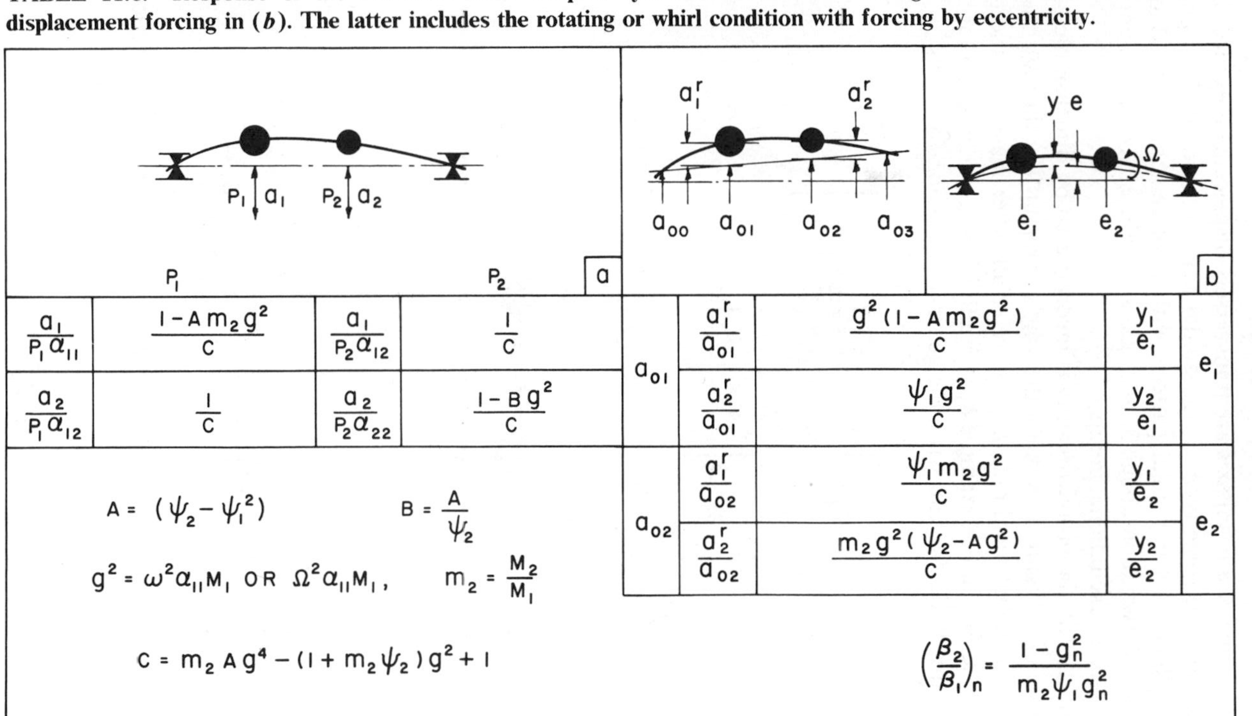

(a)

	P_1		P_2
$\dfrac{a_1}{P_1\alpha_{11}}$	$\dfrac{1-Am_2g^2}{C}$	$\dfrac{a_1}{P_2\alpha_{12}}$	$\dfrac{1}{C}$
$\dfrac{a_2}{P_1\alpha_{12}}$	$\dfrac{1}{C}$	$\dfrac{a_2}{P_2\alpha_{22}}$	$\dfrac{1-Bg^2}{C}$

$$A = (\psi_2 - \psi_1^2) \qquad B = \frac{A}{\psi_2}$$

$$g^2 = \omega^2\alpha_{11}M_1 \text{ OR } \Omega^2\alpha_{11}M_1, \qquad m_2 = \frac{M_2}{M_1}$$

$$C = m_2 A g^4 - (1 + m_2\psi_2)g^2 + 1$$

(b)

a_{01}	$\dfrac{a_1^r}{a_{01}}$	$\dfrac{g^2(1-Am_2g^2)}{C}$	$\dfrac{y_1}{e_1}$	e_1
	$\dfrac{a_2^r}{a_{01}}$	$\dfrac{\psi_1 g^2}{C}$	$\dfrac{y_2}{e_1}$	
a_{02}	$\dfrac{a_1^r}{a_{02}}$	$\dfrac{\psi_1 m_2 g^2}{C}$	$\dfrac{y_1}{e_2}$	e_2
	$\dfrac{a_2^r}{a_{02}}$	$\dfrac{m_2g^2(\psi_2-Ag^2)}{C}$	$\dfrac{y_2}{e_2}$	

$$\left(\frac{\beta_2}{\beta_1}\right)_n = \frac{1-g_n^2}{m_2\psi_1 g_n^2}$$

If two points of a beam axis are transversely displacement excited, in or out of phase, we have some type of trapezoidal pattern imposed. The special case (Figure 11.6*b*) has a triangular pattern, with displacement specified at 0. This is an angular excitation about 3, and masses farthest from the pivot tend to produce greater excitational inputs.

Quantitatively the proportionality is incorporated by dimensionless normalizing factors λ_i (Figure 11.6*c*). By drawing a straight line through the node 3 and the unit modal amplitude β_{11}, we refer all other masses to the reference coordinate at M_1. In this example $\lambda_1 = 1$ and $\lambda_2 = \frac{1}{2}$, the latter factor applying to M_2 in any mode. We can write the general expression for $\overline{m}_i$ as

$$\overline{m}_1 = \left[\frac{1 + \lambda_2 m_2 \beta_{21} + \cdots}{1 + m_2(\beta_{21})^2 + \cdots}\right]$$

$$\overline{m}_i = \left[\frac{\Sigma\lambda_q m_q(\beta_q)_{ni}}{m_{ni}}\right] \tag{11.16}$$

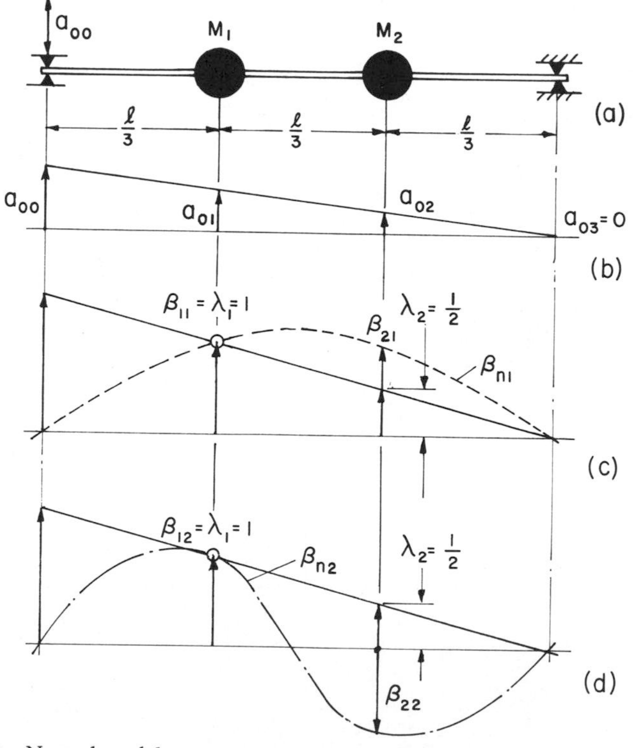

FIGURE 11.6. Normal modal component response to displacement excitation requires a linear normalization of the displacement amplitude pattern.

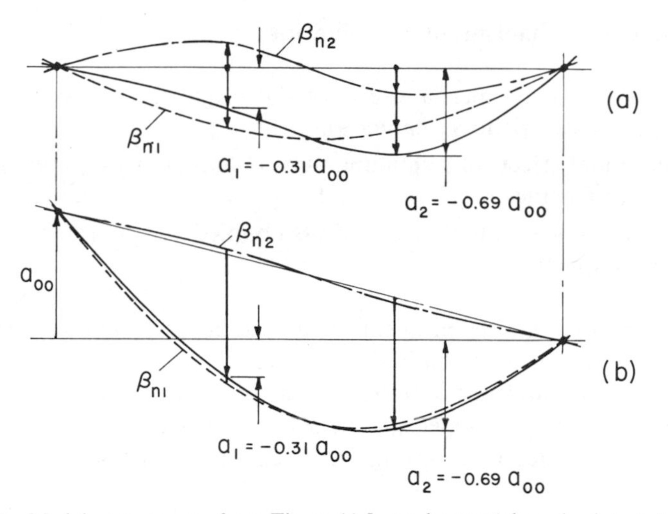

FIGURE 11.7. Modal components, from Figure 11.5, can be treated as absolute (*a*) or relative (*b*).

In this simple example we can use modal information from Table 11.4*a*, and find that $\overline{m}_1 = 0.75$ and $\overline{m}_2 = 0.25$. This verifies observations of (*c*) and (*d*) in which the λ triangle is a closer approximation to the first mode than to the second. Assuming $\mathsf{g}_1^2 = 2$, or a forcing frequency $\sqrt{2}$ times the first modal frequency, we have two alternative versions of the modal responses.

Using Equation 5.18 for *absolute* modal components, we have Figure 11.7*a*. With *relative* modal components we use Equation 5.17, and displacements are with respect to the beam axis. Adding modal and input amplitudes (Figure 11.7*b*), we find absolute amplitudes in agreement with either method.

It is instructive to note basic differences between the (*a*) and (*b*) results. The latter is a much more meaningful interpretation of the physical behavior, as beam stresses and deflections pertain to the relative displacements. Also, in (*b*) the influence of the second mode with respect to the first is much reduced; however, there are only two absolute coordinate responses, at 1 and 2, and they can be determined by either method.

In converting the linear beam displacement pattern to the normalizing λ function, coordinates are +1 at β_1, and zero at the node. The latter can be located within or beyond the system by extension of the axis. With simple translational excitation the node is at infinity.

11.12. REMARKS

In this chapter, and in Chapters 9 and 10, we have dealt with rigid masses and beam-type mounting flexibilities. We have intentionally restricted the systems to allow the use of relatively simple analytical methods but to illustrate the basic vibratory mechanisms of the following:

1. Transverse mass inertia effects of the single mass as coupled to beam slope characteristics (Chapter 9).
2. Rotational effects of a spinning mass (Chapter 10) superimposed on the mass in Chapter 9.
3. Several concentrated masses, cross-coupled on a straight elastic beam (this chapter).

Obviously, many real-life problems will not be this simple, nor as neatly packaged; nevertheless the basic concepts relative to elastic behavior, inertia effects, excitation, and response have been provided. Illustrative equilibrium equations have been developed for the selected cases. More complexly related systems can be treated by adapting the independently presented procedures as required.

EXAMPLES

EXAMPLE 11.1. A constant cross-section beam carries two masses as shown, with sinusoidal excitation at M_1 at 100 Hz, and an amplitude of 120 lb. Find:

(a) The system natural frequencies.
(b) The two normal modes.
(c) First-mode response at M_1.
(d) Second-mode response at M_1.

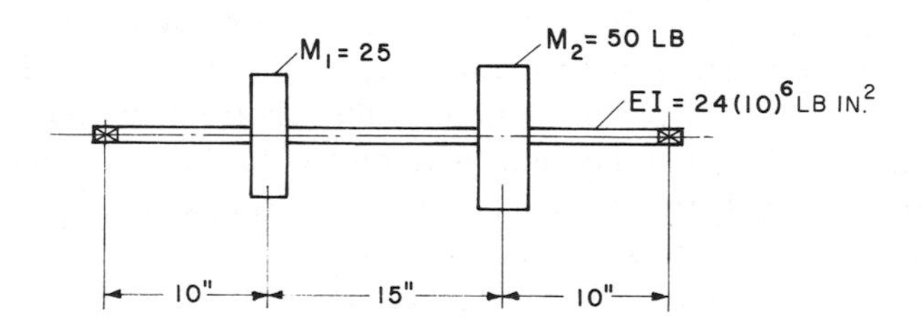

Solution. From Table 11.2*b*, the elasticity factors are

$$\alpha_{11} = 24.8(10)^{-6} \text{ in./lb}$$

$$\psi_1 = 0.82$$

$$\psi_2 = 1$$

$$(\psi_2 - \psi_1^2) = 0.328$$

(a) Using Table 11.6,

$$C_n = 0.655\mathsf{g}_n^4 - 3\mathsf{g}_n^2 + 1 = 0$$

$$\mathsf{g}_{n1}^2 = 0.362 \qquad \mathsf{g}_{n2}^2 = 4.218$$
$$\mathsf{g}_{n1} = 0.602 \qquad \mathsf{g}_{n2} = 2.054$$

$$\omega_1 \equiv \sqrt{\frac{\mathsf{g}_c}{\alpha_{11}W_1}} = \sqrt{\frac{386(10)^6}{(24.8)25}} = 789 \text{ rad/sec} = 126 \text{ Hz}$$

$$\omega_{n1} = (0.602)(126) = 76 \text{ Hz}$$

$$\omega_{n2} = (2.054)(126) = 258 \text{ Hz}$$

(b)
$$\left(\frac{\beta_2}{\beta_1}\right)_1 = \left[\frac{1 - 0.362}{(2)(0.820)(0.362)}\right] = +1.08$$

$$\left(\frac{\beta_2}{\beta_1}\right)_2 = \left[\frac{1 - 4.218}{(1.64)(4.218)}\right] = -0.47$$

(c) From Equations 4.10b, and 5.11

$$K_{n1} = \left[\frac{0.362(10)^6}{24.8}\right]\left[1 + 2(1.08)^2\right] = 48{,}300 \text{ lb/in.}$$

$$\mathsf{g}_1^2 = \left(\frac{100}{76}\right)^2 = 1.75$$

$$\mathsf{a}_{11} = \left(\frac{120}{48{,}300}\right)(-1.333) = -0.0033 \text{ in.}$$

(d)
$$\mathsf{a}_{12} = +0.0006 \text{ in.}$$

EXAMPLE 11.2. The system of Example 11.1 is excited at M_2 instead of at M_1, with the same amplitude and frequency. Find:

(a) First-mode response at M_1.

(b) First-mode response at M_2.

Solution. From Equation 5.14b,

(a)
$$\mathsf{a}_{11} = \left(\frac{P_2\bar{\beta}_{21}}{K_{n1}}\right)(-1.333) = \left[\frac{120(1.075)}{48{,}300}\right](-1.333)$$

$$= -0.0035 \text{ in.}$$

(b) $$\mathsf{a}_{21} = \beta_{21}\mathsf{a}_{11} = 1.075(-0.0035) = -0.0038 \text{ in.}$$

Note that the amplitude in Example 11.2*b* is greater than that in Example 11.1*c*, by a factor of $(\beta_{21})^2$, or 1.156.

EXAMPLE 11.3. The system of Example 11.1 is to be analyzed by the Stodola method. Find:

(a) The first natural frequency.

(b) The first normal mode.

Solution

(a) For the first trial we take $\beta_2 = \beta_1 = \omega = 1$.

$$F_1 = M_1\beta_1\omega^2 = \left(\frac{25}{386}\right)(1)(1)^2 = 0.0648 \text{ lb}$$

$$F_2 = M_2\beta_2\omega^2 = \left(\frac{50}{386}\right)(1)(1)^2 = 0.1295 \text{ lb}$$

$$y_1 = F_1\alpha_{11} + F_2\alpha_{12} = 4.236(10)^{-6} \text{ in.}$$

$$y_2 = F_1\alpha_{21} + F_2\alpha_{22} = 4.527(10)^{-6} \text{ in.}$$

$$\left(\frac{\beta_1}{y_1}\right) = \frac{(10)^6}{4.236} = 0.236(10)^6$$

$$\left(\frac{\beta_2}{y_2}\right) = \frac{(10)^6}{4.527} = 0.221(10)^6$$

For the second trial, using the new amplitudes,

$$\beta_1 = 1$$

$$\beta_2 = \left(\frac{y_2}{y_1}\right) = \left(\frac{4.527}{4.236}\right) = 1.068$$

$$F_1 = 0.0648 \text{ lb} \qquad F_2 = 0.1386 \text{ lb}$$

$$y_1 = 4.417(10)^{-6} \text{ in.} \qquad y_2 = 4.750(10)^{-6} \text{ in.}$$

$$\left(\frac{\beta_1}{y_1}\right) = 0.226(10)^6 \qquad \left(\frac{\beta_2}{y_2}\right) = 0.225(10)^6$$

For the third trial,

$$\beta_1 = 1$$

$$\beta_2 = \frac{4.750}{4.417} = 1.074$$

$$F_1 = 0.0648 \qquad F_2 = 0.1392 \text{ lb}$$

$$y_1 = 4.433(10)^{-6} \text{ in} \qquad y_2 = 4.767(10)^{-6} \text{ in.}$$

$$\left(\frac{\beta_1}{y_1}\right) = 0.226(10)^6 \qquad \left(\frac{\beta_2}{y_2}\right) = 0.226(10)^6 \quad \text{(convergence)}$$

$$\omega_{n1}^2 = \left(\frac{\beta_1}{y_1}\right) = 0.2256(10)^6$$

$$\omega_{n1} = 76 \text{ Hz}$$

(b)
$$\left(\frac{\beta_2}{\beta_1}\right) = \frac{1.075}{1.000} = +1.075$$

EXAMPLE 11.4. Analyze Example 11.1 by matrix iteration. Find:

(a) First natural frequency and mode.

(b) Second natural frequency and mode.

Solution

(a) Trial 1:

$$\beta_1 = \beta_2 = 1 \qquad (\psi_1 m_2 = (0.82)2 = 1.64)$$

$$\begin{bmatrix} \beta_1 \\ \beta_2 \end{bmatrix} = g^2 \begin{bmatrix} 1 & 1.64 \\ 0.82 & 2 \end{bmatrix} \begin{bmatrix} 1 \\ 1 \end{bmatrix} = g^2 \begin{bmatrix} 2.64 \\ 2.82 \end{bmatrix}$$

$$= 2.64\left(\frac{\omega}{\omega_1}\right)^2 \begin{bmatrix} 1.000 \\ 1.068 \end{bmatrix}$$

Trial 2:

$$\beta_1 = 1 \qquad \beta_2 = 1.068$$

$$\begin{bmatrix} \beta_1 \\ \beta_2 \end{bmatrix} = 2.64g^2 \begin{bmatrix} 1 & 1.64 \\ 0.82 & 2 \end{bmatrix} \begin{bmatrix} 1.000 \\ 1.068 \end{bmatrix} = g^2 \begin{bmatrix} 2.752 \\ 2.956 \end{bmatrix}$$

$$= 2.752\left(\frac{\omega}{\omega_1}\right)^2 \begin{bmatrix} 1.000 \\ 1.074 \end{bmatrix}$$

Trial 3:

$$\beta_1 = 1 \qquad \beta_2 = 1.074$$

$$\begin{bmatrix} \beta_1 \\ \beta_2 \end{bmatrix} = 2.752\mathrm{g}^2 \begin{bmatrix} 1 & 1.64 \\ 0.82 & 2 \end{bmatrix} \begin{bmatrix} 1.000 \\ 1.074 \end{bmatrix} = \mathrm{g}^2 \begin{bmatrix} 2.761 \\ 2.968 \end{bmatrix}$$

$$= 2.761\mathrm{g}^2 \begin{bmatrix} 1.000 \\ 1.075 \end{bmatrix}$$

$$\omega_{n1}^2 = \frac{1}{2.761}\omega_1^2 = 0.362\omega_1^2 = \mathrm{g}_{n1}^2\omega_1^2$$

$$\omega_{n1} = 0.602(125.6) = 76 \text{ Hz}$$

(b) Using orthogonal relationship for the second mode (Equation 11.13),

$$1 + (2)(1.075)\beta_{22} = 0$$

$$\beta_{22} = -0.465$$

$$\begin{bmatrix} \beta_1 \\ \beta_2 \end{bmatrix} = \mathrm{g}^2 \begin{bmatrix} 1 & 1.64 \\ 0.82 & 2 \end{bmatrix} \begin{bmatrix} 1.000 \\ -0.465 \end{bmatrix} = \mathrm{g}^2 \begin{bmatrix} 0.237 \\ -0.110 \end{bmatrix}$$

$$= 0.237\mathrm{g}^2 \begin{bmatrix} 1.000 \\ -0.465 \end{bmatrix}$$

$$\omega_{n2}^2 = \frac{1}{0.237}(\omega_1)^2 = 4.22\omega_1^2$$

$$\omega_{n2} = 2.054(125.6) = 258 \text{ Hz}$$

EXAMPLE 11.5. A two-disk system rotates at 1700 RPM with an unbalance of 1.7 in.lb at M_1. M_2 is completely balanced. Determine the whirl radius of rotation of the centroid of each disk.

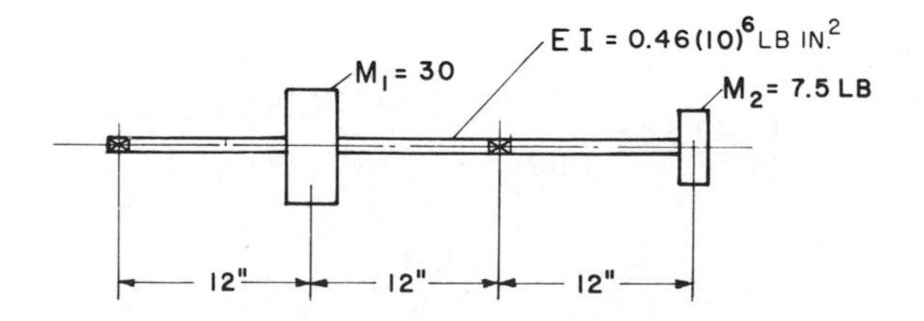

Solution. Calculating compliance factors from Table 11.3*a*,

$$\alpha_{11} = 626(10)^{-6} \text{ in./lb.} \qquad \psi_1 = -1.50$$
$$\psi_2 = 6 \qquad (\psi_2 - \psi_1^2) = 3.75$$

From Table 11.6*b*,

$$\omega_1 = \sqrt{\frac{1}{\alpha_{11} M_1}} = 143.4 \quad \text{or} \quad 22.8 \text{ Hz}$$

$$g^2 = \left[\frac{1700}{60(22.8)}\right]^2 = 1.54$$

$$e_1 = \frac{1.7}{30} = 0.0567 \text{ in.}$$

$$y_1 = \left[\frac{(0.0567)(1.54)(-0.447)}{-0.63}\right] = 0.062 \text{ in.}$$

$$r_1 = (e_1 + y_1) = 0.057 + 0.062 = 0.120 \text{ in.}$$

$$y_2 = \left[\frac{(0.0567)(-1.50)(1.54)}{-0.63}\right] = 0.208 \text{ in} = r_2$$

EXAMPLE 11.6. The system of Example 11.5 has unbalances at both masses, with $(me)_1 = 2.3$ in.lb. and $(me)_2 = 1.4$ in.lb., and lagging 1 by 90°. Calculate the first modal response to these unbalances.

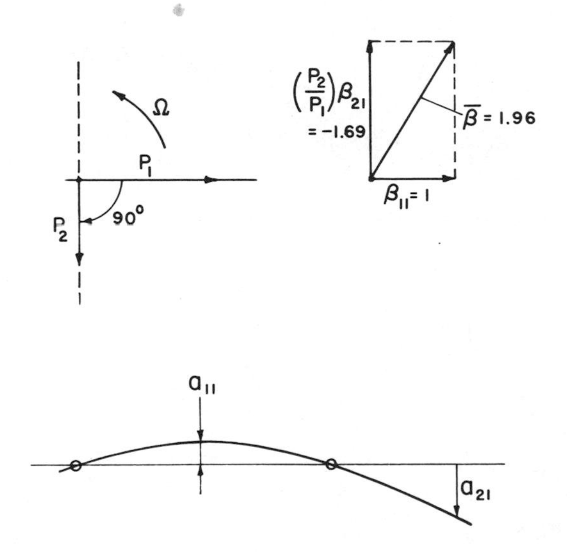

Solution. Using Table 11.6 for modal values and Equation 4.10*b* for modal spring,

$$\mathsf{g}_{n1}^2 = 0.490 \qquad \mathsf{g}_{n1} = 0.70$$
$$K_{n1} = 2290 \text{ lb/in.} \qquad \beta_{21} = -2.78$$

Excitations are obtained from unbalanced forces at operating speed.

$$P_1 = (me)_1\left(\frac{\omega^2}{\mathsf{g}_c}\right) = 2.3\left[\frac{(178)^2}{386}\right] = 189 \text{ lb}$$

$$P_2 = (me)_2\left(\frac{\omega^2}{\mathsf{g}_c}\right) = 1.4\left[\frac{(178)^2}{386}\right] = 115 \text{ lb}$$

Combining the two forces to an equivalent single force at M_1, as in Section 5.14b,

$$(P_1\beta_{11} + P_2\beta_{21}) = P_1\beta_{11}\left[1 \nrightarrow \left(\frac{P_2}{P_1}\right)\left(\frac{\beta_{21}}{\beta_{11}}\right)\right] = P_1\bar{\beta}$$

$$= 189\left[1 \nrightarrow \left(\frac{1.4}{2.3}\right)(-2.78)\right] = 189(1.96) = 370 \text{ lb}$$

$$\omega_{n1} = (22.8)(60)(0.70) = 958 \text{ RPM}$$

$$\mathsf{g}_1^2 = \left(\frac{1700}{958}\right)^2 = 3.15$$

$$\mathsf{a}_{11} = \frac{P_1\bar{\beta}}{K_{n1}}\left(\frac{1}{1 - \mathsf{g}_1^2}\right) = \frac{370}{2290}(-0.465) = -0.075 \text{ in.}$$

$$\mathsf{a}_{21} = (-0.075)(-2.78) = +0.21 \text{ in.}$$

12 UNIFORM BEAMS

In Chapter 11 concentrated masses were considered in which the supporting flexural elements had assumed negligible mass relative to the discrete masses. This chapter treats the converse, with the beams carrying no external masses. Inertia effects are due entirely to the beam, which is constantly distributed. Although not in bending, the constant prismatic bar in an axial or torsional mode is closely related and included in this category. The uniform bar or beam with various end conditions occurs frequently in mechanical systems, warranting the somewhat extensive treatment provided. Closed or classical solutions are possible because the normal modes are combinations of trigonometric and hyperbolic functions, with higher modes easily extrapolated from the lower.

12.1. AXIAL–TORSIONAL FREQUENCIES AND MODES

The clamped–free, constant cross-sectional elastic prismatic bar (Figure 12.1), can be disturbed axially or torsionally to induce free steady-state vibration. Modes and frequencies are identical, as shown in Table 12.1. It should be noted that all modes have maximum slope at a fixed end and zero slope at a free end.

The analysis can be based on partial differential equations, or quite simply using the Stodola method introduced in Chapter 11. Although Stodola is usually applied to bending systems, we adapt it to the axial case by assuming a sinusoidal normal mode (Figure 12.1*a*), which we compare with the deflection distribution that results from this loading (*d*). The normal mode is

$$\beta(x) = \sin \frac{\pi}{2}\left(\frac{x}{l}\right) \qquad (12.1)$$

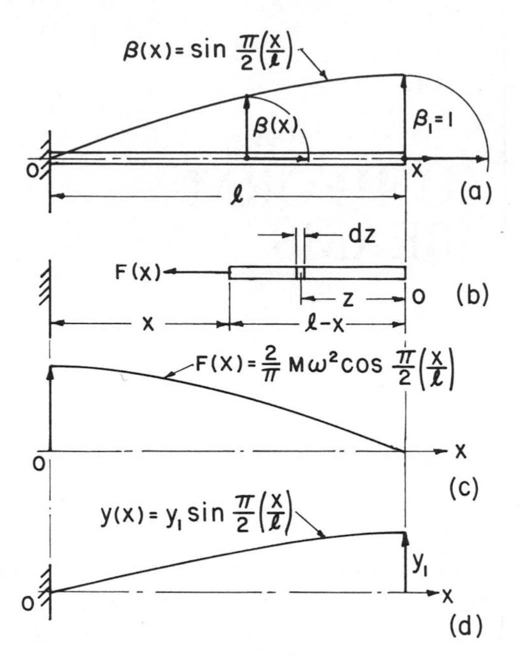

FIGURE 12.1. The axial elastic bar with sinusoidal mode (a) develops a load distribution (c) and a sinusoidal displacement (d). The axial quantities are plotted transversely, normal to the true direction.

Distributed axial loading increases from the free end, and is evaluated by integration of the differential inertia loads, introducing a coordinate z:

$$dF = -\ddot{z}\, dm = z\omega^2\mu\, dz$$

$$= \mu\omega^2 \cos\frac{\pi}{2}\left(\frac{z}{l}\right) dx$$

$$F(x) = \mu\omega^2 \int_0^{(l-x)} \cos\frac{\pi}{2}\left(\frac{z}{l}\right) dz$$

$$= \frac{2}{\pi} M\omega^2 \cos\frac{\pi}{2}\left(\frac{x}{l}\right) \tag{12.2}$$

Deflection distribution becomes

$$y(x) = \int_0^x \frac{S(x)}{E}\, dx = \frac{1}{AE}\int_0^x F(x)\, dx$$

$$= \frac{4}{\pi^2}\left(\frac{M\omega^2}{K}\right) \sin\frac{\pi}{2}\left(\frac{x}{l}\right) \tag{12.3}$$

TABLE 12.1. Natural frequencies and modal properties of the uniform bar, axially or in torsion: (*a*) clamped–free, (*b*) clamped–clamped, and (*c*) free–free. All tabulated values in (*b*) apply also to (*c*), except $f(x)$. Angular functions are in radians.

i	(a) 1	2	3	4	(b) 1	2	3	4
$f(x)$	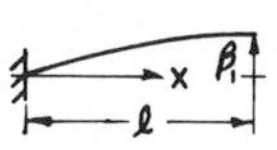	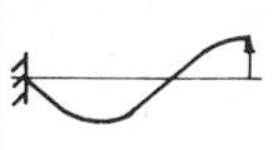	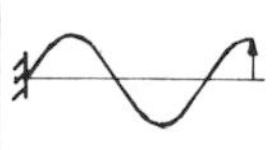	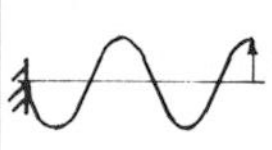	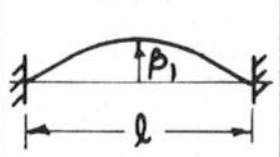	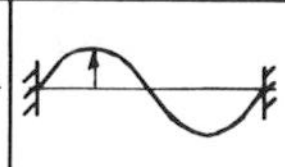	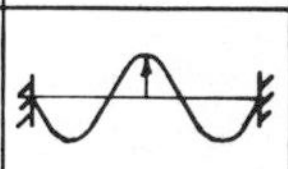	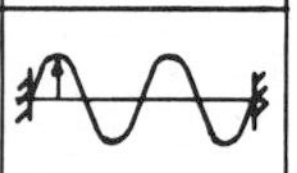
β_{ni}	$\sin \frac{\pi}{2}(\frac{x}{\ell})$	$-\sin \frac{3\pi}{2}(\frac{x}{\ell})$	$\sin \frac{5\pi}{2}(\frac{x}{\ell})$	$-\sin \frac{7\pi}{2}(\frac{x}{\ell})$	$\sin \pi (\frac{x}{\ell})$	$\sin 2\pi(\frac{x}{\ell})$	$-\sin 3\pi(\frac{x}{\ell})$	$\sin 4\pi(\frac{x}{\ell})$
g_{ni}	$\frac{\pi}{2}$	$3\left(\frac{\pi}{2}\right)$	$5\left(\frac{\pi}{2}\right)$	$7\left(\frac{\pi}{2}\right)$	π	2π	3π	4π
$\frac{M_{ni}}{M}$	$\frac{1}{2}$	$\frac{1}{2}$	$\frac{1}{2}$	$\frac{1}{2}$	$\frac{1}{2}$	$\frac{1}{2}$	$\frac{1}{2}$	$\frac{1}{2}$
$\frac{K_{ni}}{K}$	$\frac{\pi^2}{8}$	$9\left(\frac{\pi^2}{8}\right)$	$25\left(\frac{\pi^2}{8}\right)$	$49\left(\frac{\pi^2}{8}\right)$	$\frac{\pi^2}{2}$	$4\left(\frac{\pi^2}{2}\right)$	$9\left(\frac{\pi^2}{2}\right)$	$16\left(\frac{\pi^2}{2}\right)$
$\omega_{ni} = g_{ni}\,\omega_1\,, \quad \omega_1 = \frac{1}{\ell}\sqrt{\frac{E}{\rho}}$ or $\frac{1}{\ell}\sqrt{\frac{GJ}{Mr^2}}\,, \quad M = \rho A \ell$ or $\rho A \ell r^2\,, \quad K = \frac{AE}{\ell}$ or $\frac{GJ}{\ell}$								
				(c) FREE-FREE MODES	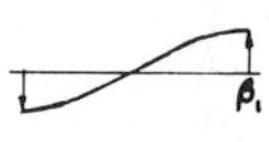	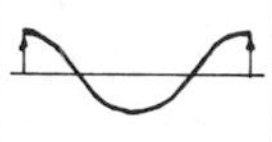	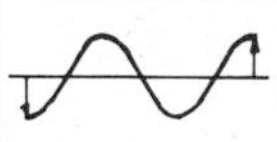	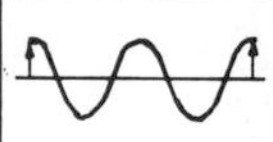

and at the free end reference point

$$y_1 = \frac{4}{\pi^2}\left(\frac{\omega^2}{K/M}\right) \tag{12.4}$$

where μ = mass per unit length
ω = circular frequency
$M = \mu l$ = total beam mass
$K = AE/l$ = reference stiffness of total bar
S = tensile stress
A = cross-sectional area
E = Young's modulus of elasticity

The mode is verified by the sine result in Equation 12.3, which agrees with Equation 12.1. We obtain the natural frequency from

$$\omega_{n1}^2 = \frac{\beta_1}{y_1} = \frac{1}{y_1} = \frac{\pi^2}{4}\left(\frac{K}{M}\right) \tag{12.5}$$

And since $K/M = (AE/l)/(\rho A l) = E/l^2\rho$, the natural frequency is

$$\omega_{n1} = g_{n1}\omega_1 = \left(\frac{\pi}{2}\right)\left(\frac{1}{l}\sqrt{\frac{E}{\rho}}\right) \tag{12.6}$$

where ρ is the mass density of the bar and the reference frequency is defined by the absolute parameters of a specific system.

Higher modes are similarly verified, including clamped–clamped and free–free bars (Table 12.1). Note that all modes are composed of the basic sinusoidal element (Equation 12.1).

Torsion, in which there is twisting deformation of a shaft or tube, is more likely physically than the axial problem, as the latter are usually much higher in natural frequency. Modes and frequencies are obtained from Table 12.1, but we must use torsional parameters, including the shear factor *GJ*. Spring rates are torsional and the mass term applies to the mass moment of inertia about the polar axis, of Mr^2.

12.2. EQUIVALENT MODAL SYSTEMS

As with discrete systems, the grounded bar can be modeled physically as an infinite number of single-degree systems (Figure 12.2). For continuously dis-

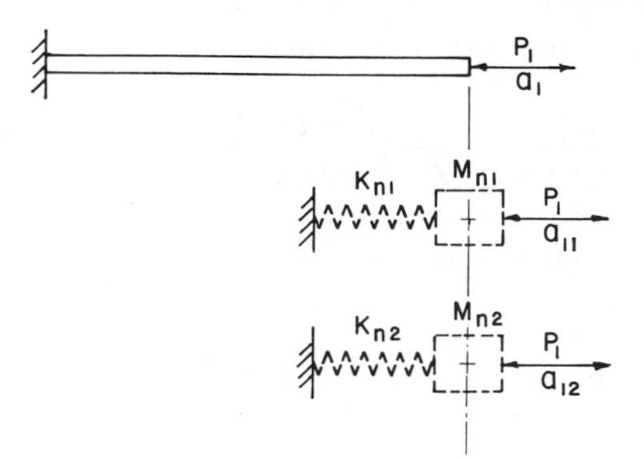

FIGURE 12.2. Equivalent modal systems, excited by P_1, respond in forced modal amplitudes.

tributed mass these expressions become

$$M_{ni} = \int_0^l \mu(x)\beta_{ni}^2(x)\,dx \tag{12.7a}$$

$$K_{ni} = \omega_{ni}^2 M_{ni} \tag{12.7b}$$

For the first axial mode of the simple bar (Equation 12.1),

$$M_{n1} = \mu \int_0^l \sin^2 \frac{\pi}{2}\left(\frac{x}{l}\right) = \frac{\mu l}{2} = \frac{M}{2} \tag{12.8a}$$

$$K_{n1} = \frac{\pi^2}{4}\left(\frac{K}{M}\right)\left(\frac{M}{2}\right) = \frac{\pi^2}{8}K \tag{12.8b}$$

Values are indicated in Table 12.1 for various modes and end conditions. Modal mass which derives from kinetic energy is $(M/2)$ for all combinations. Modal stiffness increases rapidly for higher modes, particularly for the clamped–free condition.

12.3. AXIAL–TORSIONAL FORCED RESPONSE

Forcing at the free end of the bar (Figure 12.2) excites the several modal systems:

$$\frac{\mathsf{a}_{1i}}{(P_i/K_{ni})} = \frac{1}{1 - \mathsf{g}_i^2} \tag{12.9}$$

where $\mathsf{g}_i = \omega/\omega_{ni}$. Converting to the actual bar stiffness $K = AE/l$,

$$\frac{\mathsf{a}_{1i}}{(P_1/K)} = \left(\frac{K}{K_{ni}}\right)\left(\frac{1}{1-\mathsf{g}_i^2}\right) = C_i\left(\frac{1}{1-\mathsf{g}_i^2}\right) \tag{12.10}$$

Superimposing end amplitudes,

$$\frac{\mathsf{a}_1}{P_1/K} = \Sigma C_i\left(\frac{1}{1-\mathsf{g}_i^2}\right) = \Sigma C_i A_i$$

$$= \frac{8}{\pi^2}\left(A_1 + \frac{1}{9}A_2 + \frac{1}{25}A_3 + \cdots\right) \tag{12.11}$$

Taking, for example, $g_1 = 2$, $\omega = 2\omega_{n1}$, we obtain end responses in Figure 12.3. In addition, each modal amplitude a_{1i} corresponds to a known distributed mode. We then have by the summation the total approximation Σ to three terms of the entire forced response.

As a check on the coefficients, static conditions can be observed at zero frequency, when all magnification factors revert to unity. Then

$$\frac{\mathsf{a}_1}{(P_1/K)} = \Sigma C_i = \frac{8}{\pi^2}\left(1 + \frac{1}{9} + \frac{1}{25} + \cdots\right) \tag{12.12}$$

which converges on unity. Thus the static end deflection is composed of residual vibratory effects, of which the first mode contributes $(8/\pi^2)$ or 81%. Furthermore the superposition of modes across the length will produce the linear static elongation from zero at the clamped end to (P_1/K) at the free end (Figure 12.4). Conceptually then component modal responses have physical significance even in the static situation.

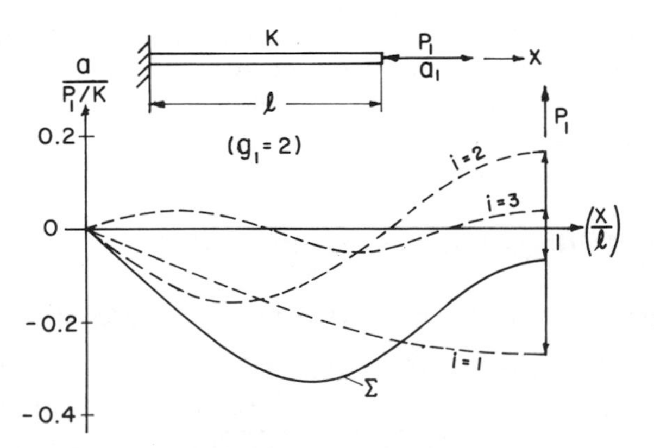

FIGURE 12.3. Forced modal components (Figure 12.2) produce amplitudes at the end that further represent the respective modal distributions along the bar.

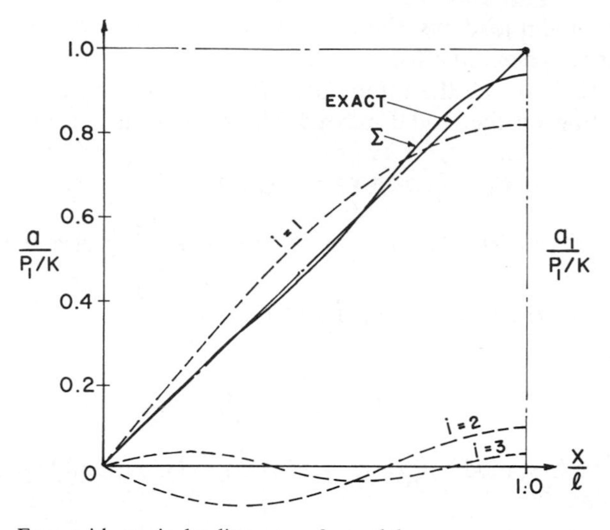

FIGURE 12.4. Even with static loading, $g_i = 0$, modal components combine to display the simple elastic elongation.

Response of the clamped–clamped bar is given in Table 12.2*b*, with sinusoidal excitation and displacement referred to the midpoint. Because of symmetry, even modes, having nodes at the center, cannot then be excited.

12.4. FORCED MODAL REACTIONS

With sinusoidal forcing at the free end (Figure 12.5), the vibratory modes create distributed inertia loads. The clamped end must carry the algebraic sum

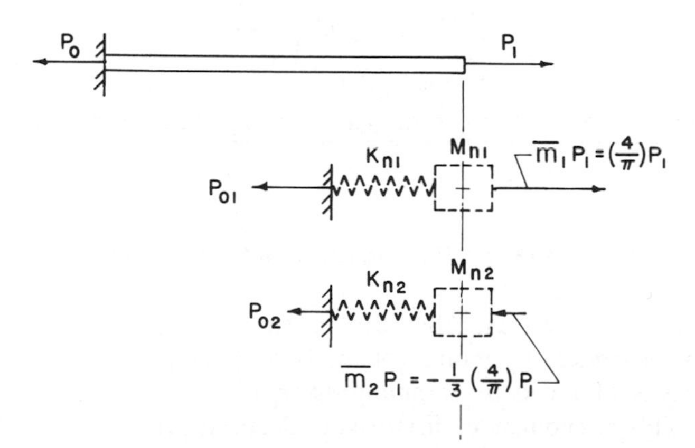

FIGURE 12.5. Vibratory reaction P_0 and axial stress at the base are determined using the equivalent modal system analog.

of P_1 and the modal loadings, the latter depending on modal distribution and forcing frequency resonant effects.

In Figure 12.5 we see the input force P_1 resolved into component modal excitations acting on the modal masses. The resolution is achieved by letting

$$P_1 = P_{11}\beta_{n1}(x) + P_{12}\beta_{n2}(x) + \cdots \tag{12.13}$$

Multiplying both sides by $\mu(x)\beta_{n1}(x)\,dx$ and integrating over the length,

$$P_1\int_0^l \mu(x)\beta_{n1}(x)\,dx = P_{11}\int_0^l \mu(x)\beta_{n1}^2\,dx + P_{12}\int_0^l \mu(x)\beta_{n1}(x)\beta_{n2}(x)\,dx + \cdots \tag{12.14}$$

Substituting the first mode, $\beta_{n1}(x) = \sin(\pi/2)(x/l)$, with all higher terms zero by orthogonality,

$$\frac{P_{11}}{P_1} = \frac{\int_0^l \mu(x)\beta_{n1}(x)\,dx}{\int_0^l \mu(x)\beta_{n1}^2(x)\,dx} = \overline{m}_1 = \frac{4}{\pi} \tag{12.15}$$

Adding single-degree effects using Table 12.2*a*,

$$\frac{-P_0}{P_1} = \Sigma\overline{m}_i\left(\frac{1}{1-\mathrm{g}_i^2}\right) \tag{12.16}$$

These are *absolute* components. In equivalent alternate form, with *relative* components (see Equations 5.17 and 5.18) we have

$$\frac{-P_0}{P_1} = 1 + \Sigma\overline{m}_i\left(\frac{\mathrm{g}_i^2}{1-\mathrm{g}_i^2}\right) \tag{12.17}$$

Stress at the base of the bar is proportional to P_0, whether an alternating axial load or torque.

12.5. MODAL RESPONSE TO DISPLACEMENT EXCITATION

The simple fixed–free elastic bar can be subjected to a sinusoidal axial or torsional displacement by motion of the base a_0 (Figure 12.6). Amplitude at the free end is obtained by summing normal modes to the degree they are developed. This phenomenon, discussed in Sections 5.10 and 11.11 for lumped masses, is now broadened to include continuously distributed mass, excitation, and normal modes.

Derivation of the modal participation factor closely parallels evaluation of harmonic coefficients in Fourier series (Chapter 6), except that we are now resolving the displacement pattern (Figure 12.6) into modal rather than sinusoidal components. We express the distributed excitation as

$$\lambda(x) = \overline{m}_1[\beta_{n1}(x)] + \overline{m}_2[\beta_{n2}(x)] + \cdots \qquad (12.18)$$

with $\overline{m}_i$ the weighting factor to be determined. This is done by multiplication of both sides, integration, and application of orthogonality, exactly as in Section 12.4, leading to a similar equation,

$$\overline{m}_i = \left[\frac{\int_0^l \lambda(x)\mu(x)\beta_{ni}(x)\,dx}{M_{ni}}\right] \qquad (12.19)$$

where $\lambda(x)\mu(x)$ is a constant. The evaluations yield $\overline{m}_1 = 4/\pi$, $m_2 = -4/3\pi$, $m_3 = 4/5\pi, \ldots$. Figure 12.6 indicates how the rectangular excitation function is approximated by adding successively higher modal terms.

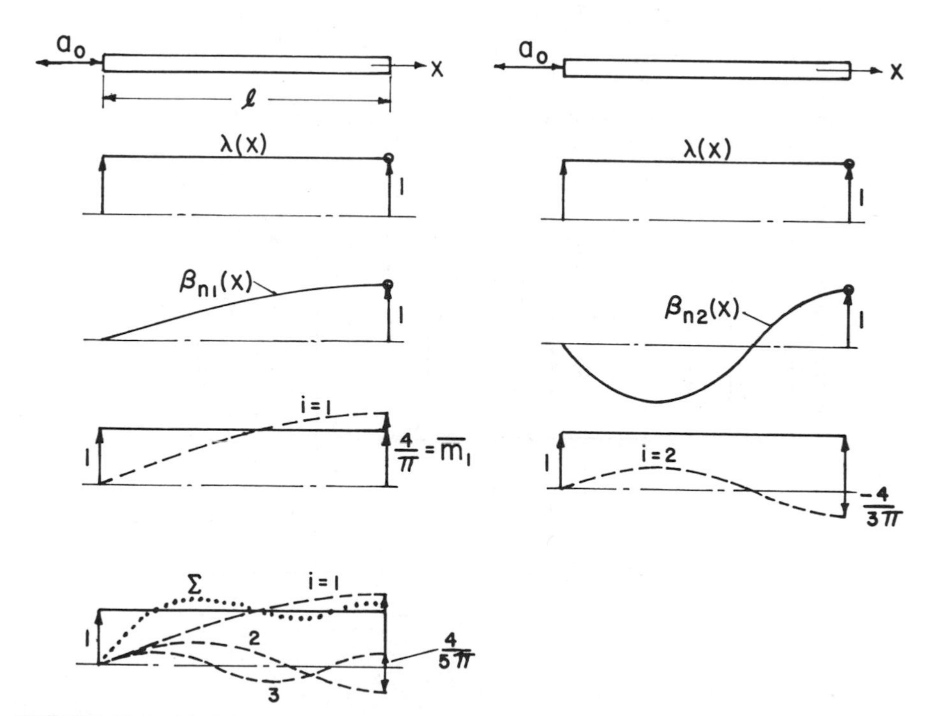

FIGURE 12.6. Modal response to base excitation requires resolution of the displacement distribution $\lambda(x)$ into component modal distributions.

TABLE 12.2. Forced modal responses of Table 12.1*a* and *b* systems. In (*c*) and (*d*), the displacement excitation is $\mathbf{a}_0$.

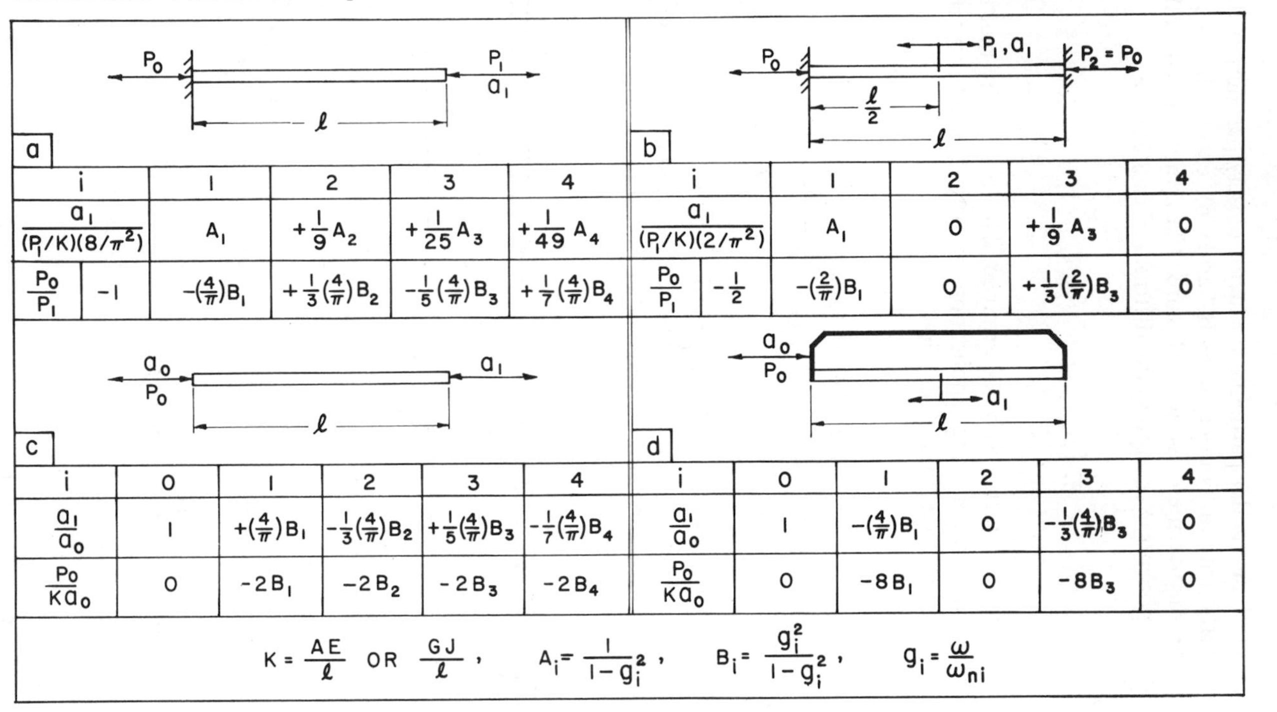

(a)

i		1	2	3	4
$\frac{a_1}{(P_1/K)(8/\pi^2)}$		A_1	$+\frac{1}{9}A_2$	$+\frac{1}{25}A_3$	$+\frac{1}{49}A_4$
$\frac{P_0}{P_1}$	-1	$-(\frac{4}{\pi})B_1$	$+\frac{1}{3}(\frac{4}{\pi})B_2$	$-\frac{1}{5}(\frac{4}{\pi})B_3$	$+\frac{1}{7}(\frac{4}{\pi})B_4$

(b)

i		1	2	3	4
$\frac{a_1}{(P_1/K)(2/\pi^2)}$		A_1	0	$+\frac{1}{9}A_3$	0
$\frac{P_0}{P_1}$	$-\frac{1}{2}$	$-(\frac{2}{\pi})B_1$	0	$+\frac{1}{3}(\frac{2}{\pi})B_3$	0

(c)

i	0	1	2	3	4
$\frac{a_1}{a_0}$	1	$+(\frac{4}{\pi})B_1$	$-\frac{1}{3}(\frac{4}{\pi})B_2$	$+\frac{1}{5}(\frac{4}{\pi})B_3$	$-\frac{1}{7}(\frac{4}{\pi})B_4$
$\frac{P_0}{Ka_0}$	0	$-2B_1$	$-2B_2$	$-2B_3$	$-2B_4$

(d)

i	0	1	2	3	4
$\frac{a_1}{a_0}$	1	$-(\frac{4}{\pi})B_1$	0	$-\frac{1}{3}(\frac{4}{\pi})B_3$	0
$\frac{P_0}{Ka_0}$	0	$-8B_1$	0	$-8B_3$	0

$$K = \frac{AE}{\ell} \text{ OR } \frac{GJ}{\ell}, \quad A_i = \frac{1}{1-g_i^2}, \quad B_i = \frac{g_i^2}{1-g_i^2}, \quad g_i = \frac{\omega}{\omega_{ni}}$$

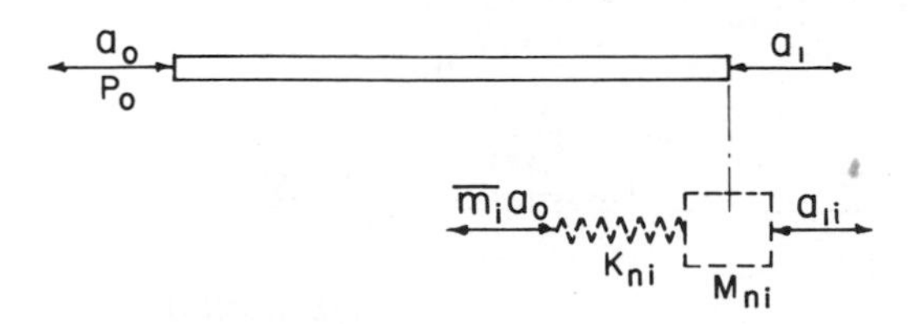

FIGURE 12.7. From Figure 12.6, the total input excitation a_0 is partitioned into component amplitudes acting on the modal springs.

From the single-degree analog (Figure 12.7), we find the modal amplitude at the end in relative and absolute form,

$$\left(\frac{a_{1i}^r}{\overline{m}_i a_0}\right) = \left(\frac{g_i^2}{1 - g_i^2}\right) \tag{12.20a}$$

$$\left(\frac{a_{1i}}{\overline{m}_i a_0}\right) = \left(\frac{1}{1 - g_i^2}\right) \tag{12.20b}$$

We have divided a_0 into modified displacements $\overline{m}_i a_0$. Collectively all modal component excitations account for the actual exciting amplitude as $\Sigma \overline{m}_i = 1.00$.

Various modal responses are given in Table 12.2*c* and *d*, including amplitudes and driving reaction.

12.6. THE SIMPLE BEAM IN BENDING

A constant beam on simply supported ends (Figure 12.8) typifies many real systems, including a shaft in self-aligning end bearings, or a uniform structural beam with negligible slope constraint at the ends. It is the most basic example of a continuous beam in bending.

Sinusoidal modes and frequencies (Table 12.3*a*), can be derived using the Stodola approach as in Section 12.1; however, determination of deflection from the assumed loading requires four successive integrations. Obviously on this basis a sinusoidal loading produces a sinusoidally distributed bending moment, and the final deflection distribution reverts to sinusoidal, with the integrations generating the constants shown in Figure 12.8. We find the first natural frequency by equating the maximum deflection to unity, as in Section 11.4.

$$y_1 = \left(\frac{l}{\pi}\right)^4 \frac{\mu\omega^2}{EI} = \beta_1 = 1$$

$$\omega_n^2 = \pi^2 \sqrt{\frac{EI}{\mu l^4}} = \pi^2 \sqrt{\frac{K}{M}} = \pi^2 \omega_1 \tag{12.21}$$

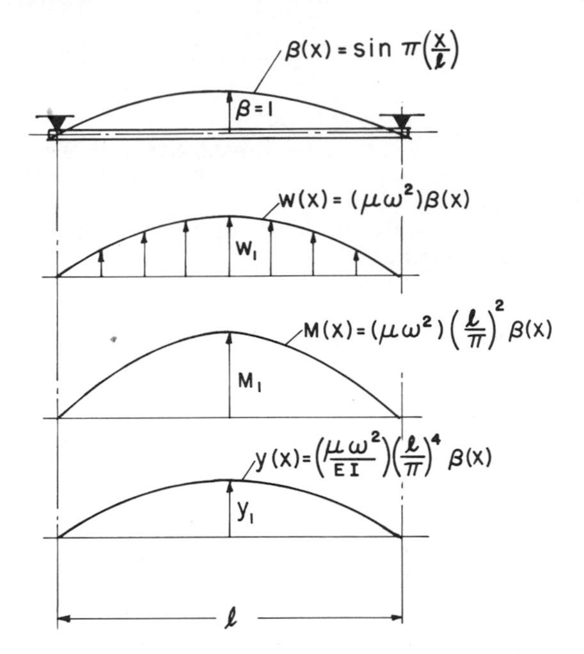

FIGURE 12.8. The Stodola method applied to a simple beam in bending confirms the sinusoidal first mode.

where $\pi^2 = g_{n1}$ = natural frequency factor for first mode
$K = EI/l^3$ = reference beam stiffness
$M = \mu l$ = total beam mass

Higher modes involve integer multiples of the first-mode length relative to the higher (Table 12.3*a*). Modal springs and masses are from Equations 12.7*a* and 12.7*b*. Table 12.4 provides modal data for the clamped–clamped and clamped–pined cases.

12.7. FORCED RESPONSE OF SIMPLE BEAM

A transverse sinusoidal force at the center of the beam (Figure 12.9) excites all odd modes to some degree, becoming maximum as the exciting frequency approaches a natural modal frequency. Magnification effects and equilibrium amplitude are applicable to the single-degree modal equivalents. Total vibratory loading consists of the concentrated exciting force and the distributed inertia loading from the vibratory modes.

TABLE 12.3. Natural frequencies and normal modal characteristics of simply supported uniform beam (*a*) and cantilever or clamped–free (*b*) in bending.

i	(a) 1	2	3	4	(b) 1	2	3	4
$f(x)$	β_1, x, ℓ				ϕ_m, β_1, ℓ			
β_{ni}	$\sin C_i\left(\frac{x}{\ell}\right)$				$\frac{1}{2}\left[\cosh C_i\left(\frac{x}{\ell}\right) - \cos C_i\left(\frac{x}{\ell}\right)\right] - D_i\left[\sinh C_i\left(\frac{x}{\ell}\right) - \sin C_i\left(\frac{x}{\ell}\right)\right]$			
g_{ni}	π^2	$4\pi^2$	$9\pi^2$	$16\pi^2$	3.52	22.0	61.7	121
$\frac{M_{ni}}{M}$	$\frac{1}{2}$	$\frac{1}{2}$	$\frac{1}{2}$	$\frac{1}{2}$	$\frac{1}{4}$	$\frac{1}{4}$	$\frac{1}{4}$	$\frac{1}{4}$
$\frac{K_{ni}}{K}$	$\frac{\pi^4}{2}$	$16\left(\frac{\pi^4}{2}\right)$	$81\left(\frac{\pi^4}{2}\right)$	$256\left(\frac{\pi^4}{2}\right)$	3.091	121.4	952	3654
C_i	π	2π	3π	4π	1.875	4.694	7.85	11.00
				D_i	0.367	0.509	0.500	0.500
				$\ell\phi_{ni}$	1.377	4.781	7.85	11.00

$$\omega_{ni} = g_{ni}\,\omega_1 \qquad \omega_1 = \sqrt{\frac{K}{M}}$$

$$K = \frac{EI}{\ell^3} \qquad M = \rho A \ell$$

TABLE 12.4. Similar to Table 12.3, but for the clamped–clamped (*a*) and the clamped–pinned beam (*b*). Frequencies for the free–free beam in bending can also be taken from (*a*). Footnote relations can be taken from Table 12.3.

i	a 1	2	3	4	b 1	2	3	4
$f(X)$	ℓ, x, β_i	b_i			ℓ, x, β_i	b_i		
β_{ni}	$E_i[\cosh C_i(\frac{x}{\ell}) - \cos C_i(\frac{x}{\ell})] - D_i[\sinh C_i(\frac{x}{\ell}) - \sin C_i(\frac{x}{\ell})]$							
g_{ni}	22.37	61.67	120.9	199.9	15.42	49.97	104.2	178.3
$\frac{M_{ni}}{M}$	0.396	$\frac{7}{16}$	$\frac{7}{16}$	$\frac{7}{16}$	$\frac{7}{16}$	$\frac{7}{16}$	$\frac{7}{16}$	$\frac{7}{16}$
$\frac{K_{ni}}{K}$	198	1660	6395	17450	104	1090	4750	13900
$\frac{b_i}{\ell}$	0.50	0.29	0.21	0.16	0.58	0.32	0.22	0.17
C	4.730	7.853	10.996	14.14	3.927	7.069	10.21	13.35
D	0.619	0.650	0.662	0.662	0.662	0.662	0.662	0.662
E	0.630	0.662	0.662	0.662	0.662	0.662	0.662	0.662

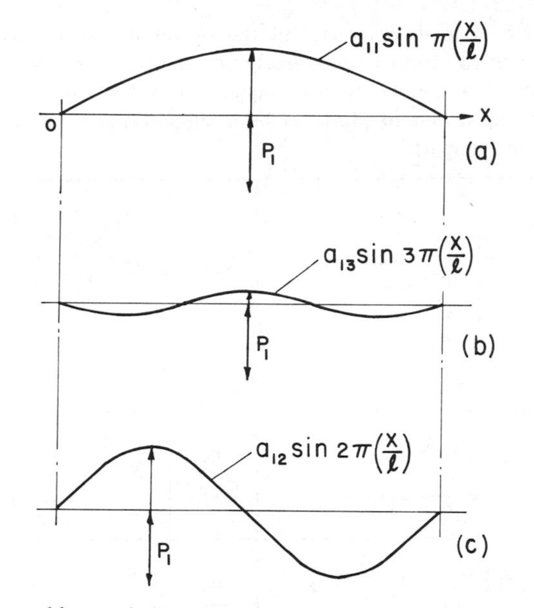

FIGURE 12.9. A central harmonic forcing function excites only odd modes of the beam (a) and (b). Even modes (c) are referred to peak modal amplitude positions when applying Equation 12.21.

For the first mode, the modal response is

$$\frac{a_{11}}{(P_1/K_{n1})} = \left(\frac{1}{1 - g_1^2}\right), \quad \frac{a_{11}}{(P_1/K)} = \left(\frac{2}{\pi^4}\right)\left(\frac{1}{1 - g_1^2}\right)$$

$$\frac{a_{1i}}{(P_1/K)} = \left(\frac{K}{K_{ni}}\right)\left(\frac{1}{1 - g_i^2}\right) \tag{12.22}$$

Total response is the sum of the center amplitudes of all the modes.

It is instructive to consider the static case, with all g_1 values zero. Then this summation is

$$\frac{a_1}{(P_1/K)} = \frac{2}{\pi^4}\left(1 + \frac{1}{81} + \frac{1}{625} + \cdots\right) \tag{12.23}$$

which approaches (1/48) in the limit, verifying the basic deflection equation. The first-mode contribution to the static deflection is major, which can now be calculated as 98.6% of the total.

As with previous systems (Equation 5.13), a force at a point different from the reference position is modified by an amplitude ratio factor, β_x. Modal effects are shown in Table 12.5a. Response is simply directly proportional to the ratio.

TABLE 12.5. Forced responses at the center of the simply supported uniform beam: (*a*) forced transversely at the center or any lateral point; (*b*) forced by a couple at one support; (*c*) transverse displacement excitations, equal and in phase at both supports; and (*d*) displacement excited at one support.

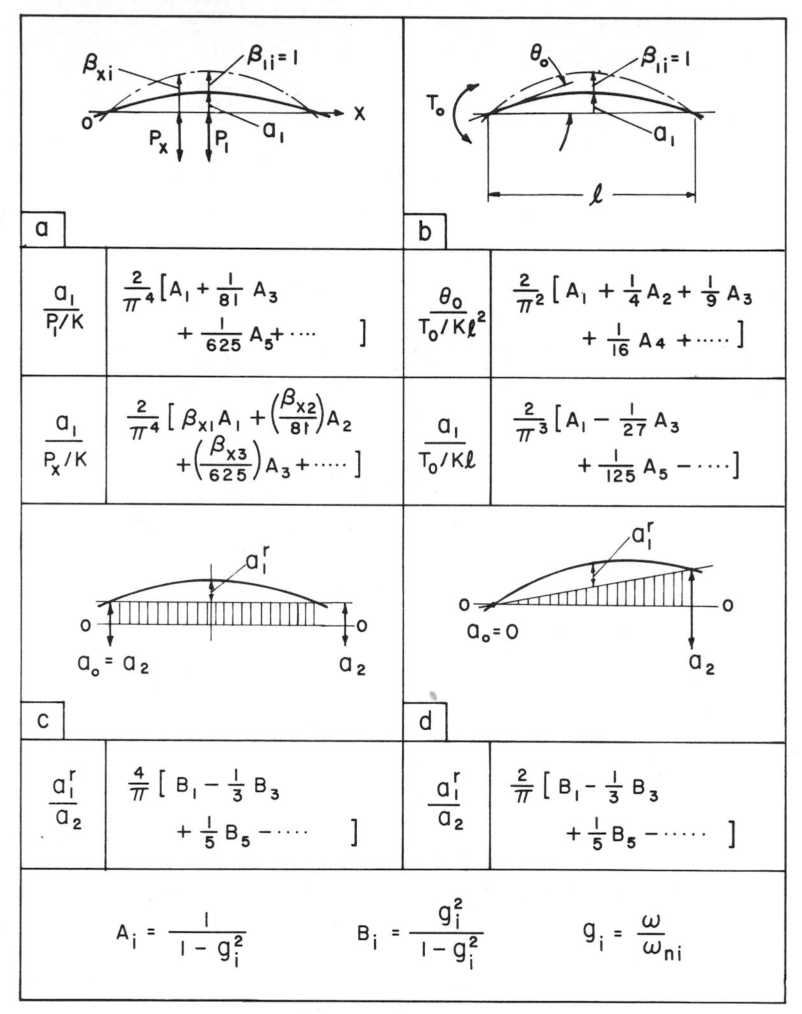

An exciting end couple (Table 12.5*b*) results in an unsymmetrical situation, with response due to all modes, including the even-numbered. These relations require the determination of the modal springs and mass at the support in an angular sense.

12.8. SUPPORT EXCITATION OF SIMPLE BEAM

Sinusoidal displacements of the beam supports cause modal responses (Tables 12.5*c* and *d*). The former pertains to pure translation and the latter to rotation about an end support. Unequal displacements at the same forcing frequency, and in or out of phase with each other, will be composed of these two component effects.

It should be noted that a_1^r is *relative* to the displaced beam axis, and the translation excites no even-numbered modes. Rotational displacement (d) excites all modes, but the even modes, though present, are nodal at the center.

End or support displacements can be orbital as well as planar. If so, the whirl response would be circular in space, with the displacement a_1^r then a radial distance.

12.9. FREQUENCIES AND MODES OF THE CANTILEVER

Table 12.3*b* summarizes free vibratory characteristics of the clamped–free uniform beam in bending for the first four modes. Normal modes for the cantilever, as for other uniform beams in bending, conform to a common modal format.[22]

$$\beta_{ni}(x) = E_i\left[\cosh C_i\left(\frac{x}{l}\right) - \cos C_i\left(\frac{x}{l}\right)\right] - D_i\left[\sinh C_i\left(\frac{x}{l}\right) - \sin C_i\left(\frac{x}{l}\right)\right] \tag{12.24}$$

where the coefficients C_i, D_i and E_i are determined by end conditions and modal number. In this and in similar tabulations, the coefficients D and E have been adjusted to produce unit amplitude at the maximum ordinates, indicated as β_1. Modal springs and masses are also referenced to β_1.

12.10. FORCED CANTILEVER RESPONSE

Table 12.6*a* and *b* provides equations for the determination of end amplitudes, and moment and shear reactions at the base when the tip is subjected to force or couple excitation. All variables exist at the common forcing frequency, and are composed of modal contributions. The ratios (P_0/P_1) and (T_0/T_1) are analogous to the transmissibility concept of Section 2.8, with damping now considered negligible.

12.11. CANTILEVER EXCITATION BY BASE DISPLACEMENT

Translation of the clamped end in a transverse direction results in modal responses and reactions (Table 12.6*c*). These relations are applicable to the complete beam, illustrated in Figure 12.10. Bending stresses at the hub are determined by T_0, and the driving force is $2P_0$. With operation exactly at a modal resonance, we would have to include damping to obtain finite values.

TABLE 12.6. Forced modal response of the clamped–free or cantilever beam to various sinusoidal excitations: (*a*) transverse force at free end; (*b*) couple at free end; (*c*) transverse displacement of the base; and (*d*) angular displacement of the base.

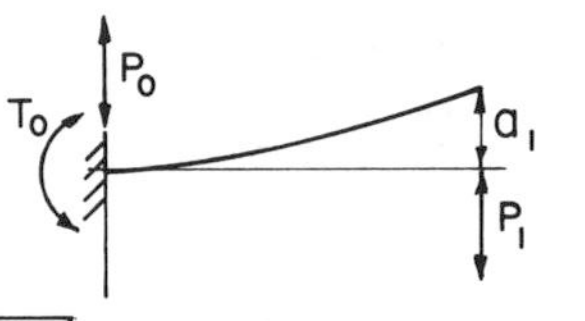

a	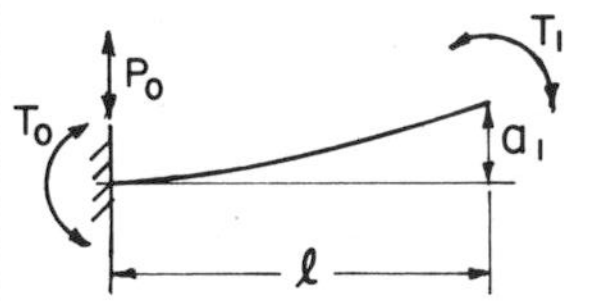	b	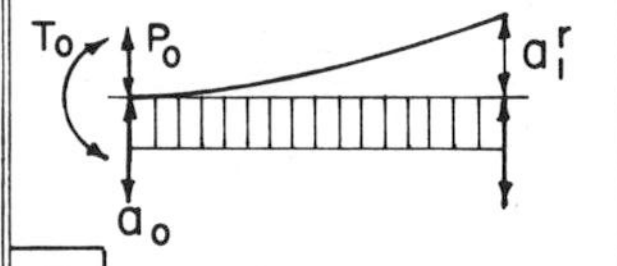	c	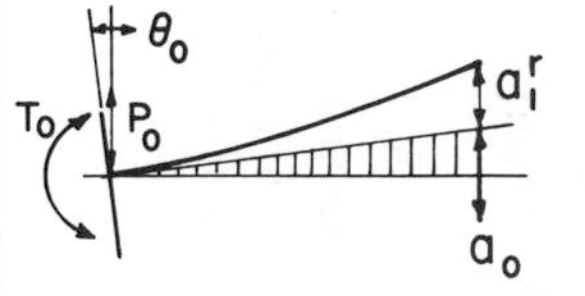	d	
$\frac{a_1}{P_1/K}$	$\sum \frac{K}{K_{ni}} A_i$	$\frac{a_1}{T_1/K\ell}$	$\sum C_i A_i$	$\frac{a_1^r}{a_o}$	$\sum \overline{m}_{iT} B_i$	$\frac{a_1^r}{a_o}$	$\sum \overline{m}_{iR} B_i$
$\frac{P_o}{P_1}$	$-1-\sum \overline{m}_{iT} B_i$	$\frac{P_o}{T_1/\ell}$	$\sum D_i B_i$	$\frac{P_o}{K a_o}$	$\sum E_i B_i$	$\frac{P_o}{K a_o}$	$\sum F_i B_i$
$\frac{T_o}{P_1 \ell}$	$-1-\sum \overline{m}_{iR} B_i$	$\frac{T_o}{T_1}$	$-1-\sum \overline{m}_{iT} B_i$	$\frac{T_o}{K \ell a_o}$	$\sum F_i B_i$	$\frac{T_o}{K \ell a_o}$	$4 \sum B_i$

i	C_i	D_i
1	0.445	-2.153
2	0.039	4.150
3	0.008	-3.974
4	0.003	3.988

$$A_i = \frac{1}{1-g_i^2} \qquad B_i = \frac{g_i^2}{1-g_i^2}$$

i	$\overline{m}_{iT}$	$\overline{m}_{iR}$
1	1.566	1.138
2	-0.868	-0.182
3.	0.509	0.065
4	-0.364	-0.033

$$K = \frac{EI}{\ell^3} \qquad g_{ni} = \frac{\omega}{\omega_{ni}}$$

i	E_i	F_i
1	7.6	5.5
2	91.4	19.1
3	247.7	31.4
4	483.6	44.0

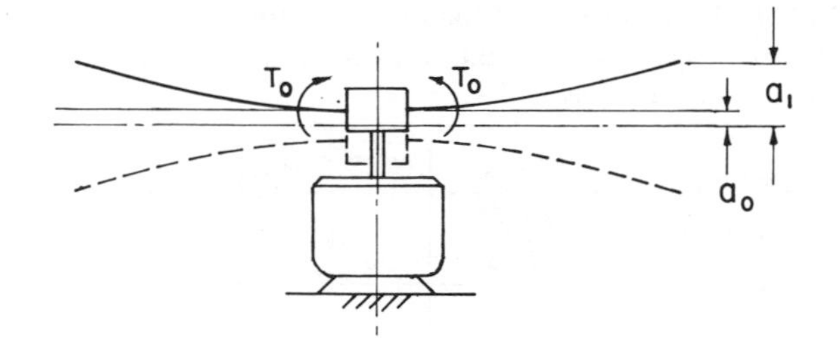

FIGURE 12.10. A typical fatigue test using an electromagnetic exciter utilizes beam resonance. The central clamp and symmetry provide moment balance at the center, and the driving force reacts both cantilevers (Table 12.6*c*).

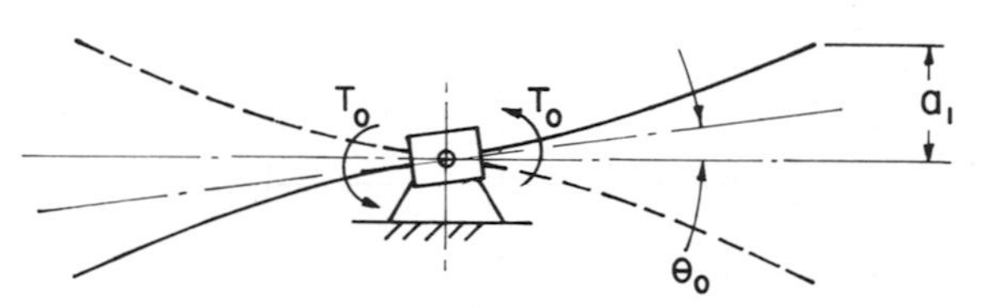

FIGURE 12.11. A complete constant beam subjected to hub excitation torsionally is equivalent to two cantilevers with a total driving couple of $(2T_0)$; see Table 12.6*d*.

In rotational response (Table 12.6*d*), the clamped end is rotated sinusoidally through a relatively small angle as shown in Figure 12.11.

The most general case involves these two displacement excitations in combination. Superposition is used if the frequencies are identical. If the excitations have a phase relation other than 0° or 180°, we would have to revert to combinations vectorially in the complex plane.

12.12. CALCULATION OF REFERENCE FREQUENCY

The frequency ω_1 is an important factor in natural frequency determination for axial, torsional, and bending modes of the uniform beam. This factor involves geometric, mass, and stiffness characteristics. For basic cross sections, these reduce to the useful relations in Table 12.7.

We note that ω_1 is directly proportional to either $\sqrt{E/\rho}$ or $\sqrt{G/\rho}$. These parameters represent the velocity of propagation of a stress wave in a given material, depending on the mass-elastic properties.

12.13. BENDING WITH SUPPORT FLEXIBILITY

An example of support flexibility relative to a uniform beam is indicated in Figure 12.12. Although the combination is fairly simple, several additional

TABLE 12.7. Evaluation of the reference frequency ω_1 for various basic cross sections in axial, torsional, or bending modes, as required in the previous tables.

CROSS SECTION	AXIAL $\frac{\omega_1 \ell}{\sqrt{E/\rho}}$	TORSIONAL $\frac{\omega_1 \ell}{\sqrt{G/\rho}}$	BENDING $\frac{\omega_1 \ell}{\sqrt{E/\rho}}$
Solid circle, D	1	1	$0.25\left(\frac{D}{\ell}\right)$
Hollow circle, d, D	1	1	$0.25\left(\frac{D}{\ell}\right)\sqrt{1+\left(\frac{d}{D}\right)^2}$
Thin-walled circle, D	1	1	$0.35\left(\frac{D}{\ell}\right)^2$
Solid square, b	1	0.92	$0.29\left(\frac{b}{\ell}\right)$
Thin-walled square, b	1	1.08	$0.41\left(\frac{b}{\ell}\right)$
Equilateral triangle (60°), b	1	0.78	$0.24\left(\frac{b}{\ell}\right)$
FOR STEEL & ALUMINUM ALLOY	$\sqrt{E/\rho} = 0.195(10)^6$ IPS OR 5000 m/s $\sqrt{G/\rho} = 0.125(10)^6$ IPS OR 3200 m/s		

concepts are required to determine the two lowest translational natural frequencies. In the extreme position shown, the springs are subjected to inertia loadings due to both translation and flexure, with resultant values of

$$F_1 = (\mu \omega^2 \beta_1) l = M\omega^2 \beta_1$$

$$F_2 = (\mu \omega^2 \beta_2)\left(\frac{2}{\pi}\right) l = \frac{2}{\pi}(M\omega^2 \beta_2)$$

producing equal *spring deflections* of

$$y_1 = \frac{F_1 + F_2}{K_1} = \frac{M\omega^2}{K_1}\left(\beta_1 + \frac{2}{\pi}\beta_2\right) \tag{12.25}$$

Central beam deflection relative to the translation of the ends is the result of the two distributed loadings. Anticipating sinusoidal flexural behavior, we take the sinusoidal component of the translational loading. As with the rectangle

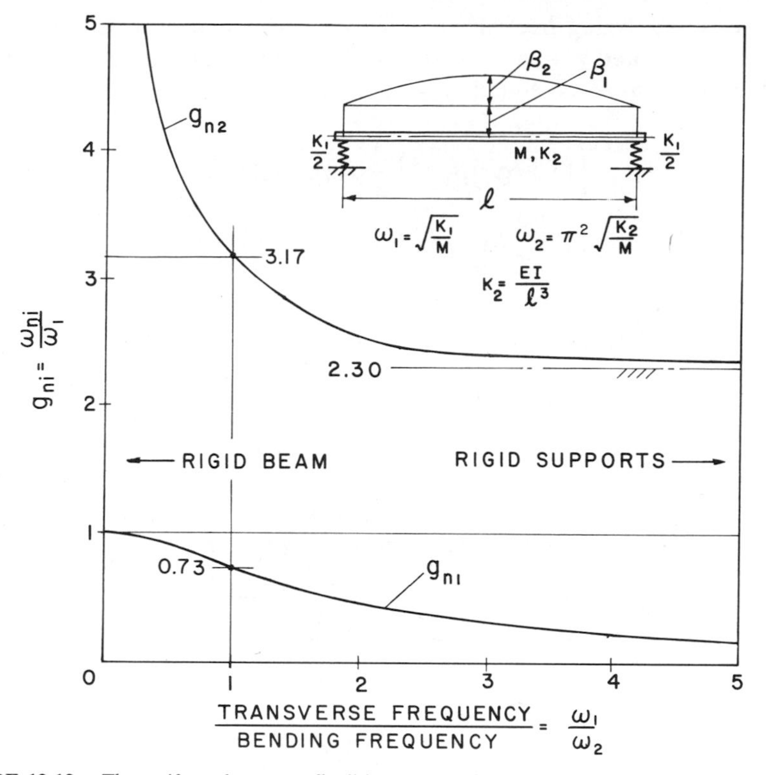

FIGURE 12.12. The uniform beam on flexible supports has two translational modes related to the first beam bending mode. Reference frequency is rigid beam on springs.

(Figure 12.6), the normalization factor is $4/\pi$, and

$$w_1(x) = \frac{4}{\pi}\left[\mu\omega^2 \sin\pi\left(\frac{x}{l}\right)\right]\beta_1 \tag{12.26}$$

Additional inertia loading due to beam flexure is

$$w_2(x) = \left[\mu\omega^2 \sin\pi\left(\frac{x}{l}\right)\right]\beta_2 \tag{12.27}$$

for a combined sinusoidal loading of

$$w(x) = \left[\mu\omega^2 \sin\pi\left(\frac{x}{l}\right)\right]\left(\frac{4}{\pi}\beta_1 + \beta_2\right) \tag{12.28}$$

This then converts to a resulting central *beam deflection* from Figure 12.8 of

$$y_2 = \frac{1}{EI}\left(\frac{l}{\pi}\right)^4 w(x) = K_2\left(\frac{l}{\pi^4}\right)w(x) \tag{12.29}$$

Since we are analyzing free vibration, ω must be a natural frequency, and in Stodola, $y_1 = \beta_1$ and $y_2 = \beta_2$. Equations 12.25 and 12.29 can be combined into simultaneous normalized equations:

$$\begin{cases} (1 - \mathsf{g}_n^2)\beta_1 - \left(\dfrac{2}{\pi}\mathsf{g}_n^2\right)\beta_2 = 0 \\ \left(-\dfrac{4}{\pi}C\mathsf{g}_n^2\right)\beta_1 + (1 - C\mathsf{g}_n^2)\beta_2 = 0 \end{cases} \tag{12.30}$$

where $\mathsf{g}_n = \omega_n/\omega_1 = \omega_n/\sqrt{K_1/M}$

$$C = \left(\frac{1}{\pi^4}\right)(K_1/K_2) = (\omega_1/\omega_2)^2$$

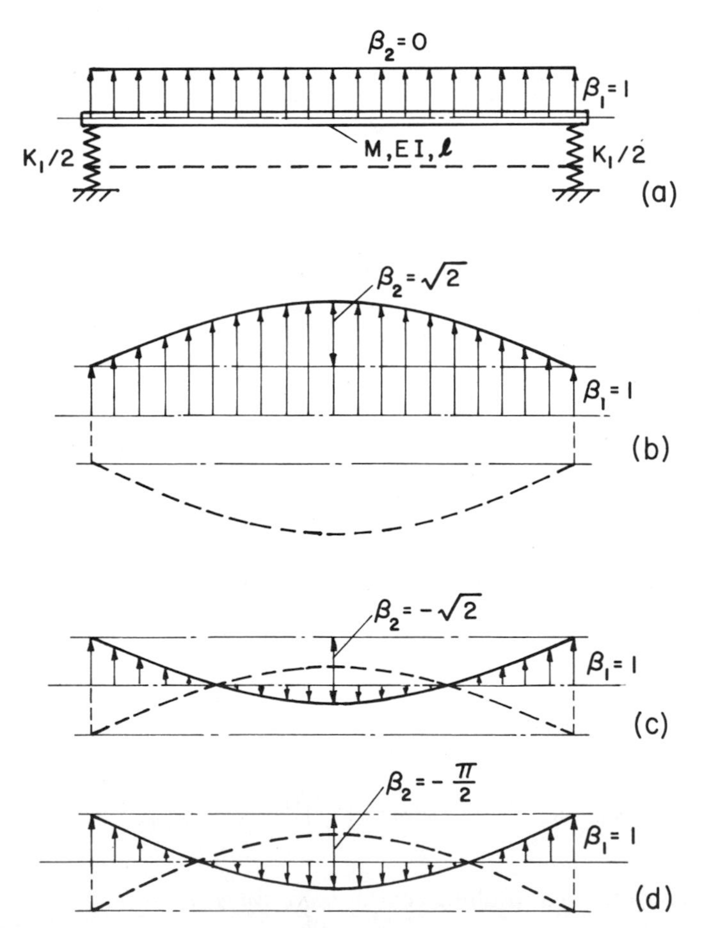

FIGURE 12.13. Normal modes for special frequency ratios: (*a*) rigid beam; (*b*) lower, $\omega_1 = \omega_2$; (*c*) higher, $\omega_1 = \omega_2$; (*d*) free–free, $\omega_1/\omega_2 = 0$, $\omega_n/\omega_2 = 2.30$.

Equating the denominator of the determinant to zero, we obtain the natural frequency quadratic:

$$A\mathrm{g}_n^4 - B\mathrm{g}_n^2 + 1 = 0 \tag{12.31}$$

where $A = \left(\dfrac{1}{\pi^4}\right)(1-8/\pi^2)(K_1/K_2) = (1-8/\pi^2)(\omega_1/\omega_2)^2$

$B = 1 + \left(\dfrac{1}{\pi^4}\right)(K_1/K_2) = 1 + (\omega_1/\omega_2)^2$

$\omega_1 = \sqrt{K_1/M}$ = translational natural frequency, beam rigid

$\omega_2 = \pi^2\sqrt{K_2/M}$ = beam natural frequency, supports rigid

From the first expression (Equation 12.30), normal modes are

$$(\beta_2)_n = \frac{\pi}{2}\left(\frac{1-\mathrm{g}_n^2}{\mathrm{g}_n^2}\right) \tag{12.32}$$

TABLE 12.8. First two translational frequencies and normal modes of the uniform beam on equally flexible end supports.

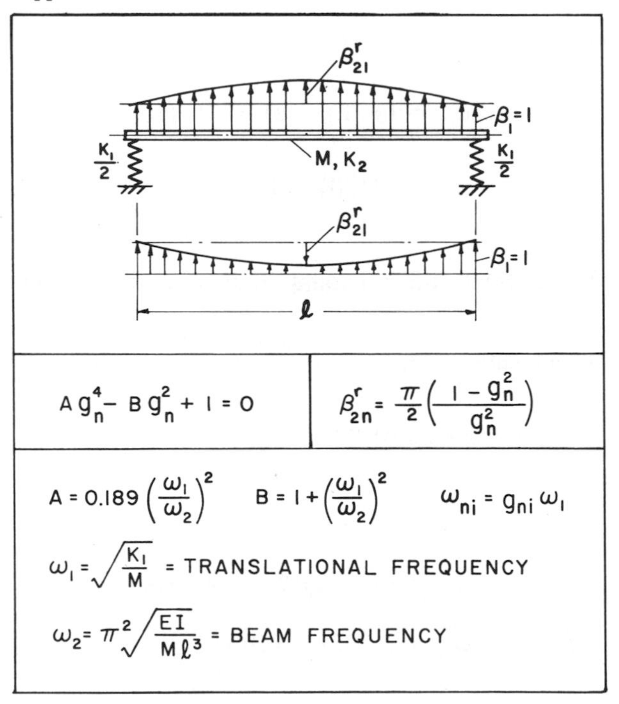

Generalized frequency behavior can be observed in Figure 12.12, as a function of the frequency ratio ω_1/ω_2. If these should be identical, the system lower frequency is decreased to 0.73 of either frequency because of the combined flexibility; however, the higher fundamental mode is 3.17 times the reference frequencies. Also, the higher mode frequency factor has a minimum asymptote of 2.30.

Behavior of the normal modes is interesting. In Figure 12.13 the rigid beam case is shown in (*a*), with $\beta_2 = 0$. For equal component frequencies, the relative modal amplitude of the beam is $+\sqrt{2}$ and $-\sqrt{2}$ for the lower and higher modes (*b*) and (*c*). As the springs approach zero, the beam becomes free–free, with a natural frequency of 2.30 ω_2, (Figure 12.13*d*).

Summarized results for this system are given in Table 12.8.

EXAMPLES

EXAMPLE 12.1. A square aluminum alloy bar, 6 × 6 cm and 2.4 m long, is clamped at one end and free at the other. Find the first two natural frequencies in the axial mode.

Solution. From Table 12.7,

$$\omega_1 = 5000/2.4 = 2082 \text{ r/s} = 332 \text{ Hz}$$

From Table 12.1,

$$\omega_{n1} = (\pi/2)(332) = 520 \text{ Hz}$$

$$\omega_{n2} = 3(520) = 1560 \text{ Hz}$$

EXAMPLE 12.2. Calculate the torsional frequencies for the bar of Example 12.1.

Solution. From Table 12.7,

$$\omega_1 = (3200/2.4)(0.92) = 1230 \text{ r/s} = 195 \text{ Hz}$$

From Table 12.1,

$$\omega_{n1} = (\pi/2)(195) = 307 \text{ Hz}$$

$$\omega_{n2} = 3(307) = 920 \text{ Hz}$$

EXAMPLE 12.3. Repeat Example 12.1 for bending about either principal axis.

Solution. From Table 12.6,

$$\omega_1 = 0.29(6/240)(5000/2.4) = 15.1 \text{ r/s} = 2.40 \text{ Hz}$$

From Table 12.3*b*,

$$\omega_{n1} = 3.52(2.40) = 8.5 \text{ Hz}$$

$$\omega_{n2} = 22.0(2.40) = 53 \text{ Hz}$$

EXAMPLE 12.4. A steel shaft 2.25 in. in diameter is rigidly clamped at both ends. What is the minimum distance between the ends if the first natural torsional frequency of the shaft must be above 500 Hz?

Solution. From Tables 12.7 and 12.1*b*,

$$\omega_{n1} = \pi\left[\frac{1}{l}(0.125)(10)^6\right] = 3142 = 500(2\pi)$$

$$l = 125 \text{ in.} \quad \text{(minimum)}$$

EXAMPLE 12.5. The bar of Example 12.1 has an axial base excitation of 0.10 cm at 900 Hz. What is the amplitude of the driving force?

Solution. From Table 12.2*c*,

$$K = (0.06)(0.06)\left[\frac{70(10)^9}{2.4}\right]$$

$$= 105(10)^6 \text{ N/m}$$

$$g_1 = 900/520 = 1.73 \qquad g_2 = 900/1560 = 0.58$$

$$g_3 = 900/2605 = 0.35 \qquad g_4 = 900/3640 = 0.25$$

$$B_1 = -1.50 \qquad B_2 = +0.50$$

$$B_3 = +0.14 \qquad B_4 = +0.065$$

$$P_0 = 105(10)^6(0.001)(-2)\Sigma B = -210(10)^3(-0.80)$$

$$= 170{,}000 \text{ N}$$

The positive result indicates that P_0 is *in phase* with a_0.

EXAMPLE 12.6. A 3 × 3 in. square steel tube has a wall of 1/8 in. and is 8 ft long. Find the first and second modal frequencies if the ends are supported on springs each 960 lb/in.

Solution. In Table 12.8

$$M = (1.44)(0.29)(96) = 40 \text{ lb}$$

$$K_2 = \left(\frac{EI}{l^3}\right) = \left[\frac{30(10)^6(1.98)}{96^3}\right] = 67 \text{ lb/in.}$$

$$\omega_1 = \sqrt{\frac{2(960)386}{40}} = 136 \text{ r/s} = 21.7 \text{ Hz}$$

$$\omega_2 = \pi^2\sqrt{\frac{(67)(386)}{40}} = 251 \text{ r/s} = 39.9 \text{ Hz}$$

$$A = (0.189)(0.295) = 0.055$$

$$B = 1 + 0.295 = 1.295$$

$$0.055\mathsf{g}_n^4 - 1.295\mathsf{g}_n^2 + 1 = 0$$

$$\mathsf{g}_{n1} = 0.89$$

$$\mathsf{g}_{n2} = 4.76$$

$$\omega_{n1} = (0.89)(21.7) = 19.4 \text{ Hz}$$

$$\omega_{n2} = 103 \text{ Hz}$$

Note that the frequency of 19.4 Hz represents a reduction from the beam frequency of 39.9 Hz caused by the end springs.

EXAMPLE 12.7. In the tube of Example 12.6 the beam is simply supported at both ends and at the center. Calculate the first two modal frequencies in bending.

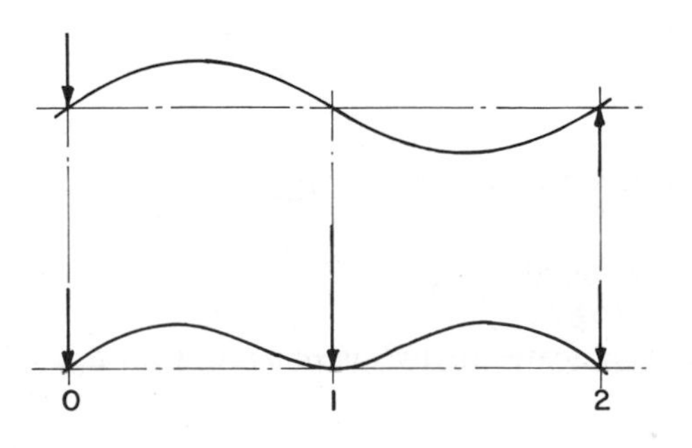

Solution. Considering the second mode in Table 12.3*a*, the center point is a node with or without a center support, and

$$\omega_{n1} = \left(\frac{4\pi^2}{\pi^2}\right)\omega_2 = 4(39.9) = 160 \text{ Hz}$$

From symmetry the second beam mode has zero slope at the center supports, providing an effective fixed end for both spans. For the clamped–pinned case (Table 12.4*b*), we consider the half-beam, with $l = 48$ in. and $M = 20$ lb.

$$K = \left(\frac{EI}{l^3}\right) = \left[\frac{30(10)^6(1.98)}{48^3}\right] = 537 \text{ lb/in.}$$

$$\omega_{n2} = 15.42\sqrt{\frac{(537)(386)}{20}} = 1570 \text{ r/s} = 250 \text{ Hz}$$

13 WHIRLING OF FLEXIBLE ROTORS

Analysis of a whirling rotor in bearings with distributed mass and stiffness is an involved subject. In addition to the basic natural frequencies, modes, and responses there are complications arising from bearing effects, rotational inertia, and numerous other factors. An extensive treatment of this subject will not be attempted here; however, the basic concepts of whirl response and balancing procedures are presented in generalized form. The rotor is really an extension of previously discussed systems; those of Chapters 11 and 12 can be construed as rotors with discrete or uniformly distributed mass, respectively. Rotor excitation is typically due to unbalance distributed throughout the length, and the behavior is related to simple whirl (Chapter 10).

In Sections 8.11–8.13, dynamic or two-plane unbalance was discussed for the *rigid* rotor, which is the usual case. For the fixed geometry this rotor can be completely balanced at all speeds by means of two corrections in different planes determined by measurements at any convenient balancing speed; however, if a rotor operates at a speed greater than approximately one-half of its first critical speed, it is usually considered to be *flexible*. Distortions of the axis in whirl cause the response to vary with speed.

This problem is now addressed on the basis of modal response and modal corrections. It will be shown that any given mode can be balanced by means of a single correction.[10]

13.1. NATURAL FREQUENCIES AND NORMAL MODES

The distributed rotor with varying cross section is shown in Figure 13.1, with mass and stiffness distributions indicated. With vibration in the plane shown, it acts as a beam subjected to harmonically varying transverse inertia loading, as in Chapter 12. If modeled as a series of concentrated masses connected by springs of different bending stiffness, the methods of Chapter 11 apply, including Stodola and matrix iteration. Several modes and frequencies are usually of interest; the nature of the first two are shown in Figure 13.1. Reasonably accurate data of this nature are obviously necessary for the modal methods to follow.

Although we view the rotor problem in Figure 13.1 as planar bending modes, whirl is three-dimensional, as in Chapter 10. The modal amplitudes become radii in rotation and the modal frequencies are more correctly *critical speeds*.

In Figure 13.2, single-degree modal systems are shown. Again, in whirl equal modal springs and masses must exist in both the vertical and horizontal directions.

13.2. PLANAR RESPONSE TO SINGLE UNBALANCE

If we have a concentrated unbalance excitation on the beam of $(me)_c$ (see Figure 13.2a), a vertical harmonic forcing function is developed as shown with the unbalance revolving with an angular velocity ω about the stationary beam.

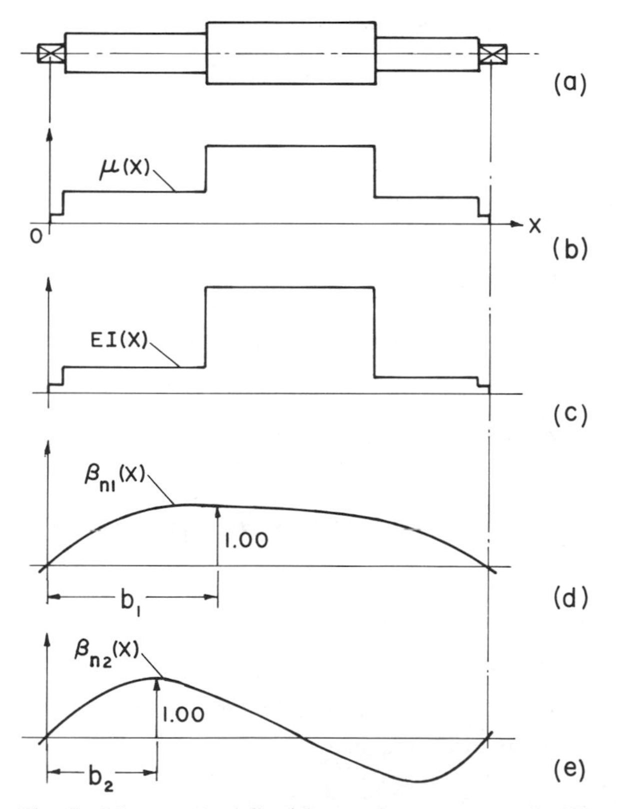

FIGURE 13.1. The flexible rotor is defined by continuous mass and stiffness distributions. Normal modes are maximum at different axial positions.

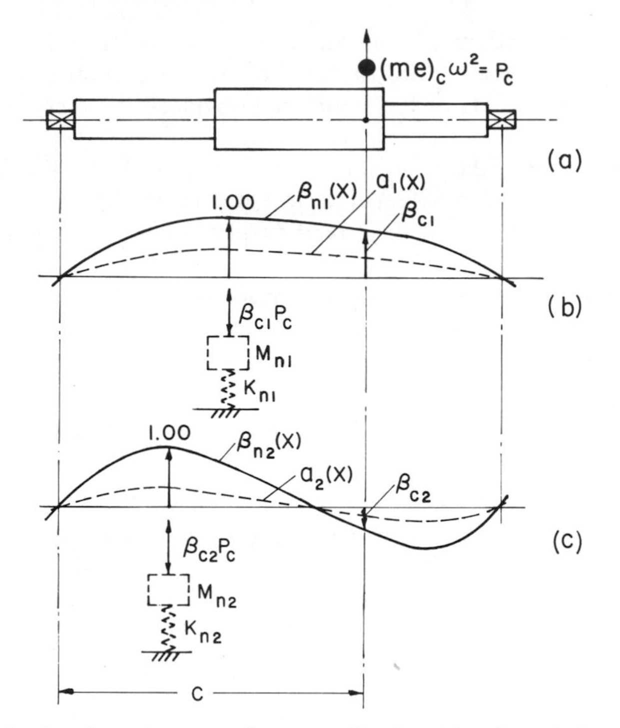

FIGURE 13.2. In the planar beams modes a centrifugal exciting force P_c is referred to the respective modal systems using the modal ratios at the force.

Response in the first mode requires referral of the force to the modal normalizing point, or multiplication by the factor β_{c1} (Figure 13.2*b*). Then,

$$a_1 = \left[\frac{(me)_c \omega^2 \beta_{c1}}{K_{n1}}\right]\left(\frac{1}{1 - g_1^2}\right) \tag{13.1}$$

which in generalized form becomes

$$a_i = \left[\frac{(me)_c \beta_{ci}}{M_{ni}}\right]\left(\frac{g_i^2}{1 - g_i^2}\right) \tag{13.2}$$

where $M_{ni} = \int_0^l \mu(x)\beta_{ni}^2(x)\,dx$ = modal mass

$g_i = (\omega/\omega_{ni})$ = forced modal frequency ratio

ω = rotational velocity of the unbalance

ω_{ni} = modal natural frequency

Thus Equation 13.2 allows us to calculate the response amplitude in a given mode at the normalizing point due to centrifugal forcing at any axial position.

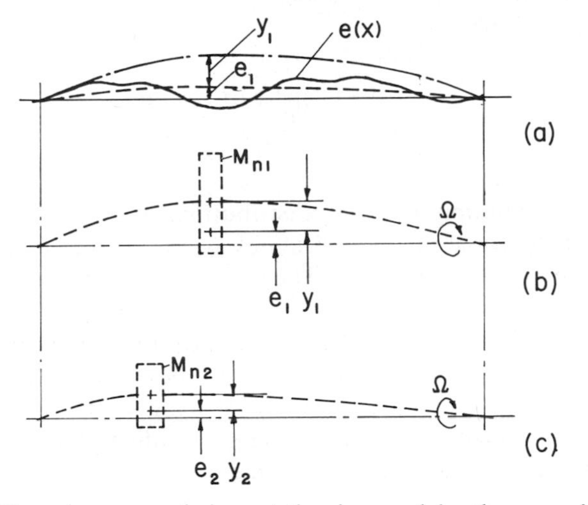

FIGURE 13.3. Flexural response during rotation is caused by the sum of all the modal unbalances.

This single coordinate in turn determines the complete normal modal distribution. We then have

$$\mathsf{a}(x) = \mathsf{a}_1\beta_{n1}(x) + \mathsf{a}_2\beta_{n2}(x) + \cdots \tag{13.3}$$

providing by summation the complete coordinate distribution of the forced beam response, including as many modes as necessary. In Figure 13.2*b* and *c* the sum of the two dashed curves provides a two-term approximation.

Rotation of the beam containing a fixed single unbalance effectively converts the beam to a rotor, with the rotor velocity Ω replacing ω and y_i replacing a_i. Now y_i is the radial elastic deflection of the axis at the normalizing position, obtained from Equation 13.2, as seen in Figure 13.3.

13.3. RESPONSE TO DISTRIBUTED UNBALANCE

A rotor is typically unbalanced to some degree throughout its length. Actual unbalance is the product at any point of centroidal eccentricity and the differential mass at the point. In this approach the rotor mass effect (Figure 13.1*b*) is incorporated in the solution, and the unbalance is specified by only the eccentricity, $e(x)$, as shown in Figure 13.3*a*. The irregular solid curve shows an arbitrary distribution in the plane which can be resolved into modal components.

$$e(x) = e_1\beta_{n1}(x) + e_2\beta_{n2}(x) + \cdots \tag{13.4}$$

Evaluation of the coefficients e_i requires an integration similar to Equation

12.19, using the relation

$$e_i = \frac{\int_0^l e(x)\mu(x)\beta_{ni}(x)\,dx}{M_{ni}} \tag{13.5}$$

To solve this we multiply the $e(x)$ distribution, if known, by the other two functions, plot the triple product ordinates against x, and determine the area under the curve as a full scale value. The integral dimensionally will involve the product of mass and distance.

We note that Equation 12.19 is similar to 13.5, and they are related; however, $\overline{m}_i$ in the former is dimensionless because the $\lambda(x)$ function is normalized to unity and is in turn dimensionless. But in the latter $e(x)$ represents absolute centroidal displacement, providing absolute values of e_i.

By multiplication, Equation 13.5 becomes

$$M_{ni}e_i = \int_0^l e(x)\mu(x)\beta_{ni}(x)\,dx \tag{13.6}$$

and we interpret the left term as *modal unbalance* (Figure 13.3). The distributed centroidal eccentricity has been converted to an equivalent unbalance which excites *only the mode for which it is calculated.*

The modal response in Equation 13.2 now simplifies to an equivalent of Equation 10.2, becoming

$$\left(\frac{y_i}{e_i}\right) = \left(\frac{\mathrm{g}_i^2}{1 - \mathrm{g}_i^2}\right) \tag{13.7}$$

again verifying the modal single-degree analog.

Complete distributed response is obtained by the superposition of the several modal responses in the plane. Additionally, in a rotor there will be a different eccentricity distribution in a perpendicular plane containing the rotor axis, and a given mode at the normalizing position has two perpendicular components of e_i, which are calculated independently. The position of the resultant determines the planar direction in space of the modal response, and these planes will be in different directions for different modes (Figure 13.4).

13.4. ANALYSIS OF DISTRIBUTED RESPONSE

We have seen how distributed rotor eccentricity generates radial modal response; however, this is the reverse of the real problem. Usually we must determine the unbalance and corrections from the total distributed deflection developed by the unbalance during rotation. As with dynamic balancing, rotation provides the source of information regarding the latent unbalance.

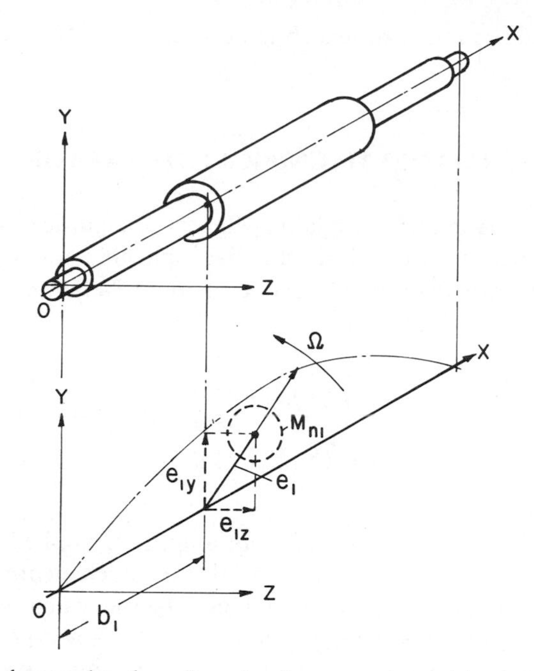

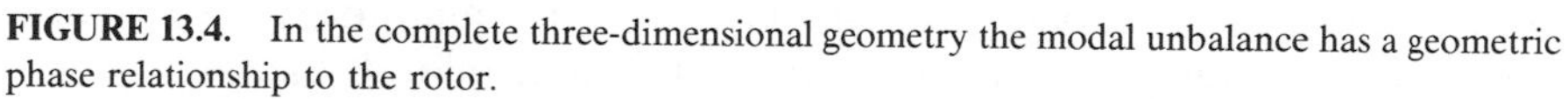

FIGURE 13.4. In the complete three-dimensional geometry the modal unbalance has a geometric phase relationship to the rotor.

We now require some type of transducer to traverse the rotor and associated instrumentation to record the change in orbital amplitudes in going from zero speed to a predetermined balancing speed. The static or rollover readings become the reference. Obviously the transducer cannot contact the moving surface and must measure an air gap.

We must also have some means for making measurements at selected phase positions, 90° from each other, and of relating phase direction to the rotor when it is stopped for locating corrections.

With these data available, for simplicity we revert to the single xy plane and resolve the $y(x)$ measured deflection function into modal components,

$$y(x) = y_1\beta_{n1}(x) + y_2\beta_{n2}(x) + \cdots \tag{13.8}$$

from which we extract the component coefficients, as in Equation 12.19.

$$y_i = \frac{\int_0^l y(x)\mu(x)\beta_{ni}(x)\,dx}{M_{ni}} \tag{13.9}$$

From Equation 13.7 we evaluate e_i:

$$e_i = \left(\frac{1 - \mathsf{g}_i^2}{\mathsf{g}_i^2}\right) y_i \tag{13.10}$$

The modal unbalance at the normalizing point is then $(M_{ni}e_i)$, as Figure 13.3 indicates.

13.5. THE GENERALIZED MAGNIFICATION FACTOR

Equation 13.7 is a basic form of relative vibratory magnification effect similar to the basic system of Table 2.1*c*, and is applicable to any mode of the distributed rotor with distributed excitation, or unbalance. We can further express this ratio using Equations 13.9 and 13.5,

$$\left(\frac{y_i}{e_i}\right) = \frac{\int_0^l y(x)\mu(x)\beta_{ni}(x)\,dx}{\int_0^l e(x)\mu(x)\beta_{ni}(x)\,dx} = \left(\frac{g_i^2}{1 - g_i^2}\right) \tag{13.11}$$

Thus the ratio of relative response to input amplitude applied to any mode at its reference axial position conforms to the most elementary of vibratory concepts. Forced frequency ratio determines a factor which, when applied to the static case, (e_i) predicts the dynamic, (y_i). Now, however, it is two entire normal modal distributions, $y_i(x)$ and $e_i(x)$, whose amplitudes are compared. The ratio of the integrals is required to extract the amplitude factors.

13.6. BALANCING SPEEDS

We note the importance of the balancing speed on the accuracy of the data in Equation 13.7. By balancing at a speed close to a critical speed, y_i for that mode approaches infinity for a given e_i. But this in turn can create inaccuracy because of the sharpness of the resonant peak and abrupt shifts in phase angle. If possible it seems advisable to operate near the criticals, taking several sets of data for several modes if necessary, in order to take advantage of some amplification of the modal unbalances.

13.7. UNBALANCE CORRECTION

Knowing the first modal unbalance (Figure 13.4), we can correct the rotor in the perpendicular plane a distance b_1 from the end by removing material in the e_1 direction or adding unbalance in the opposite direction.

$$(me)_1 = -(M_{n1}e_1) \tag{13.12}$$

However, it may not be feasible to make corrections in the reference plane. Using a different position requires a larger correction, modifying by the modal amplitude ratio as in Figure 13.2:

$$(me)_{d1} = -\left(\frac{M_{n1}e_1}{\beta_{d1}}\right) \tag{13.13}$$

If the second mode is also of concern, the first-mode correction can complicate the problem by possibly increasing the second-mode unbalance. One means of avoiding this is to locate the first modal correction at the node of the second mode, providing complete independence with respect to the second mode.

In the general case of balancing two modes, a simultaneous solution is required for corrections in two balancing planes, requiring analysis in the xy and xz planes (Figure 13.4); that is, we must deal with planar components algebraically rather than with single resultant modal corrections.

To solve simultaneously, assume we have obtained the modal unbalances $(M_{n1}e_1)$ and $(M_{n2}e_2)$ and they are in the same xy plane (Figure 13.3). In Figure 13.5, we will use correction planes at A and B. The equations for balance of first and second modes are, respectively,

$$\begin{cases} (me)_A\beta_{A1} + (me)_B\beta_{B1} = -M_{n1}e_1 \\ (me)_A\beta_{A2} + (me)_B\beta_{B2} = -M_{n2}e_2 \end{cases} \tag{13.14}$$

where $(me)_A$ and $(me)_B$ are solved for as the two unbalances to be added in the A and B planes.

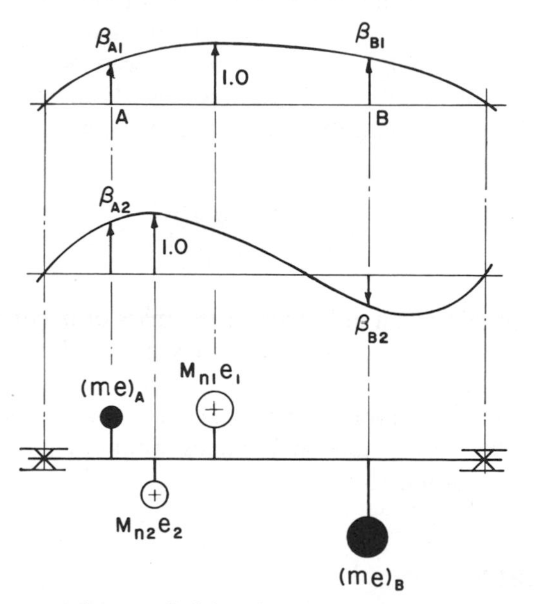

FIGURE 13.5. Two corrections applied in arbitrary planes A and B simultaneously correct unbalance at the first two critical speeds.

As the number of modes to be balanced increases, the number of required correction planes increases, equaling the number of modes to be balanced. In the simultaneous approach no plane should be selected at a node.

13.8. RESPONSE TO BEARING EXCITATION

A rotor can be excited externally at the supporting bearings, either of which can provide sinusoidal displacement. These can be in an axial plane or orbital. As in the previous balancing discussion and in Section 11.11, we then have a trapezoidal or triangular excitation pattern rather than the arbitrary unknown distribution $e(x)$ (Figure 13.3a) due to the effect of one bearing. The cases are similar, with modal responses predicted from Equations 13.5 and 13.7.

One important difference is that the rotating unbalance is correctable as shown, whereas the transverse vibration is not.

13.9. BALANCING OF THE DISCRETE SYSTEM

Multiple masses on an elastic shaft (Chapter 11) comprise another family of flexible rotors, and the continuous rotor is sometimes reduced to this model. Unbalance can then occur only in the rotational planes of the masses. We will then have as many unbalances and corrections as disks to achieve complete correction for all modes, and the result is to move the centroid of each disk to the axis of rotation. This requires running the rotor and measuring amplitudes at as many different balancing speeds as there are disks.

Actually if only a first mode or critical speed is of importance, one correction in one plane will suffice and we use Equation 13.9 to find the component modal deflection after calculating the first natural frequency and normal mode (Figure 13.6).

$$y_{b1} = \frac{\Sigma y_q M_q \beta_{q1}}{\Sigma M_q \beta_{q1}^2} \tag{13.15}$$

where y_{b1} is the elastic radial deflection at the mass of maximum amplitude in the first mode. This is assumed to be the mass in which the single correction will be made.

Evaluating e_b from Equation 13.10 we have the required modal correction of $(me)_b = M_{n1}e_b$. Multiple planes and modes would require the methods of Section 13.7 and Equation 13.14.

13.10. REMARKS

Methods indicated provide a rational approach to flexible rotor balancing and a means for eliminating a trial-and-error procedure in the shop or field.

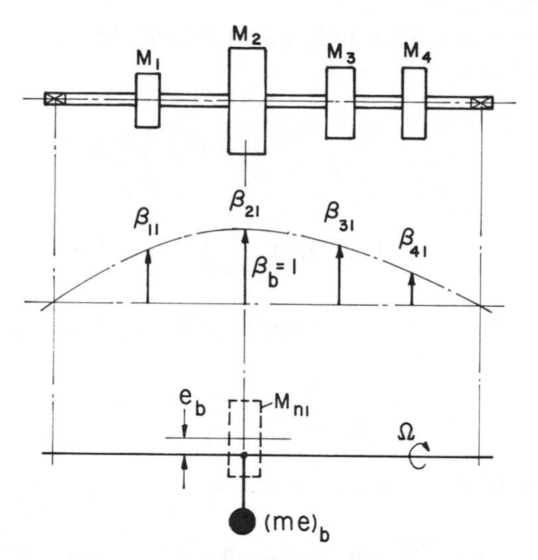

FIGURE 13.6. Correction of the first critical speed in a discrete mass rotor compensates for the first modal unbalance.

Approximate balancing is essential if an operating speed is above or near a critical speed. The greater the accuracy of balancing, the less deflection and roughness will be experienced in operation; however, no degree of balancing can ensure stability at a critical speed. The elastic resistance then exactly balances the inertia loading and increases at the same rate with deflection, and the radial position of the rotor is undefined.

Although distortions caused by unbalance are generally undesirable, in steady state no fatigue stresses are developed, as indicated in Chapter 10. This is because the rotor maintains a fixed deflection with constant flexure, characteristic of all excitations with $N = +1$. Fatigue failure and structural damage can occur in passing through a critical speed because this is not steady state and can involve large transient activity related to phase shifts and tendencies to large radii.

EXAMPLES

EXAMPLE 13.1. A straight steel shaft operated at 80% of the first critical speed has a measured deflection along its length as shown. Amplitudes are in mils, or 0.001 in. What first-mode correction should be added at the center of the span?

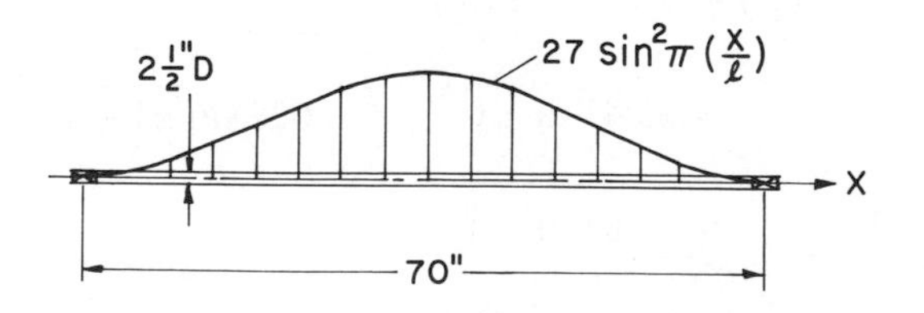

Solution. From Equation 13.9 and Table 12.3a we find the first modal amplitude component:

$$y_1 = \frac{(0.027)\mu l}{(\mu l/2)} \int_0^l \sin^3\pi\left(\frac{x}{l}\right) dx$$

$$= (0.027)(2)\left(\frac{4}{3\pi}\right) = 0.023 \text{ in.}$$

Note that this is 85% of the total amplitude. Using Equation 13.10,

$$e_1 = \left(\frac{1 - 0.64}{0.64}\right)(0.023) = 0.013 \text{ in.}$$

Modal unbalance is the required correction.

$$M_{n1}e_1 = \left(\frac{\rho A l}{2}\right)(0.013) = 49.8(0.013) = 0.65 \text{ in. lb}$$

EXAMPLE 13.2. The shaft analyzed in Example 13.1 is perfectly homogeneous, but initially bent to the shape shown. Calculate the first modal response when rotating at 80% of the first critical speed.

Solution. The curve now represents the eccentricity distribution $e(x)$. From Equation 13.5

$$e_1 = \left[\frac{\mu l \int_0^l \sin^3\pi(x/l)\, dx}{\mu l/2}\right] = 0.023 \text{ in.}$$

With Equation 13.7, we have

$$y_1 = \left(\frac{0.64}{1 - 0.64}\right)(0.023) = 1.777(0.023) = 0.041 \text{ in.}$$

EXAMPLE 13.3. Referring to Example 13.2, what is the approximate total response distribution?

Solution. By superimposing the first modal deformation on the initial eccentricity distribution,

$$y(x) \approx y_1(x) + e(x)$$

$$\approx 0.041 \sin \pi\left(\frac{x}{l}\right) + 0.027 \sin^2\pi\left(\frac{x}{l}\right)$$

$$\approx 0.068 \text{ in. at the center}$$

There is no second-mode contribution because the integration of a symmetrical $e(x)$ and an antisymmetrical $\beta_{n2}(x)$ is zero.

EXAMPLE 13.4. A two-mass system on a shaft of negligible mass has the normal modes shown. Mass eccentricity of M_1 is 0.90 cm and of M_2 is -0.13 cm. Verify that corrections of $-0.90M_1$ and $+0.13M_2$ will completely balance the system using the modal approach.

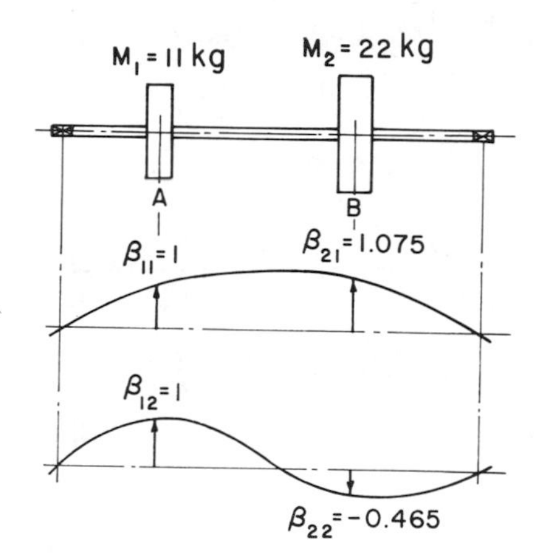

Solution. From Equation 13.6 we determine the two modal unbalances at the reference position A.

$$M_{n1}e_{A1} = 11(0.90) + 22(-0.13)(1.075) = +6.83 \text{ kg} \cdot \text{cm}$$

$$M_{n2}e_{A2} = 11(0.90) + 22(-0.13)(-0.47) = +11.23 \text{ kg} \cdot \text{cm}$$

Then the simultaneous solutions become, from Equation 13.14,

$$\begin{cases} (me)_A(1) + (me)_B(1.075) = -6.83 \\ (me)_A(1) + (me)_B(-0.47) = -11.23 \end{cases}$$

$$(me)_A = -9.90 \quad \text{and} \quad (me)_B = 2.86 \text{ kg} \cdot \text{cm}$$

These corrections are equal and opposite to the specified mass unbalances.

EXAMPLE 13.5. A steel rotor is conical in one span and a slender shaft in the other. The cone is considered rigid and the shaft massless. Eccentricity in the cone varies linearly in a plane from 0.020 in. at 0 to zero at 1. Find the whirl radius at 1 if the rotor operates at 125% of the critical speed.

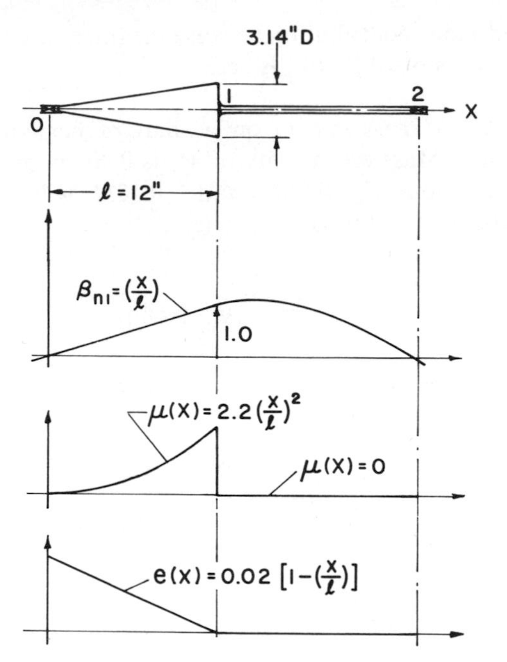

Solution. In the cone, mass rate increases as the diameter squared, or as $(x/l)^2$, and is a maximum of 2.20 lb/in. at 1. The modal mass is

$$M_{n1} = \int_0^l \mu(x)\beta_{n1}^2(x)\,dx = 2.2l\int_0^1 \left(\frac{x}{l}\right)^2\left(\frac{x}{l}\right)^2 \frac{dx}{l}$$

$$= 2.2l\left[\frac{1}{5}\left(\frac{x}{l}\right)^5\right]_0^1 = 2.2(12)\left(\frac{1}{5}\right) = 5.28 \text{ lb}$$

From Equation 13.5,

$$e_1 = \frac{12\int_0^1 (0.020)[1-(x/l)]\left[2.2(x/l)^2\right](x/l)\,dx/l}{5.28} = 0.005 \text{ in.}$$

$$\left(\frac{y_1}{e_1}\right) = \left[\frac{(1.25)^2}{1-(1.25)^2}\right] = -2.78 \quad \text{and} \quad y_1 = -0.014 \text{ in.}$$

Since there is no eccentricity at 1, this is also the total radius.

14 SPATIAL COORDINATES

Many systems are well defined in terms of degrees of freedom and in the directions in which the various masses oscillate in free vibratory modes. Translational, rotational, and distributed cases have been discussed in terms of the obviously related directional motions, but thus far there has been no need to determine space relationships for more complex elastic suspensions. In this chapter emphasis is on relatively simple systems having a single mass, with the mass of the elastic structure neglected. These are also limited to three-degree or three-coordinate cases, but this could rise to six; however, the cases developed illustrate basic concepts and probably cover the majority of coordinate coupled systems.

Although vibratory aspects of systems are the principal concern, all solutions evolve from elastic factors, and any attempt to develop such procedures would be prohibitively lengthy. Methods are available in strength of materials and structural theory for deflection computations. If a structure is in existence, accurate experimental measurements may be possible.

Tabular results presented are often illustrated by simple spherical masses to emphasize generality; however, the mass can have any geometric form, as can the elastic member or members that connect the mass to ground.

14.1. DIRECTIONAL CHARACTERISTICS OF THE CONSTANT BEAM

The most simple shape for a beam cross section is the circle. This geometry is completely symmetrical in the sense that equal bending stiffness exists in any angular direction in a longitudinal plane containing the beam axis. Similarly, other geometries with a high degree of symmetry can deflect and vibrate in any axial plane (Figure 14.1*a*). These include all equipolygons, solid or hollow. Although shown with a simple end mass with centroid coincident with the central axis of the beam, uniform beams of Chapter 12 would behave similarly.

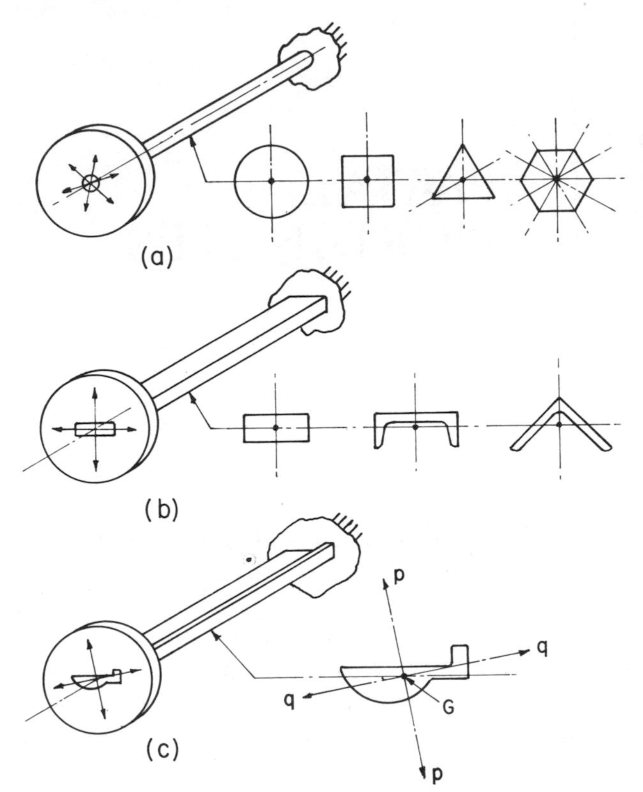

FIGURE 14.1. In (a) there is no directional preference and only one natural frequency. In (b) and (c) modes exist in two perpendicular directions.

Thus we have a two-degree-of-freedom system with respect to the coordinates of the centroid in the plane of the mass, but only a single or common frequency.

A more general case of symmetry about two perpendicular axes is shown in Figure 14.1*b*. The basic rectangle restricts bending modes to vertical and horizontal, or *flatwise* and *edgewise*, with the former corresponding to the lower natural frequency. We see that the (a) geometries have a single natural frequency with undefined modal directions, and the (b) sections have two frequencies and perpendicularly defined modal directions.

In Figure 14.1*c* a constant arbitrary beam section is indicated. Several integrations are necessary to calculate the center of gravity and the principal axes. The system has two bending modes related to these directional axes. At the lower the beam bends in the minimum stiffness plane, p–p about the neutral axes in the q–q direction, and vice versa. Absolute values of I_p and I_q in turn determine the bending frequency values.

The general end mass is not concentrated, but has three orthogonal principal axes. With conventional shapes, however, one body axis is often aligned with the beam axis, and the other two with the principal axes of the beam (Figure 14.1*b*) There is one torsional and two bending modes in each of the respective perpendicular planes, the latter discussed in Chapter 9. Because of the geometric orthogonality the torsional mode and the paired bending modes are uncoupled.

14.2. ORTHOGONAL PLANAR ELASTICITY

As illustrated by the planar bent beam (Figure 14.2), a force in the plane at an arbitrary direction produces a deflection with magnitude and direction governed by the geometry and the elasticity of the structure. There are, however, two perpendicular directions in which a load produces a deflection vector coincident with the force vector, and these are the planar principal axes p and q. There will incidentally be a third principal axis, z, perpendicular to the p–q plane. These three axes are completely uncoupled with respect to linear displacements caused by forces in these directions; that is, F_p causes no q or z displacement.

Restricting our attention to the p and q axes, these also represent directions of maximum and minimum deflection produced by a given load acting in any angular direction. Initially we will not be able to identify these directions, but will analyze in arbitrary coordinates, say x and y (Figure 14.2). Using methods

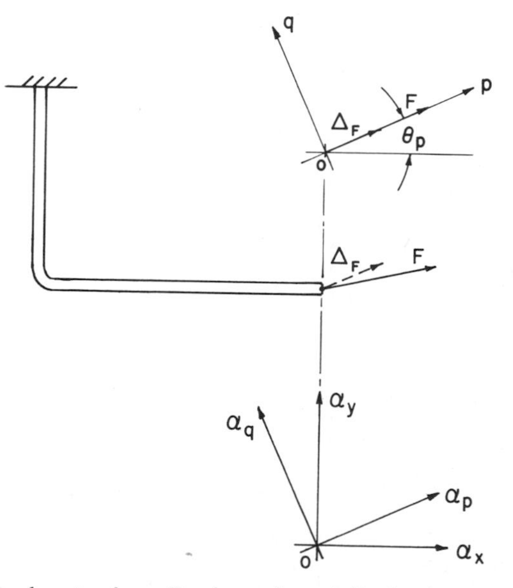

FIGURE 14.2. A general vector force F only produces deflection in a coincident direction if applied along a principal axis p or q.

from structural analysis we obtain the basic influence factors:

$\alpha_x = X_x/F_x =$ deflection in the x direction caused by unit force in the x direction

$\alpha_y = Y_y/F_y =$ deflection in the y direction caused by unit force in the y direction

$\alpha_{yx} = Y_x/F_x =$ deflection in the y direction caused by unit force in the x direction

$\alpha_{xy} = X_y/F_y =$ same as α_{yx}, reversed, and equal

Because of the reciprocity of the cross-factors there are three independent calculations required to define the coordinate stiffness. The two direct factors will always be positive, but the cross-factors carry algebraic sign. Following previous practice we generalize these numerical results by normalization with respect to the reference compliance α_x.

$$\psi_1 = \frac{\alpha_{yx}}{\alpha_x} = \frac{\alpha_{xy}}{\alpha_x} \qquad \psi_2 = \frac{\alpha_y}{\alpha_x}$$

transforming from x–y to p–q coordinates we have

$$\psi_p = \frac{\alpha_p}{\alpha_x} \qquad \psi_q = \frac{\alpha_q}{\alpha_x}$$

Table 14.1 summarizes the notation and indicates the equations for this

TABLE 14.1. Elastic factors for the constant section two-leg bent. Relations for the *p* and *q* principal compliances from *xy* data apply to any planar structure.

ψ_1	$\frac{\alpha_{yx}}{\alpha_x}$	$\frac{3}{2}\left(\frac{b}{l}\right)$	$\tan 2\theta_p$	$\frac{2\psi_1}{1-\psi_2}$	$\alpha_p = \frac{\Delta p}{F_p}$
ψ_2	$\frac{\alpha_y}{\alpha_x}$	$\left(\frac{b}{l}\right)^2\left(3+\frac{b}{l}\right)$	θ_q	$\theta_p + \frac{\pi}{2}$	$\alpha_q = \frac{\Delta q}{F_q}$
ψ_p	$\frac{\alpha_p}{\alpha_x}$	$1+\psi_1 \tan\theta_p$	α_x	$\frac{l^3}{3EI}$	$\alpha_{pq} = \alpha_{qp} = 0$
ψ_q	$\frac{\alpha_q}{\alpha_x}$	$1+\psi_1 \tan\theta_q$			

TABLE 14.2. Steady-state response of the concentrated mass to forcing in arbitrary perpendicular directions x and y (a), and to x or y coordinate displacement excitation (b).

	P_x		P_y (a)		a_{ox}		a_{oy} (b)
$\frac{a_{xx}}{P_x \alpha_x}$	$\frac{1-Ag^2}{C}$	$\frac{a_{xy}}{P_y \alpha_{xy}}$	$\frac{1}{C}$	$\frac{a_{xx}}{a_{ox}}$	$\frac{1-\psi_2 g^2}{C}$	$\frac{a_{xy}}{a_{oy}}$	$\frac{\psi_1 g^2}{C}$
$\frac{a_{yx}}{P_x \alpha_{xy}}$	$\frac{1}{C}$	$\frac{a_{yy}}{P_y \alpha_x}$	$\frac{\psi_2 - Ag^2}{C}$	$\frac{a_{yx}}{a_{ox}}$	$\frac{\psi_1 g^2}{C}$	$\frac{a_{yy}}{a_{oy}}$	$\frac{1-g^2}{C}$
$\tan \theta_x$	$\frac{\psi_1}{1-Ag^2}$	$\tan \theta_y$	$\frac{\psi_2 - Ag^2}{\psi_2}$	$\tan \theta_x$	$\frac{\psi_1 g^2}{1-\psi_2 g^2}$	$\tan \theta_y$	$\frac{1-g^2}{\psi_1 g^2}$

$C = Ag^4 - (1+\psi_2)g^2 + 1$ $\quad A = (\psi_2 - \psi_1^2)$ $\quad g = \frac{\omega}{\omega_1} = \omega\sqrt{\alpha_x M}$ $\quad \psi_2 = \frac{\alpha_y}{\alpha_x}$ $\quad \psi_1 = \frac{\alpha_{xy}}{\alpha_x}$ $\quad \left(\frac{\beta_y}{\beta_x}\right)_n = \tan \phi_n = \frac{1-g_n^2}{\psi_1 g_n^2}$

conversion for any planar system, with specific values α_x, ψ_1, and ψ_2 given in Table 14.2 for the simple bent beam. We note that $\alpha_{pq} = \alpha_{qp} = 0$, as there is no elastic coupling relative to the principal axes.

14.3. FREE PLANAR VIBRATION OF CONCENTRATED MASS

A concentrated mass mounted at the coordinate origin (Figure 14.2) has two degrees of freedom. Modal amplitudes lie along the p and q axes, and the system behaves like two independent single-degree cases. The two natural frequencies are

$$\omega_{n1} = \sqrt{\frac{1}{\alpha_p M}} \quad \text{and} \quad \omega_{n2} = \sqrt{\frac{1}{\alpha_q M}} \tag{14.1}$$

and if directionally forced along p or q we apply the basic equations of Table 2.1.

Analyzing in x–y coordinates we have for free vibration

$$\begin{cases} \mathrm{a}_x = \alpha_x\left(M\omega^2 \mathrm{a}_x\right) + \alpha_{xy}\left(M\omega^2 \mathrm{a}_y\right) & (14.2\text{a}) \\ \mathrm{a}_y = \alpha_{yx}\left(M\omega^2 \mathrm{a}_x\right) + \alpha_y\left(M\omega^2 \mathrm{a}_y\right) & (14.2\text{b}) \end{cases}$$

which normalize to

$$\begin{cases} a_x(1 - g^2) - a_y\psi_1 g^2 = 0 & (14.3a) \\ -a_x(\psi_1 g^2) + a_y(1 - \psi_2 g^2) = 0 & (14.3b) \end{cases}$$

Resulting frequencies and modes are given in Table 14.2 with $C = C_n = 0$ for the frequency quadratic. These should verify values obtained based on principal axes considerations.

14.4. PLANAR FORCED CONCENTRATED MASS

Sinusoidal forcing applied to a mass in any angular direction can be converted to in-phase perpendicular components P_x and P_y (Table 14.2*a*). Each force in turn will produce x and y components of response which can then be superimposed. These relations result from Equation 14.2, for instance by adding $\alpha_x P_x$ and $\alpha_{yx} P_x$ to 14.2a and 14.2b, respectively. If the base of an elastic structure is oscillated we have the displacement excited absolute amplitudes of Table 14.2*b*.

It should be emphasized that Table 14.2 represents a single concentrated mass elastically supported by a planar structure not shown. It could be the bent, or any other planar configuration for which the compliances are calculated.

14.5. THREE-COORDINATE ELASTICITY

Proceeding to an additional degree of freedom we have several fundamental situations on which we will focus (Figure 14.3 and Tables 14.3 and 14.4). These cases are as follows:

a. Displacements of a point G in space with x, y, and z components. The several spans are subjected to bending, torsion, or both (Figure 14.3*a*).
b. Displacements only in the xy plane, which now include the slope coordinate θ_z. The latter pertains to the mass moment of inertia about the z axis, perpendicular to the xy plane (Figure 14.3*b*).
c. Displacements for motion generally transverse to the xy plane, consisting of a linear z deflection, and rotations in the yz and xz planes (Figure 14.3*c*).

Although cases b and c represent the same system for the bent, they are not coupled to each other. Also we typically use coordinates centered on the mass centroid and aligned with the principal axes of the mass and the orthogonal

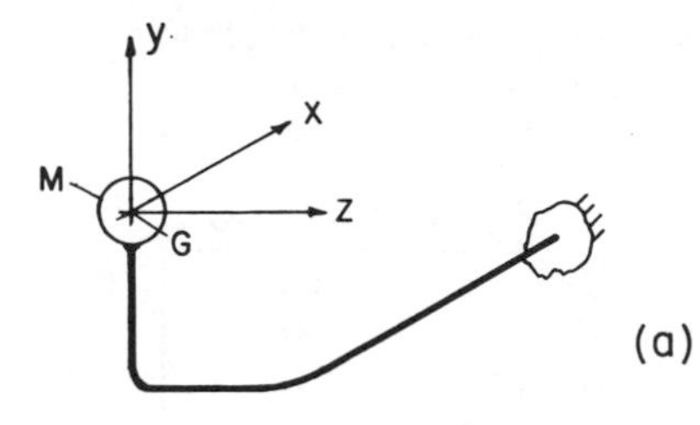

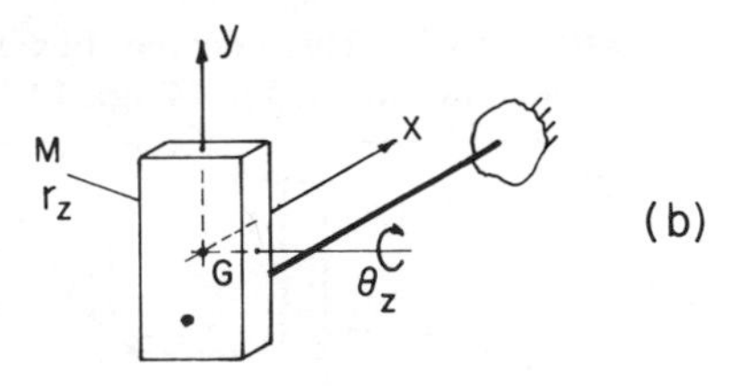

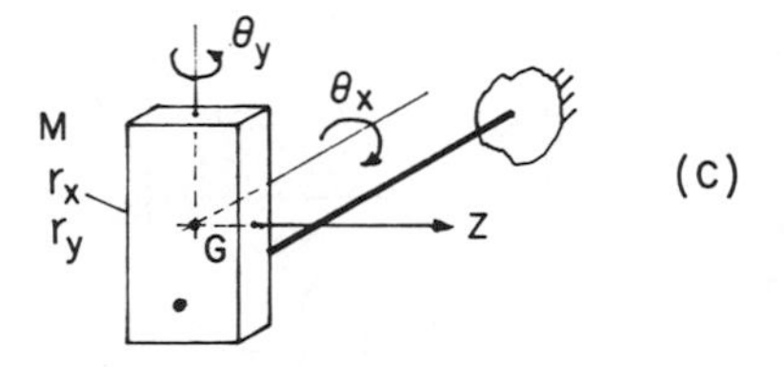

FIGURE 14.3. The basic examples of three-coordinate coupling are as follows: (*a*) three perpendicular linear coordinates with a concentrated mass; (*b*) displacement and rotation in the *xy* plane; and (*c*) out-of-plane action with *z* motion of the centroid and two related angular displacements.

TABLE 14.3. Expressions for bent influence factors. Cases (*a*), (*b*), and (*c*), (Table 14.4). Only (*c*) involves torsion.

a (ℓ, b, Y, Z, X)		b (ℓ, EI, GJ, b, Y, θ_z, X)		c (ℓ, b, θ_y, Z, θ_x)	
ψ_1	$\frac{3}{2}\left(\frac{b}{\ell}\right)$	ψ_1	$\frac{3}{2}\left(\frac{b}{\ell}\right)$	$\ell\psi_1$	$\frac{3}{2A}$
ψ_2	$\left(\frac{b}{\ell}\right)^2\left(3+\frac{b}{\ell}\right)$	ψ_2	$\left(\frac{b}{\ell}\right)^2\left(3+\frac{b}{\ell}\right)$	$\ell^2\psi_2$	$\frac{3}{A}\left(1+\rho\frac{b}{\ell}\right)$
ψ_3	0	$\ell\psi_3$	$3\left(\frac{b}{\ell}\right)\left(1+\frac{b}{\ell}\right)$	$\ell^2\psi_3$	0
ψ_4	A	$\ell^2\psi_4$	$3\left(1+\frac{b}{\ell}\right)$	$\ell^2\psi_4$	$\frac{3}{A}\left(\rho+\frac{b}{\ell}\right)$
ψ_5	0	$\ell\psi_5$	$\frac{3}{2}$	$\ell\psi_5$	$\frac{3}{A}\left(\frac{b}{\ell}\right)\left(\rho+\frac{1}{2}\frac{b}{\ell}\right)$
α_x	$\frac{\ell^3}{3EI}$	α_x	$\frac{\ell^3}{3EI}$	α_z	$\frac{A\ell^3}{3EI}$
$A=\left[1+3\rho\left(\frac{b}{\ell}\right)^2+\left(\frac{b}{\ell}\right)^3\right]$ $\quad \rho=\left(\frac{EI}{GJ}\right)$					

TABLE 14.4. Three-degree flexibility notation for the basic coupling combinations outlined in Figure 14.3, including normalized factors ψ.

F_x		F_y		F_z	a	ψ_1	$\frac{\alpha_{yx}}{\alpha_x}$	$\frac{\alpha_{xy}}{\alpha_x}$
						ψ_2	$\frac{\alpha_y}{\alpha_x}$	
α_x	$\frac{X_x}{F_x}$	α_y	$\frac{Y_y}{F_y}$	α_z	$\frac{Z_z}{F_z}$	ψ_3	$\frac{\alpha_{yz}}{\alpha_x}$	$\frac{\alpha_{zy}}{\alpha_x}$
α_{yx}	$\frac{Y_x}{F_x}$	α_{xy}	$\frac{X_y}{F_y}$	α_{xz}	$\frac{X_z}{F_z}$	ψ_4	$\frac{\alpha_z}{\alpha_x}$	
α_{zx}	$\frac{Z_x}{F_x}$	α_{zy}	$\frac{Z_y}{F_y}$	α_{yz}	$\frac{Y_z}{F_z}$	ψ_5	$\frac{\alpha_{zx}}{\alpha_x}$	$\frac{\alpha_{xz}}{\alpha_x}$
F_x		F_y		T_z	b	ψ_1	$\frac{\alpha_{yx}}{\alpha_x}$	$\frac{\alpha_{xy}}{\alpha_x}$
						ψ_2	$\frac{\alpha_y}{\alpha_x}$	
α_x	$\frac{X_x}{F_x}$	α_y	$\frac{Y_y}{F_y}$	γ_z	$\frac{\theta_z}{T_z}$	ψ_3	$\frac{\beta_{yz}}{\alpha_x}$	$\frac{\beta_{zy}}{\alpha_x}$
α_{yx}	$\frac{Y_x}{F_x}$	α_{xy}	$\frac{X_y}{F_y}$	β_{xz}	$\frac{X_z}{T_z}$	ψ_4	$\frac{\gamma_z}{\alpha_x}$	
β_{zx}	$\frac{\theta_x}{F_x}$	β_{zy}	$\frac{\theta_y}{F_y}$	β_{yz}	$\frac{Y_z}{T_z}$	ψ_5	$\frac{\beta_{zx}}{\alpha_x}$	$\frac{\beta_{xz}}{\alpha_x}$
F_z		T_x		T_y	c	ψ_1	$\frac{\beta_{xz}}{\alpha_z}$	$\frac{\beta_{zx}}{\alpha_z}$
						ψ_2	$\frac{\gamma_x}{\alpha_z}$	
α_z	$\frac{Z_z}{F_z}$	γ_x	$\frac{\theta_x}{T_x}$	γ_y	$\frac{\theta_y}{T_y}$	ψ_3	$\frac{\gamma_{xy}}{\alpha_z}$	$\frac{\gamma_{yx}}{\alpha_z}$
β_{xz}	$\frac{\theta_{xz}}{F_z}$	γ_{yx}	$\frac{\theta_{yx}}{T_x}$	γ_{xy}	$\frac{\theta_{xy}}{T_y}$	ψ_4	$\frac{\gamma_y}{\alpha_z}$	
β_{yz}	$\frac{\theta_{yz}}{F_z}$	β_{zx}	$\frac{Z_x}{T_x}$	β_{zy}	$\frac{Z_y}{T_y}$	ψ_5	$\frac{\beta_{yz}}{\alpha_z}$	$\frac{\beta_{zy}}{\alpha_z}$

directions of the structure. The angular terms now require rotational compliance factors β and γ, introduced in Chapter 9.

Table 14.4 indicates the complete elasticity notation applicable to the three cases. The ψ ratios of Chapter 9 are now expanded from two to five. Specific values for the simple bent are given in Table 14.3 with the ψ factors carrying dimensions and signs as necessary.

14.6. EQUATIONS FOR CASE (a)

As seen in Table 14.3*a*, the bent is uncoupled elastically with respect to the *z* direction, as $\psi_3 = \psi_5 = 0$, but the Figure 14.3*a* structure involves all coupling possibilities, as will generally more complex geometries. The equilibrium of the unforced concentrated mass with respect to three arbitrarily directed perpendicular axes, as for Equation 14.3, becomes

$$\begin{cases} \mathsf{a}_x(1 - g^2) - \mathsf{a}_y(\psi_1 \mathsf{g}^2) - \mathsf{a}_z(\psi_5 \mathsf{g}^2) = 0 & (14.4\text{a}) \\ -\mathsf{a}_x(\psi_1 \mathsf{g}^2) + \mathsf{a}_y(1 - \psi_2 \mathsf{g}^2) - \mathsf{a}_z(\psi_3 \mathsf{g}^2) = 0 & (14.4\text{b}) \\ -\mathsf{a}_x(\psi_5 \mathsf{g}^2) - \mathsf{a}_y(\psi_3 \mathsf{g}^2) + \mathsf{a}_z(1 - \psi_4 \mathsf{g}^2) = 0 & (14.4\text{c}) \end{cases}$$

These equations lead to the natural frequency cubic, $C_a = 0$, and the normal modal amplitudes β_y and β_z in Table 14.5. From the literal concept of orthogonality of modes the directions of the three resultant modal vectors in space will be mutually perpendicular, and will define the three principal elastic axes.

Introduction of forcing terms in Equation 14.4 provides the responses in Table 14.5*a*. Sinusoidal base excitation leads to the six possible magnification factors in Table 14.6*a*. All amplitudes are absolute, as seen by letting g approach zero. The three direct ratios then approach unity, or $\mathsf{a}_x = \mathsf{a}_{0x}$, or the motion of the mass equals the imposed displacements in a given direction.

14.7. TOTAL RESPONSES FOR CASES (b) AND (c)

Derivations and results for cases *b* and *c* parallel those for case *a*. Natural frequency cubics are also given in Table 14.5, and modal amplitudes which also shows the three possible responses to the three possible applied excitations. Reciprocity reduces the total relations from nine to six.

Displacement excitation results are presented in Table 14.6 for the three coordinate excitations.

Incidentally, for case *b* in Table 14.5*b*, the natural frequency cubic with $C_b = 0$ by definition for the three-degree system should lead to three natural frequencies and to three planar modes. However, if the elastic supporting structure is a simple bent (Table 14.3*b*), *only two modes exist, and the third cannot occur*. This relates to the coefficient A_1 in Table 14.5, which will always be negative for the particular coupling terms in Table 14.3*b*. This results in a positive coefficient for g_n^6. The violation of the correspondence of the number of modes to the number of degrees of freedom apparently is caused by the restrictive elastic properties of the bent, which cannot accommodate a third combination of the deflection coordinates. On this basis the bent degenerates to the beam situation of Chapter 9 (see Figure 9.2), and only two free vibratory modes are possible.

TABLE 14.5. **Amplitude responses for coupling combinations (*a*), (*b*), and (*c*), to sinusoidal forcing at the mass in the three respective coordinate directions. Perpendicular coordinate axes coincide with the centroid and principal axes of the mass.**

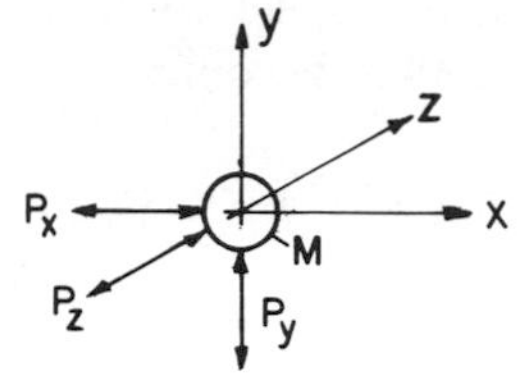

(a)

$\frac{a_x}{P_x \alpha_x}$	$\frac{A_1 g^4 - (B_1 + B_2) g^2 + 1}{C_a}$	$\frac{a_{yx}}{P_x \alpha_x}$	$\frac{a_{xy}}{P_y \alpha_x}$	$\frac{D_1 g^2 + \psi_1}{C_a}$
$\frac{a_y}{P_y \alpha_x}$	$\frac{A_1 g^4 - (B_3 + B_2) g^2 + \psi_2}{C_a}$	$\frac{a_{zx}}{P_x \alpha_x}$	$\frac{a_{xz}}{P_z \alpha_x}$	$\frac{D_2 g^2 + \psi_5}{C_a}$
$\frac{a_z}{P_z \alpha_x}$	$\frac{A_1 g^4 - (B_3 + B_1) g^2 + \psi_4}{C_a}$	$\frac{a_{zy}}{P_y \alpha_x}$	$\frac{a_{yz}}{P_z \alpha_x}$	$\frac{D_3 g^2 + \psi_3}{C_a}$
$\left(\frac{\beta_y}{\beta_x}\right)_n$	$\left(\frac{\psi_3 + D_3 g_n^2}{\psi_5 + D_2 g_n^2}\right)$	$\left(\frac{\beta_z}{\beta_x}\right)_n$	$\frac{1}{\psi_5}\left[\frac{1}{g_n^2} - 1 - \psi_1 \left(\frac{\beta_y}{\beta_x}\right)_n\right]$	

$$A_1 = (\psi_{24} + 2\psi_{135}) - (\psi_2 \psi_5^2 + \psi_1^2 \psi_4 + \psi_3^2)$$

$$C_a = -A_1 g^6 + (B_1 + B_2 + B_3) g^4 - (\psi_2 + \psi_4 + 1) g^2 + 1$$

$$B_1 = (\psi_4 - \psi_5^2) \qquad D_1 = (\psi_{35} - \psi_{14})$$

$$B_2 = (\psi_2 - \psi_1^2) \qquad D_2 = (\psi_{13} - \psi_{25}) \qquad \psi_{ij} = (\psi_i)(\psi_j)$$

$$B_3 = (\psi_{24} - \psi_3^2) \qquad D_3 = (\psi_{15} - \psi_3) \qquad \psi_{ijk} = (\psi_i)(\psi_j)(\psi_k)$$

$$g_a^2 = g_b^2 = \omega^2 (\alpha_x M), \quad g_c^2 = \omega^2 (\alpha_z M)$$

$$C_b = -r_z^2 A_1 g^6 + [r_z^2 (B_1 + B_3) + B_2] g^4 - (r_z^2 \psi_4 + \psi_2 + 1) g^2 + 1$$

$$C_c = -r_y^2 r_x^2 A_1 g^6 + [r_y^2 r_x^2 B_3 + r_y^2 B_1 + r_x^2 B_2] g^4 - (r_y^2 \psi_4 + r_x^2 \psi_2 + 1) g^2 + 1$$

Table 14.5. Continued.

b				
$\dfrac{a_x}{P_x \alpha_x}$	$\dfrac{r_z^2 A_1 g^4 - (r_z^2 B_1 + B_2) g^2 + 1}{C_b}$	$\dfrac{a_{yx}}{P_x \alpha_x}$	$\dfrac{a_{xy}}{P_y \alpha_x}$	$\dfrac{r_z^2 D_1 g^2 + \psi_1}{C_b}$
$\dfrac{a_y}{P_y \alpha_x}$	$\dfrac{r_z^2 A_1 g^4 - (r_z^2 B_3 + B_2) g^2 + \psi_2}{C_b}$	$\dfrac{\theta_{zx}}{P_x \alpha_x}$	$\dfrac{a_{xz}}{T_z \alpha_x}$	$\dfrac{D_2 g^2 + \psi_5}{C_b}$
$\dfrac{\theta_z}{T_z \alpha_x}$	$\dfrac{A_1 g^4 - (B_1 + B_3) g^2 + \psi_4}{C_b}$	$\dfrac{\theta_{zy}}{P_y \alpha_x}$	$\dfrac{a_{yz}}{T_z \alpha_x}$	$\dfrac{D_3 g^2 + \psi_3}{C_b}$
$\left(\dfrac{\beta_y}{\beta_x}\right)_n$	$\left(\dfrac{\psi_3 + D_3 g_n^2}{\psi_5 + D_2 g_n^2}\right)$	$\left(\dfrac{r_z \phi_z}{\beta_x}\right)_n$	$\dfrac{1}{r_z \psi_5}\left[\dfrac{1}{g_n^2} - 1 - \psi_1 \left(\dfrac{\beta_y}{\beta_x}\right)_n\right]$	

c				
$\dfrac{a_z}{P_z \alpha_z}$	$\dfrac{r_y^2 r_x^2 A_1 g^4 - (r_y^2 B_1 + r_x^2 B_2) g^2 + 1}{C_c}$	$\dfrac{\theta_{xz}}{P_z \alpha_z}$	$\dfrac{a_{zx}}{T_x \alpha_z}$	$\dfrac{r_y^2 D_1 g^2 + \psi_1}{C_c}$
$\dfrac{\theta_x}{T_x \alpha_z}$	$\dfrac{r_y^2 A_1 g^4 - (r_y^2 B_3 + B_2) g^2 + \psi_2}{C_c}$	$\dfrac{\theta_{yz}}{P_z \alpha_z}$	$\dfrac{a_{zy}}{T_y \alpha_z}$	$\dfrac{r_x^2 D_2 g^2 + \psi_5}{C_c}$
$\dfrac{\theta_y}{T_y \alpha_z}$	$\dfrac{r_x^2 A_1 g^4 - (r_x^2 B_3 + B_1) g^2 + \psi_4}{C_c}$	$\dfrac{\theta_{xy}}{T_y \alpha_z}$	$\dfrac{\theta_{yx}}{T_x \alpha_z}$	$\dfrac{D_3 g^2 + \psi_3}{C_c}$
$\left(\dfrac{r_x \phi_x}{\beta_z}\right)_n$	$\dfrac{1}{r_x}\left(\dfrac{\psi_3 + D_3 g_n^2}{\psi_5 + D_2 g_n^2}\right)$	$\left(\dfrac{r_y \phi_y}{\beta_z}\right)_n$	$\dfrac{1}{r_y \psi_5}\left[\dfrac{1}{g_n^2} - 1 - r_x \psi_1 \left(\dfrac{r_x \phi_x}{\beta_z}\right)_n\right]$	

TABLE 14.6. Similar to Table 14.5, but with impressed sinusoidal displacement of the supporting structure. Definitions of the coefficients are given in Table 14.5.

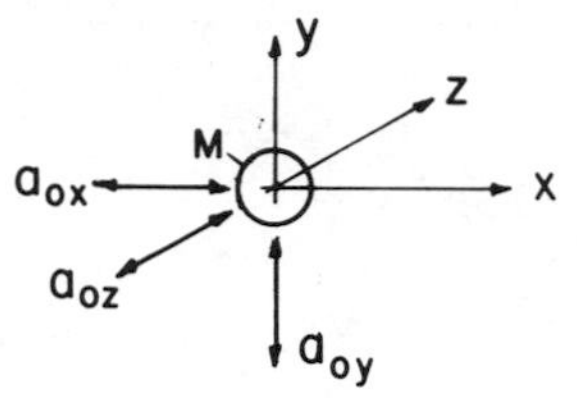

(a)

$\frac{a_x}{a_{ox}}$	$\frac{B_3 g^4 - (\psi_2 + \psi_4) g^2 + 1}{C_a}$	$\frac{a_{yx}}{a_{ox}}$	$\frac{a_{xy}}{a_{oy}}$	$\frac{g^2 (D_1 g^2 + \psi_1)}{C_a}$
$\frac{a_y}{a_{oy}}$	$\frac{B_1 g^4 - (\psi_4 + 1) g^2 + 1}{C_a}$	$\frac{a_{zx}}{a_{ox}}$	$\frac{a_{xz}}{a_{oz}}$	$\frac{g^2 (D_2 g^2 + \psi_5)}{C_a}$
$\frac{a_z}{a_{oz}}$	$\frac{B_2 g^4 - (\psi_2 + 1) g^2 + 1}{C_a}$	$\frac{a_{zy}}{a_{oy}}$	$\frac{a_{yz}}{a_{oz}}$	$\frac{g^2 (D_3 g^2 + \psi_3)}{C_a}$

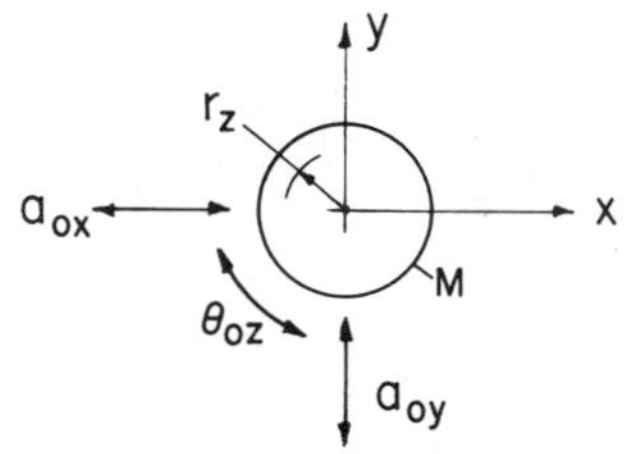

(b)

$\frac{a_x}{a_{ox}}$	$\frac{r_z^2 B_3 g^4 - (\psi_2 + r_z^2 \psi_4) g^2 + 1}{C_b}$	$\frac{a_{yx}}{a_{ox}}$	$\frac{a_{xy}}{a_{oy}}$	$\frac{g^2 (r_z^2 D_1 g^2 + \psi_1)}{C_b}$
$\frac{a_y}{a_{oy}}$	$\frac{r_z^2 B_1 g^4 - (r_z^2 \psi_4 + 1) g^2 + 1}{C_b}$	$\frac{\theta_{zx}}{a_{ox}}$	$\frac{a_{xz}}{\theta_{oz}}$	$\frac{g^2 (r_z^2 D_2 g^2 + r_z^2 \psi_5)}{C_b}$
$\frac{\theta_z}{\theta_{oz}}$	$\frac{B_2 g^4 - (\psi_2 + 1) g^2 + 1}{C_b}$	$\frac{\theta_{zy}}{a_{oy}}$	$\frac{a_{yz}}{\theta_{oz}}$	$\frac{g^2 (D_3 g^2 + \psi_3)}{C_b}$

Table 14.6. Continued.

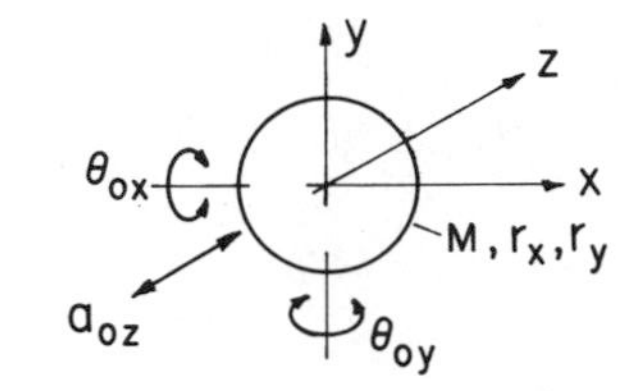

				c
$\frac{a_z}{a_{oz}}$	$\frac{r_x^2 r_y^2 B_3 g^4 - (r_x^2 \psi_2 \;\; r_y^2 \psi_4) g^2 + 1}{C_c}$	$\frac{a_{zx}}{r_x^2 \theta_{ox}}$	$\frac{\theta_{xz}}{a_{oz}}$	$\frac{g^2 (r_y^2 D_1 g^2 + \psi_1)}{C_c}$
$\frac{\theta_x}{\theta_{ox}}$	$\frac{r_y^2 B_1 g^4 - (r_y^2 \psi_4 + 1) g^2 + 1}{C_c}$	$\frac{a_{zy}}{r_y^2 \theta_{oy}}$	$\frac{\theta_{yz}}{a_{oz}}$	$\frac{g^2 (r_x^2 D_2 g^2 + \psi_5)}{C_c}$
$\frac{\theta_y}{\theta_{oy}}$	$\frac{r_x^2 B_2 g^4 - (r_x^2 \psi_2 + 1) g^2 + 1}{C_c}$	$\frac{\theta_{xy}}{r_y^2 \theta_{oy}}$	$\frac{\theta_{yx}}{r_x^2 \theta_{ox}}$	$\frac{g^2 (D_3 g^2 + \psi_3)}{C_c}$

14.8. MODAL SPRINGS AND PARTICIPATION FACTORS

We have considered the total vibratory response specifically for coupling cases *a*, *b*, and *c*, but response can be approached on the basis of modal components. A given three-degree system can be replaced by three equivalent modal springs and masses, each related to a particular coordinate. Applying kinetic energy and frequency equivalence, as in Section 4.6, we have, for case *b*,

$$\tfrac{1}{2} M_{nx} (\beta_x \omega)_n^2 = \tfrac{1}{2} M (\beta_x \omega)_n^2 + \tfrac{1}{2} M (\beta_y \omega)_n^2 + \tfrac{1}{2} M r_z^2 (\phi_z \omega)_n^2$$

$$\left(\frac{M_{nx}}{M} \right) = \left[1 + \left(\frac{\beta_y}{\beta_x} \right)_n^2 + \left(\frac{r_z \phi_z}{\beta_x} \right)_n^2 \right] \tag{14.5}$$

$$K_{nx} = \omega_n^2 M_{nx} = \left[1 + \left(\frac{\beta_y}{\beta_x} \right)_n^2 + \left(\frac{r_z \phi_z}{\beta_x} \right)_n^2 \right] M \omega_n^2$$

$$\left(\frac{K_{nx}}{K_x} \right) = \left[1 + \left(\frac{\beta_y}{\beta_x} \right)_n^2 + \left(\frac{r_z \phi_z}{\beta_x} \right)_n^2 \right] g_n^2 \tag{14.6}$$

Use of the spring rate *K* rather than the reciprocal is more convenient when considering the single-degree analog.

Participation, or $\overline{m}$ factors, for a single mass become

$$\overline{m}_i = \left(\frac{\Sigma M \beta}{\Sigma M \beta^2} \right) = \left(\frac{M}{M_{ni}} \right) = \left(\frac{1}{m_{ni}} \right) \tag{14.7}$$

Results for cases a, b, and c are summarized in Table 14.7. As with other systems, we can check numerical values of $\overline{m}$ if all three modes are calculated because

$$\overline{m}_{x1} + \overline{m}_{x2} + \overline{m}_{x3} = 1$$

Or, in general terms,

$$\sum_{1}^{3} \overline{m}_{qi} = 1 \tag{14.8}$$

All terms in brackets in Table 14.7 represent normalized modal mass, $(M_{ni}/M) = m_{ni}$. All terms in this table are properties of a system representing dimensionless modal characteristics, deriving from the inertia–elasticity parameters. Nine terms are necessary to describe completely the dynamic responses of a three-degree system in this context.

14.9. FORCED MODAL RESPONSE—CASE (b)

We now apply the modal parameters of Table 14.7 to actual responses, taking the case b system to illustrate the procedure. In Table 14.8 the three coordinate excitations, P_x, P_y, and T_z, are indicated. Taking the first case with P_x, the single-degree modal system includes K_{nx} and M. The horizontal amplitude of the mass in any of the three modes is given by

$$\frac{\mathsf{a}_{nx}}{(P_x/K_{nx})} = \left(\frac{1}{1 - \mathsf{g}_i^2}\right) \tag{14.9}$$

and the total response by superposition of modes is

$$\mathsf{a}_x = \left(\frac{P_x}{K_{n1x}}\right)\left(\frac{1}{1 - \mathsf{g}_1^2}\right) + \left(\frac{P_x}{K_{n2x}}\right)\left(\frac{1}{1 - \mathsf{g}_2^2}\right) + \left(\frac{P_x}{K_{n3x}}\right)\left(\frac{1}{1 - \mathsf{g}_3^2}\right) \tag{14.10}$$

This is similar to Equation 5.13 without the rigid-mode term. Each magnification factor in brackets relates to the forced modal frequency ratio, $\mathsf{g}_i = (\omega/\omega_{ni})$.

But we are dealing with a coupled system, and in a given principal mode a_{nx} has coupled displacements

$$\frac{\mathsf{a}_{ny}}{\mathsf{a}_{nx}} = \left(\frac{\beta_y}{\beta_x}\right)_n \qquad \frac{\theta_{nz}}{\mathsf{a}_{nx}/r_z} = \left(\frac{r_z\phi_z}{\beta_x}\right) \tag{14.11}$$

TABLE 14.7. Modal springs for cases (*a*), (*b*), and (*c*) are calculated from modal and frequency data. Also shown are $\overline{m}$ factors for modal response to displacement excitation.

a (β_x, β_y, β_z)		b (β_x, β_y, ϕ_z)		c (ϕ_x, ϕ_y, β_z)	
$\frac{K_{nx}}{K_x}$	$\left[1+\left(\frac{\beta_y}{\beta_x}\right)_n^2+\left(\frac{\beta_z}{\beta_x}\right)_n^2\right]g_n^2$	$\frac{K_{nx}}{K_x}$	$\left[1+\left(\frac{\beta_y}{\beta_x}\right)_n^2+\left(\frac{r_z\phi_z}{\beta_x}\right)_n^2\right]g_n^2$	$\frac{K_{nz}}{K_z}$	$\left[1+\left(\frac{r_x\phi_x}{\beta_z}\right)_n^2+\left(\frac{r_y\phi_y}{\beta_z}\right)_n^2\right]g_n^2$
$\frac{K_{ny}}{K_x}$	$\left[1+\left(\frac{\beta_x}{\beta_y}\right)_n^2+\left(\frac{\beta_z}{\beta_y}\right)_n^2\right]g_n^2$	$\frac{K_{ny}}{K_x}$	$\left[1+\left(\frac{\beta_x}{\beta_y}\right)_n^2+\left(\frac{r_z\phi_z}{\beta_y}\right)_n^2\right]g_n^2$	$\frac{K_{n\theta x}}{K_z r_x^2}$	$\left[1+\left(\frac{\beta_z}{r_x\phi_x}\right)_n^2+\left(\frac{r_y\phi_y}{r_x\phi_x}\right)_n^2\right]g_n^2$
$\frac{K_{nz}}{K_x}$	$\left[1+\left(\frac{\beta_x}{\beta_z}\right)_n^2+\left(\frac{\beta_y}{\beta_z}\right)_n^2\right]g_n^2$	$\frac{K_{n\theta z}}{K_x}$	$\left[1+\left(\frac{\beta_x}{r_z\phi_z}\right)_n^2+\left(\frac{\beta_y}{r_z\phi_z}\right)_n^2\right]g_n^2$	$\frac{K_{n\theta y}}{K_z r_y^2}$	$\left[1+\left(\frac{\beta_z}{r_y\phi_y}\right)_n^2+\left(\frac{r_x\phi_x}{r_y\phi_y}\right)_n^2\right]g_n^2$
$\frac{1}{\overline{m}_x}$	$1+\left(\frac{\beta_y}{\beta_x}\right)_n^2+\left(\frac{\beta_z}{\beta_x}\right)_n^2$	$\frac{1}{\overline{m}_x}$	$1+\left(\frac{\beta_y}{\beta_x}\right)_n^2+\left(\frac{r_z\phi_z}{\beta_x}\right)_n^2$	$\frac{1}{\overline{m}_z}$	$1+\left(\frac{r_x\phi_x}{\beta_z}\right)_n^2+\left(\frac{r_y\phi_y}{\beta_z}\right)_n^2$
$\frac{1}{\overline{m}_y}$	$1+\left(\frac{\beta_x}{\beta_y}\right)_n^2+\left(\frac{\beta_z}{\beta_y}\right)_n^2$	$\frac{1}{\overline{m}_y}$	$1+\left(\frac{\beta_x}{\beta_y}\right)_n^2+\left(\frac{r_z\phi_z}{\beta_y}\right)_n^2$	$\frac{1}{\overline{m}_{\theta x}}$	$1+\left(\frac{\beta_z}{r_x\phi_x}\right)_n^2+\left(\frac{r_y\phi_y}{r_x\phi_x}\right)_n^2$
$\frac{1}{\overline{m}_z}$	$1+\left(\frac{\beta_x}{\beta_z}\right)_n^2+\left(\frac{\beta_y}{\beta_z}\right)_n^2$	$\frac{1}{\overline{m}_{\theta z}}$	$1+\left(\frac{\beta_x}{r_z\phi_z}\right)_n^2+\left(\frac{\beta_y}{r_z\phi_z}\right)_n^2$	$\frac{1}{\overline{m}_{\theta y}}$	$1+\left(\frac{\beta_z}{r_y\phi_y}\right)_n^2+\left(\frac{r_x\phi_x}{r_y\phi_y}\right)_n^2$

TABLE 14.8. Case (*b*) three-degree planar modal response requires determination in the sense of the forcing, with other coupled components following from the related normal modes.

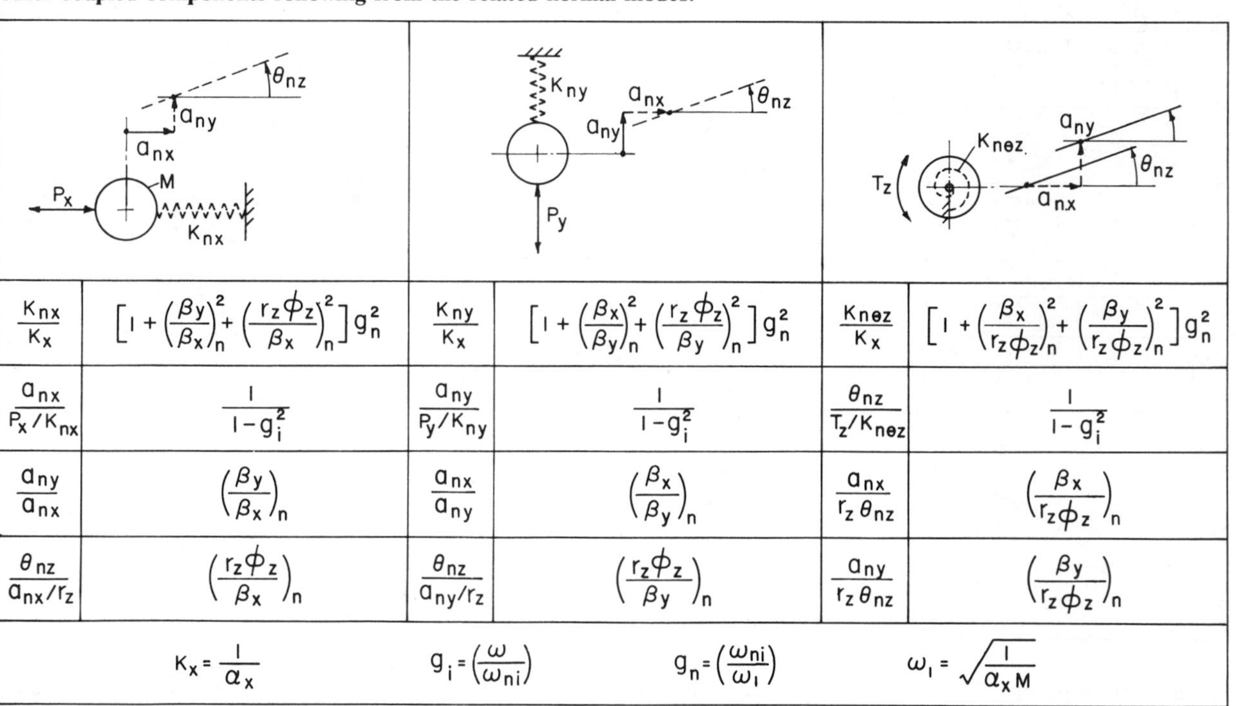

$\frac{K_{nx}}{K_x}$	$\left[1+\left(\frac{\beta_y}{\beta_x}\right)_n^2+\left(\frac{r_z\phi_z}{\beta_x}\right)_n^2\right]g_n^2$	$\frac{K_{ny}}{K_x}$	$\left[1+\left(\frac{\beta_x}{\beta_y}\right)_n^2+\left(\frac{r_z\phi_z}{\beta_y}\right)_n^2\right]g_n^2$	$\frac{K_{n\theta z}}{K_x}$	$\left[1+\left(\frac{\beta_x}{r_z\phi_z}\right)_n^2+\left(\frac{\beta_y}{r_z\phi_z}\right)_n^2\right]g_n^2$
$\frac{a_{nx}}{P_x/K_{nx}}$	$\frac{1}{1-g_i^2}$	$\frac{a_{ny}}{P_y/K_{ny}}$	$\frac{1}{1-g_i^2}$	$\frac{\theta_{nz}}{T_z/K_{n\theta z}}$	$\frac{1}{1-g_i^2}$
$\frac{a_{ny}}{a_{nx}}$	$\left(\frac{\beta_y}{\beta_x}\right)_n$	$\frac{a_{nx}}{a_{ny}}$	$\left(\frac{\beta_x}{\beta_y}\right)_n$	$\frac{a_{nx}}{r_z\theta_{nz}}$	$\left(\frac{\beta_x}{r_z\phi_z}\right)_n$
$\frac{\theta_{nz}}{a_{nx}/r_z}$	$\left(\frac{r_z\phi_z}{\beta_x}\right)_n$	$\frac{\theta_{nz}}{a_{ny}/r_z}$	$\left(\frac{r_z\phi_z}{\beta_y}\right)_n$	$\frac{a_{ny}}{r_z\theta_{nz}}$	$\left(\frac{\beta_y}{r_z\phi_z}\right)_n$

$K_x=\frac{1}{\alpha_x}$ $\quad g_i=\left(\frac{\omega}{\omega_{ni}}\right)$ $\quad g_n=\left(\frac{\omega_{ni}}{\omega_1}\right)$ $\quad \omega_1=\sqrt{\frac{1}{\alpha_x M}}$

TABLE 14.9. Case (*b*) modal response to displacement excitation.

Case					
Excitation a_{ox} (diagram: mass M, centre G, axes x, y, rotation θ_z)		Excitation a_{oy} (diagram: axes x, y, rotation θ_z)		Excitation θ_o (diagram: axes x, y, rotation θ_z)	
$\frac{1}{\overline{m}_x}$	$1+\left(\frac{\beta_y}{\beta_x}\right)_n^2+\left(\frac{r_z\phi_z}{\beta_x}\right)_n^2$	$\frac{1}{\overline{m}_y}$	$1+\left(\frac{\beta_x}{\beta_y}\right)_n^2+\left(\frac{r_z\phi_z}{\beta_y}\right)_n^2$	$\frac{1}{\overline{m}_{\theta z}}$	$1+\left(\frac{\beta_x}{r_z\phi_z}\right)_n^2+\left(\frac{\beta_y}{r_z\phi_z}\right)_n^2$
$\frac{a_{nx}}{\overline{m}_x a_{ox}}$	$\frac{1}{1-g_i^2}$	$\frac{a_{ny}}{\overline{m}_y a_{oy}}$	$\frac{1}{1-g_i^2}$	$\frac{\theta_{nz}}{\overline{m}_{\theta z}\theta_o}$	$\frac{1}{1-g_i^2}$
$\frac{a_{ny}}{a_{nx}}$	$\left(\frac{\beta_y}{\beta_x}\right)_n$	$\frac{a_{nx}}{a_{ny}}$	$\left(\frac{\beta_x}{\beta_y}\right)_n$	$\frac{a_{nx}}{r_z\theta_{nz}}$	$\left(\frac{\beta_x}{r_z\phi_z}\right)_n$
$\frac{\theta_{nz}}{a_{nx}/r_z}$	$\left(\frac{r_z\phi_z}{\beta_x}\right)_n$	$\frac{\theta_{nz}}{a_{ny}/r_z}$	$\left(\frac{r_z\phi_z}{\beta_y}\right)_n$	$\frac{a_{ny}}{r_z\theta_{nz}}$	$\left(\frac{\beta_y}{r_z\phi_z}\right)_n$
$g_i=\left(\frac{\omega}{\omega_{ni}}\right)$			$\omega_1=\sqrt{\frac{1}{\alpha_x M}}$		

Thus we first determine the primary amplitude for which there is a physical interpretation of (P_x/K_{nx}) as a static deflection or equilibrium amplitude. The coupled responses then follow from the normal modes.

This same analysis is applicable to P_y and T_z, with each applied to the corresponding modal spring statically to establish the reference amplitude to which the magnification factor is applied.

For a given coordinate excitation there are a total of three modes to be added for each of the three coordinate responses, or a total of nine components associated with a given excitation. Practically we may only be concerned with proximity to one modal resonance, which reduces the response determination to the three coordinate amplitudes at the forcing frequency and mode of interest.

14.10. MODAL RESPONSE TO DISPLACEMENT FORCING

Again using the case *b* example, we can apply the single-degree analog as in Equation 5.18. Given a horizontal base excitation a_{0x}, absolute modal response of the mass in the horizontal direction is

$$\frac{\mathsf{a}_{nx}}{\mathsf{a}_{0x}} = \overline{m}_x \left(\frac{1}{1 - \mathsf{g}_i^2} \right) \tag{14.12}$$

where $\overline{m}_x$ is understood to be a factor related to a particular principal mode, although not subscripted by n.

In Table 14.9 the modal response equations are given for the primary and for the coupled responses that follow. The table yields absolute amplitudes, but we can solve on the basis of relative amplitudes with respect to the excitational motion using the alternative magnification factor form,

$$\frac{\mathsf{a}_{nx}^r}{\mathsf{a}_{0x}} = \overline{m}_x \left(\frac{\mathsf{g}_i^2}{1 - \mathsf{g}_i^2} \right) \tag{14.13}$$

EXAMPLES

EXAMPLE 14.1. The two-bar truss is pin-connected and carries a cylindrical mass. The 38 kg. mass has polar inertia, but is decoupled by the bearing mounting. Link 1–2 has twice the cross-sectional area of 0–1. Calculate the natural frequencies and modal directions if

$$\alpha_x = \alpha_y = 0.75(10)^{-6} \quad \text{and} \quad \alpha_{xy} = 0.25(10)^{-6}\ \text{m/N}$$

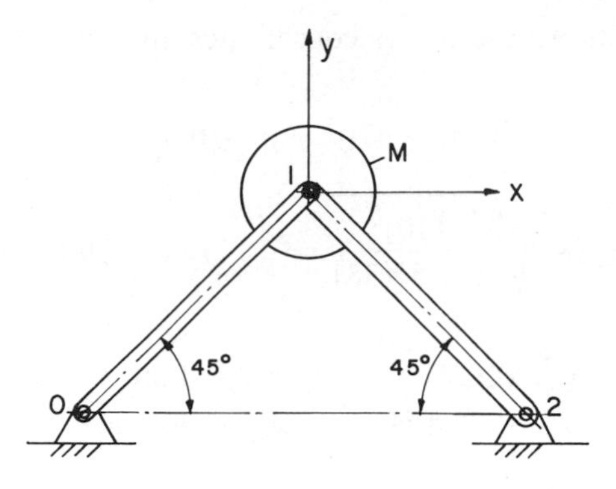

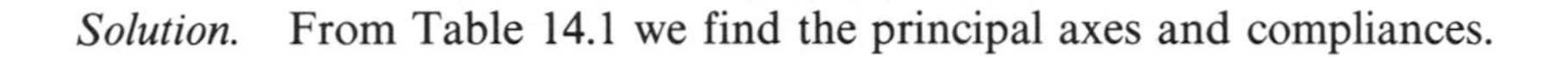

Solution. From Table 14.1 we find the principal axes and compliances.

$$\psi_1 = \left(\frac{0.25}{0.75}\right) = \frac{1}{3}$$

$$\psi_2 = \left(\frac{0.75}{0.75}\right) = 1$$

$$\tan 2\theta_p = \left(\frac{2/3}{0}\right) = \infty$$

$$\theta_p = 45°$$

$$\theta_q = 135°$$

Principal axes are in the direction of the links.

$$\left(\frac{\alpha_p}{\alpha_x}\right) = \left(1 + \frac{1}{3}\right) = 4/3 \qquad \left(\frac{\alpha_q}{\alpha_x}\right) = \left(1 - \frac{1}{3}\right) = \frac{2}{3}$$

As uncoupled or independent single-degree systems, the natural frequencies are

$$\omega_{np} = \sqrt{\frac{(10)^6}{(0.75)(4/3)(38)}} = 162 \text{ r/s} = 25.8 \text{ Hz}$$

$$\omega_{nq} = \sqrt{\frac{(10)^6}{(0.75)(2/3)(38)}} = 229 \text{ r/s} = 36.5 \text{ Hz}$$

Or, with Table 14.2*a* we use the *xy* coordinates directly.

$$C = 0.889\mathsf{g}^4 - 2\mathsf{g}^2 + 1 = 0$$

$$\omega_1 = \sqrt{\frac{(10)^6}{(0.75)(38)}} = 187 \text{ r/s} = 29.8 \text{ Hz}$$

$$\mathsf{g}_{n1}^2 = 0.75 \qquad \mathsf{g}_{n2}^2 = 1.50$$
$$\omega_{n1} = 25.8 \text{ Hz} \qquad \omega_{n2} = 36.5 \text{ Hz}$$

$$\left(\frac{\beta_y}{\beta_x}\right)_{n1} = \left[\frac{1 - 0.75}{(1/3)(0.75)}\right] = \left(\frac{0.25}{0.25}\right) = +1$$

$$\left(\frac{\beta_y}{\beta_x}\right)_{n2} = \left[\frac{1 - 1.50}{(1/3)(1.5)}\right] = -1$$

These directions correspond to $+45°$ and $-45°$ respectively, and agree with the previous results.

EXAMPLE 14.2. A constant section cantilevered beam supports a concentrated mass which is excited horizontally. Neglect the mass of the structure, and find

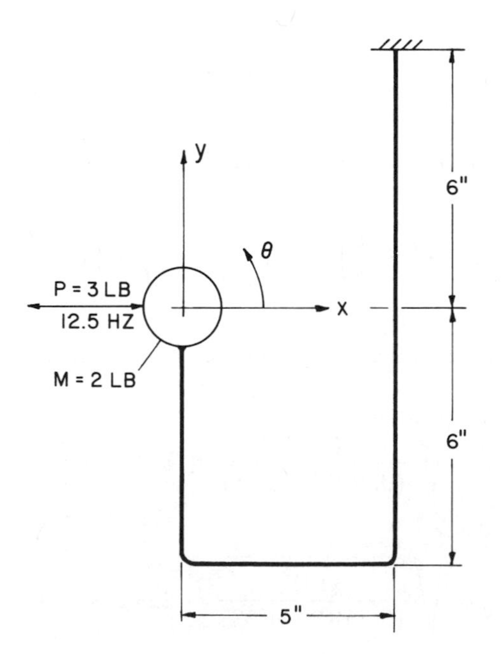

the magnitude and direction of the steady-state response if

$$\alpha_x = +0.10$$

$$\alpha_y = +0.0864$$

$$\alpha_{xy} = +0.0189 \text{ in./lb}$$

Solution. From Table 14.2*a*, with $\psi_1 = 0.189$ and $\psi_2 = 0.864$,

$$\omega_1 = 43.9 \text{ r/s} = 7.0 \text{ Hz} \qquad \mathsf{g}^2 = \left(\frac{12.5}{7}\right)^2 = 3.20$$

$$\left(\frac{\mathsf{a}_{xx}}{P_x\alpha_x}\right) = \left[\frac{1-(0.828)(3.2)}{3.51}\right] = -0.47$$

$$\mathsf{a}_{xx} = (3)(0.10)(-0.47) = -0.141 \text{ in.}$$

$$\frac{\mathsf{a}_{yx}}{(P_x\alpha_{xy})} = \left(\frac{1}{C}\right) = +0.285$$

$$\mathsf{a}_{yx} = (3)(0.189)(0.285) = 0.016 \text{ in.}$$

EXAMPLE 14.3. In Example 14.2 the radius of gyration of the mass is 2 in. Find the three natural frequencies and modes in the plane with elastic properties of

$$\alpha_x = \frac{396}{EI} \qquad \alpha_y = \frac{342}{EI} \qquad \alpha_{xy} = \frac{75}{EI} \text{ in./lb}$$

$$\beta_{yz} = \frac{72.5}{EI} \qquad \beta_{zx} = \frac{48}{EI} \text{ lb}^{-1} \qquad \gamma_z = \frac{23}{EI} \text{ (in. lb)}^{-1}$$

Solution. This is case 2 (Table 14.7*b*), with C_b from Table 14.5. Normalized parameters, defined in Table 14.4*b*, are

$$\psi_1 = 0.189 \qquad \psi_2 = 0.864$$
$$\psi_3 = 0.183 \qquad \psi_5 = 0.121 \text{ in.}^{-1}$$
$$\psi_4 = 0.058 \text{ in.}^{-2}$$
$$B_1 = 0.0435 \qquad B_2 = 0.828$$
$$B_3 = 0.0167 \text{ in.}^{-2}$$

The frequency cubic becomes

$$-0.0412\mathsf{g}_n^6 + 1.068\mathsf{g}_n^4 - 2.096\mathsf{g}_n^2 + 1 = 0$$

for which the roots are

$$\begin{array}{lll} \mathsf{g}_{n1}^2 = 0.77 & \mathsf{g}_{n2}^2 = 1.32 & \mathsf{g}_{n3}^2 = 23.8 \\ \mathsf{g}_{n1} = 0.88 & \mathsf{g}_{n2} = 1.15 & \mathsf{g}_{n3} = 4.88 \end{array}$$

If EI is not specified, the general reference frequency is, also for any mass,

$$\omega_1 = \sqrt{\frac{EI}{396M}} \quad \text{and} \quad \omega_{ni} = \mathsf{g}_{ni}\omega_1$$

Using modal equations from Table 14.7*b* for the first mode,

$$\begin{array}{llccc} & & i = 1 & i = 2 & i = 3 \\ \left(\dfrac{\beta_y}{\beta_x}\right)_{n1} = \left[\dfrac{0.183 - 0.16(0.77)}{0.121 - 0.07(0.77)}\right] & = & +0.89 & -1.00 & +2.35 \\ \left(\dfrac{r_z\phi_z}{\beta_x}\right)_{n1} = \dfrac{1}{2(0.121)}(0.131) & = & +0.54 & -0.23 & -5.8 \end{array}$$

Orthogonality can be verified between any two modal combinations using the relation

$$1 + \left(\frac{\beta_y}{\beta_x}\right)_{ni}\left(\frac{\beta_y}{\beta_x}\right)_{nj} + \left(\frac{r_z\phi_z}{\beta_x}\right)_{ni}\left(\frac{r_z\phi_z}{\beta_x}\right)_{nj} = 0$$

Or, for first and second modes,

$$1 + (0.89)(-1.00) + (0.54)(-0.23) = -0.014$$

where the remainder is due to round-off error.

EXAMPLE 14.4. The structure of Example 14.3 has a beam stiffness of $EI = 0.77(10)^6$ and a mass of $M = 5$ lb. It is excited at the base vertically with a harmonic displacement of 0.09 in. at 80 Hz. Determine the first modal component responses.

Solution. Using first-mode data from Example 14.3, and Table 14.9 for a_{0y} excitation,

$$\overline{m}_y = \left[\frac{1}{1 + (1/0.89)^2 + (0.54/0.89)^2}\right] = 0.38$$

$$\omega_{n1} = (0.88)\sqrt{\frac{386(0.77)(10)^6}{396(5)}} = 341 \text{ r/s} = 54.3 \text{ Hz}$$

$$\mathsf{g}_1 = \left(\frac{80}{54.3}\right) = 1.47$$

$$\left(\frac{\mathsf{a}_{ny}}{\overline{m}_y \mathsf{a}_{0y}}\right) = \left(\frac{1}{1 - 2.174}\right) = -0.85$$

$$\mathsf{a}_{ny} = (-0.85)(0.38)(0.09) = -0.029 \text{ in.}$$

$$\mathsf{a}_{nx} = -0.029\left(\frac{1}{0.89}\right) = -0.033 \text{ in.}$$

$$\theta_{nz} = \frac{-0.029}{2}\left(\frac{0.54}{0.89}\right) = -0.0088 \text{ rad} = 0.50°$$

15 CIRCULAR RINGS

The circular ring or arc is a common structural element, and related frequencies and modes are often of engineering importance. A complete ring represents a continuous beam with the ends connected, and in bending or torsion relates to the concepts of Chapter 12. In- and out-of-plane modes are possible, and since bending and torsion of the elastic elements are involved, the most basic form of the beam or bar is a round cross section. Most of the following analysis concerns lumped masses, single or disposed symmetrically, with the mass of the circular beam assumed negligible and sufficiently slender so curved beam nonlinearities are avoided. Modal frequencies for the uniform ring without masses are given in a number of references indicated; however, numerical results are presented here only for the free planar modes in bending.

It is impossible to summarize the many variations of geometries and modes for ring-mounted masses. Those presented are some of the more elementary, but the concepts underlying the analytical methods will provide insights to similar cases and more complex systems. Elastic coupling techniques in three coordinates, developed in previous chapters, are now illustrated for circular geometries in which coupling effects are prevalent.

15.1. SEMICIRCULAR PIN-CONNECTED BEAMS

Table 15.1 indicates several elementary planar examples in which the beams are pinned semicircular segments. All masses can have significant mass moment of inertia about the polar axis perpendicular to the plane of the ring, but this factor only enters the frequency calculation of the angular modes (g), (h), and (i). The frequency factors indicated apply to the single mode indicated.

Obviously the bending flexibilities of the circular beams must be evaluated to arrive at these results, but no attempt is made here to develop the required subject matter. These have been determined by elastic energy considerations

TABLE 15.1. Pinned semicircular beam elements for selected symmetrical combinations in translation, (*a*) to (*f*), and in torsion, (*g*) to (*i*), are uncoupled and have single modal frequencies. *M* is for a single mass.

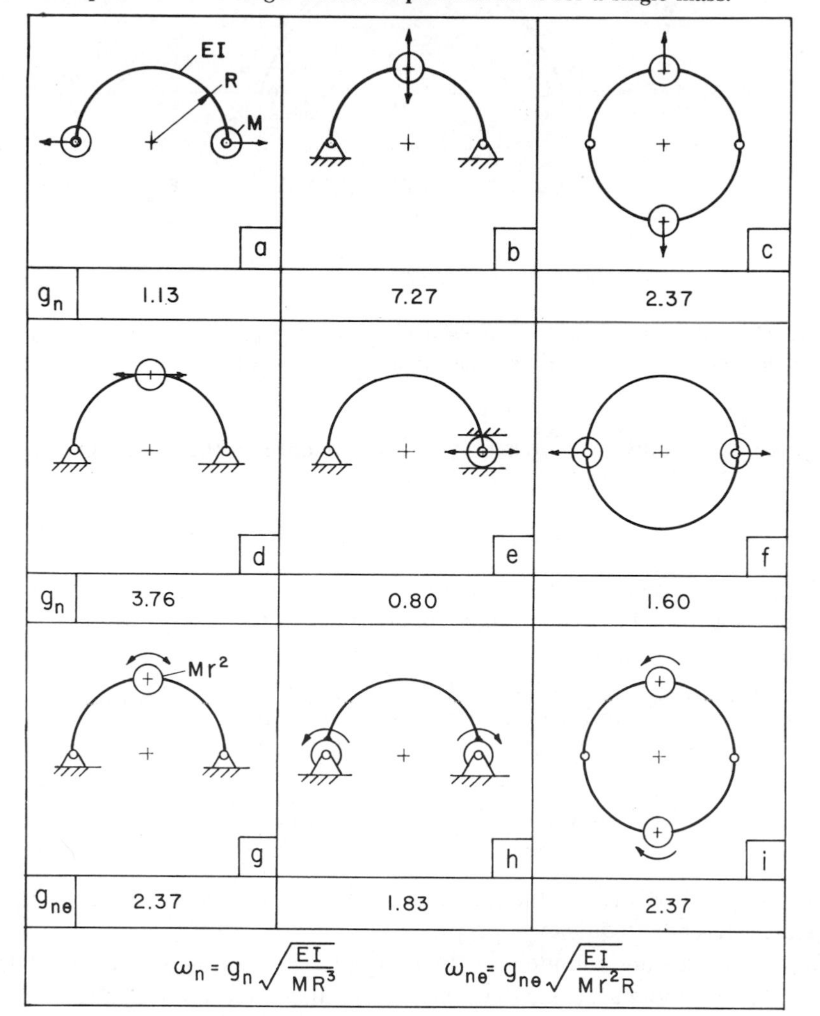

for the statically determinate structures, and later for the indeterminate cases. Results given are based entirely upon bending elasticity, neglecting direct stress and beam shear, which are usually negligible.

Taking the basic system (Table 15.1*a*), the two masses vibrate in opposition, with the total center of gravity unchanged and no connection to ground. Inertia forces are in equilibrium, as in (*c*), (*f*), and (*i*). These self-contained systems are termed *reactionless modes*.

The nature of static diametral compliance for Table 15.1*a* is shown in Figure 15.1. With opposed tensile loads the curved beam behaves as a spring,

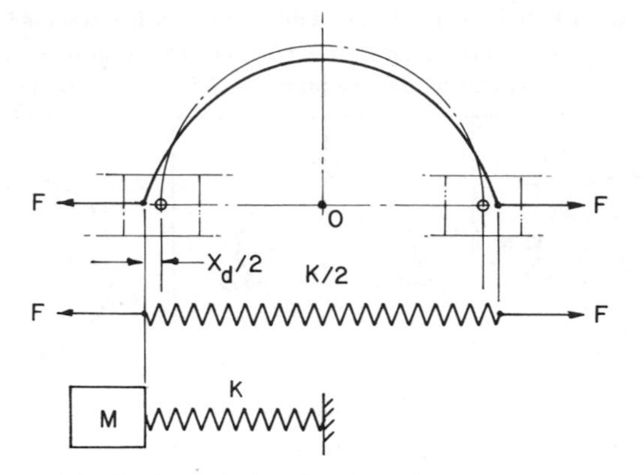

FIGURE 15.1. The semicircular beam in bending is equivalent to a diametral spring and to a radial spring grounded at the center for natural frequency purposes.

and in the free condition there is no vertical deflection at the forces, and

$$x_d = \frac{\pi}{2}\left(\frac{FR^3}{EI}\right) \tag{15.1}$$

where x_d is the relative deflection between the two ends. Adding the two masses, concentrated or pinned, there is no effect of the change in slope at the ends and we have a two-degree free–free system. With the node at the center it reduces to a single-mass case with a spring rate twice the total, and $\omega_n = \sqrt{K/M}$. Using the influence or compliance factor notation, or the reciprocal of K, the system frequency is

$$\omega_n = \sqrt{\frac{EI}{0.7854\, MR^3}} = 1.128\sqrt{\frac{EI}{MR^3}} \tag{15.2}$$

where $\alpha = (\pi/4)(\mathrm{R}^3/\mathrm{EI})$ or the radial deflection per unit load, and $\omega_n = \mathsf{g}_n\, \omega_1$ using the reference frequency concept.

Tabular values are rounded to three significant figures as numerical frequencies are no more accurate than the factors in ω_1, and the modulus is never known precisely. Furthermore, practical considerations of natural frequencies do not generally require a high degree of accuracy.

Examples of torsional modes with the semicircular ring are shown in Table 15.1*g*, *h*, and *i*. For (*g*) the compliance factor at the mass relative to an applied couple is

$$\frac{\gamma}{R/EI} = \left(\frac{3}{8}\pi - 1\right) = 0.178$$

with a natural frequency of

$$\omega_{n\theta} = \sqrt{\frac{1}{\gamma Mr^2}} = \sqrt{\frac{EI}{0.178Mr^2R}} = 2.37\sqrt{\frac{EI}{Mr^2R}} \tag{15.3}$$

15.2. PLANAR MODES OF THE CONTINUOUS RING

Table 15.2*a*, *b*, and *c* indicates frequencies for the constant ring with the only possible radial modes for the free–free systems. Comparing Table 15.1*c* with Table 15.2*a*, we see that the latter ring is appreciably stiffer horizontally because of the continuity of the beam at the masses. The Table 15.2*c* system is much stiffer than that shown in Table 15.2*b* because of the nature of the flexure in the two modes.

In Table 15.2*d* and *g* there are no radial displacements coupled to the angular motions of the masses. We note in Figure 15.2 that the angular–planar

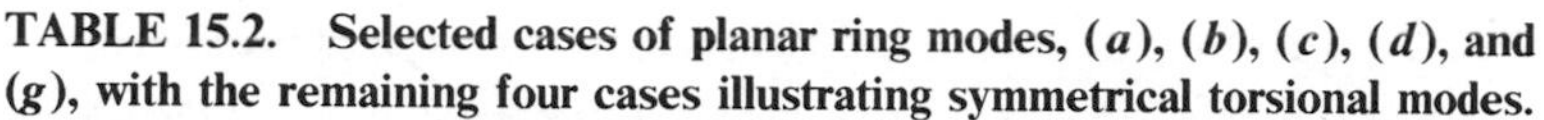

TABLE 15.2. Selected cases of planar ring modes, (*a*), (*b*), (*c*), (*d*), and (*g*), with the remaining four cases illustrating symmetrical torsional modes.

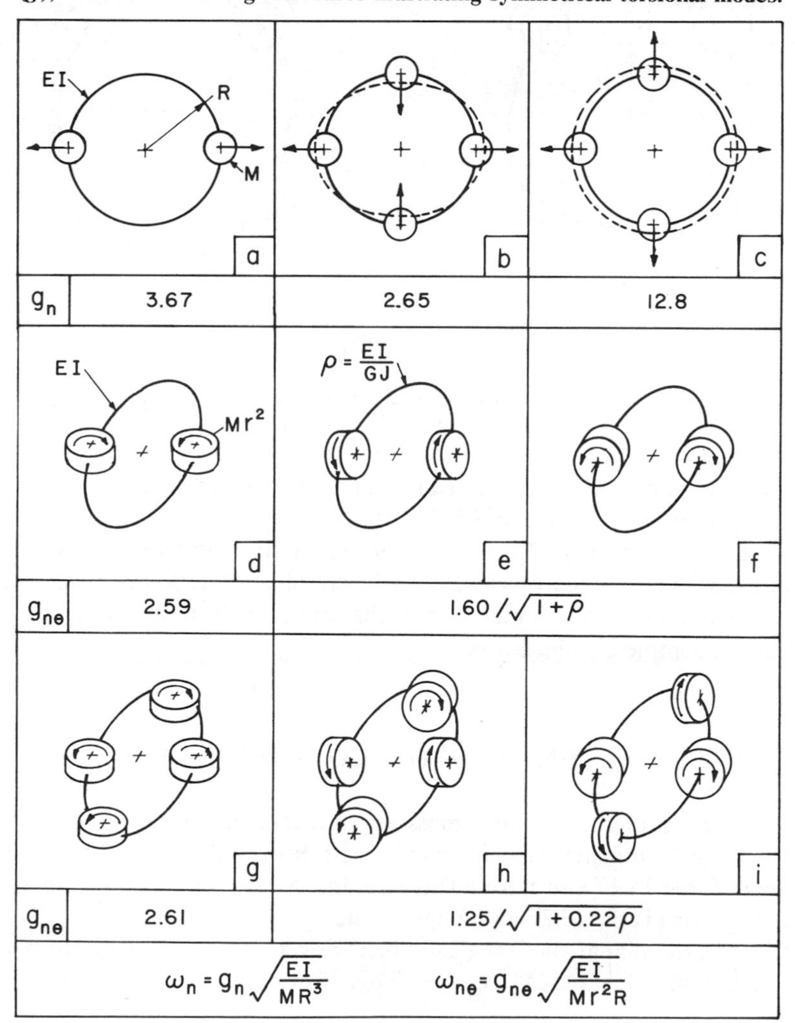

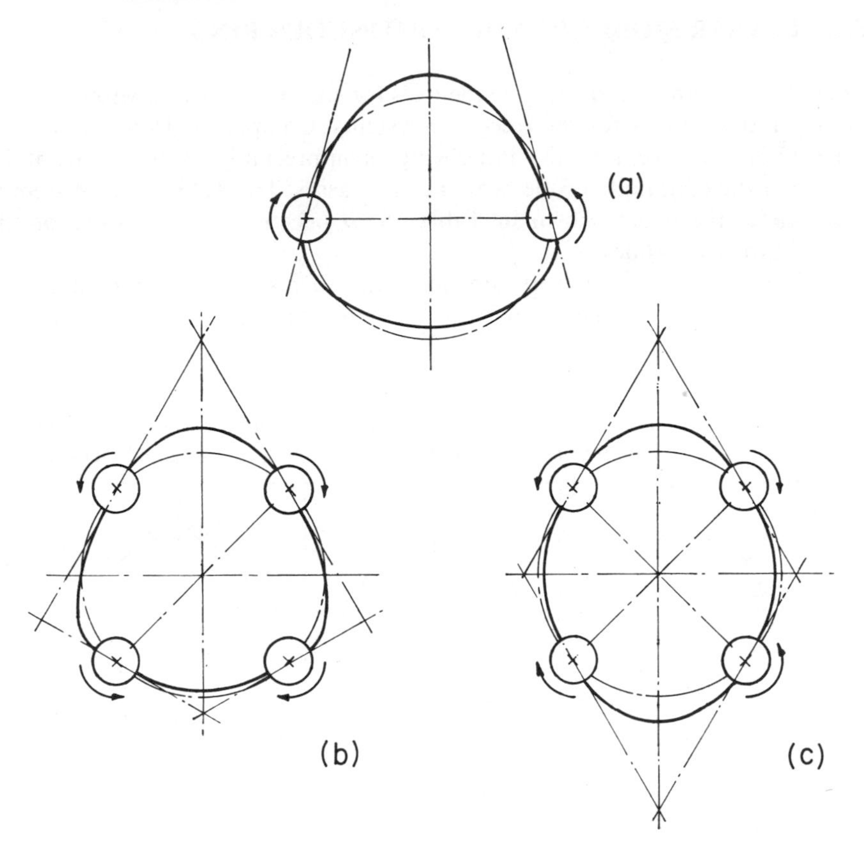

FIGURE 15.2. Reactionless planar modes can be symmetrical (*a*) and (*b*), or bisymmetrical (*c*).

combinations can be symmetrical, (*a*) and (*b*), or bisymmetrical, (*c*); the latter represents a different mode from Table 15.2*g*.

As we see in Figure 15.2*a*, the lack of symmetry about the horizontal axis accounts for the absence of radial elastic coupling. The lower ring tends to increase the horizontal diameter, but the upper half tends to decrease it, resulting in complete cancelation.

15.3. TORSIONAL MODES OF THE CONTINUOUS RING

Rotational or rolling action of a mass about a tangent to a ring can similarly involve simple symmetry or bisymmetry, with potentially related translational coupling (Table 15.2*f* and *i*, and Table 15.3*b*). Masses can also rotate about a radial ring axis (Table 15.2*e* and *h* and Table 15.3*a*).

Except in the planar modes, elastic behavior of a ring entails both bending and torsion in all stressed sections. This introduces the torsional stiffness

TABLE 15.3. Frequency equations for solid or hollow round beams, $\rho = EI/GJ = 1.25$; (*b*) and (*c*) involve two coupled modes in translation and rotation.

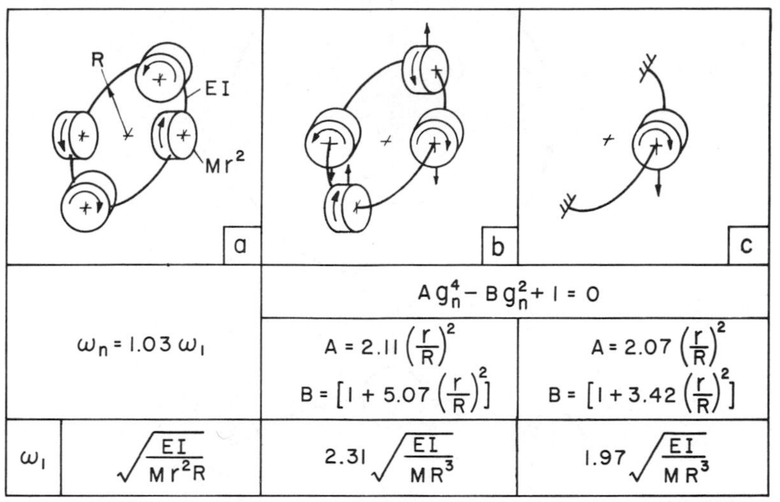

	a	b	c
		$Ag_n^4 - Bg_n^2 + 1 = 0$	
	$\omega_n = 1.03\,\omega_1$	$A = 2.11\left(\frac{r}{R}\right)^2$ $B = \left[1 + 5.07\left(\frac{r}{R}\right)^2\right]$	$A = 2.07\left(\frac{r}{R}\right)^2$ $B = \left[1 + 3.42\left(\frac{r}{R}\right)^2\right]$
ω_1	$\sqrt{\frac{EI}{Mr^2R}}$	$2.31\sqrt{\frac{EI}{MR^3}}$	$1.97\sqrt{\frac{EI}{MR^3}}$

parameter GJ in addition to the bending parameter EI, where

E = Young's modulus of elasticity in direct stress
I = cross-sectional area moment of inertia about a transverse neutral axis at the centroid
G = shear modulus of elasticity
J = polar area moment of inertia about the central axis

For a solid round cross section we have a dimensionless ratio in conventional units of

$$\rho = \frac{EI}{GJ} = \frac{30(10)^6(\pi d^4/64)}{12(10)^6(\pi d^4/32)} = 1.25$$

The 1.25 factor also applies to a hollow round cross section, for most common metals, and in SI units. For noncircular sections the cross section is reduced to an equivalent J factor.

15.4. OUT-OF-PLANE RING MODES

If concentrated masses on a ring vibrate transversely to the plane of the ring, we can have with symmetry a reactionless mode (Figure 15.3); however, *this cannot happen with less than four masses*. In the figure there is complete

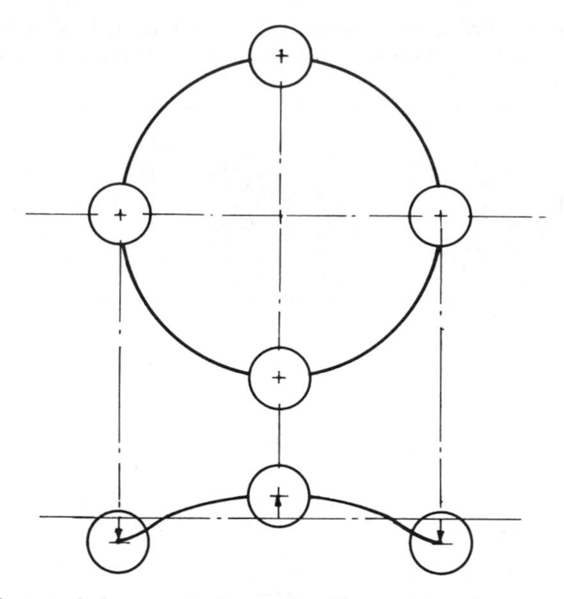

FIGURE 15.3. The out-of-plane reactionless mode with concentrated masses has the uncoupled reference frequency of Table 15.3*b*.

equilibrium of inertia forces and moments from this loading, and the deflection rate from the central plane is

$$\frac{y}{F} = \frac{R^3}{4}\left[\left(\frac{\frac{\pi}{2} - 1}{EI}\right) + \left(\frac{\pi - 3}{GJ}\right)\right]$$

$$\frac{\alpha}{R^3/EI} = 0.1427(1 + 0.2481\rho)$$

If $\rho = 1.25$,

$$\frac{\alpha}{R^3/EI} = 0.187$$

$$\omega_1 = \sqrt{\frac{1}{\alpha_m}} = 2.313\sqrt{\frac{EI}{MR^3}} \qquad (15.4)$$

This result is shown in Table 15.3*b* as the reference frequency for four concentrated masses at $r = 0$. With polar inertia of the masses the mode shown is equivalent to the coupled cases for each mass (Figure 9.2), and the resulting natural frequency quadratic, Equation 9.4.

Similarly, in Table 15.3*c* the semicircular clamped ring has a single reference frequency with concentrated masses and two coupled frequencies when the inertia is significant.

15.5. THE CIRCULAR PLANAR CANTILEVER

The uniform slender beam with constant curvature is fixed to ground (Table 15.4), with deflection data for x and y loading and torsional loading about the perpendicular z axis at the free end. This corresponds to the coupling case b in Chapter 14. Expressions are given for all compliance factors as a function of the central angle. There are three additional factors related by reciprocity, with notation explained in Table 14.4*b*. Specific numerical values for deflection determination are shown in Table 15.5*a*. These data can be used for static structural behavior; however, for vibratory purposes we will apply a mass at the end of the beam, the centroid coinciding with the origin of the axes (Figure 15.4).

In Table 14.4*b* we normalize the deflection factors with respect to α_x, and these dimensionless ψ_i terms are shown in Table 15.5*b*. Further combined coefficients deriving from the elastic behavior are calculated in Table 15.6. The A terms in Table 15.6*a* apply to the natural frequency cubic shown. The B and D coefficients are used for the normal mode and the mass response if forced (Table 14.5*b*). With sinusoidal forcing at the base of the beam we use Table 14.6*b* for coordinate responses of the mass.

TABLE 15.4. Equations for all the planar direct and cross-deflection coefficients of the circular cantilever, similar to Table 14.3*b*.

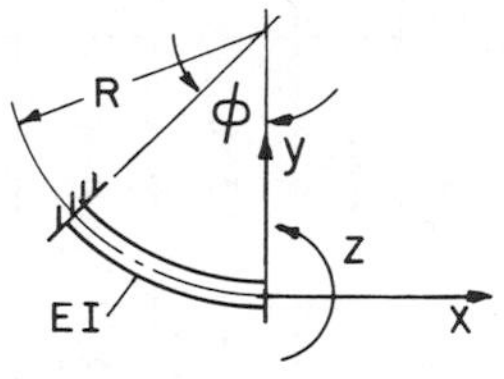

$\frac{\alpha_x}{R^3/EI}$	$\frac{3}{2}\phi - 2\sin\phi + \frac{1}{4}\sin 2\phi$	$\frac{\beta_{yz}}{R^2/EI}$	$1 - \cos\phi$
$\frac{\alpha_{xy}}{R^3/EI}$	$\frac{1}{2} - \cos\phi + \frac{1}{2}\cos^2\phi$	$\frac{\gamma_z}{R/EI}$	ϕ
$\frac{\alpha_y}{R^3/EI}$	$\frac{1}{2}\phi - \frac{1}{4}\sin 2\phi$	$\frac{\beta_{xz}}{R^2/EI}$	$\phi - \sin\phi$

TABLE 15.5. Absolute compliance coefficients (*a*), and normalized ratios (*b*), for various beam angles (case (*b*) Chapter 14).

a

ϕ	$\pi/2$	π	$3\pi/2$	2π
$\frac{\alpha_x}{R^3/EI}$	0.356	4.712	9.069	9.425
$\frac{\alpha_{xy}}{R^3/EI}$	0.500	2.000	0.500	0
$\frac{\alpha_y}{R^3/EI}$	0.785	1.571	2.356	3.142
$\frac{\beta_{yz}}{R^2/EI}$	1.000	2.000	1.000	0
$\frac{\gamma_z}{R/EI}$	1.571	3.142	4.712	6.283
$\frac{\beta_{zx}}{R^2/EI}$	0.571	3.142	5.712	6.283

b

ϕ	$\pi/2$	π	$3\pi/2$	2π
ψ_1	1.404	0.424	0.055	0
ψ_2	2.205	0.333	0.260	0.333
$R\psi_3$	2.807	0.424	0.110	0
$R^2\psi_4$	4.410	0.667	0.520	0.667
$R\psi_5$	1.603	0.667	0.630	0.667

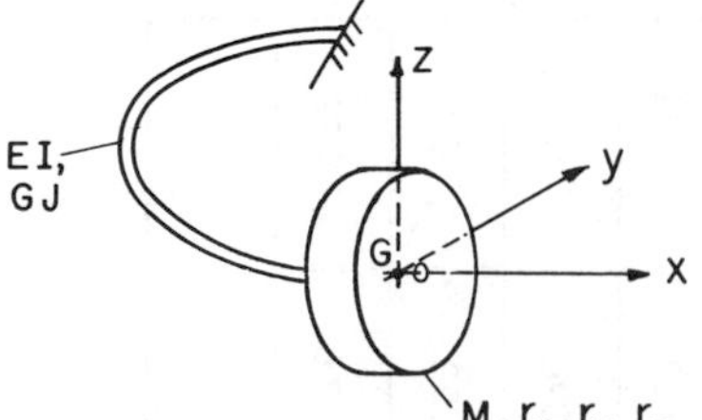

FIGURE 15.4. A centrally supported mass is analyzed using elasticity factors from Tables 15.4–15.8.

15.6. TRANSVERSE MODES OF THE CIRCULAR CANTILEVER

Whereas the case *b* coupling just discussed involves only planar motion, the end mass can vibrate perpendicularly to the plane with coupled rotations (case *c*; see Table 14.4*c*). We now have z, θ_x, and θ_y displacements coupled elastically, and independently of the x, y, and θ_z coupling combination.

Case *c* elasticity factors are given in Tables 15.7 and 15.8. These introduce the torsional stiffness factor GJ, or the ratio $\rho = EI/GJ$, which complicates the expressions somewhat. In Tables 15.7*a* and 15.8 any ρ ratio can be used. In the latter the compliance factors result from summing A and B. For instance, with a 90° bar,

$$\alpha_z = \frac{R^3}{EI}(0.785 + \rho 0.356)$$

$$\beta_{yz} = \beta_{zy} = \frac{0.500R^2}{EI}(1 + \rho)$$

Actual calculations can be extended numerically as required for any case *c* equations in Tables 14.5, 14.6, and 14.7, for frequency, modal, or response results in three coordinates.

15.7. ELASTIC COEFFICIENTS FOR AN EXTENDED BODY

Elastic data and corresponding frequencies in the previous sections are limited to planar circular beams with a mass centered at the free end, and to the planar and out-of-plane combinations. We now broaden the use of Tables 15.4–15.8 to any end-mounted rigid body, provided the centroid coincides with the plane of the beam.

In Figure 15.5 the body extends beyond the beam a distance b and below a distance c. For case *b* (Chapter 14), all deflections are in the xy plane and a horizontal force at G applies a direct load and a couple to the beam at O, or F_x and cF_x respectively. As seen in Table 14.4*b* we have deflections at O due to F_x of X_x, Y_x, and θ_x. The couple, or T_z, produces θ_z, X_z, and Y_z deflections.

TABLE 15.6. Calculated coefficients for circular cantilever elements for Tables 14.5*b* and 14.6*b*.

R φ y z x

a

ϕ	$\pi/2$	π	$3\pi/2$	2π
R^2A_1	0.121	0.014	0.026	0.074
R^2A_2	3.684	0.264	0.246	0.444
A_3	0.235	0.153	0.257	0.333
R^2A_4	4.410	0.666	0.520	0.666
A_5	3.205	1.333	1.260	1.333

$$C_b = r_z^2A_1g^6 + [r_z^2A_2 + A_3]g^4 - [r_z^2A_4 + A_5]g^2 + 1$$

b

ϕ	$\pi/2$	π	$3\pi/2$	2π
R^2B_1	1.842	0.222	0.123	0.222
B_2	0.235	0.153	0.257	0.333
R^2B_3	1.842	0.042	0.123	0.222
R^2D_1	−1.691	0	0.042	0
$R\ D_2$	0.407	−0.042	−0.158	0
$R\ D_3$	−0.558	0.142	−0.076	0.222

TABLE 15.7. Three-coordinate compliance coefficients for case (*c*) in terms of arc angle (*a*) and numerical terms for round cross sections (*b*).

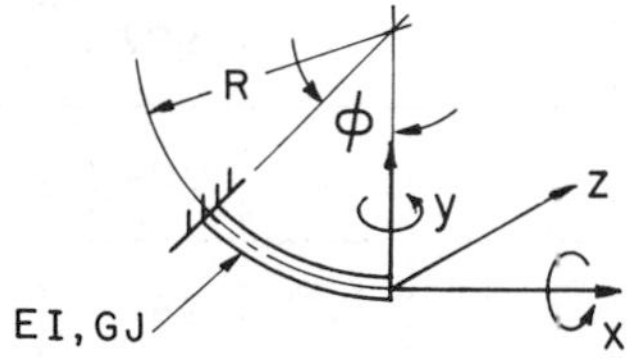

(a)

			(b) $\rho = \frac{EI}{GJ} \approx 1.25$ COMMON METALS, SOLID OR HOLLOW CIRCULAR SECTION				
$\frac{\alpha_z}{R^3/EI}$	$\left(\frac{1}{2}\phi - \frac{1}{4}\sin 2\phi\right)$	$+\rho\left(\frac{3}{2}\phi - 2\sin\phi + \frac{1}{4}\sin 2\phi\right)$	ϕ	$\pi/2$	π	$3\pi/2$	2π
$\frac{\beta_{xz}}{R^2/EI}$		$+\rho\left(\frac{1}{2}\phi - \sin\phi + \frac{1}{4}\sin 2\phi\right)$	$R\psi_1$	0.42	0.47	0.47	0.47
$\frac{\gamma_x}{R/EI}$		$+\rho\left(\frac{1}{2}\phi + \frac{1}{4}\sin 2\phi\right)$	$R^2\psi_2$	1.44	0.47	0.37	0.47
$\frac{\gamma_{xy}}{R/EI}$	$\frac{1}{2}\sin^2\phi$	$-\rho\left(\frac{1}{2}\sin^2\phi\right)$	$R^2\psi_3$	-0.10	0	0	0
$\frac{\gamma_y}{R/EI}$	$\left(\frac{1}{2}\phi - \frac{1}{4}\sin 2\phi\right)$	$+\rho\left(\frac{1}{2}\phi - \frac{1}{4}\sin 2\phi\right)$	$R^2\psi_4$	1.44	0.47	0.39	0.47
$\frac{\beta_{yz}}{R^2/EI}$	$\frac{1}{2}\sin^2\phi$	$-\rho\left(\cos\phi + \frac{1}{2}\sin^2\phi\right)$	$R\psi_5$	0.91	0.34	0.08	0

TABLE 15.8. Deflection coefficients for case (*c*) coordinates for any *EI*/*GJ* ratio, and for the ratio $\rho = 1.25$, *C*.

$$\frac{\alpha}{R^3/EI} \text{ OR } \frac{\beta}{R^2/EI} \text{ OR } \frac{\gamma}{R/EI} = (A + \rho B)$$

$$\rho = \frac{EI}{GJ}, \quad C = (A + 1.25B) \text{ FOR ROUND STEEL}$$

ϕ	$\pi/2$			π			$3\pi/2$			2π		
	A	B	C	A	B	C	A	B	C	A	B	C
$\frac{\alpha_z}{R^3/EI}$	0.785	0.356	1.23	1.571	4.712	7.46	2.356	9.069	13.7	3.142	9.425	14.9
$\frac{\beta_{xz}}{R^2/EI}$	0.785	-0.215	0.52	1.571	1.571	3.54	2.356	3.356	6.55	3.142	3.142	7.07
$\frac{\gamma_x}{R/EI}$	0.785	0.785	1.77	1.571	1.571	3.54	2.356	2.356	5.30	3.142	3.142	7.07
$\frac{\gamma_{xy}}{R/EI}$	0.500	-0.500	-0.13	0	0	0	0.500	-0.500	-0.13	0	0	0
$\frac{\gamma_y}{R/EI}$	0.785	0.785	1.77	1.571	1.571	3.54	2.356	2.356	5.30	3.142	3.142	7.07
$\frac{\beta_{yz}}{R^2/EI}$	0.500	0.500	1.13	0	2.000	2.50	0.500	0.500	1.13	0	0	0

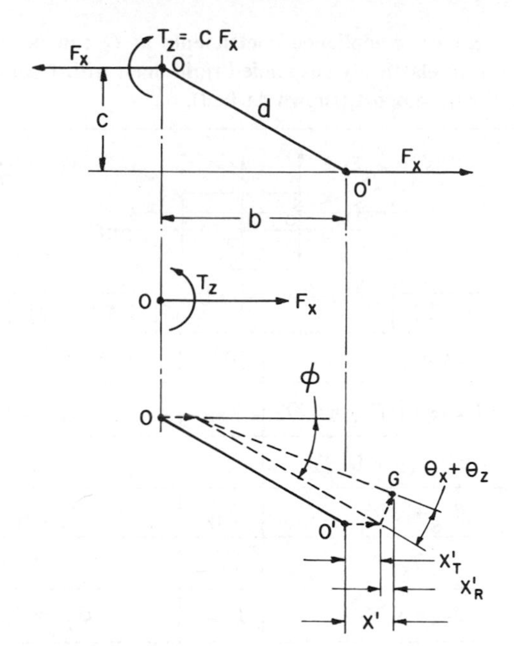

FIGURE 15.5. Geometry leading to Equation 15.5, showing the point of attachment of the rigid body to the beam at O. Point O' is the initial centroid and G is the final deflected position.

Translational deflection at O horizontally is

$$X'_T = \alpha_x(F_x) + \beta_{xz}(cF_x)$$

transmitted from O to O', or to G, by horizontal displacement of the body, which also has rotation about O of $\theta_x = \beta_{zx}F_x$ and $\theta_z = \gamma_z(cF_x)$. These cause tangential deflections about O of $(OG)\theta_x$ and $(OG)\theta_z$, with a total horizontal component of

$$X'_R = (\theta_x + \theta_z)\, d \sin\phi$$

$$= c[(\beta_{zx}F_x) + \gamma_z(cF_x)]$$

Summing all four components,

$$\frac{X}{F_x} = \alpha_x = \alpha_x + 2c\beta_{2x} + c^2\gamma_z \tag{15.5}$$

Similar developments can be applied to all three in-plane and all three out-of-plane loadings, with results in Table 15.9 enabling us to transpose all elastic factors from O to O'. Although illustrated in connection with the

TABLE 15.9. Known compliance coefficients at O can be converted to factors at O' for an elastically suspended rigid mass with G in the plane of any xy planar elastic support (shown dashed).

CASE (b)		CASE (c)	
α'_x	$\alpha_x + 2c\beta_{zx} + c^2\gamma_z$	α'_z	$\alpha_z + 2c\beta_{zy} + c^2\gamma_y$
α'_{xy}	$\alpha_{xy} + c\beta_{xz} + b\beta_{yz} + c^2\gamma_z$	β'_{xz}	$\beta_{xz} + c\gamma_x$
α'_y	$\alpha_y + 2b\beta_{yz} + b^2\gamma_z$	γ'_x	γ_x
β'_{yz}	$\beta_{yz} + b\gamma_z$	γ'_{xy}	γ_{xy}
γ'_z	γ_z	γ'_y	γ_y
β'_{zx}	$\beta_{zx} + c\gamma_z$	β'_{yz}	$\beta_{yz} + b\gamma_y$

TABLE 15.10. Reactionless planar bending modes of the uniform ring, with N the number of nodes.

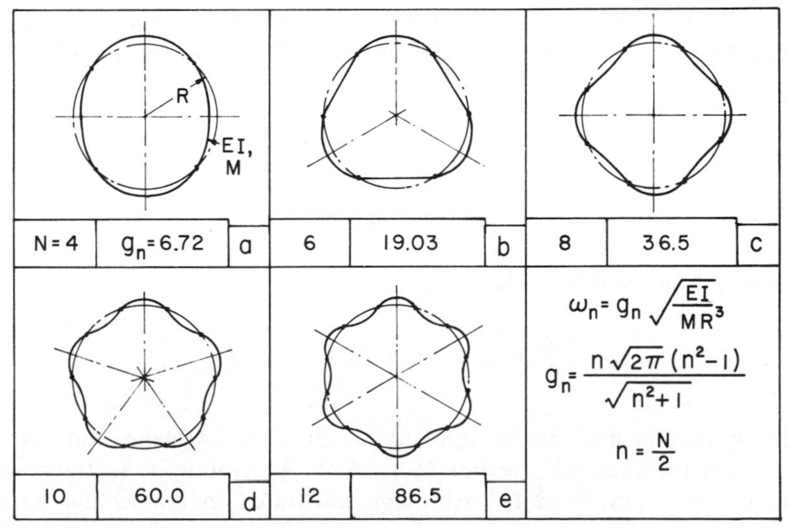

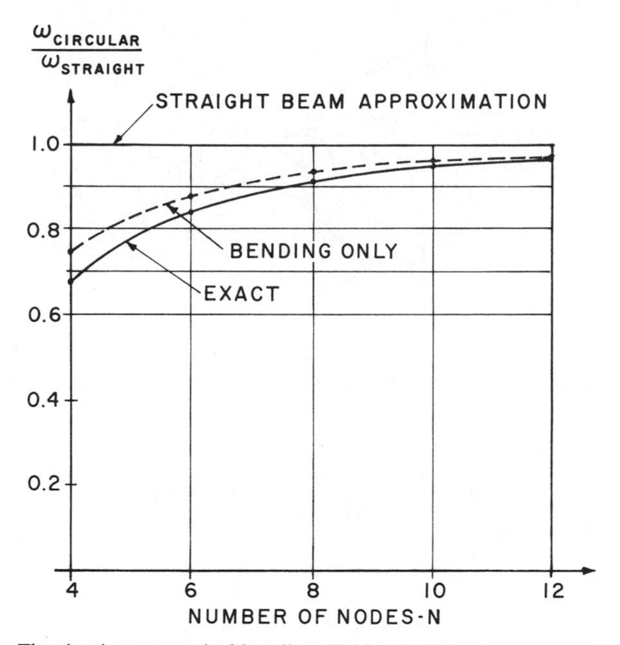

FIGURE 15.6. The ring in symmetrical bending (Table 15.10) has lower natural frequencies than calculated using straight beam equations for the same node-to-node length.

circular cantilever, this transformation is applicable to any planar cantilever-type suspension.

15.8. THE UNIFORM CIRCULAR BAR IN PLANAR BENDING

The complete circle is a special case of the uniform beam discussed in Chapter 12 with one end joined to the other (Table 15.10). Reactionless ring frequencies in bending are affected by the circular geometry with respect to the radial inertia loading. Additionally, there are tangential modal displacements.[21] Both effects tend to reduce the ring frequency relative to a corresponding straight beam of the same length. As shown in Figure 15.6, the deviation is pronounced at the lower modes, but the straight beam is an excellent approximation at the higher modes. This figure does not really represent continuous curves because the number of nodes must be an even integer greater than 2; however, the curves are drawn to show the basic trend.

EXAMPLES

Example 15.1. A pinned–pinned semicircular ring is made of round aluminum tubing, 1.50 × 1.25 in. The mass is cylindrical and centrally mounted. Find all the planar natural frequencies neglecting beam mass.

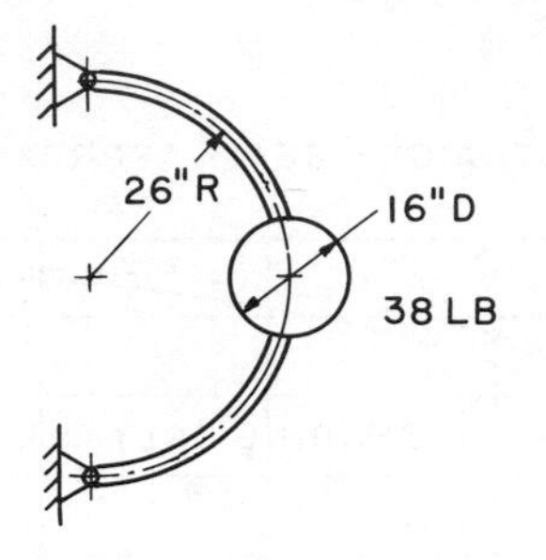

Solution. The modes correspond to Table 15.1*b*, *d*, and *g*, and are uncoupled.

$$I = \frac{\pi(D^4 - d^4)}{64} = 0.129 \text{ in.}^4$$

Table 15.1*b*:

$$\omega_1 = \sqrt{\frac{10(10)^6(0.129)(386)}{38(26)^3}}$$

$$= 27.3 \text{ rad/sec} = \frac{27.3}{2\pi} \text{ or } 4.34 \text{ Hz}$$

$$\omega_n = 7.27(4.34) = 31.6 \text{ Hz}$$

Table 15.1*d*:

$$\omega_n = 3.76(4.34) = 16.3 \text{ Hz}$$

Table 15.1*g*:

$$\omega_1 = \sqrt{\frac{10(10)^6(0.129)(386)}{38(26)(32)}} \qquad r^2 = \frac{1}{2}(8)^2 = 32 \text{ in.}^2$$

$$= 125.5 \text{ or } 20 \text{ Hz}$$

$$\omega_n = 2.37(20) = 47.3 \text{ Hz}$$

EXAMPLE 15.2. Mass–elastic data from Example 15.1 apply to the system in Table 15.3*c*. Find the natural frequencies for the modes shown.

Solution. For a concentrated mass in Table 15.3*c*,

$$\omega_1 = 1.97\sqrt{\frac{1.29(10)^6(386)}{38(26)^3}} = 53.8 \text{ or } 8.56 \text{ Hz}$$

Including moment of inertia effects, with $r^2 = 32$,

$$A = 2.07\frac{32}{(26)^2} = 0.980$$

$$B = 1 + 3.42\left[\frac{32}{(26)^2}\right] = 1.162$$

$$C_n = 0.098\mathrm{g}_n^4 - 1.162\mathrm{g}_n^2 + 1 = 0$$

$$\omega_{n1} = (0.967)(8.56) = 8.27 \text{ Hz}$$

$$\omega_{n2} = (3.30)(8.56) = 28.3 \text{ Hz}$$

EXAMPLE 15.3. A slender circular cantilever with a central angle of 60° supports a concentrated end mass. Find the angular directions of the planar modes.

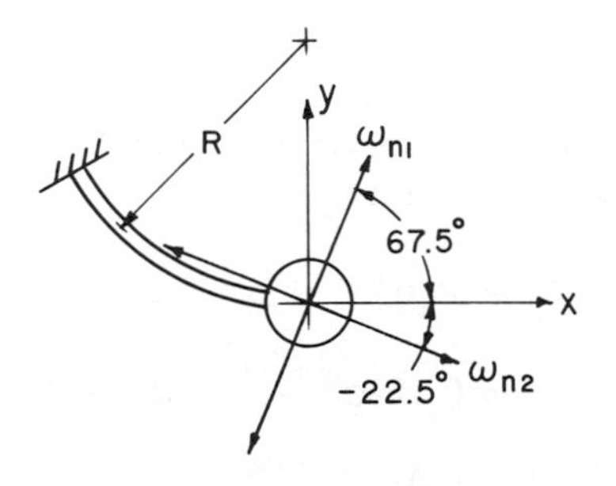

Solution. Determining the directions of the principal axes from Table 14.1 requires the calculation of ψ_1 and ψ_2. Elastic data in Table 15.4 provide the necessary α terms as a function of $\phi = 60°$.

$$\frac{\alpha_x}{R_3/EI} = 0.055 \qquad \tan 2\theta_p = \frac{2(2.27)}{1 - 5.58} = -0.991$$

$$\frac{\alpha_{xy}}{R^3/EI} = 0.125 \qquad \theta_p = 67.5°$$

$$\frac{\alpha_y}{R^3/EI} = 0.307 \qquad \theta_q = 157.5°$$

$$\psi_1 = \frac{0.125}{0.055} = 2.27 \qquad \psi_2 = \frac{0.307}{0.055} = 5.58$$

EXAMPLE 15.4. The semicircular bar is round and steel, and supports an end load with vertical excitation. Considering the mass concentrated and neglecting the mass of the bar, find:

(a) The planar natural frequencies.
(b) The forced vertical amplitude.
(c) The forcing frequency for zero vertical amplitude.

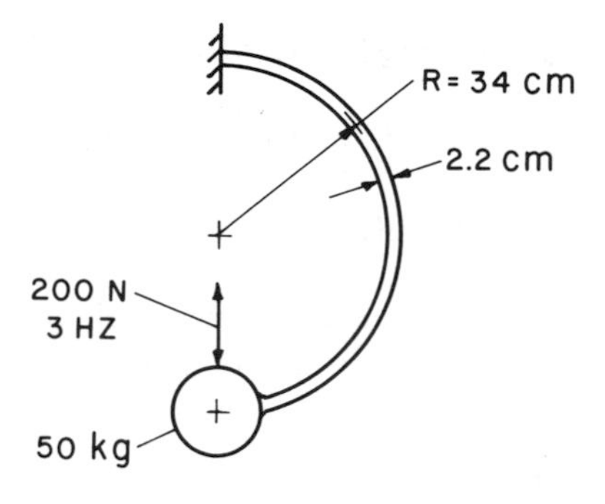

Solution

(a) From Table 15.6*a* with $\phi = \pi$, terms in the frequency cubic involving r_z will vanish, leaving the quadratic

$$C_b = A_3 g_n^4 - A_5 g_n^2 + 1 = 0$$

$$= 0.153 g_n^4 - 1.333 g_n^2 + 1 = 0$$

$$g_{n1} = 0.91$$

$$g_{n2} = 2.81$$

For the beam section, using cm units, and Table 15.5,

$$EI = 20(10)^6 \left[\frac{\pi(2.2)^4}{64}\right] = 23(10)^6 \text{ N} \cdot \text{cm}^2$$

$$\alpha_x = 4.712 \frac{R^3}{EI} = 0.0081 \text{ cm/N}$$

The reference frequency is

$$\omega_1 = \sqrt{\frac{100}{(0.0081)(50)}} = 15.8 \text{ r/s or } 2.50 \text{ Hz}$$

$$\omega_{n1} = \mathsf{g}_{n1}\omega_1 = 2.28 \text{ Hz}$$

$$\omega_{n2} = \mathsf{g}_{n2}\omega_1 = 7.0 \text{ Hz}$$

(b) Response equations are in Table 14.5b.

$$\left(\frac{a_y}{P_y \alpha_x}\right) = \left(\frac{-B_2 \mathsf{g}^2 + \psi_2}{C_b}\right) \qquad \mathsf{g} = \left(\frac{3}{2.5}\right) = 1.2$$

From Table 15.6*b*, $B_2 = 0.153$. From Table 15.5*b*, $\psi_2 = 0.333$.

$$C_b = 0.153(1.2)^4 - 1.333(1.2)^2 + 1 = -0.603$$

$$\mathsf{a}_y = (200)(0.0081)(-0.187) = -0.30 \text{ cm}$$

(c) The numerator is zero if $B_2\mathsf{g}^2 = \psi_2$, $g = 1.48$, and $\omega = 3.7$ Hz.

EXAMPLE 15.5. Using Example 15.4, calculate the out-of-plane or case *c* natural frequencies.

Solution. With no radius of gyration, $r_y = r_x = 0$, and there is only one uncoupled transverse frequency. The cubic in Table 14.5 reduces to

$$C_c = -\mathsf{g}_n^2 + 1 = 0$$

$$\mathsf{g}_n = 1 = \omega_n \sqrt{\alpha_z M}$$

For α_z we take $C = 7.46$ from Table 15.8.

$$\omega_n = 10\sqrt{\frac{EI}{7.46R^3M}} = 12.5 \text{ r/s or } 2.0 \text{ Hz}$$

EXAMPLE 15.6. The bar of the previous two examples now carries an extended rigid mass, as shown, with the centroid in the plane of the ring. Find the three static coordinate deflections due to the weight in the vertical position shown.

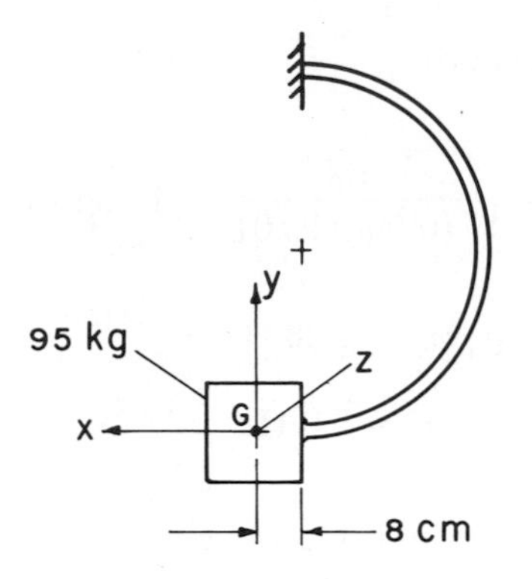

Solution. For case b, Table 15.9, $c = 0$ and $b = 8$ cm. Referring to Table 14.4b, for a vertical load F_y we require the factors α'_y, α'_{xy}, and β'_{zy}. These in turn require α_{xy}, β_{yz}, α_y, and γ_z from Table 15.5a.

$$\alpha_{xy} = 2\frac{R^3}{EI} = 3400(10)^{-6} \text{ cm/N}$$

$$\beta_{yz} = 2\frac{R^2}{EI} = 101(10)^{-6} \text{ N}^{-1}$$

$$\alpha_y = 1.571\frac{R^3}{EI} = 2700(10)^{-6} \text{ cm/N}$$

$$\gamma_z = 3.142\frac{R}{EI} = 4.64 \text{ (N} \cdot \text{cm)}^{-1}$$

Converting from the end of the bar to the centroid,

$$\alpha'_{xy} = \alpha_{xy} + 8\beta_{yz} = 4200(10)^{-6} \text{ cm/N}$$

$$\alpha'_y = \alpha_y + 2(8)\beta_{yz} + (8)^2\gamma_z = 4600(10)^{-6} \text{ cm/N}$$

$$\beta'_{zy} = \beta'_{yz} = \beta_{yz} + 8\gamma_z = 138(10)^6 \text{ N}^{-1}$$

Final deflections are due to the weight, and

$$Mg_c = 95(9.80) = 931 \text{ N}$$

$$X_y = \alpha'_{xy}F_y = 3.9 \text{ cm to the right}$$

$$Y_y = \alpha'_y F_y = 4.3 \text{ cm down}$$

$$\theta_y = \beta'_{zy}F_y = 0.13 \text{ r or } 7.4° \text{ counterclockwise}$$

Sign conventions are affected by the downward weight, which is negative, and by the reflection of the bar about the vertical diameter.

16 EXPERIMENTAL METHODS

The preceding chapters are purely analytical and idealized with respect to mass, spring, and dashpot characteristics. Systems are defined independent of the usually complex real systems from which they would normally be derived. Obviously, meaningful analytical results hinge upon meaningful modeling of the actual system, whether in physical existence or as a design concept. In the former case we can apply experimental techniques in the laboratory or in the field to supplement and verify analysis. Analysis is important, however, in visualizing the nature of a vibratory problem, which can often be solved deductively without exact quantitative information.

Experimentally we can determine many system parameters including natural frequencies, mode shapes, stiffnesses, masses, damping, excitation, and stress, and these results are useful to refine analytical procedures. Field testing is often required to ensure that operational vibratory conditions do not violate specified limits of amplitude or stress. Many of the experimental procedures are directly applicable to fatigue testing, as it involves exciters, transducers, and control of variables.

There are many commercially available test devices and systems to excite and measure vibratory and other dynamic phenomena. Some are extremely sophisticated with respect to excitational programs and the processing of multiple signals obtained; however, here, as in previous chapters, we approach the subject on a simple conceptual basis.

Data are interpreted with respect to steady-state and decay behavior as already developed, particularly in Chapters 2 and 3. Exciters are treated in terms of fundamental types and function. Transducers are discussed as elementary vibratory systems and on the vibratory basis for the signals generated.

16.1. DETERMINATION OF STIFFNESS STATICALLY

The simple static load-deflection test is often significant in determining the stiffness of an elastic structure. Load can be applied along a geometric axis or in any desired three-dimensional direction. Angular loads, torque or moment, can also be applied but may require a pair of equal and opposite forces to provide a couple. Loads can be applied and measured by testing machines, scales, weights, or load cells.

It is desirable to apply loads and measure deflections incrementally so the data when plotted indicate any zero error or nonlinearity. Slope, or dF/dx, is usually the most important result.

Dial gages are a convenient and accurate means of measuring deflection, although there are many options. Sometimes we might wish to measure a relative deflection, but in most cases it is important that the indicator be mounted rigidly so it will sense the true absolute deflection.

16.2. DETERMINATION OF MASS PARAMETERS

Although very basic, weighing a part is sometimes necessary, particularly when involving complex geometry difficult to calculate. In the case of models a scale factor is used with respect to size or density.

Center of gravity is often important, For an actual part we simply suspend it in a horizontal plane by a wire or cord at a balance point located by trial and error. Also, balancing on a horizontal fulcrum in two different perpendicular successive directions will establish the coordinates of the centroid in that plane. This procedure can be repeated for various directional planes of interest.

When mass moment of inertia is to be determined for an element or assembly about a transverse or polar axis, pendulum natural frequency is the usual factor measured. The body can be suspended and oscillated as a compound pendulum (see Chapter 8 and Examples 8.1 and 8.2). Radius of gyration follows from Equation 8.2, which can be reduced to

$$r = \sqrt{l(9.78T^2 - l)} \tag{16.1}$$

where r = radius of gyration about centroidal axis

l = distance from fulcrum to center of gravity

T = period of the pendulum about the fulcrum (seconds per cycle)

All terms apply to the plane of oscillation and to axes perpendicular to this plane. Small amplitudes should be used.

16.3. NATURAL FREQUENCIES

This basic characteristic of vibratory systems can be observed simply by applying a disturbance and measuring the transient decay. For normal damping there will be sufficient cycles to record as a function of time to determine the frequency with good accuracy. In a multimass system, all modes tend to be excited by an impact, but the first mode tends to persist during the decay.

Resonances are identified more systematically by applying some type of exciter to induce steady state with an adjustable exciting frequency. Then resonant peaks can often be located by visual observation, otherwise by transducers. In steady state, stroboscopic lighting near forcing frequency helps to display the modes if the amplitudes are sufficient.

16.4. EQUIVALENT VISCOUS DAMPING

There are two basic methods of studying system damping. The most obvious is to interpret the rate of decay (Figure 3.7). By recording the transient, or by memory display on an oscilloscope, successive amplitude ratios are an index of equivalent viscous damping. In more convenient form, Equation 3.22 for low damping reduces to

$$\zeta \approx \frac{-\log_e(x_{n+1}/x_n)}{2\pi} \tag{16.2}$$

or we read ζ from Figure 3.7.

With an exciter having controlled constant force input over the frequency range, we can measure single-degree amplitudes similar to Figure 2.12 with $N = 1$. With excitation by a rotating unbalance the curve is that of Figure 2.12 with $N = \mathsf{g}^2$. In Table 2.1a and b, at resonance, $\mathsf{g} = 1$ and the MF factors are

$$\frac{\mathsf{a}_1}{(P_1/K_1)} = \frac{\mathsf{a}_1}{(me/M_1)} = \left(\frac{1}{jB}\right) = \left(\frac{1}{j2\zeta}\right) \tag{16.3}$$

Thus, knowing the ratio of output amplitude to input function we can solve for ζ from the resonant factor using absolute values,

$$\zeta = \frac{1}{2}\left|\frac{P_1/K_1}{\mathsf{a}_1}\right| = \frac{1}{2}\left|\frac{me}{M_1\mathsf{a}_1}\right| \tag{16.4}$$

For the multimode systems we can analyze using modal springs, masses, and damping, as in Chapter 5.

From Equation 16.3 resonance is further identified by a 90° phase difference between a_1 and P_1 as seen by the j operator. If the phase is measured

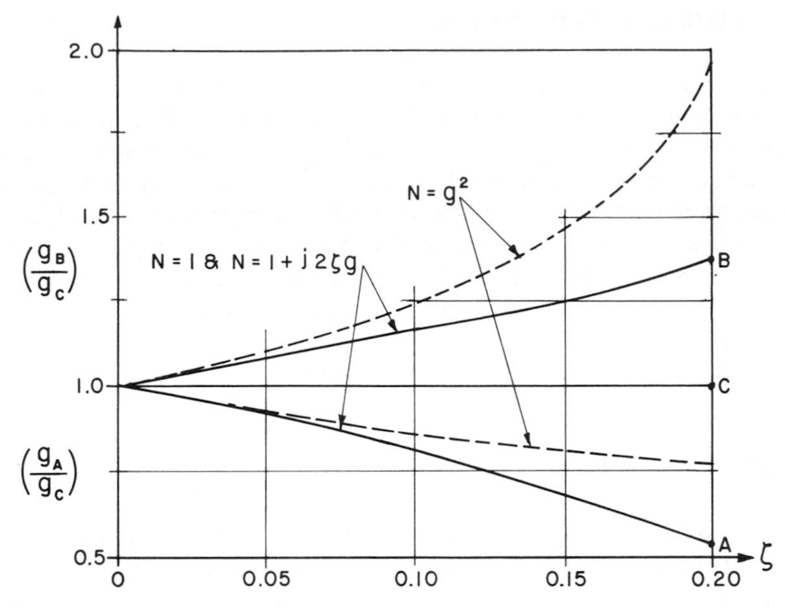

FIGURE 16.1. Viscous damping ratio is determined from response peak data (Figure 2.12) at the half-peak ordinates. g_A and g_B are the corresponding frequency ratio abscissas below and above $g_C = 1$ at resonance.

accurately it is a useful index of the resonant frequency, and at this condition Equation 16.4 applies exactly as $g = 1$.

With centrifugal forcing, $N = g^2$ in Table 2.1*b*, responses have different characteristics, shown dashed in Figure 16.1. The other possibility is displacement excitation, $N = (1 + j2\zeta g)$ in Table 2.1*c*. Although not exact, this latter behavior is very close to the $N = 1$ case and the average of these two cases is plotted as a solid curve in Figure 16.1. This figure then incorporates results for any of three types of excitation.

16.5. DAMPING DETERMINATION FROM PEAK GEOMETRY

In Figure 16.2 we have a family of response curves with various damping ratios and constant forcing, $N = 1$ in Table 2.1*a*. These are all rectified to a peak of unity and a peak frequency of unity, though peak heights and frequencies in terms of actual magnification factor and the g ratio are all different. The common rectified peak is *C*.

At half-height a horizontal line intersects particular frequencies relative to the peak frequency g_C. Since the *AB* frequency range increases with damping, it is a quantitative indication of damping. For instance, if $\zeta = 0.20$, at *A* the forcing frequency is 51% of the peak frequency. Above resonance $g_B = 1.31$.

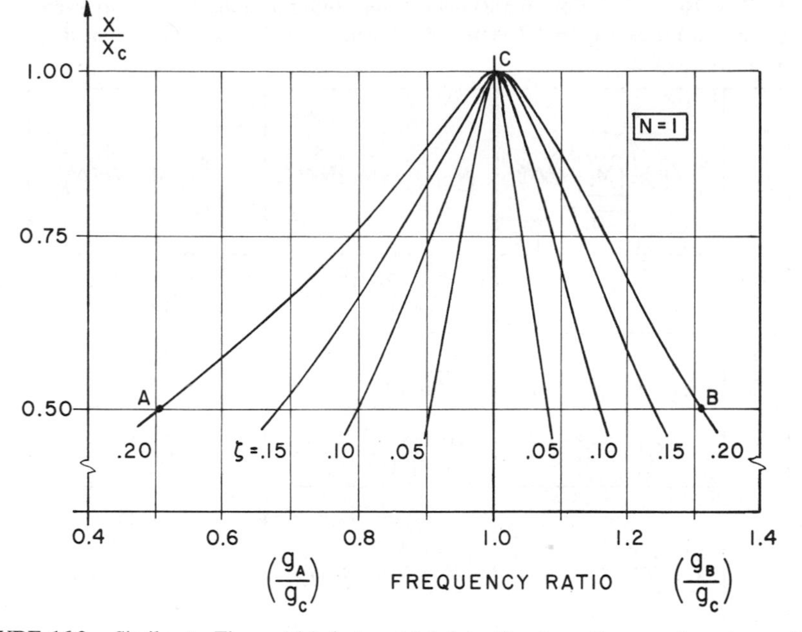

FIGURE 16.2. Similar to Figure 16.1, but restricted to $N = 1$, or the mass-forced case (Table 2.1*a*).

If in an actual response diagram the coordinates are measured at half maximum peak to be

$$\omega_A = 24.5 \text{ Hz} \qquad \omega_C = 34 \text{ Hz} \qquad \omega_B = 41 \text{ Hz}$$

the ratios are

$$\left(\frac{g_A}{g_C}\right) = 0.72 \qquad \left(\frac{g_B}{g_C}\right) = 1.21$$

Locating these ratios on the horizontal line in Figure 16.2, ζ interpolates to 0.13. If our experimental curve is not this well behaved because of erroneous data or nonlinear damping, ζ_A may not agree with ζ_B. An average can be taken.

16.6. EXCITER FUNDAMENTALS

Sinusoidal forcing of a single-degree main system can be achieved by the three basic methods in Table 2.1. These approaches are shown in Table 16.1, with the corresponding devices (*a*), (*c*), and (*d*). If the exciter does not react directly to ground but instead reacts against the exciter mass, we have somewhat different behavior (Table 16.1*b* and *e*).

TABLE 16.1. Relation of various exciter configurations to the force–displacement ratios required. Relative displacement within the exciter unit is e, or a stroke of $2e$.

	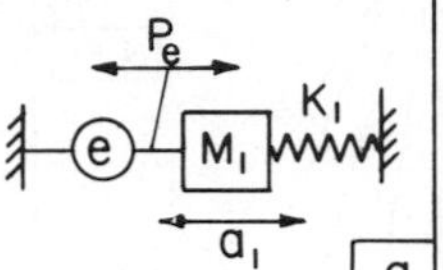 P_e, e, M_1, K_1, a_1 — a	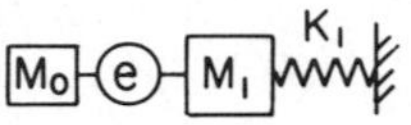 M_0, e, M_1, K_1 — b	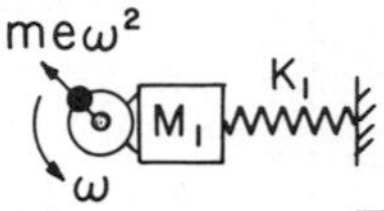 $me\omega^2$, ω, M_1, K_1 — c
$\frac{a_1}{e}$	1	$\frac{m_0 g^2}{1-g^2(1+m_0)}$	$\left(\frac{m}{M_1}\right)\left(\frac{g^2}{1-g^2}\right)$
$\frac{a_1}{P_e}$	$\left(\frac{1}{K_1}\right)\left(\frac{1}{1-g^2}\right)$		
$\frac{P_e}{e}$	$K_1(1-g^2)$	$\frac{K_1 m_0 g^2(1-g^2)}{1-g^2(1+m_0)}$	$K_1\left(\frac{m}{M_1}\right)g^2$
	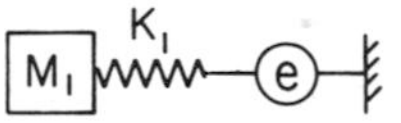 M_1, K_1, e — d	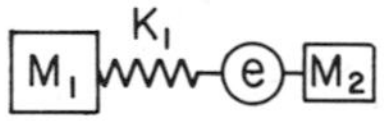 M_1, K_1, e, M_2 — e	$g = \frac{\omega}{\omega_1}$ $\omega_1 = \sqrt{K_1/M_1}$ $m_0 = \frac{M_0}{M_1}$ $m_2 = \frac{M_2}{M_1}$ K_1, M_1 = SYSTEM UNDER TEST
$\frac{a_1}{e}$	$\frac{1}{1-g^2}$	$\frac{m_2}{1+m_2(1-g^2)}$	
$\frac{a_1}{P_e}$	$\frac{1}{K_1 g^2}$		
$\frac{P_e}{e}$	$\frac{K_1 g^2}{1-g^2}$	$\frac{K_1 m_2 g^2}{1+m_2(1-g^2)}$	

The excitation can be electromagnetic or hydraulic with adjustable stroke and frequency. A positive mechanism (Figure 2.10) has adjustable frequency by means of speed, but the amplitude is usually fixed during a given test. The rotating unbalance or inertia exciter (Table 16.1c) is also not adjustable during test, but provides a known force at a given speed.

In Table 16.1 three characteristic ratios are shown as a function of the forced frequency ratio g, with (a_1/e) the output, or usable amplitude per unit exciter amplitude. The maximum exciter amplitude is one-half the design stroke or is the radial eccentricity.

Another parameter is (a_1/P_e), or the test amplitude per unit driving force amplitude, indicating the force required of the exciter to oscillate the mass M_1 at a given amplitude as a function of g.

Although the idealized single-degree main system is much simpler than most that are excited, we can reduce multidegree systems conceptually and analytically to simple modal equivalents, as in Section 4.6.

16.7. THE FORCE-DISPLACEMENT GENERATOR

A basic driving function generator is the electromagnetic shaker (Figure 12.10). The body is essentially a permanent magnet within which a coil is attached to the output stem. The coil carries an alternating current and reacts with the magnet to oscillate the stem, similar in principle to an audio speaker. Frequency and amplitude of the driving stem are easily varied by control of the oscillator circuit to the coil, and the excitation can be adjusted manually or by a programmed schedule.

The driving unit can similarly be a cylinder supplied with cyclically controled hydraulic fluid as the source or actuation. Because of the valving problem it is difficult to design for high frequencies and to achieve a completely sinusoidal output, but the hydraulic system tends to be capable of high forces for studying large structures.

Combining into a parameter applying to the exciter only (Table 16.1), we have (P_e/e), showing the variation with frequency of the vibratory load on the exciter per unit exciter amplitude.

Obviously we can attenuate the electromagnetic shaker to reduce e if necessary, and to thereby also reduce both the exciter load and the output amplitude a_1, but the three tabulated ratios are design parameters applicable to a given exciter and define the required capability.

In Table 16.1*a*, with a mechanical drive, resonance could not be observed as an amplitude peak of a_1, since this is constant; however, load sensing would indicate a minimum force P_e at resonance.

With the suspended exciter (Table 16.1*b*), we have a reduced system natural frequency which requires correction in the results, but the vibratory behavior is typically single degree. In Table 16.1*c* we similarly affect the natural frequency if the mass of the exciter attached to M_1 is appreciable.

The cases shown in Table 16.1*d* and *e* will exhibit resonant behavior, but with reduced frequency in (*e*). If a_1 approaches resonance, so must the exciter force, P_e. The several combinations of variables are not plotted, but this is easily done if necessary for a particular application.

16.8. UNBALANCE EXCITERS

In Table 2.1*b* a rotating unbalance excites the single-degree system with M_1 constrained to move horizontally. The horizontal component of the centrifugal force provides the harmonically varying excitation. Rotative speed determines the frequency, and the force increases as the square of the speed. The unbalance can readily be made adjustable, but normally this change is only made with the rotor at rest.

Although the unbalance must be mounted directly on the test system and provided with some type of flexible power drive to the shaft, we have the advantage of simplicity, a known exciting force, and a source for a phase signal as a point on the rotating disk passes a stationary transducer.

With a rotating unbalance (Figure 16.3*a*), we have available polar or whirl-type excitation, enabling us to excite simultaneously in two directions. Thus a vibratory motion can be detected *in any compass direction in a plane*. For instance, in Figure 16.4*a* the exciter sweeps the *xy* plane, and will detect any resonances that have linear modal amplitudes in the bending plane of the structure.

If the rotor is turned to spin about the *x* axis (Figure 16.4*b*), we search the *yz* plane. A further possibility is to align the spin axis with the *y* axis, to sweep

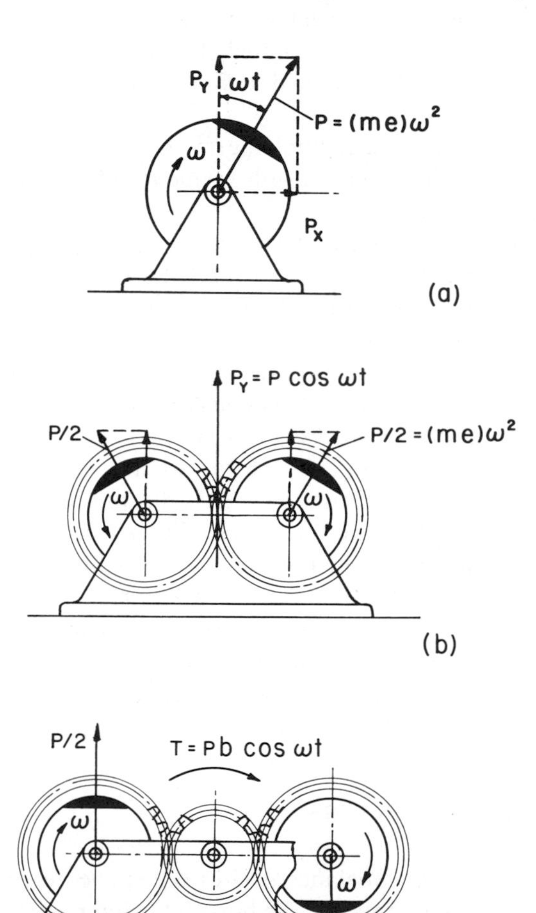

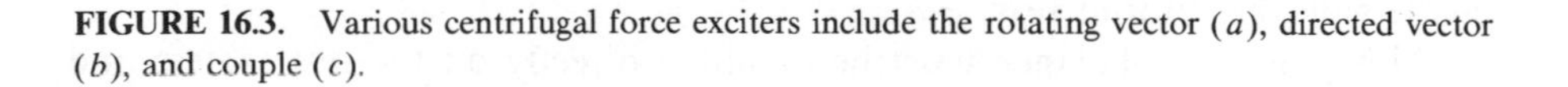

FIGURE 16.3. Various centrifugal force exciters include the rotating vector (*a*), directed vector (*b*), and couple (*c*).

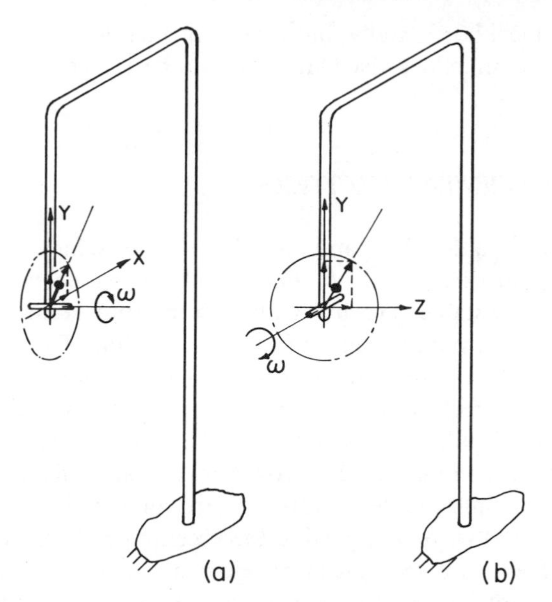

FIGURE 16.4. A single rotating unbalance exciter can search the xy plane (a), or the yz plane (b). With rotation about the y axis, the xz plane is searched for resonance.

the xz plane, which is another indication of the versatility of the rotating single disk in the study of two- or three-dimensional systems.

16.9. GEARED UNBALANCE EXCITERS

By combining two equal unbalances that rotate in opposite directions we develop a harmonically varying force perpendicular to the centerline of two gears as shown in Figure 16.3*b*. For the phase relations shown, the horizontal components of the centrifugal forces cancel, but the vertical components add. Then we can select an exact space direction for applying an excitation in three dimensions by proper positioning of the unit. The magnitude is also predictable.

This device is used on certain engines as a *secondary balancer* or *Lanchester balancer* with the resultant sinusoidal force at the mesh used to counteract undesirable shaking forces. In these applications the forces to be canceled are caused by the inertia of the reciprocating parts which also increase as the square of the speed.

A further alternative provides a sinusoidally varying pure couple for torsional excitation (Figure 16.3*c*). Using an idler between the two unbalanced gears, we rotate the gears and unbalances in the same direction. In the position shown the net torque is maximum and clockwise. After 90° of rotation the

($P/2$) forces cancel horizontally, but after a further 90° the resultant torque is maximum and counterclockwise. Any gear can be driven.

16.10. DISPLACEMENT EXCITERS

Base excitation (Tables 2.1*c* and 16.1*d*) can be achieved by means of a reciprocating platform, vertical or horizontal, for which both frequency and amplitude are adjustable, Driving with a crank and connecting rod (Figure 16.5) provides an amplitude equal to the crank radius. This approach subjects the entire test system to directional inertia-coupled loading. There are obvious limitations with respect to frequency and size, but at low frequencies relatively large defined amplitudes can be obtained. The connecting rod is usually long with respect to the crank radius, providing a practically undistorted simple harmonic motion. Shake tables of this type are commercially available.

There are several ways of adjusting the eccentric radius, with one shown in Figure 16.6. There are two eccentric elements, and by rotating the inner element with respect to the outer one we can vary the radius continuously from zero to $2R$. After setting, the two are clamped for operation.

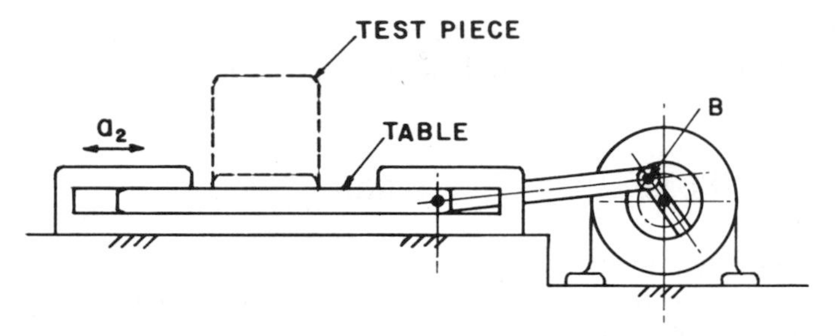

FIGURE 16.5. A slider–crank mechanism provides positive sinusoidal base excitation for a specimen mounted on the table.

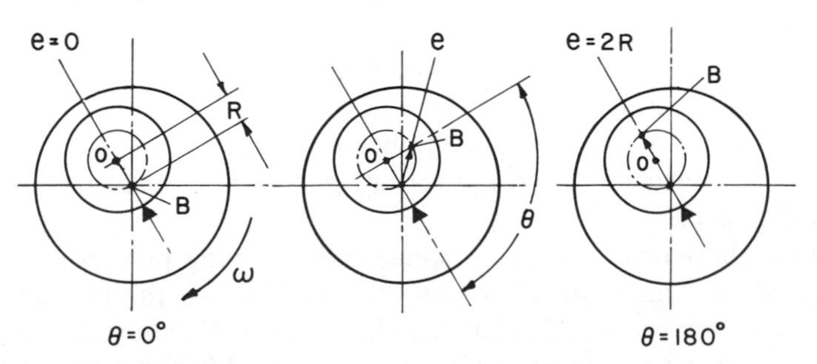

FIGURE 16.6. The crank radius in Figure 16.5 or Figure 16.7 is infinitely adjustable to zero if a double eccentric head is used.

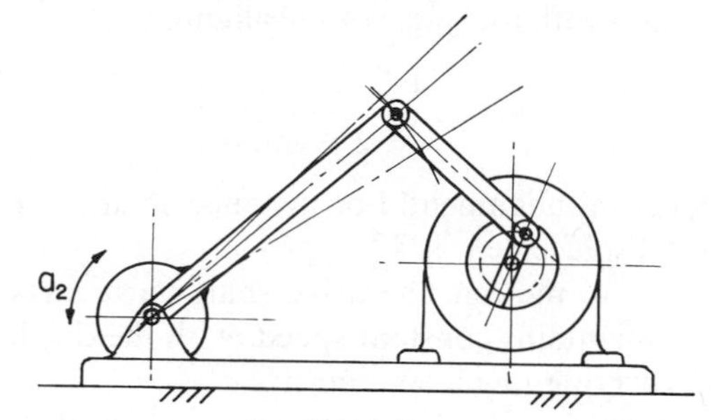

FIGURE 16.7. A four-bar linkage with adjustable crank produces torsional displacement excitation.

In the nonrotative case an excitational angular displacement can be obtained by applying the connecting rod to a crank, as shown in Figure 16.7. If the pivot bearing at 2 is integral with the exciter, the output shaft provides the desired displacement without any transverse forces transmitted by the mechanism.

16.11. TORSIONAL EXCITATION DURING ROTATION

It is difficult to excite a rotating system, but torsional effects can be of interest. The universal, or Hooke's, joint is a mechanism that enables us to accomplish this. Normally, universal joints are used in pairs. With proper angular orientation of one to the other, speed variations developed by one are canceled by the other. The result is usually a constant speed output.

By using only one joint (Figure 16.8), there is a vibratory relative displacement generated between the shafts which is predominantly double rotative

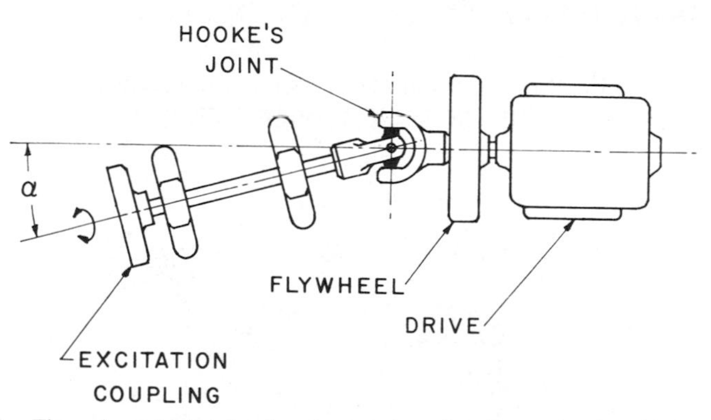

FIGURE 16.8. The universal joint develops known second-order torsional excitation in a rotating specimen.

frequency and increases with the angle of misalignment.

$$a_{12} = \frac{1}{2}\left(\frac{\sin^2\alpha}{2 - \sin^2\alpha}\right) \tag{16.5}$$

where α is the angular misalignment. For instance, if this is 15°, the amplitude of the second-order displacement is 1°.

Since this is a relative motion, the drive shaft must carry high polar mass inertia, so this shaft maintains constant speed, with the displacement imparted to the driven shaft carrying much less inertia.

Because the frequency is second order we cannot vary the frequency and the speed independently, and the maximum available frequency will be limited to twice the maximum possible speed.

This arrangement is particularly useful for the calibration of torsional vibration transducers, in which case the device is relatively small.

16.12. DISPLACEMENT TRANSDUCERS

Vibratory motion can be measured directly as a displacement if a ground point is available. In Figure 16.9 the motion a_2 produces stresses in the spring elements which can be sensed by means of conventional resistance strain gages, and static calibration is possible. In these examples, stress, strain, force, and bending moment are all proportional to the displacement. The spring must have low stiffness in order not to have a significant influence on the test system, and it should have a reasonable stress; however, the spring must not be so flexible as to have vibratory responses of its own at the frequencies involved.

There are a number of other methods for detecting displacement directly, including inductance bridges, air gap sensing, and reflected microwaves.

16.13. SEISMIC DISPLACEMENT TRANSDUCERS

The principle of a flexibly mounted mass, as used in earthquake seismology, is employed on a small scale in many commercially manufactured vibration

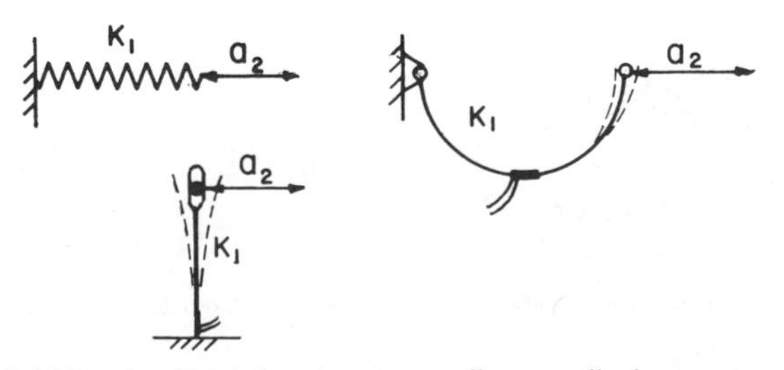

FIGURE 16.9. A grounded spring senses vibratory displacement as stress.

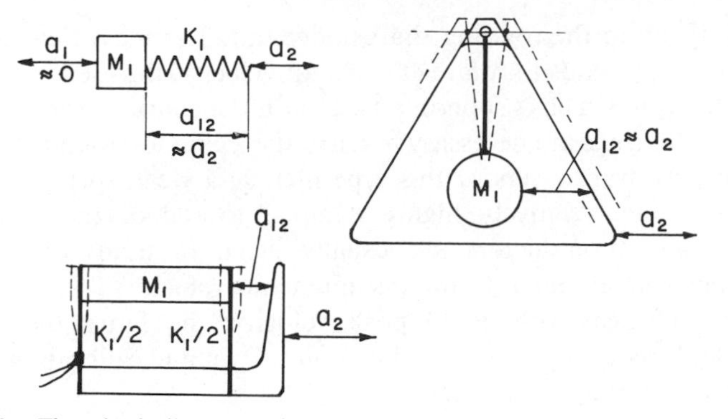

FIGURE 16.10. The seismically mounted mass M_1 simulates ground in order that relative motion a_{12} approximates the measured motion a_2.

transducers. In Figure 16.10 a low natural frequency is provided relative to the frequency to be measured a_2. From Table 2.1*c* we see that the ratio a_{12}/a_2 then approaches unity with considerable accuracy at forced frequency ratios of $g = 3$ or greater (Figure 2.9). We have in effect used the inertia of the mass M_1 to establish a reference ground point from which to measure motion.

Relative motion between 1 and 2 can be used to generate an electrical signal using a coil and a permanent magnetic field. The magnet usually is part of the seismic mass, with a small coil rigidly driven by the input a_2 (Figure 16.10). If the coil cuts the magnetic flux path with sinusoidal amplitude, a voltage is generated proportional to the velocity of the motion. Since the sinusoidal velocity is in turn proportional to the product of the frequency and amplitude, the voltage is proportional to the product, or ωa_2. This device is therefore often termed a *velocity* pickup.

This concept is illustrated in Figure 16.11 for the torsional velocity transducer, or torsiograph. The small flywheel is the seismic element with the coil

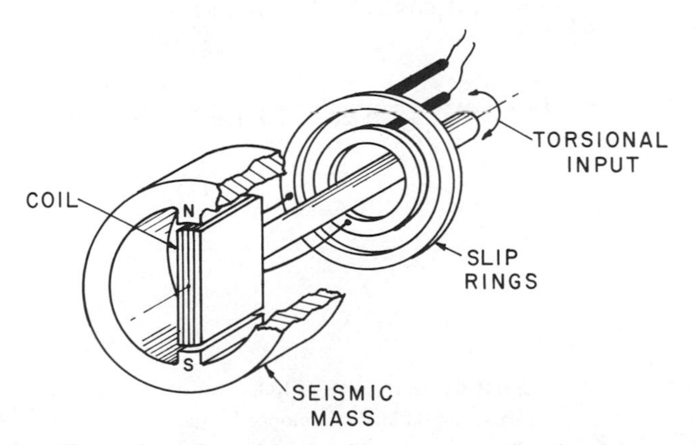

FIGURE 16.11. The torsiograph used in rotating systems typifies the electromagnetic pickup schematic for a velocity transducer.

connected rigidly to the rotating shaft under test. The mass that incorporates the magnetic poles tends to rotate at constant velocity, thus sensing the relative motion of the coil, which is superimposed upon the constant rotational motion of the shaft. Slip rings are necessary to carry the signal to ground.

Most velocity transducers of this type include a weak spring for centering the coil. They are relatively highly damped to reduce the persistence of transients. Calibration factors are usually given in terms of the ratio of instantaneous voltage signal to the instantaneous velocity. They are also given as the ratio of peak voltage to peak velocity, the latter the product of amplitude and frequency in appropriate units. Dynamic calibration is required over the specified frequency range.

16.14. ACCELEROMETERS

Subjecting an essentially rigid auxiliary system to a vibratory acceleration field causes inertia loads in the spring that supports the mass (Figure 16.12). With K_1 essentially stiff, a sinusoidal base motion a_2 creates an inertia force at M_1 of $M_1(\omega^2\mathsf{a}_1) \approx M_1(\omega^2\mathsf{a}_2)$, where the term in parentheses represents the maximum sinusoidal acceleration. In the general transient case, $M_1\ddot{x}_1 \approx M_1\ddot{x}_2$.

The inertia load at M_1 creates a bending stress that can be sensed, shown in the figure by strain gages. Actual sensing in the accelerometer is often provided by piezoelectric crystal elements which generate a small voltage when stressed.

This scheme is the inverse of the seismic. Here M_1 experiences the motion to be measured. Previously M_1 was intended to remain stationary.

With a complex repetitive displacement function, or Fourier expression,

$$f(\omega t) = A_1 \sin \omega t + A_2 \sin 2\omega t + \cdots \tag{16.6a}$$

$$\frac{df(\omega t)}{dt} = \omega(A_1 \cos \omega t + 2A_2 \cos 2\omega t + \cdots) \tag{16.6b}$$

$$\frac{d^2f(\omega t)}{dt^2} = -\omega^2(A_1 \sin \omega t + 4A_2 \sin 2\omega t + \cdots) \tag{16.6c}$$

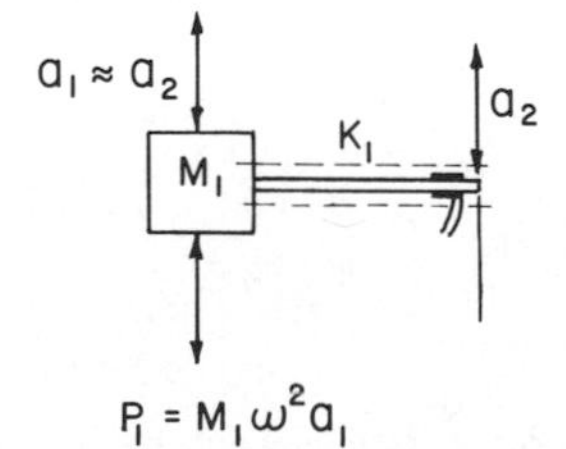

FIGURE 16.12. Accelerometers use relatively stiff suspensions to sense inertia loads developed by the instrument mass M_1. Input is a_2.

In Equation 16.6*a* with displacement sensing, the signal indicates components in direct proportion to actual amplitude. In Equation 16.6*b*, with velocity sensing, higher harmonics are amplified by the harmonic number. With accelerometers (Equation 16.6*c*) there is amplification as the square of the harmonic number, or 1, 4, 9, 16,.... When actual acceleration is important, we can record the signal in Equation 16.6*c* directly, but true displacements from an accelerometer can be obtained by a double-integrating circuit. Similarly, a velocity signal can be single-integrated.

16.15. PHASE INDICATION

It is often desirable to generate a signal to indicate phase of a vibratory signal with respect to a rotating shaft. For example, in an engine test a contactor pulse that is recorded with other vibratory data gives an accurate indication of speed and expedites order identification, or cycles per revolution. With dynamic balancing (Chapters 8 and 13), phase indications are essential to the angular orientation of corrections when the rotor is stopped.

As indicated, any short electrical pulse or pip introduced during rotation and superimposed on a vibratory signal serves this purpose. There are also stroboscopic methods in which a strobe light is triggered by the vibratory signal, enabling a direct observation of phase in a rotating part.

Phase relative to a shaft can also be provided by a sinusoidal signal generated by a two-pole tachometer alternator. If the pole element is movable, the reference signal position can be shifted if desired.

Phase data can also be provided by simultaneous signals. In Table 2.1*a*, a force signal from P_1 and an amplitude signal from a_1 will indicate resonance when P_1 leads a_1 by a time phase angle of 90°.

16.16. OBSERVATION OF SERVICE PATTERNS

Data are often available from machines or mechanisms experiencing vibratory problems. In the worst scenario, fatigue failures occur; however, these serve to identify the nature of stress distribution in a mode or to an operating speed that corresponds to resonance. Similarly, wear patterns are often recorded on surfaces. If carefully interpreted these patterns can focus attention on the basic nature of a problem by means of frequency and directional information. This technique can be refined by an amplified recording of the wear surfaces.

Although sophisticated equipment is required in many cases, some of the simple observations indicated can be especially significant in defining or quantifying a problem.

16.17. REMARKS

Successful solution of vibration problems requires a blend of analytical and experimental information, often obtained on an iterative basis. When conflicts arise we must favor the latter, as proper testing yields the ultimate answers; however, analysis provides insights necessary for system visualization or modeling. In turn this allows us to select the most meaningful modes and frequencies, and to utilize test time and equipment effectively.

The most powerful single tool in vibratory sleuthing is frequency, usually the easiest parameter to compute and to measure.

Operational mechanisms consisting of several interacting components can obscure a problem. It is important to identify the one that is the source of a difficulty. As with crime, there are often many suspects but one criminal, and we should consider all components potentially guilty until a final judgment is proven. Otherwise we may prematurely try to solve a problem by blaming the wrong element or mechanism.

BIBLIOGRAPHY

1. Blevins, R. D., *Formulas for Natural Frequency and Mode Shape*, Van Nostrand-Reinhold, New York, 1979.
2. Crossley, F. R. E., *Dynamics in Machines*, Ronald, New York, 1954.
3. Den Hartog, J. P., *Mechanical Vibrations*, 4th ed., McGraw-Hill, New York, 1955.
4. Gorman, D. J., *Free Vibration Analysis of Beams and Shafts*, Wiley, New York, 1975.
5. Grover, G. K. and R. J. Harker, A Study of the Overhung Damped Dynamic Vibration Absorber with Tuning Optimization, *Proceedings of the 9th International MTDR Conference*, Pergamon Press, Elmsford, N.Y., 1968.
6. Harker, R. J., "Theory of the Centrifugally Tuned Vibration Absorber," *J. Aeronaut. Sci.*, **11** (3): 197–204, July 1944.
7. Harker, R. J., Vibratory Response of Rigid Rotors to Dynamic Unbalance, *Proceedings of the 1st U.S. Congress of Applied Mechanics*, Edwards Brothers, Ann Arbor, Mich., 1951, pp. 119–124.
8. Harris, C. M. and C. E. Crede, *Shock and Vibration Handbook*, 2nd ed., McGraw-Hill, New York, 1976.
9. Hudson, G. E., "A Method of Estimating Equivalent Static Loads in Simple Elastic Structures," *Proc. Soc. Exp. Stress Anal.* **6** (2): 72–79, 1948.
10. Hundal, M. S. and R. J. Harker, "Balancing of Flexible Rotors Having Arbitrary Mass and Stiffness Distribution," *ASME J. Eng. Ind.*, **88** (2): 217–223, May 1966.
11. Ker Wilson, W., *Practical Solution of Torsional Vibration Problems*, 2nd ed., Wiley, New York, 1948.
12. Lewis, R. C. and K. Unholz, "A Simplified Method for the Design of Vibration-Isolating Suspensions," *Trans. ASME*, **69** (8), 813–820, November 1947.
13. O'Connor, B. E., "Viscous Torsional Vibration Damper," *SAE Trans.*, **1** (1): 87–97, January 1947.
14. Pitloun, R. *Vibrating Beams*, Veb Verlag für Bauwesen, Berlin, 1971.
15. Ricciardiello, A. M., "Finding Vibration Mode Shapes Quickly," *Machine Design*, October 26, 1978, pp. 113–114.
16. Roark, R. J. and W. C. Young, *Formulas for Stress and Strain*, 5th ed., McGraw-Hill, New York, 1975.
17. Seireg, A., *Mechanical System Analysis*, International Textbook, Scranton, Pa., 1969.
18. Seto, W. W., *Theory and Problems of Mechanical Vibrations*, Schaum's Outline Series, New York, 1964.

19. Thomson, W. T., *Theory of Vibration with Applications*, 2nd ed., Prentice-Hall, Englewood Cliffs, N.J., 1981.
20. Thureau, P. and D. Lecler, *An Introduction to the Principles of Vibrations of Linear Systems*, Halsted–Wiley, New York, 1981.
21. Timoshenko, S. *Vibration Problems in Engineering*, 3rd ed., Van Nostrand, New York, 1955.
22. Young, D. and R. P. Felgar, Jr., *Tables of Characteristic Functions Representing Normal Modes of Vibration of a Beam*, University of Texas Publication No. 4913, July 1949.

PROBLEMS

Chapter 1

1.1. A mass vibrates with simple harmonic motion at an amplitude of 0.30 cm and a frequency of 25 Hz. Find:

(a) The maximum vibratory velocity.

(b) The maximum vibratory acceleration.

(c) The circular frequency.

(d) The period of the vibration.

1.2. In Problem 1.1 when $t = 0$, $x = +0.22$ cm, and the velocity is negative. What is the elapsed time for the mass to reach maximum positive displacement?

1.3. Plot the displacement, velocity, and acceleration vectors in the complex plane for the initial conditions of Problem 1.2.

1.4. A disk vibrates as a torsional pendulum with an amplitude of 2° and a frequency of 18 Hz. Calculate:

(a) The maximum vibratory velocity.

(b) The maximum vibratory acceleration.

1.5. If $\mathsf{a}_1 = (4 + j2)$ and $\mathsf{a}_2 = (-3 + j8)$,

(a) By what phase angle does a_1 lead a_2?

(b) What is $(\mathsf{a}_1 - \mathsf{a}_2)$ as a complex number?

(c) What is the absolute ratio $\mathsf{a}_2/\mathsf{a}_1$?

(d) Determine the angle and radius of $(\mathsf{a}_1 + \mathsf{a}_2)$.

1.6. Considering the bar stiff and neglecting its mass, determine:

(a) The static deflection of the cylindrical disk.

(b) The torsional spring rate about 0.

(c) The equivalent concentrated mass at the spring so the natural frequency will be the same as for the actual system.

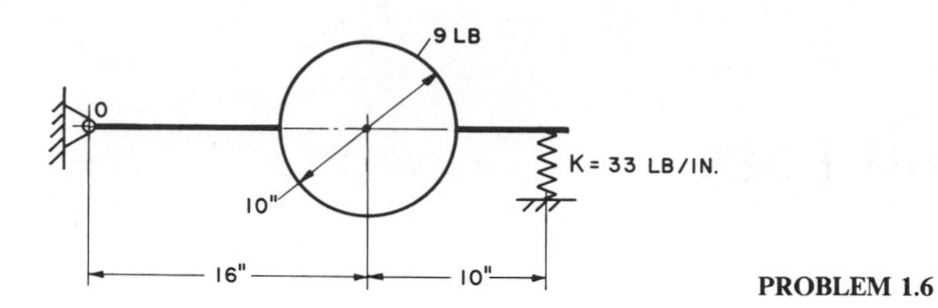

PROBLEM 1.6

1.7. The initial unloaded dimensions of the two helical springs are shown. If the springs are pulled together and hooked, find the axial tensile preload in each spring.

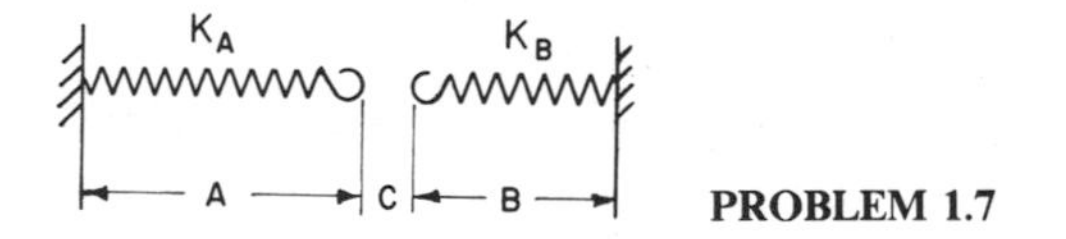

PROBLEM 1.7

1.8. During an oscillation with simple harmonic motion the instantaneous conditions are

$$x = 0.24 \text{ cm} \qquad \dot{x} = 145 \text{ cm/s}$$

Calculate the instantaneous magnitude of the external force F.

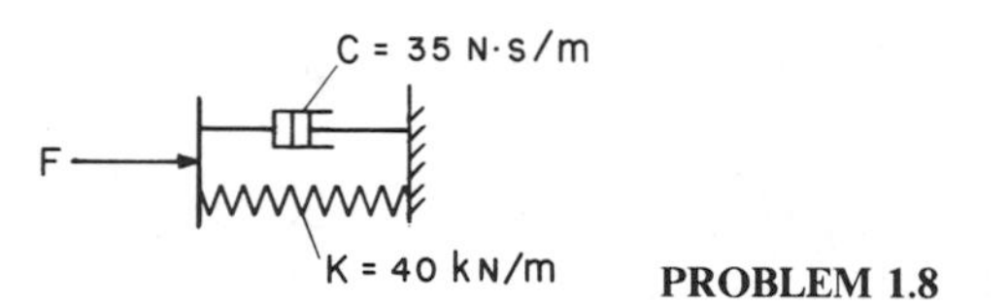

PROBLEM 1.8

Chapter 2

2.1. A turbine is directly connected to a generator with the armature assumed to have infinite rotational inertia with respect to the turbine. The turbine rotor weighs 8 lb with a radius of gyration of 3.60 in. If the connecting shaft is 1.50 in. in diameter and 30 in. long, calculate the natural frequency in torsion.

2.2. The steel shaft supported in self aligning bearings at each end, is 4.5 cm in diameter. Mass of the disk is 55 kg. Considering the mass concentrated, and neglecting the mass of the shaft, what is the natural frequency in bending?

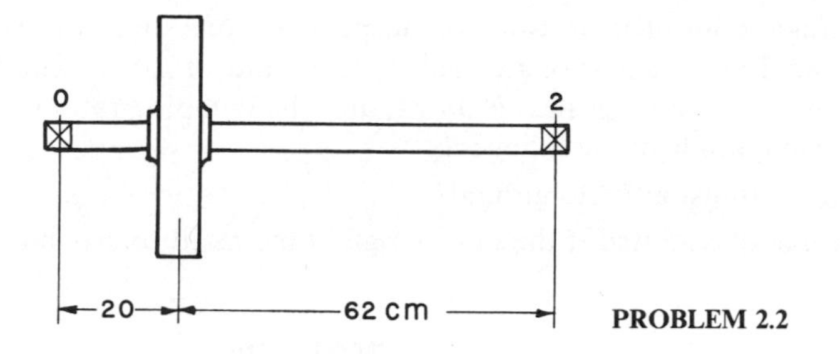

PROBLEM 2.2

2.3. In Problem 2.2 the polar radius of gyration of the disk is 16 cm. Find the torsional natural frequency if the shaft is:

(a) Clamped at 0 and free at 2.

(b) Clamped at 2 and free at 0.

(c) Clamped at both 0 and 2.

2.4. Find the natural frequency of the system neglecting beam and spring masses. All helical springs are equal.

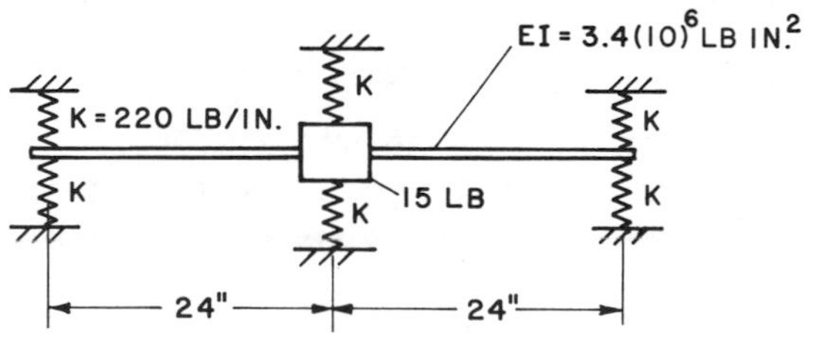

PROBLEM 2.4

2.5. A steel cantilever beam 36 in. long carries a concentrated end mass of 5 lb. Determine the minimum size of a square bar that will provide a natural frequency of at least 15 Hz:

(a) Neglecting the beam mass.

(b) Including an effective beam mass at the end of 0.25 of the total beam weight. (Density of steel is 0.29 lb/in.^{3})

2.6. A basic forced system (Table 2.1*a*), has the following specifications:

$$K_1 = 340 \text{ lb/in.} \qquad \zeta_1 = 0.12$$
$$M_1 = 18.5 \text{ lb} \qquad P_1 = 1.9 \text{ lb at } 10 \text{ Hz}$$

Find:

(a) Vibratory amplitude of the mass.

(b) Amplitude of the reaction force.

(c) Steady-state rate of energy dissipation.

2.7. The centrifugal pump has an isolation suspension consisting of a heavy sub-base and four equal springs, each with a rate of 10 kN/m. The weight of the entire sprung mass is 30 kg, and the damping ratio is 0.05. What is the amplitude of the following:

(a) The force transmitted to ground?

(b) The force transmitted if the unit is rigidly mounted to ground?

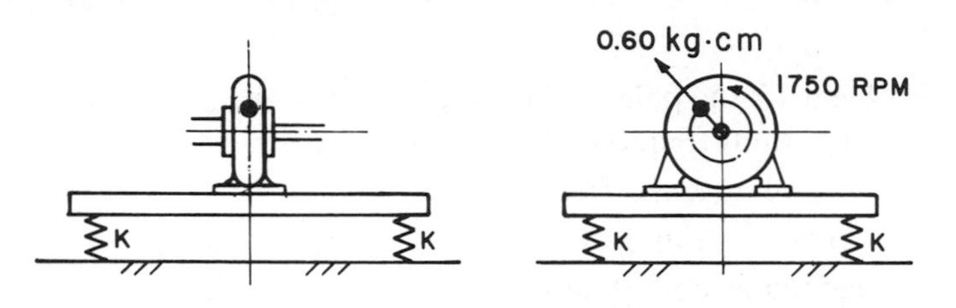

PROBLEM 2.7

2.8. A mass is sinusoidally excited at a spring, as in Table 2.3*c*. Given the following,

$$K_1 = 44 \text{ N/cm} \qquad C_0 = 0.55 \text{ N s/cm}$$
$$M_1 = 12 \text{ kg} \qquad P_2 = 2.1 \text{ N at } 4.5 \text{ Hz}$$

calculate:

(a) The amplitude a_2.

(b) The amplitude of the dashpot force.

(c) The phase angle by which a_2 leads a_1.

2.9. The two-mass system is excited sinusoidally as shown. Find:

(a) The absolute amplitude of each mass.

(b) The required amplitude of the driving force. (Hint: Use the results of part *a* for part *b*.)

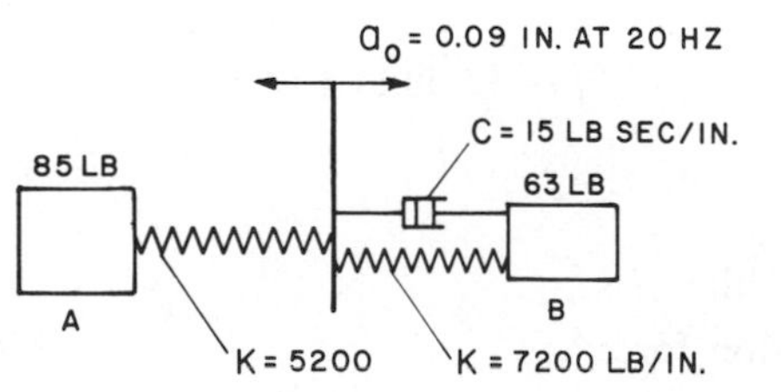

PROBLEM 2.9

Chapter 3

3.1. A spring–mass–dashpot system with a damping ratio of 0.30 is released from an initial displacement of 1.30 in. If the natural frequency is 11 Hz, determine:

(a) The displacement after $1\frac{1}{2}$ complete cycles.

(b) The acceleration after $1\frac{1}{2}$ complete cycles.

(c) The maximum absolute velocity during this interval.

(d) The elapsed time to the several instants of zero displacement.

3.2. The damped pendulum has a recorded time-displacement behavior shown by the curve. Determine:

(a) The damping ratio.

(b) The natural frequency.

(c) The initial velocity.

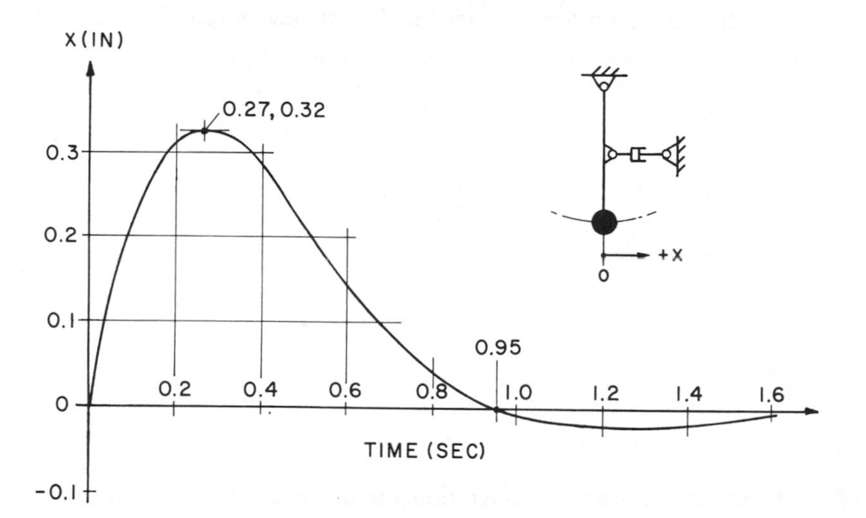

PROBLEM 3.2

3.3. The truck of Example 3.4 rebounds to a constant free reversed velocity after impact. Calculate this terminal velocity.

3.4. A single-degree system having a natural frequency of 15 Hz has critical viscous damping. If the mass has an initial displacement of 0.28 in. and an initial velocity of +40 in./sec, find:

(a) The maximum positive velocity after release.

(b) The maximum negative velocity after release.

(c) The time to achieve the maximum negative velocity.

3.5. A flywheel weighing 7.3 kg and supported in sleeve bearings coasts from 1000 to 700 RPM in 40 sec. The radius of gyration about the rotational axis is 16.1 cm. Find:

(a) The equivalent torsional damping constant of the bearings.

(b) The total number of revolutions during the measured interval.

3.6. A damped single-degree system, $\zeta = 0.50$, sustains a constant force pulse on the mass for one-half of an undamped period. If the force is 4.7 lb and the spring is 12.5 lb/in., determine:

(a) The displacement at the instant of pulse removal.

(b) The maximum total displacement during the entire transient response.

3.7. The disk is subjected to a variable torque pulse, as shown by the curve, for 3 sec, with the torque in in. lb. Find:

(a) The maximum torque in the spring during the pulse.

(b) The maximum torque after the disturbance is removed.

(c) Total elapsed time to reach the maximum torque.

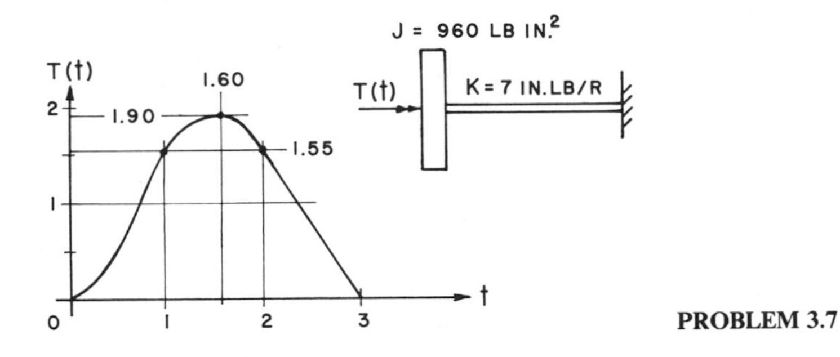

PROBLEM 3.7

3.8. The car has constant acceleration for 8 sec after starting from rest. It then maintains a constant terminal velocity of 30 mph. Determine the maximum dynamic force in the hitch.

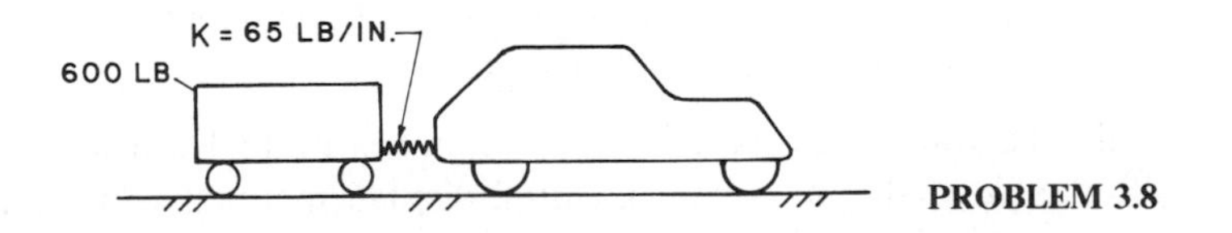

PROBLEM 3.8

Chapter 4

4.1. A normalized two-mass system (Table 4.1*b*), has parameters of $m_2 = 0.6$ and $k_2 = 0.4$. It is desired to raise the first-mode natural frequency by 15%.

(a) What is the new value of k_2?
(b) What change in m_2 only will accomplish the same result?

4.2. Verify the Example 4.4 solution using a single Holzer normalized calculation.

4.3. For the four-mass torsional system,

$$\begin{aligned} J_1 &= 750 \text{ lb in.}^2 & K_1 &= 0.6(10)^6 \text{ in. lb/rad} \\ J_2 &= 1000 & K_2 &= 1.3(10)^6 \\ J_3 &= 800 & K_3 &= 1.1(10)^6 \\ J_4 &= 550 \end{aligned}$$

(a) Determine the first and second natural frequencies.
(b) Check the modes using orthogonality.

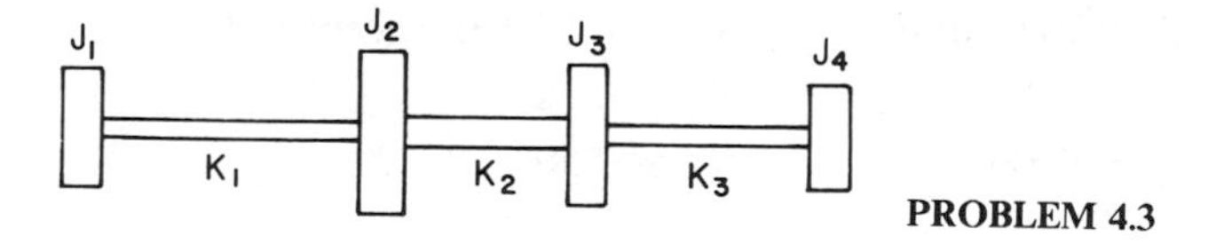

PROBLEM 4.3

4.4. Calculate the first and second natural translational frequencies of the four-mass vibratory system if

$$\begin{aligned} M_1 &= 7.5 \text{ kg} & K_1 &= 600 \text{ N/cm} \\ M_2 &= 10 & K_2 &= 1300 \\ M_3 &= 8 & K_3 &= 1100 \\ M_4 &= 5.5 \end{aligned}$$

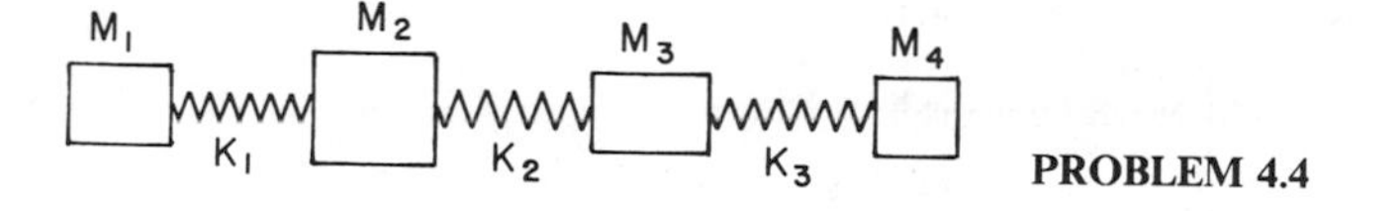

PROBLEM 4.4

4.5. Mass–elastic data for the geared system are

$$\begin{aligned} J_1 &= 7000 \text{ lb in.}^2 & K_1 &= 1.5(10)^6 \text{ in. lb/rad} \\ J_2 &= 8000 & K_2 &= 1.8(10)^6 \\ J_3 &= 8600 & K_5 &= 1.1(10)^6 \end{aligned}$$

Find the first natural torsional frequency using the Holzer method, working as follows:

(a) From 1 to 5.

(b) From 1 to the mesh, and from 5 to the mesh.

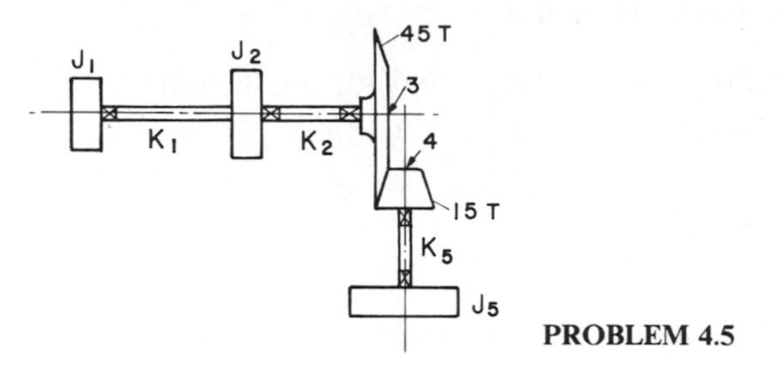

PROBLEM 4.5

4.6. For the three-branch geared system,

$$J_1 = 2150 \text{ lb in.}^2 \qquad K_1 = 0.85(10)^6 \text{ in. lb/rad}$$
$$J_4 = 3400 \qquad K_4 = 0.70$$
$$J_6 = 4000 \qquad K_6 = 0.50$$

(a) Find the lowest modal frequency from the complete results using the single-degree analogy.

(b) Plot the normal mode.

(c) Verify by applying orthogonality to the first and zero modes.

(d) Check results summing torques at 2.

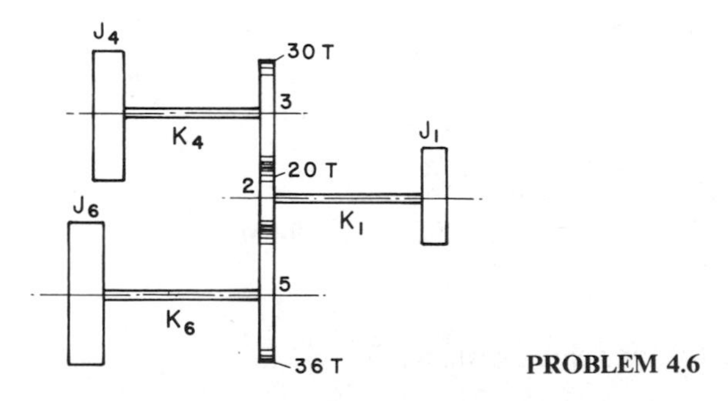

PROBLEM 4.6

4.7. A ski lift has dual chairs located every 65 ft along its length. When loaded, the mass suspended at each of these points, including cable mass, averages 350 lb. Supporting sheaves at the towers provide a cable span of

260 ft, and the counterweight applies a cable tension of 10,000 lb. Find the first natural frequency (or frequencies) of the suspended system.

Chapter 5

5.1. A two-mass system is excited horizontally by a rotating unbalance of 1.80 in. lb at 200 RPM. Find the steady-state response of each mass.

$$M_1 = 70 \text{ lb} \qquad K_0 = K_1 = K_2 = 150 \text{ lb/in.}$$
$$M_2 = 100 \text{ lb}$$

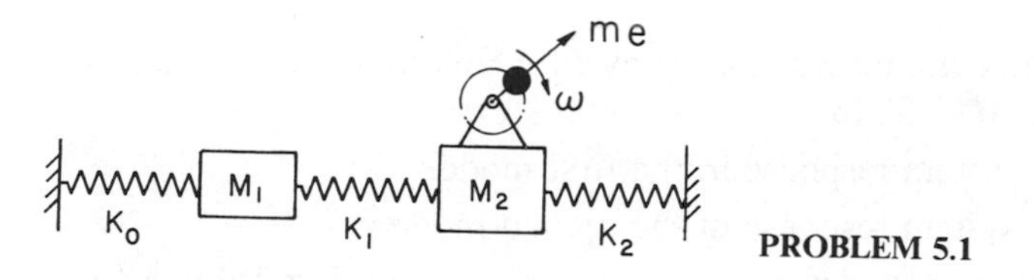

PROBLEM 5.1

5.2. Repeat Problem 5.1 with an additional equal and opposite unbalance applied simultaneously at M_1, counterrotating at the same speed.

(a) Find a_1 and a_2.

(b) Will $\mathsf{a}_1 = 0$ at any frequency? Explain.

5.3. The damped two-mass system is excited at system resonance at one mass by a sinusoidal force. Find

(a) The amplitude of the vibratory force in the spring.

(b) The amplitude of the dashpot force.

(c) The amplitude of the masses in the rigid mode.

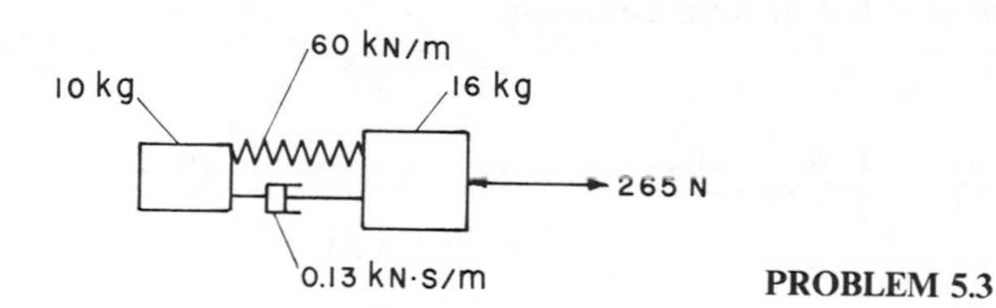

PROBLEM 5.3

5.4. A sinusoidal torque T_0 acts on the torsional system at a frequency of 55 Hz. Spring data are not available but $J_1 = 1400$ and $J_2 = J_3 = 1800$ lb in.^{2} Determine:

(a) The spring constant K_2.

(b) The amplitude T_0.

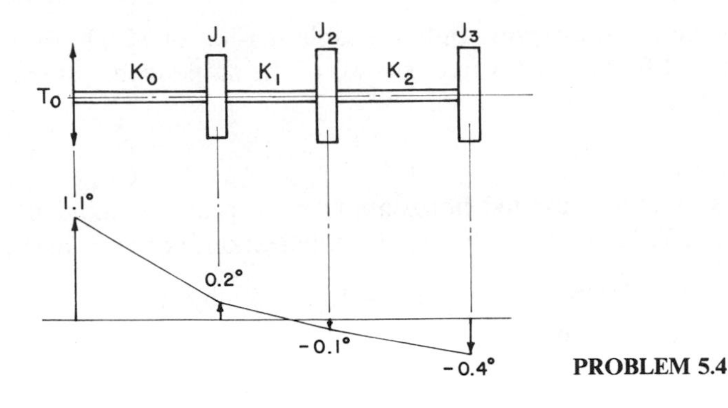

PROBLEM 5.4

5.5. The intermediate mass is forced by $P_2 = 260$ lb at 10.5 Hz, and $K = 190$ lb/in. and $M = 30$ lb.

(a) Find system response in the first mode.

(b) Find system response in the second mode.

(c) Verify amplitudes from the total response using Table 5.1*a*.

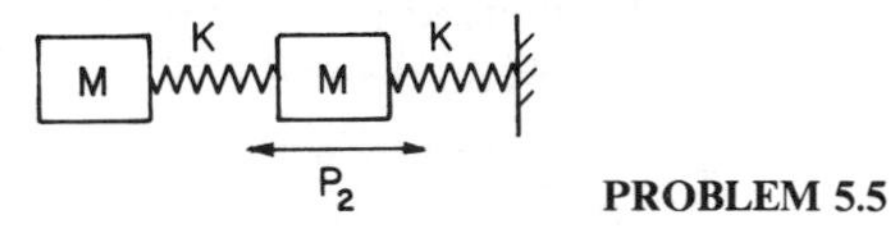

PROBLEM 5.5

5.6. A spring–mass system is excited by a sinusoidal displacement a_3 of 0.20 in. at 12 Hz. $M = 350$ lb and $K = 8000$ lb/in. Using the forced Holzer tabular solution, find:

(a) The total response of each mass.

(b) The amplitude of the exciting force.

(c) The amplitude of the ground reaction.

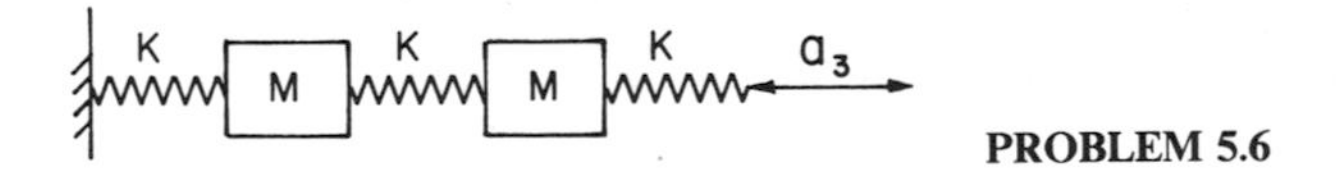

PROBLEM 5.6

5.7. The torsional vibratory system is displacement-excited as shown. If $\mathsf{g}_{n1} = 0.429$, calculate for the first modal response the amplitude of each disk:

(a) On an absolute basis.

(b) Relative to the rigid mode.

(c) Plot these distributions.

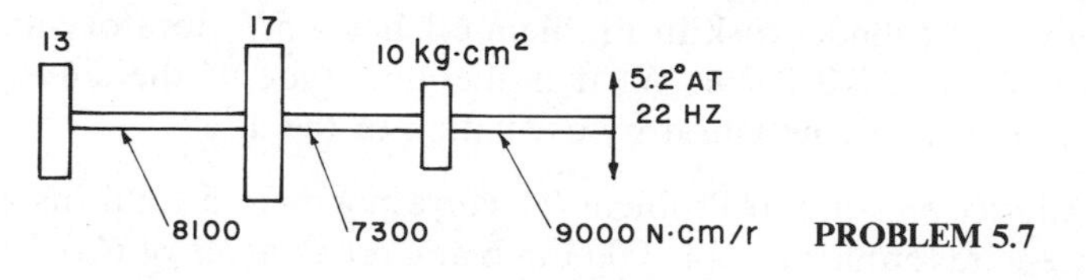

PROBLEM 5.7

5.8. A system similar to that of Table 5.4*c* has the following specifications:

$M_1 = 50$ lb $\quad K_1 = 1000$ lb/in. $\quad C_1 = 8$ lb sec/in.
$M_2 = 70$ $\quad K_2 = 800$ $\quad C_2 = 0$
$a_3 = 0.75$ in. at 10 Hz

Determine the total steady-state response using the Holzer system, including the following:

(a) Absolute amplitude of each mass.
(b) Amplitude of the exciting force.

5.9. A double pendulum (Table 5.5) consists of two equal-length links, each 55 cm center to center. Respective masses are $M_1 = 40$ kg and $M_2 = 70$ kg.

(a) Find two natural frequencies.
(b) Find two normal modes.
(c) Verify the orthogonality in part b.

Chapter 6

6.1. A four-cycle 10×10.75 in. (bore and stroke) engine has calculated values for each cylinder as follows:

Polar mass moment of inertia of crankshaft about rotational axis = 582 lb in.^{2} per cylinder
Length of connecting rod = 19.5 in.
Weight of connecting rod = 18.7 lb
Weight of piston assembly = 13.6 lb
Distance from wristpin to rod centroid = 12.3 in.
Radius of gyration of rod about wristpin = 13.4 in.

Determine:

(a) The equivalent inertia of the disk to represent each cylinder for modeling purposes.
(b) The amplitude of the $2N$ inertia torque developed at one cylinder by reciprocating parts at an engine speed of 3000 RPM.
(c) Total inertia torque in part b at a crank angle of 90°.

6.2. The equivalent cylinder disk in Problem 6.1 has a $5N$ vibratory amplitude of 0.38° at 2740 RPM. What is the amplitude of the sinusoidal torque developed by the vibratory oscillation of the disk?

6.3. What cylinder pressure in Problem 6.1 corresponds to a total instantaneous pressure torque of $-145{,}000$ in. lb at a crank angle of 675°? (The angle is measured from top dead center on the power stroke.)

6.4. A large diesel engine crankshaft is made of nodular cast iron. Shear modulus of the material is found to vary $\pm 5\%$ from the average value in the torsional frequency computation.

(a) What deviation from the calculated frequency will be expected as a result? Explain.

(b) If this engine mode had damping equivalent to 2% of critical, and operation ranges over a resonant peak, what corresponding variation in vibratory magnification factor could occur at the nominal resonant speed?

6.5. A function varies periodically as shown. Find the amplitude of the fundamental sine component of the Fourier series and the constant term.

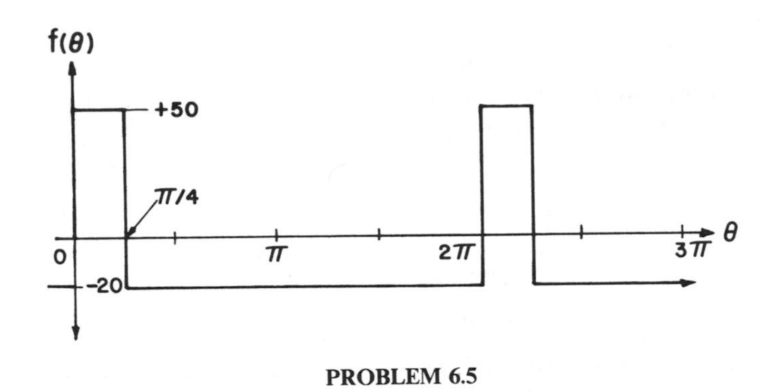

PROBLEM 6.5

6.6. A repetitive sawtooth function is given. Calculate the Fourier components A_1, B_0, and B_1.

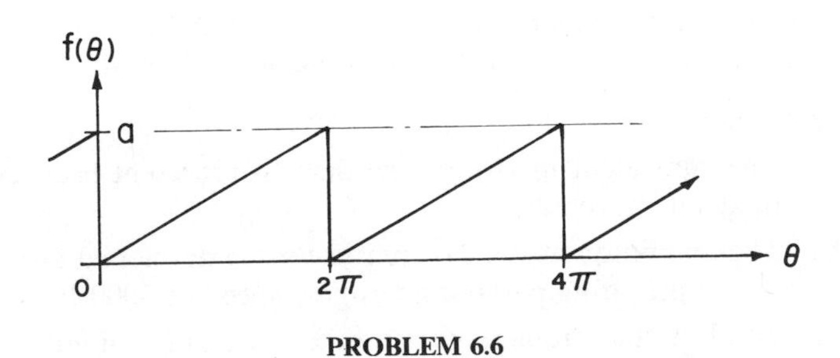

PROBLEM 6.6

6.7. On torsiograph test an engine develops a peak amplitude. When the speed is increased 8% beyond the peak speed, crankshaft amplitude in this mode falls to 60% of the maximum. What is the resonant magnification factor, assuming viscous damping?

6.8. A four-cylinder, two-cycle engine has a first-mode natural frequency of 7202 CPM, and firing order 1, 4, 2, 3, 1,... . Mass–elastic data are as follows:

$$K_1 = K_2 = K_3 = 0.92(10)^6 \qquad K_4 = 0.69(10)^6 \text{ in. lb/rad}$$

$$J_1 = J_2 = J_3 = J_4 = 85; \qquad J_5 = 1060 \text{ lb in.}^2$$

(a) What are the major orders of excitation?
(b) What is $\bar{\beta}$ for $2N$ excitation?
(c) If the first-mode vibratory amplitude of J_1 is 0.73° due to a $3N$ excitation, calculate the first-mode vibratory torque in the span of maximum torque.
(d) Find the first modal mass and spring.
(e) At what operational speed will $6N$ excitation cause a first-mode resonance?

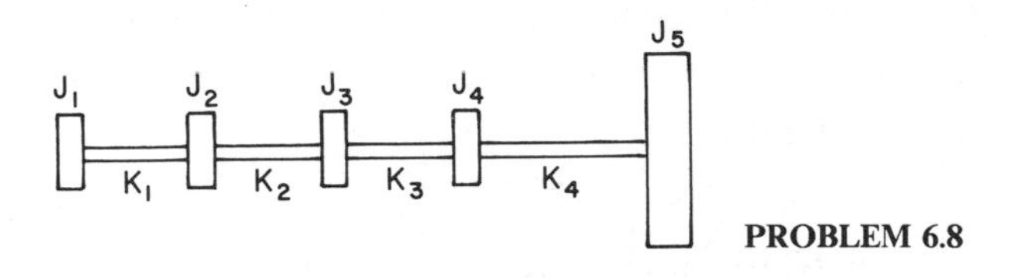

PROBLEM 6.8

6.9. A three-cylinder, two-cycle engine has a $3N$ resonance with end amplitude at J_1 of 1.75°. The following data apply:

Damping is 3% of critical.
Modal amplitudes are 1.00, 0.72, 0.22, and −0.27.
Modal spring is $0.77(10)^6$ in. lb/rad.

(a) Calculate the equilibrium amplitude in degrees.
(b) Determine the amplitude of the $3N$ exciting torque per cylinder.

6.10. The six-cylinder diesel engine-generator has the following specifications:

Bore and stroke = $20 \times 21\frac{1}{2}$ in.
Rating = 6000 HP at 400 RPM

Firing order = 1, 6, 2, 4, 3, 5, 1,...
Main bearings = $19\frac{1}{2}$ in. O. D. × 13 in. I. D.
Connecting rod bearings = 16 in. O. D. × 10 in. I. D.
Reciprocating weight = 1574 lb/cylinder
Connecting rod length = 51 in.

$$J_1 = 432{,}807 \text{ lb in.}^2$$

$$J_2 = J_3 = J_4 = J_5 = J_6 = 370{,}024 \text{ lb in.}^2$$

$$J_7 = 505{,}056$$

$$J_8 = 812{,}315$$

$$J_9 = 33{,}700{,}000 \text{ lb. in.}^2$$

$$K_1 = K_2 = K_3 = K_4 = K_5 = 1300(10)^6 \text{ in. lb/rad}$$

$$K_6 = 2121(10)^6$$

$$K_7 = 3839(10)^6$$

$$K_8 = 8097(10)^6 \text{ in. lb/rad}$$

Calculate the first natural frequency of the torsional system, which can be normalized with respect to J_2.

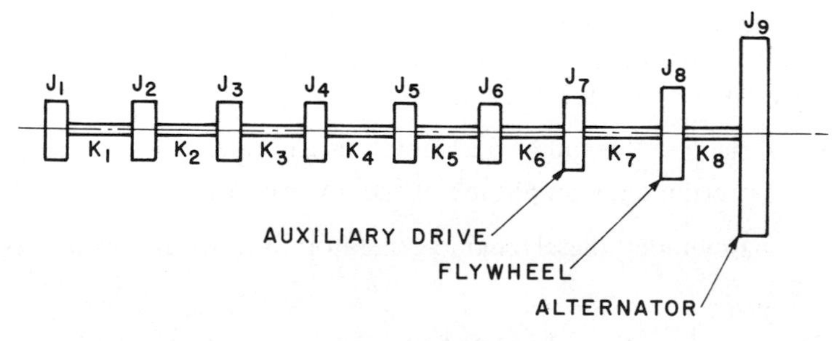

PROBLEM 6.10

6.11. Pressure versus crank angle data for the six-cylinder engine are as follows:

Crank Angle	Pressure (psi)	Crank Angle	Pressure (psi)	Crank Angle	Pressure (psi)
0	1188	95	131	275	50
5	1275	100	125	280	56
10	1256	105	106	285	69
15	1125	110	88	290	75
20	975	115	75	295	88
25	825	120	56	300	106
30	687	125	44	305	125
35	575	130	31	310	150
40	481	135	19	315	187
45	406	140	12	320	231
50	350	145	6	325	281
55	300	150–235	0	330	350
60	256	240	12	335	425
65	219	245	25	340	513
70	188	250	32	345	600
75	169	255	38	350	706
80	150	260	38	355	950
85	144	265	44	360	1188
90	138	270	50		

By harmonic analysis determine the Fourier components of the torque due to cylinder pressure for orders 6 through 12 inclusive.

6.12. Make a complete torsional response analysis of the six-cylinder engine, assuming a resonant magnification factor limited by system damping to 25, including the following:

(a) Star diagrams for first mode.

(b) A frequency diagram similar to Figure 6.7.

(c) Tabulate resonant speeds, vibratory resonant amplitudes of J_1, and corresponding resonant stresses for the order excitations $6N$ through $12N$.

(d) Plot approximate predicted modal stresses versus speed at the span of maximum torque, similar to Figure 6.13.

Chapter 7

7.1. A Frahm vibration absorber is applied to a main system with a mass of 3.8 kg and a natural frequency of 34 Hz. The objectional exciting frequency is 30 Hz. If the absorber mass is to be 0.90 kg, find:

(a) The proper absorber spring constant.

(b) The vibratory amplitude of the absorber at the null condition if the amplitude of the exciting force is 100 *N*.

(c) The two natural frequencies.

7.2. A dynamic damper has a mass ratio of 0.30 and optimum tuning and damping. Calculate and plot the amplitude of the main system as a function of frequency, with g varying from 0 to 1.5 using increments of 0.10.

7.3. If a dynamic damper has the following specifications,

	Main System	Auxiliary System
Mass	80 lb	16 lb
Spring	2200 lb/in.	500 lb/in.
Dashpot	0	1.2 lb sec/in.

find the coordinates of the independent intersection point Q in the response curves; that is, find the frequency ratio and the magnification factor at Q.

7.4. The bar is stiff and assumed weightless, and carries two masses as shown.

(a) What frequency of T will result in a stationary bar?

(b) When the bar is then stationary, what is the vibratory amplitude of the 4.4 lb mass?

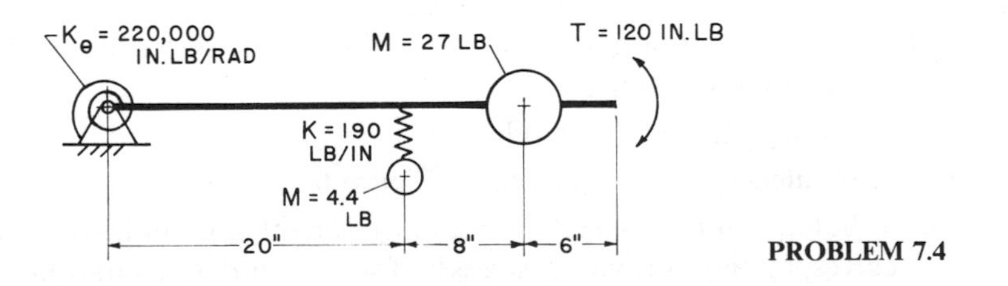

PROBLEM 7.4

7.5. We wish to convert the small spring–mass system of Problem 7.4 into a dynamic damper by adding damping and optimizing the tuning, keeping the same mass. What is the required new value of the spring?

7.6. A dynamic damper is to limit the maximum vibratory magnification factor to 4.0 over the complete range of exciting frequencies. Determine:

(a) The optimum damping ratio.

(b) The frequencies at which the main system peaks with the damper operative if the original main-system resonance is at 41 Hz.

7.7. If a dynamic damper has a mass ratio of 0.12 and a main-system natural frequency of 45 Hz, calculate:

(a) The natural frequency of the auxiliary system for optimum tuning.

(b) Optimum damping ratio for the damper.

(c) Maximum main-system magnification factor.

(d) Frequencies at which the peaks occur.

(e) Main-system magnification factor at 54 Hz.

7.8. A viscous damper acts upon a torsional pendulum with the damper case directly attached below the disk. Polar mass moment of inertia of the disk, including the damper case, is 0.040 $kg \cdot m^2$, and the auxiliary inertia is 0.012 $kg \cdot m^2$. Stiffness of the vertical shaft is 5200 $N \cdot m/r$, and sinusoidal torque on the disk is 75 $N \cdot m$ at 3200 CPM. For optimum viscous damping, calculate:

(a) The damping constant for the damper, indicating proper units.

(b) Amplitude of the main disk in degrees.

(c) Main-disk response in degrees with the damper disk locked to the case.

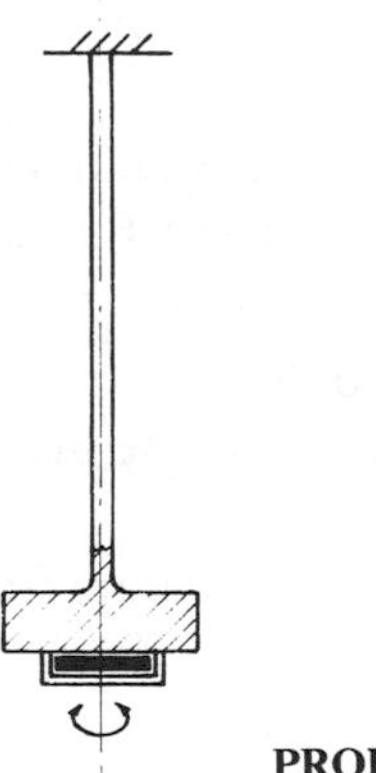

PROBLEM 7.8

7.9. A viscous coupled damper on a single-degree torsional system has the following parameters:

Main mass = 880 lb in.^{2}

Damper mass = 90 lb in.^{2}

Main-system natural frequency = 36 Hz
Damping = optimum
Exciting torque = 750 in. lb at 55 Hz

Find the following:

(a) The amplitude of the main mass.
(b) Absolute amplitude of the damper mass.
(c) Relative amplitude of the damper with respect to the main mass.
(d) Amplitude of the main mass if the damper is removed.

7.10. A bifilar pendulum absorber is tuned to $6N$, with the radius of rotation to the center of gravity of the suspended mass 11.60 in. Find the diameter of the pins if the diameter of the holes is 1.500 in.

7.11. A centrifugal pendulum absorber has a pendulum length of 0.410 in. and a radius to the pivot point of 7.00 in. Pendulum mass is 2.80 lb, mass ratio is 0.09, and excitation is $4N$.

(a) What is the moment of inertia of the main system about the rotational axis?
(b) Will there be infinite response?
(c) If so there is infinite response, at what rotational speed relative to the original $4N$ resonance without the pendulum?

7.12. A bifilar pendulum with a mass ratio of 0.11 is to shift a $5N$ resonance from an operational speed of 2800 RPM to 4000, thereby moving it above the running range. Find the diameter of the pins if hole diameters are 1.350 in. and the radius to the pendulum centroid is 14 in.

Chapter 8

8.1. A bronze sphere 12 in. in diameter is suspended as a pendulum with a distance of 11 in. from the center of the sphere to the fulcrum. Determine:

(a) The natural frequency as a compound pendulum.
(b) The dynamically equivalent two-mass system, with one mass at the fulcrum.
(c) The natural frequency of the equivalent simple pendulum from part b.

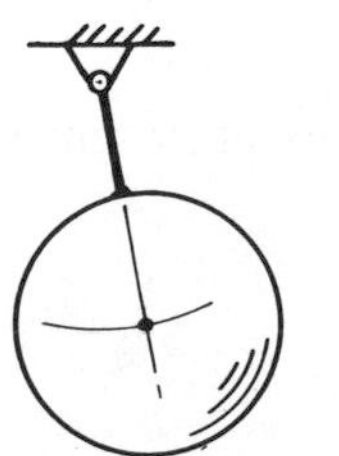

PROBLEM 8.1

8.2. A slender rigid bar, 80 cm long, is to be suspended by a pivot on the bar centerline, between the ends. Total weight is 8.5 kg. Find the end distance x for a pendular period of 2.00 sec.

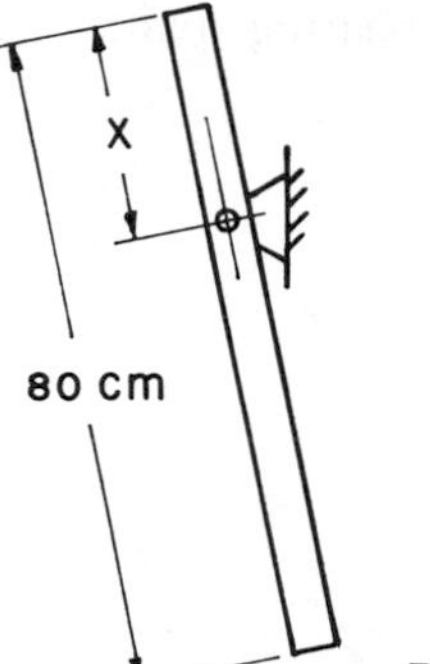

PROBLEM 8.2

8.3. The bar of Problem 8.2 is excited horizontally by a harmonic displacement of 3.30 cm at 1.40 Hz. Find the amplitude of the angular response.

8.4. For the connecting rod of Problem 6.1, calculate:

(a) The distance from the wristpin of M_2.

(b) The value of M_2.

8.5. The rigid system is constrained to oscillate in the horizontal plane shown, with the centroid at C. Consider the end mass to be concentrated at C.

(a) Find the natural frequency or frequencies.

(b) Determine the normal mode or modes, and plot to scale.

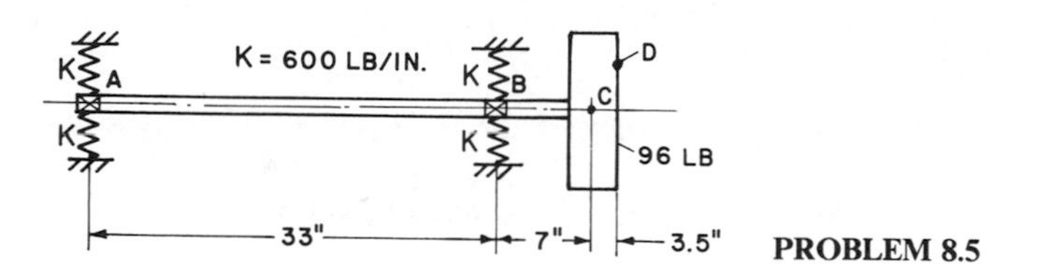

PROBLEM 8.5

8.6. Repeat Problem 8.5, assuming that the radius of gyration of the end mass is 10.72 in. about a perpendicular axis through C. Also verify orthogonality of modes.

8.7. The system of Problem 8.5 is excited transversely at D by an unbalance at D of 15.04 lb in. as the system rotates at 200 RPM. Find the forced amplitudes at A and B.

Chapter 9

9.1. A simply supported shaft carries a central mass. Find:

(a) The natural frequency factor, g_n, if the mass is considered concentrated.

(b) The natural frequency factors if the radius of gyration of the disk about a transverse axis is 30% of the total bearing span.

(c) The normal modes corresponding to part b.

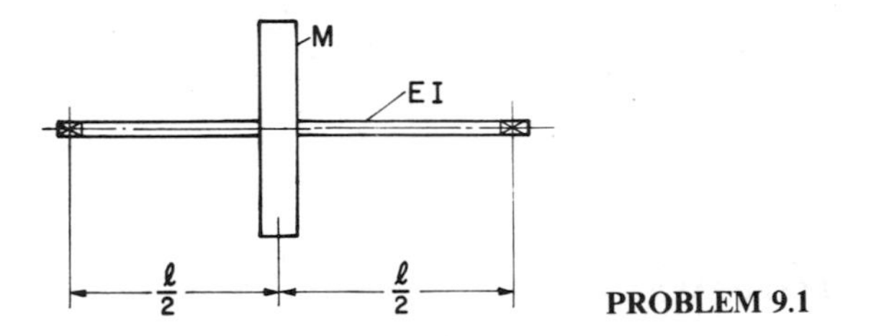

PROBLEM 9.1

9.2. A uniform steel shaft, 3.80 cm in diameter, carries a 6.80 kg disk for which the radius of gyration about a transverse diameter is 30.5 cm. Find the natural frequencies and modes, neglecting the mass of the shaft.

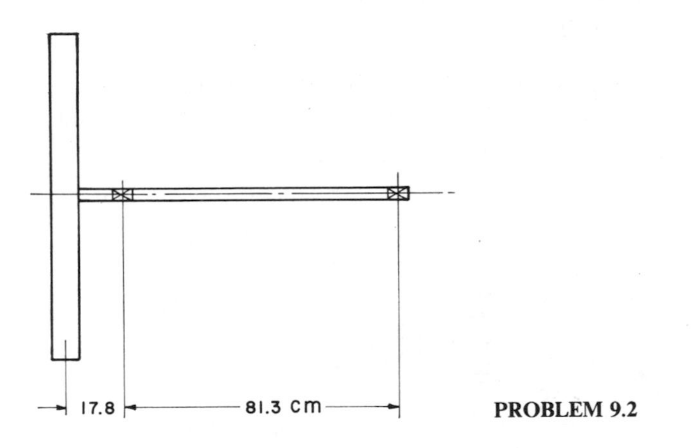

PROBLEM 9.2

9.3. Calculate the first modal springs for the system of Problem 9.2.

9.4. Find the natural frequencies of the Problem 9.2 system if each bearing support has a transverse flexibility of $30(10)^6$ N/m.

9.5. A steel bar is 4 × 4 in. in cross section. Determine the natural frequencies and normal modes in the plane.

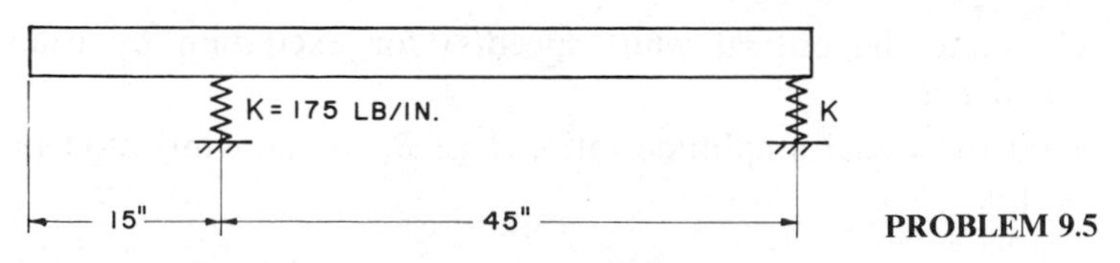

PROBLEM 9.5

Chapter 10

10.1. An 8 lb disk is centrally supported on a $\frac{3}{4}$ in. shaft, 12 in. between bearing centers. Mass eccentricity is 0.003 in. Calculate the orbital radius of the mass center at the following:

(a) 1750 RPM.

(b) 3500 RPM.

(c) 5250 RPM.

(d) 7500 RPM.

10.2. A thin disk is cantilever-mounted to a larger shaft, assumed fixed. Determine:

(a) The nonrotative natural frequencies.

(b) The rotative natural frequencies relative to unbalance excitation.

(c) Normal modes in part b.

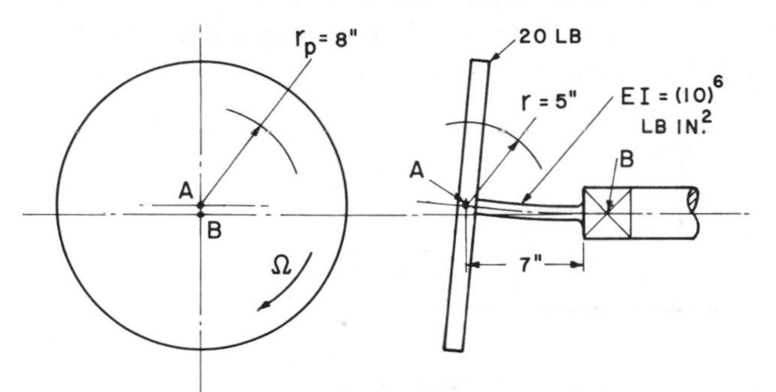

PROBLEM 10.2

10.3. The bearing at B in Problem 10.2 orbits at three times the rotational velocity, and in the opposite (counterclockwise) sense. Find the resonant whirl speed(s) and the normal mode(s).

10.4. The rotor is mounted in relatively flexible bearings, with equal stiffness in all directions at each end. System constants are as follows:

$$K = 550 \text{ lb/in.} \qquad I_G = 6970 \text{ lb in.}^2$$
$$M = 67 \text{ lb} \qquad J = 17{,}800 \text{ lb in.}^2$$

(a) Find the natural frequency if the mass is considered concentrated at G.

(b) Find the planar frequencies for the nonrotative case considering the rigid inertia.

(c) Calculate the critical whirl speed(s) for excitation by rotating unbalance.

(d) Find the radial amplitude ratio, A to B, in the whirl mode(s) in part c.

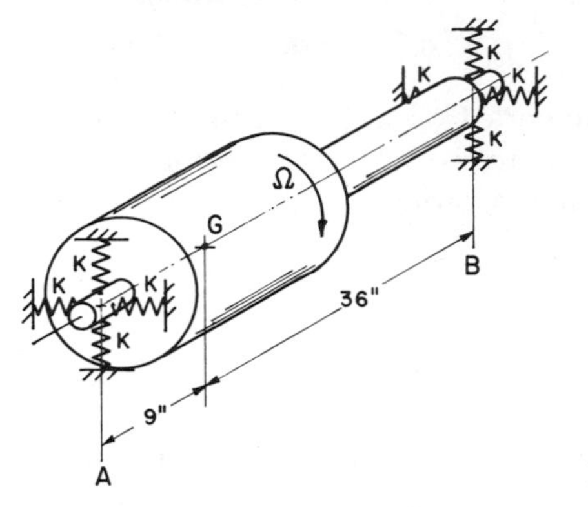

PROBLEM 10.4

10.5. In Problem 10.4 the static unbalance in the G plane is 8.5 in. lb. Find the radial displacements at A and B (elastic) at 800 RPM.

10.6. The steel rotor is assumed rigid between B and C, and we neglect mass effects of the end shafts. Find the natural frequencies and modes if the rotor is excited by rotational unbalance.

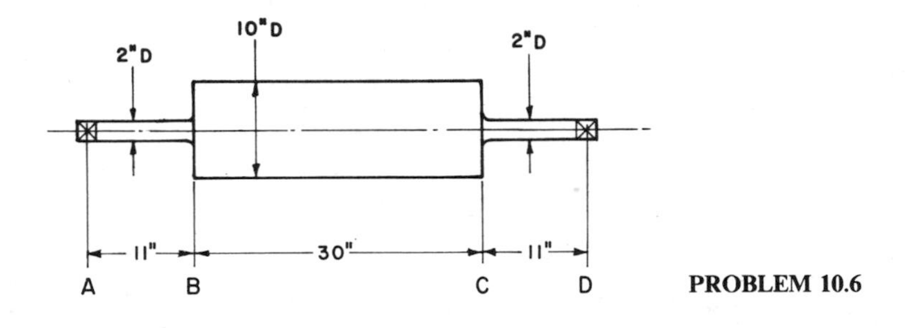

PROBLEM 10.6

Chapter 11

11.1. A two-mass flexural system, checked for flexibility statically, has a transverse load applied at B of 40 lb, which causes a deflection of 0.106 in. at B and 0.068 in. at D. A load at D of 40 lb produces a deflection at D of 0.125 in. Neglecting the mass of the shaft, calculate:

(a) The two lateral frequencies in CPM.

(b) The natural frequency if the mass at D is removed, leaving only the mass at B.

(c) The natural frequency if the mass at B is removed, leaving only the mass at D.

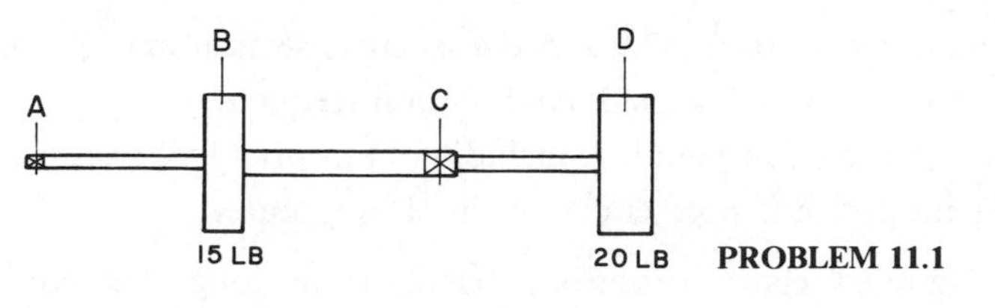

PROBLEM 11.1

11.2. Using Rayleigh's method, find the approximate first-mode natural frequency in bending, and the corresponding normal mode.

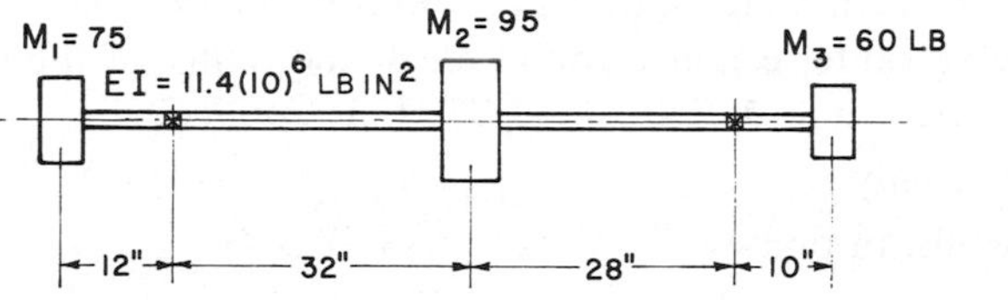

PROBLEM 11.2

11.3. Repeat Problem 11.2 using matrix iteration.

11.4. Calculate the natural frequencies and the modal springs and masses at M_1 for the two-mass case, neglecting beam mass, with masses concentrated.

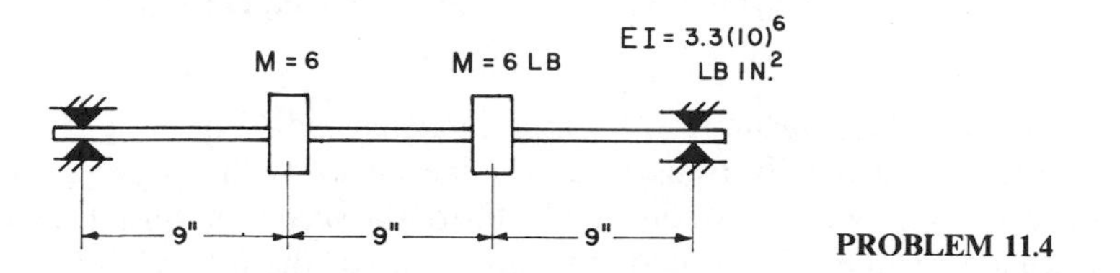

PROBLEM 11.4

11.5. Verify the first natural frequency in Problem 11.4 using the Stodola method.

11.6. In-phase sinusoidal displacements are applied to the supports in the plane at 150 Hz, with amplitudes of 0.060 in. and 0.120 in. at the left and right ends respectively. Compute the total absolute response amplitudes of the masses using modal components.

11.7. Repeat Problem 11.6 for equal in-phase bearing amplitudes of 0.13 in. at 364.4 Hz.

Chapter 12

12.1. Calculate normal modal amplitudes for the first cantilever mode of a uniform beam in bending.

12.2. A rectangular steel bar is 3.5 × 5 cm in cross section and 3.1 m long.

(a) Find the first and second axial natural frequencies.

(b) Find the same in bending, including both principal axes.

(c) Repeat part b if both ends are simply supported.

12.3. A steel clamped–clamped uniform bar is 55 in. long. The equilateral triangular cross section is 2.00 in. on a side. Find the first and second natural frequencies:

(a) In torsion.

(b) In bending.

12.4. An aluminum I beam weighing 14 kg is excited in the plane shown. Determine the absolute response amplitude at the center of the beam if sinusoidal end excitation of 0.40 cm at 75 Hz is applied:

(a) At one end only.

(b) At both ends, in phase.

(c) At both ends, out of phase.

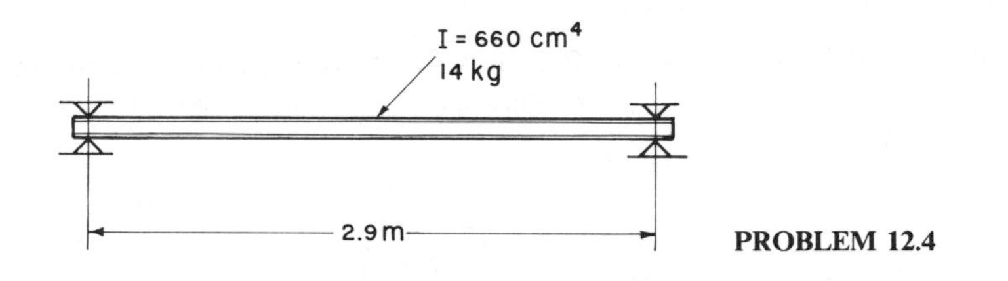

PROBLEM 12.4

12.5. A steel cantilever beam is 1.50 in. in diameter and 42 in. long. The base is forced sinusoidally transversely to the beam, with an amplitude of 0.008 in. If the exciting frequency is 70% of the second natural frequency, calculate for the second mode the amplitude of the following:

(a) The shear force at the base.

(b) The bending moment at the base.

(c) The bending stress at the base.

12.6. Repeat Problem 12.5, but the excitation is now a sinusoidal couple at the free end of 500 in. lb.

Chapter 13

13.1. The rotor of Problem 10.6 has a concentrated unbalance in plane *B* of 35 in. lb.

(a) What is the first modal response at 3600 RPM?

(b) What unbalance must be added in plane *C* to completely eliminate first-mode response?

(c) In the central plane of the rotor?

(d) With respect to possible second-mode problems, which is preferable?

13.2. A steel rotor consists of two constant-diameter sections, the larger essentially rigid with respect to the smaller. The approximate bending mode is shown, including equations using dimensionless abscissae. In the flexural section a parabolic elastic curve is assumed. Calculate the modal mass at 1.

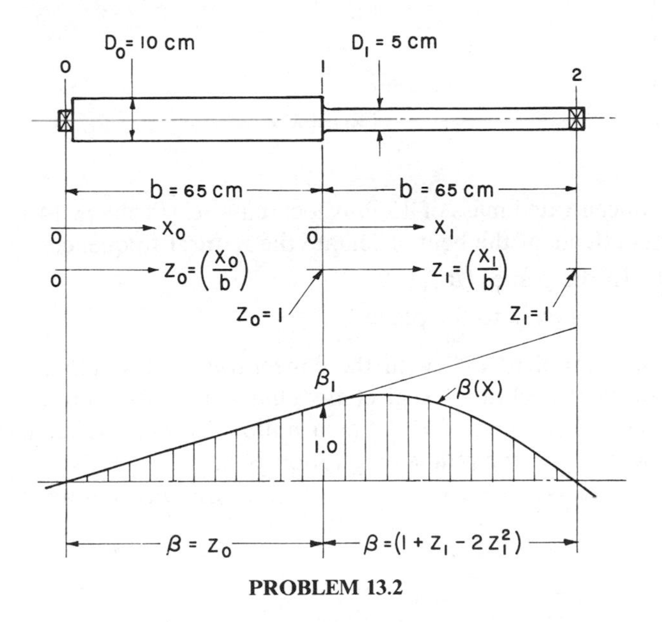

PROBLEM 13.2

13.3. In Problem 13.2 the entire rotor orbits translationally at a radius of 0.34 cm due to in-phase eccentricities in the machining of the bearing journals. What correction at 1 is required for first modal balance?

13.4. In Problem 13.3, what is the correction if the correction is located:

(a) At the center of the larger span?

(b) At the center of the smaller span?

13.5. Neglecting gyroscopic effects and the mass of the smaller section, approximate the critical speed in Problem, 13.2. (Hint: See Table 8.9.)

Chapter 14

14.1. A steel tube $\frac{7}{8} \times \frac{5}{8}$ in. in diameter is formed into a bent, *ABC*. Find:

(a) The maximum static compliance factor in the plane of the bent at *C* and the angular direction, θ.

(b) Repeat for the minimum compliance.

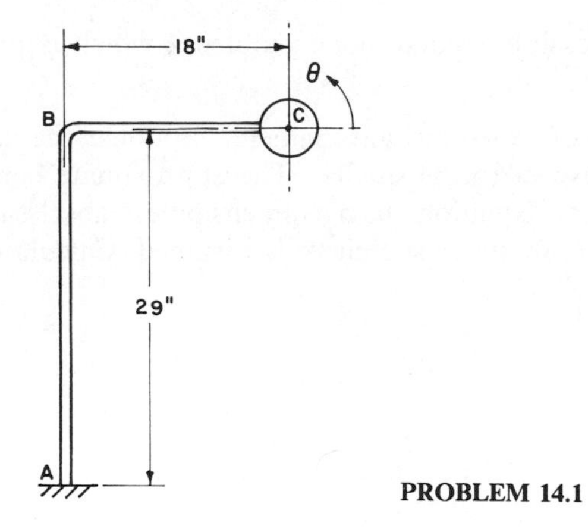

PROBLEM 14.1

14.2. A concentrated mass of 15 lb is located at C (Problem 14.1). Neglecting mass effects of the bent, calculate the natural frequencies:

(a) In the plane ABC.

(b) Transverse to the plane.

14.3. Repeat Problem 14.2 with the concentrated mass replaced by a large cylindrical steel tube. Neglect any change in stiffness of the bent due to a slight shortening of BC. For a hollow cylinder about the transverse principal axis, the radius of gyration is

$$0.289\sqrt{3(R^2 + r^2) + L^2}$$

PROBLEM 14.3

14.4. The Problem 14.3 system is excited vertically in the AB direction by a sinusoidal displacement of 0.13 in. at 6 Hz. Calculate the resultant or total translational amplitude of C.

Chapter 15

15.1. A planar circular steel cantilever has a central angle of 90°. The section is a circular tube 3 × 2.75 cm in diameter, and the mean radius is 44 cm. Calculate the three out-of-plane primary compliance factors (case c).

15.2. Repeat Problem 15.1 for a central angle of 135°.

15.3. A rigid body has a center of gravity above the two uniform beams upon which it is mounted, resulting in angular elastic coupling effects.

(a) Find expressions for the six elastic compliance factors with respect to the centroid from Tables 9.1 and 15.9.

(b) Determine the five ψ ratios.

(c) Find the coefficients of g^2 in the denominator, C_b, in Table 14.5, neglecting beam mass.

(d) How many modal frequencies exist? Explain.

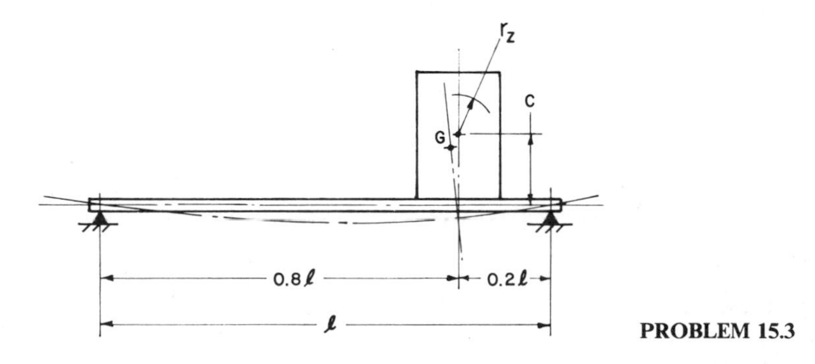

PROBLEM 15.3

15.4. A rotational agitator is mounted radially in a cylindrical steel tank. Consider the problem to be two-dimensional in the plan view shown. If the tank is excited angularly by unbalance of the impeller, find the potential resonant agitator speeds between 800 and 1100 RPM.

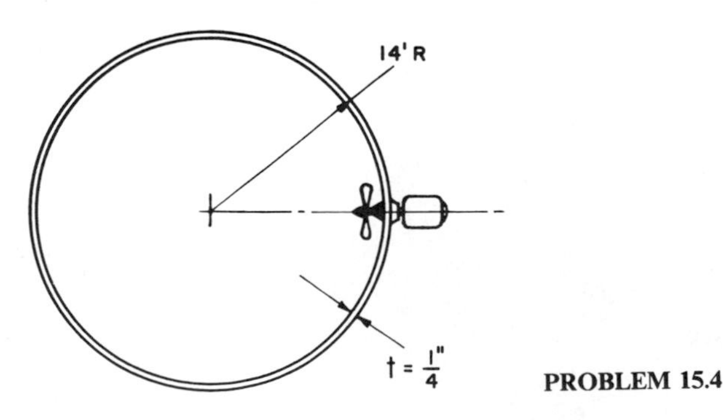

PROBLEM 15.4

ANSWERS TO PROBLEMS

1.1(a) 47.1 cm/s
(b) 7400 cm/s^2
(c) 157 r/s
(d) 0.04 s

1.2 0.0352 s

1.4(a) 3.95 r/s
(b) 447 r/s^2

1.5(a) 276°
(b) $7 - j6$
(c) 1.91
(d) 10.05 at 84.3°

1.6(a) 0.103 in.
(b) 22,300 in.lb/rad
(c) 3.57 lb

1.7 $F_B = c\left(\dfrac{K_A K_B}{K_A + K_B}\right)$

1.8 96 + 50.8 = 146.8 N

2.1 137 Hz

2.2 54.4 Hz

2.3(a) 53.8 Hz
(b) 30.6 Hz
(c) 61.9 Hz

2.4 25.4 Hz

2.5(a) 0.92 in.
(b) 1.02 in.

2.6(a) 0.0117 in.
(b) 4.04 lb
(c) 0.26 in.lb/sec

2.7(a) 9.32 N
(b) 202 N

2.8(a) 0.027 cm
(b) 0.34 N
(c) 163°

2.9(a) 0.27 in. (A) and 0.134 in. (B)
(b) 1290 lb

3.2(a) 0.63
(b) 0.68 Hz
(c) 2.78 in./sec

3.3 2.96 ft/sec

3.5(a) 0.0017 N · m · s
(b) 560 revolutions

3.7(a) 3.1 in. lb
(b) 2.9 in. lb
(c) 4.2 sec

3.8 137 lb

4.1(a) $k_2 = 0.57$
(b) $m_2 = 0$

4.4 $\mathsf{g}_{n1}^2 = 0.985$, $\mathsf{g}_{n2}^2 = 2.685$

4.5 30.3 Hz

4.7 0.46 Hz

5.1 0.0165 in. (1) and 0.0242 in. (2)

5.3(a) 476 N
(b) 102 N
(c) 0.104 cm

5.5(a) $\mathsf{a}_{11} = -0.437$ in. and $\mathsf{a}_{21} = -0.270$ in.
(b) $\mathsf{a}_{12} = -0.731$ in. and $\mathsf{a}_{22} = +1.182$ in.
(c) $\mathsf{a}_1 = -1.17$ in. and $\mathsf{a}_2 = +0.911$ in.

5.7(a) $\mathsf{a}_{11} = -9.32°$ and $\mathsf{a}_{21} = -7.6°$
(b) $\mathsf{a}_{11}^r = -15.5°$

5.9(a) 0.53 and 1.07 Hz
(b) 0.38 and -1.52

6.1(a) $M_R = 11.8$ and $M_T = 6.9$ lb
$J_e = 1220$ lb in.^{2}
(b) 75,700 in.lb
(c) 41,700 in.lb

6.3 402 psi

6.5 $A_1 = 6.53$

6.6 $A_1 = -\dfrac{a}{\pi}$, $B_1 = 0$

6.7 7.62

6.9(a)	0.105°
(b)	727 in.lb
7.1(a)	32000 N/m
(b)	0.31 cm
(c)	25.2 and 40.4 Hz
7.3	1.192 and 1.41
7.4(a)	20.5 Hz
(b)	0.032 in.
7.6(a)	0.21
7.8(a)	3.55 N · m · s
(b)	6.3°
7.10	1.186 in.
7.12	0.701 in.
8.1(a)	0.89 Hz
(b)	12.3 in.
(c)	0.89 Hz
8.2	34.3 cm
8.3	2.18°
8.5(a)	8.98 Hz
8.6	8.58 and 43.9 Hz
8.7	a_A = 0.0054 in.
9.1(a)	1.00
(b)	1.00 and 1.73
(c)	0 and ∞
9.2	3.87 and 266 Hz
9.3	K_{n1} = 21,000 N/cm and $K_{n\theta}$ = 4.6(10)^6 N · cm/r
9.5	3.2 and 5.0 Hz
10.1(b)	0.0038 in. at 3500 RPM
10.3	685 and 2950 RPM
10.4(a)	920 RPM
(c)	950 RPM
(d)	+ 2.85
10.6	39 and 122 Hz
11.1(a)	650 and 1340 CPM
11.2	13.1 Hz
11.4	94 and 364 Hz $M_{n1} = M_{n2}$ = 12 lb
11.6	− 0.70 in.
12.2(a)	815 and 1630 Hz
(b)	18.8 and 51.8 Hz flatwise 26.8 and 74 Hz edgewise
(c)	8.5 and 34 Hz; 12 and 49 Hz

12.3(a) 890 Hz
(b) 96 Hz

12.5(a) 70 lb
(b) 625 in. lb
(c) 1900 lb/in.^{2}

13.1(a) 0.089 in.

13.2 21.8 kg

13.3 9.8 kg · cm

13.5 4200 RPM

14.1(a) 0.0273 in./lb
(b) 0.00314 in./lb

14.2(a) 4.9 and 14.4 Hz
(b) 4.37 Hz

14.3 (a) 4.76 (b) 13.45 Hz

15.1 $\alpha_z = 0.0045$ cm/N
$\gamma_x = 3.33(10)^{-6}$ r/N · cm
$\gamma_y = 3.33(10)^{-6}$ r/N · cm

15.3(b) $\psi_1 = 1, \psi_4 = \left(\frac{1}{c^2}\right), \psi_5 = -\left(\frac{1}{c}\right)$

(c) $C_b = +\left(\frac{r_z}{c}\right)^2\left[\left(\frac{\beta_{yz}}{\gamma_z}\right) + 1\right]^2$

INDEX